Mitochondria in Health and Disease

OXIDATIVE STRESS AND DISEASE

Series Editors

LESTER PACKER, PH.D.
ENRIQUE CADENAS, M.D., PH.D.
University of Southern California School of Pharmacy
Los Angeles, California

Mitochondria in Health and Disease

edited by

Carolyn D. Berdanier

CRC Press
Taylor & Francis Group
Boca Raton   London   New York

CRC Press is an imprint of the
Taylor & Francis Group, an **informa** business

CRC Press
Taylor & Francis Group
6000 Broken Sound Parkway NW
Boca Raton, FL 33487-2742

First issued in paperback 2019

© 2005 by Taylor & Francis Group
CRC Press is an imprint of Taylor & Francis Group, an Informa business

No claim to original U.S. Government works

ISBN-13: 978-0-8247-5442-6 (hbk)
ISBN-13: 978-0-367-39269-7 (pbk)

Library of Congress Card Number 2004064967

Library of Congress Cataloging-in-Publication Data

Mitochondria in health and disease / Carolyn D. Berdanier, editor.
    p. cm.
Includes bibliographical references and index.
ISBN 0-8247-5442-5 (alk. paper)
1. Mitochondrial pathology. 2. Mitochondria. 3. Oxidative stress. I. Oxidative stress and disease.

RB147.5.M57 2005
616.07--dc22   2004064967

Visit the Taylor & Francis Web site at
http://www.taylorandfrancis.com

and the CRC Press Web site at
http://www.crcpress.com

Series Introduction

Oxygen is a dangerous friend. Through evolution, oxygen — itself a free radical — was chosen as the terminal electron acceptor for respiration. The two unpaired electrons of oxygen spin in the same direction; thus, oxygen is a biradical. Other oxygen-derived free radicals, such as superoxide anion or hydroxyl radicals, formed during metabolism or by ionizing radiation are stronger oxidants, i.e., endowed with a higher chemical reactivity. Oxygen-derived free radicals are generated during oxidative metabolism and energy production in the body and are involved in regulation of signal transduction and gene expression, activation of receptors and nuclear transcription factors, oxidative damage to cell components, the antimicrobial and cytotoxic action of immune system cells, neutrophils and macrophages, as well as in aging and age-related degenerative diseases. Overwhelming evidence indicates that oxidative stress can lead to cell and tissue injury. However, the same free radicals that are generated during oxidative stress are produced during normal metabolism and, as a corollary, are involved in both human health and disease.

In addition to reactive oxygen species, research on reactive nitrogen species has been gathering momentum to develop an area of enormous importance in biology and medicine. Nitric oxide or nitrogen monoxide (NO) is a free radical generated by nitric oxide synthase (NOS). This enzyme modulates physiological responses in the circulation such as vasodilation (eNOS) or signaling in the brain (nNOS). However, during inflammation, a third isoenzyme is induced, iNOS, resulting in the overproduction of NO and causing damage to targeted infectious organisms and to healthy tissues in the vicinity. More worrisome, however, is the fact that NO can react with superoxide anion to yield a strong oxidant, peroxynitrite. Oxidation of lipids, proteins, and DNA by peroxynitrite increases the likelihood of tissue injury.

Both reactive oxygen and nitrogen species are involved in the redox regulation of cell functions. Oxidative stress is increasingly viewed as a major upstream component in the signaling cascade involved in inflammatory responses and stimulation of adhesion molecule and chemoattractant production. Hydrogen peroxide decomposes in the presence of transition metals to the highly reactive hydroxyl radical, which by two major reactions — hydrogen abstraction and addition — accounts for most of the oxidative damage to proteins, lipids, sugars, and nucleic acids. Hydrogen peroxide is also an important signaling molecule that, among others, can activate NF-kB, an important transcription factor involved in inflammatory responses. At low concentrations, hydrogen peroxide regulates cell signaling and stimulates cell proliferation; at higher concentrations it triggers apoptosis and, at even higher levels, necrosis.

Virtually all diseases thus far examined involve free radicals. In most cases, free radicals are secondary to the disease process, but in some instances free radicals are causal. Thus, there is a delicate balance between oxidants and antioxidants in health and disease. Their proper balance is essential for ensuring healthy aging.

The term oxidative stress indicates that the antioxidant status of cells and tissues is altered by exposure to oxidants. The redox status is thus dependent on the degree to which a cell's components are in the oxidized state. In general, the reducing environment inside cells helps to prevent oxidative damage. In this reducing environment, disulfide bonds (S–S) do not spontaneously form because sulfhydryl groups are maintained in the reduced state (SH), thus preventing protein misfolding or aggregation. This reducing environment is maintained by oxidative metabolism and by the action of antioxidant enzymes and substances, such as glutathione, thioredoxin, vitamins E and C, and enzymes such as superoxide dismutases, catalase, and the selenium-dependent glutathione reductase and glutathione and thioredoxin hydroperoxidases, which serve to remove reactive oxygen species (hydroperoxides).

Changes in the redox status and depletion of antioxidants occur during oxidative stress. The thiol redox status is a useful index of oxidative stress mainly because metabolism and NADPH-dependent enzymes maintain cell glutathione (GSH) almost completely in its reduced state. Oxidized glutathione (glutathione disulfide, GSSG) accumulates under conditions of oxidant exposure and this changes the ratio GSSG/GSH; an increased ratio is usually taken as indicating oxidative stress. Other oxidative stress indicators are ratios of redox couples such as NADPH/NADP, NADH/NAD, thioredoxin$_{reduced}$/thioredoxin$_{oxidized}$, dihydrolipoic acid/α-lipoic acid, and lactate/pyruvate. Changes in these ratios affects the energy status of the cell, largely determined by the ratio ATP/ADP + AMP. Many tissues contain large amounts of glutathione, 2–4 mM in erythrocytes or neural tissues and up to 8 mM in hepatic tissues. Reactive oxygen and nitrogen species can oxidize glutathione, thus lowering the levels of the most abundant nonprotein thiol, sometimes designated as the cell's primary preventative antioxidant.

Current hypotheses favor the idea that lowering oxidative stress can have a health benefit. Free radicals can be overproduced or the natural antioxidant system defenses weakened, first resulting in oxidative stress, and then leading to oxidative injury and disease. Examples of this process include heart disease, cancer, and neurodegenerative disorders. Oxidation of human low-density lipoproteins is considered an early step in the progression and eventual development of atherosclerosis, thus leading to cardiovascular disease. Oxidative DNA damage may initiate carcinogenesis. Environmental sources of reactive oxygen species are also important in relation to oxidative stress and disease. A few examples: UV radiation, ozone, cigarette smoke, and others are significant sources of oxidative stress.

Compelling support for the involvement of free radicals in disease development originates from epidemiological studies showing that an enhanced antioxidant status is associated with reduced risk of several diseases. Vitamins C and E and prevention of cardiovascular disease are a notable example. Elevated antioxidant status is also associated with decreased incidence of cataracts, cancer, and neurodegenerative disorders. Some recent reports have suggested an inverse correlation between antioxidant status and the occurrence of rheumatoid arthritis and diabetes mellitus. Indeed, the number of indications in which antioxidants may be useful in the prevention and/or the treatment of disease is increasing.

Oxidative stress, rather than being the primary cause of disease, is more often a secondary complication in many disorders. Oxidative stress diseases include inflammatory bowel diseases, retinal ischemia, cardiovascular disease and restenosis, AIDS, adult respiratory distress syndrome, and neurodegenerative diseases such as stroke, Parkinson's disease, and Alzheimer's disease. Such indications may prove amenable to antioxidant treatment (in combination with conventional therapies) because there is a clear involvement of oxidative injury in these disorders.

In this series of books, the importance of oxidative stress and disease associated with organ systems of the body is highlighted by exploring the scientific evidence and the medical applications of this knowledge. The series also highlights the major natural antioxidant enzymes and antioxidant substances such as vitamins E, A, and C, flavonoids, polyphenols, carotenoids, lipoic acid, coenzyme Q_{10}, carnitine, and other micronutrients present in food and beverages. Oxidative stress is an underlying factor in health and disease. More and more evidence indicates that a proper balance between oxidants and antioxidants is involved in maintaining health and longevity and that altering this balance in favor of oxidants may result in pathophysiological responses causing functional disorders and disease. This series is intended for researchers in the basic biomedical sciences and clinicians. The potential of such knowledge for healthy aging and disease prevention warrants further knowledge about how oxidants and antioxidants modulate cell and tissue function.

Lester Packer

Enrique Cadenas

Preface

The mitochondria serve as the powerhouses of the cells. Interest in these organelles (especially their DNA) has blossomed over the last few decades. Mitochondrial diseases have been identified. At first, it was thought that these diseases were rare and that only a few people were affected. Then, as our knowledge expanded, we began to realize that several chronic diseases were associated with malfunctioning mitochondria. Several of these malfunctions were found to be nuclear in origin. Indeed, some of the degenerative chronic diseases were found to have an adverse effect on mitochondrial number, mitogenesis, and the base sequence of the mitochondrial DNA. This idea — that disease can affect the DNA and that mutated mitochondrial DNA can result in disease — has become a puzzle. Diabetes, in particular, can be found to be associated with DNA mutation and can cause mutation.

Mitochondria in Health and Disease is a collection of chapters written by experts in particular areas of mitochondrial investigations. It is hoped that you, the reader, will find it of sufficient breadth and depth to meet your needs to learn more about the function of this organelle and what happens when it malfunctions. I am grateful to the authors who have taken the time to share their expertise; in some cases, it was a difficult task, but each carried it to completion in an admirable way. Thank you.

Carolyn D. Berdanier

About the Editor

Carolyn D. Berdanier is a professor emerita of the University of Georgia. She earned her B.S. from Pennsylvania State University and her M.S. and Ph.D. in nutrition/biochemistry from Rutgers University. After a postdoctoral year with Dr. Paul Griminger, she served as a research nutritionist with the Human Nutrition Research division of the Agricultural Research Service, USDA, and concurrently held a faculty position with the University of Maryland. After 7 years, Dr. Berdanier moved to the College of Medicine of the University of Nebraska in Omaha, where she again was a full-time researcher with some teaching assignments. Later, she moved to the University of Georgia, where she headed the Department of Nutrition for 11 years, followed by 11 years as a teacher and researcher in the genetic aspects of nutritional response in diabetes. Her research was supported by grants from NIH, USDA, several commodity groups, and USDC Sea Grant Program.

Dr. Berdanier is a member of the American Society for Nutrition Science, The American Physiology Society, The American Diabetes Association, and the Society of Experimental Biology and Medicine. She has served on a number of editorial boards, including those of the *Journal of Nutrition*, the *FASEB Journal*, *Nutrition Research*, and *Biochemistry Archives*. She has published 150 research articles, authored/edited 12 books, and authored 29 invited reviews in peer-reviewed scientific publications. Dr. Berdanier has also authored 40 book chapters and a variety of other short reviews and articles for lay readers.

Contributors

Carolyn D. Berdanier
University of Georgia
Watkinsville, Georgia

Deepa Bhatt
Department of Biochemistry
  and Molecular Biology
University of Florida
Gainesville, Florida

Brian D. Cain
Department of Biochemistry
  and Molecular Biology
University of Florida
Gainesville, Florida

Tammy Bohannon Grabar
Department of Biochemistry
  and Molecular Biology
University of Florida
Gainesville, Florida

Moon-Jeong Chang Kim
Department of Nutrition
Kook-Min University
Seoul, Korea

Hong Kyu Lee
Department of Internal
  Medicine
Institute of Endocrinology,
  Nutrition, and Metabolism
Seoul National University
  College of Medicine
Seoul, Korea

Hsin-Chen Lee
Department of Pharmacology
School of Medicine
National Yang-Ming University
Taipei, Taiwan, Republic of
  China

Carl A. Pinkert
Department of Pathology and
  Laboratory Medicine
Center for Aging and
  Developmental Biology
University of Rochester Medical
  Center
Rochester, New York

John M. Shoffner
Horizon Molecular Medicine
  and
Georgia State University
Atlanta, Georgia

Jan-Willem Taanman
University Department of
  Clinical Neurosciences
Royal Free and University
  College Medical School
University College London
London, United Kingdom

Nobuakira Takeda
Department of General
  Medicine
Aoto Hospital, Jikei University
  School of Medicine
Tokyo, Japan

Ian A. Trounce
Centre for Neuroscience
University of Melbourne
Victoria, Australia

Richard L. Veech
Laboratory of Metabolic
  Control, NIAAA
Bethesda, Maryland

Yau-Huei Wei
Department of Biochemistry
  and Molecular Biology
School of Medicine
National Yang-Ming University
Taipei, Taiwan, Republic of
  China

Siôn Llewelyn Williams
University Department of
  Clinical Neurosciences
Royal Free and University
  College Medical School
University College London
London, United Kingdom

Contents

Chapter 13
BHE/Cdb Rats as Tools to Study Mitochondrial
Carolyn D. Berdanier and Moon-Jeong Chang Kim

1

Introduction to Mitochondria

CAROLYN D. BERDANIER

CONTENTS

INTRODUCTION

The mitochondria are organelles suspended in the cytoplasm that are the powerhouses or power plants of the cell. It is this organelle that traps the energy released by oxidative reactions in the high-energy bond of ATP. This trapping occurs through the process of coupling the synthesis of water via the respiratory chain to the synthesis of ATP via the $F_1F_0ATPase$ in a process called oxidative phosphorylation (OXPHOS). The mitochondria have other functions as well. Fatty acid oxidation and the citric acid cycle occur here and gluconeogenesis and urea synthesis begin in this compartment. The mitochondria play a role in apoptosis and have the components for the synthesis of a few proteins of OXPHOS.

The mitochondria serve as the central integrators of intermediary metabolism. Redox and phosphorylation states are managed here. The energy "coinage" (high-energy phosphate bonds) of the cell is manufactured here and, through this manufacture, the cell is able to conserve energy and store it for use in times of need. Throughout metabolism, the flux through many of the pathways is determined by the ratio of ATP to ADP, the availability of ATP, and the ratio of reduced to oxidized metabolic intermediates. It is the purpose of this chapter to provide the framework for subsequent chapters designed to describe mitochondrial function as well as specific aspects of mitochondrial DNA in health and disease.

ANATOMY AND PHYSIOLOGY OF THE MITOCHONDRION

Structural Components

Morphologically, the mitochondria are remarkably similar between species in terms of structure and function (1). Cells vary in the number of mitochondria in the cytoplasm and in the shape of these organelles. The liver cell contains ~800 mitochondria, while bone cells contain fewer than 400 and erythrocytes contain none. In contrast, the ovum contains 2000–20000. Mitochondria are distributed throughout the cytoplasm in a nonrandom manner by cytoskeleton motors. High-energy dependent cells tend to have their mitochondria located close to where energy is needed. For example, muscle cells have mitochondria located in close proximity to the contracting muscle fibers. Muscle contraction requires considerable energy provided by creatine phosphate and ATP. Both of these high-energy compounds are synthesized in the mitochondrial compartment. Cells that are rapidly dividing have mitochondria close to the nucleus and to ribosomes because *de novo* protein synthesis is highly dependent on the energy provided by ATP. Nerve conduction is dependent on ATP; thus, nerve tissue has many mitochondria located close to where conduction takes place. Any shortfall in ATP production and availability will affect the function of the central and peripheral nervous systems. Sensory perception, movement, and the responses to a variety of stimuli will be affected.

The typical mitochondrion is about 0.5 μm in diameter and from 0.5 μm to several microns long. A typical mitochondrion is about the size of a typical bacterium. The shape of the mitochondrion varies with the cell type and with the metabolic activity of the cell. In the liver, the typical mitochondrion is oblong or sausage shaped; however, spindle or rod shapes can also be found in this cell type. The mitochondria in the adipose cells are generally round, but other shapes can be found as well.

Table 1.1 Components of Mitochondria and Their Functions

Component	Function
Outer membrane	Oxidation of neuroactive aromatic amines
	Cardiolipin synthesis
	Import of cytoplasmic synthesized proteins
	Ion transfer
Intermembrane space	Maintenance of adenine nucleotide balance
	Creatine kinase and adenylate kinase located here
	Electron transfer from complex III to complex IV of the respiratory chain
	Processing of proteins imported into the mitochondrion
Inner membrane	Oxidative phosphorylation (OXPHOS)
	Transport of pyridine nucleotides
	Calcium ion transport
	Transport of metabolites (pyruvate, $H_2PO_4^-/OH^-$, dicarboxylates, citrate/malate, carnitine/acylcarnitine)
Matrix	Oxidation of pyruvate to acetyl CoA (pyruvate dehydrogenase complex)
	Oxidation of ketone bodies
	Oxidation of amino acids
	Initiation of the urea cycle
	Fatty acid oxidation, citric acid cycle occurs
	Suppress free radical damage through Mn-SOD
	Process imported proteins
	Synthesize 13 components of OXPHOS
	Heme synthesis

The mitochondrion consists of an outer membrane, an inner membrane, a space between the two membranes, and a matrix within the inner membrane. Each of these components has a specific yet different function, as listed in Table 1.1. It is possible to fractionate the cell and isolate the organelles by differential centrifugation so as to study their function. Furthermore, the mitochondria can be subfractionated so as to study the individual components and their associated functions (see Appendix 1 for methods). Enzymes typical of the matrix, the inner membrane, the intermembrane space, or the outer membrane can be isolated and studied.

Table 1.2 lists some of the enzymes found in each of the mitochondrial components. For example, all of the enzymes of the citric acid cycle and for fatty acid oxidation can be found in the mitochondrial matrix. The intermembrane space contains several enzymes that use ATP, including creatine kinase and adenylate kinase. The F_1F_0ATPase has its base (the F_0 portion) embedded in the inner membrane and the head protrudes out into the matrix. The flux through the citric acid cycle or the fatty acid oxidation pathway or the F_1F_0ATPase can be studied in detail. The activity of specific enzymes as well as the changes in substrate levels of these segments of metabolism can be determined.

The matrix also contains a small genome (~16.4 kilobases in mammals) and the machinery necessary to transcribe and translate the 13 OXPHOS proteins encoded by this DNA. The size of the genome is species specific. Shown in Table 1.3 are some species and their mitochondrial DNA size. Regardless of size, the map of the genome is fairly consistent from species to species. That is, within the genome the location of each of the structural genes of the ribosomes and each of the tRNAs are found in roughly the same place. Most of the proteins of OXPHOS are encoded by the nuclear genome, synthesized in the cytoplasm, and imported into the mitochondria for insertion into their appropriate locations within the OXPHOS system. Chapter 3 of this volume describes this DNA in detail — its transcription, translation, and role in mitochondrial protein synthesis and its replication. Subsequent chapters address specific disorders associated with mitochondrial DNA mutation.

Membranes

The composition and shape of the inner and outer membranes differ significantly. Although these membranes are composed primarily of protein and phospholipid, the proteins and phospholipids differ — particularly the composition of the fatty acids in the phospholipids. The phospholipid fatty acids of the

Table 1.2 Enzymes in the Mitochondria

E.C. #	Enzyme	Comment
	Outer Membrane	
2.7.1.1	Hexokinase	Component of the glycolytic sequence attached to the outer surface of this membrane
2.7.1.11	Phosphofructokinase	Component of the glycolytic sequence attached to the outer surface of this membrane
1.2.1.12	Glyceraldehyde-phosphate dehydrogenase	Component of the glycolytic sequence attached to the outer surface of this membrane
2.7.1.40	Pyruvate kinase	Component of the glycolytic sequence attached to the outer surface of this membrane
1.1.1.27	L-lactate dehydrogenase	Component of the glycolytic sequence attached to the outer surface of this membrane
6.2.1.3	Fatty acyl CoA synthetase (ATP)	Synthesizes long chain fatty acid CoA
1.4.3.4	Monoamine oxidase	
1.14.13.9	Kyneurenine hydroxylase (mono-oxygenase)	
2.7.8.2	Choline phosphotransferase	Enzyme in phospholipid synthesis
2.3.1.15	Glycerolphosphate acyl transferase	Enzyme in phospholipid synthesis
2.7.7.41	CDP-diglyceride pyrophosphorylase	Enzyme in phospholipid synthesis
2.7.8.5	Glycerolphosphate phosphatidyl transferase	Enzyme in phospholipid synthesis
3.1.3.27	Phosphatidylglycero-phosphatase	Enzyme in phospholipid synthesis
3.1.3.4	Phosphatidate phosphatase	Enzyme in phospholipid synthesis
1.6.2.2	Cytochrome b_5 reductase	
	Intermembrane Space	
1.8.2.1	Sulfite dehydrogenase	
2.7.4.6	Nucleoside diphosphokinase	
2.7.4.3	Adenylate kinase	
4.6.1.1	Adenylate cyclase	
2.7.3.2	Creatine kinase	

(continued)

Table 1.2 (continued)   Enzymes in the Mitochondria

E.C. #	Enzyme	Comment
	Inner Membrane	
2.3.1.21	Carnitine acyltransferase	Part of fatty acid oxidation
1.3.99.3	Acyl CoA dehydrogenase (long chain)	Part of fatty acid oxidation
1.3.99.2	Acyl CoA dehydrogenase (short chain)	Part of fatty acid oxidation
4.2.1.17	Enoyl CoA hydratase	Part of fatty acid oxidation
1.1.1.35	L-Hydroxyacyl CoA dehydrogenase	Part of fatty acid oxidation
2.3.1.16	3-Oxoacyl CoA thiolase	Part of fatty acid oxidation
5.3.3.8	*cis*-3,*trans*-2-Enoyl CoA isomerase	Part of fatty acid oxidation
5.1.2.3	3-Hydroxyacyl CoA epimerase	Part of fatty acid oxidation
1.3.99.1	Succinate dehydrogenase	Part of the citric acid cycle
1.1.1.30	Hydroxybutyrate dehydrogenase	Important in ketone body metabolism
1.6.99.3	NADH dehydrogenase	
1.6.1.1	NAD(P) transhydrogenase	
1.6.2.4	NADPH-cytochrome reductase	
1.1.99.5	Glycerol phosphate dehydrogenase	
1.5.99.1	Sarcosine dehydrogenase	
1.5.99.2	Dimethylglycine dehydrogenase	
1.3.99.2	Acyl CoA dehydrogenase	Part of fatty acid oxidation
1.1.99.1	Choline dehydrogenase	
1.1.2.3	Lactate dehydrogenase (cytochrome b_2)	
1.10.2.2	Ubiquinone-cytochrome c reductase	
1.9.3	Cytochrome oxidase	
3.6.1.3	F_1F_0ATP synthetase	
3.6.1.1	Pyrophosphatase	
2.1.2.10	Glycine synthase	
1.14.15.4	Steroid 11β-mono-oxygenase	
4.99.1.1	Ferrochelatase	

(continued)

Table 1.2 (continued) Enzymes in the Mitochondria

E.C. #	Enzyme	Comment
	Matrix	
2.7.1.99	Pyruvate dehydrogenase kinase	Part of the citric acid cycle
3.1.3.43	Pyruvate dehydrogenase phosphatase	Part of the citric acid cycle
4.1.3.7	Citrate synthase	Part of the citric acid cycle
4.2.1.3	Aconitase	Part of the citric acid cycle
1.1.1.41	Isocitrate dehydrogenase (NAD)	Part of the citric acid cycle
1.1.1.42	Isocitrate dehydrogenase (NADP)	Part of the citric acid cycle
1.2.4.2	Oxoglutarate dehydrogenase complex	Part of the citric acid cycle
2.3.1.61	Oxoglutarate dehydrogenase complex	Part of the citric acid cycle
1.6.4.3	Oxoglutarate dehydrogenase complex	Part of the citric acid cycle
6.2.1.4	Succinate thiokinase (GTP)	Part of the citric acid cycle
6.2.1.5	Succinate thiokinase (ATP)	Part of the citric acid cycle
4.2.1.2	Fumarase (fumerate hydratase)	Part of the citric acid cycle
1.1.1.37	Malate dehydrogenase	Part of the citric acid cycle
1.1.1.40	Malic enzyme (NADP) (oxaloacetate decarboxylating)	
6.4.1.1	Pyruvate carboxylate (ATP)	
4.1.1.32	Phosphoenolpyruvate carboxykinase	Location in the matrix is species specific
6.2.1.2	Fatty acyl CoA synthetase (ATP)	Activates medium chain fatty acids
6.2.1.10	Fatty acyl CoA synthetase (GDP)	Important in fatty acid activation
6.2.1.1	Acetyl CoA synthetase	Important in fatty acid activation
6.4.1.3	Propionyl CoA carboxylase	Important to propionyl CoA metabolism
5.1.99.1	Methylmalonyl CoA racemase	Important to propionyl CoA metabolism
5.4.99.2	Methylmalonyl CoA mutase	Important to propionyl CoA metabolism

(continued)

Table 1.2 (continued) Enzymes in the Mitochondria

E.C. #	Enzyme	Comment
2.3.1.9	Acetoacetyl CoA thiolase	Important in ketone metabolism
4.1.3.5	Hydroxymethylglutaryl CoA synthase	Important in ketone metabolism
4.1.3.4	Hydroxymethylglutaryl CoA lyase	Important in ketone metabolism
2.8.3.5	Succinyl CoA transferase	Not in liver mitochondrial matrix
1.3.1.8	Enoyl CoA reductase	Important to fatty acid elongation
1.4.1.3	Glutamate dehydrogenase	Important to amino acid metabolism
2.6.1.1	Aspartate transaminase	
3.5.1.2	Glutaminase	
6.3.4.16	Carbamyl phosphate synthetase	First committed step in urea synthesis
1.5.1.12	Pyrroline carboxylate dehydrogenase	
2.1.2.1	Serine hydroxymethyl transferase	
2.3.1.37	δ-Aminolaevulinate synthase	Important to heme synthesis
2.7.7.7	DNA polymerase	Important to mitochondrial protein synthesis
3.1.21.1	Deoxyribonuclease	Important to mitochondrial protein synthesis
2.7.7.6	RNA polymerase	Important to mitochondrial protein synthesis
1.15.1.1	Superoxide dismutase	Important in free radical suppression

inner membrane are highly unsaturated. Cardiolipin is abundant in the inner mitochondrial membrane and less so in the outer membrane. The inner membrane contains 19% of its phospholipids as cardiolipin and the outer membrane contains 4%. Cardiolipin is used as a marker for the identification of the mitochondrial membrane fraction when the membranes are studied. The inner membrane has very little (5.1 µg/mg protein) cholesterol. The outer membrane has some (30.1 µg/mg protein) cholesterol, but not as much as the plasma membrane.

Table 1.3 Mitochondrial DNA in a Variety of Species

Species	No. Bases	Remarks	Ref.
Yeast	~50 kb	Some may be linear	2
Red alga	25.8 kb	51 genes	3
Amoeboid protozoan	41.6 kb	Contains open reading frames	4
Pythuim (fungus)	50–150 kb	Linear and circular forms	5
Xenopus laevis	17.553 kb		6
Japanese pond frog	18.4–19.1 kb		7,8
Honey bee		Paternal DNA present	9
Sheep	~16.5 kb	Wild and domestic sheep differ in control region length	10
Cows	16.338 kb	Many polymorphisms reported	11–13
Horse	~16.5	Very diverse	14,15
Donkey	16.67 kb	Length can vary	16
Rat	16.298		17
Gibbon	16.472 kb	Four genes lack complete stop codons	18
Man	16.569 kb	Polymorphisms are common	19,20

The plasma membrane has no cardiolipin but contains phosphatidylserine; the mitochondrial membranes have none of this phospholipid. Phosphatidylcholine (42% in the inner, 54% in the outer membrane) and phosphatidylethanolamine (32% in the inner and 26% in the outer membrane) are found in both mitochondrial membranes. The inner membrane contains 4% of its phospholipids as phosphatidylinositol and the outer membrane contains 14%. The inner membrane contains ubiquinone and the outer membrane has none. Enzymes within the mitochondrial compartment play a role in the *in situ* production of some of these phospholipids (see Table 1.2). However, some of the phospholipids are synthesized outside the mitochondrion and imported (22).

The outer membrane surrounds the organelle and serves to maintain its shape (23). It is smooth and contains about 30 to 40% lipid and 60 to 70% protein. The protein in the outer membrane is rich in porin-a protein that contains many β-sheets. These sheets form large channels across the membrane, thus explaining its relative permeability. The channels permit the free diffusion of molecules of less than 10,000 in molecular weight. The channels are voltage-dependent anion channels (VDAC) permeable to solutes up to 5 kDa. Small molecules such as ADP or ATP pass easily through the outer membrane, as do certain nonelectrolytes such as inulin. Ion movement through the outer membrane depends largely on the charge of the ion. NAD and NADH+ cannot pass into the mitochondria through these membranes.

The inner membrane has many folds into the matrix that serve to increase its surface area for the metabolic activities that take place there. The wrinkles of the inner membrane form lamellar structures called cristae. On the inner membrane (facing into the matrix) are numerous small spheres on stalks, which are the structures associated with the multi-subunit ATP synthetase or the $F_1F_0ATPase$. The inner membrane is permeable only to water, oxygen, and carbon dioxide and is impermeable to a wide variety of molecules and ions that, in turn, require specific transporters for entry and exit. A number of transport systems and enzymes are in the inner membrane, thus accounting for the fact that this membrane consists of ~80% by weight protein and only 20% lipid.

Located within the inner membrane are the many proteins that together comprise the respiratory chain. One of these, cytochrome c, in addition to being important to the synthesis of water via the respiratory chain, also plays an important role in apoptosis. In this role it is part of the apoptosis-inducing factor (AIF). About 90% of cytochrome c is stored in vesicles created by the in-foldings or wrinkles of the inner mitochondrial membrane. The remaining 10% is located in the intermembrane space. Cytochrome c functions in respiration and also as an electron transfer protein in many different redox reactions.

As described earlier, the inner and outer membranes have a large percentage of their structure as lipid. The lipids in these membranes are phospholipids that form lipid bilayers. The phospholipids are oriented so that their hydrophobic portions, the fatty acids at positions 1 and 2 of the phospholipid, face into the center of the membrane, thus leaving the

Intermembrane space

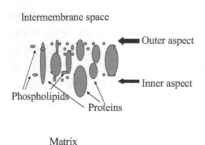

Matrix

Figure 1.1 Schematic representation of the inner mitochondrial membrane. The phospholipids are represented as having the hydrophilic head group on the outside (small circle) with the hydrophobic fatty acid tails extending in towards the center of the bilayer. Proteins are shown in several positions within the bilayer.

hydrophilic portion (the phosphorylated substituent) of their structure facing outward. Embedded in these lipid bilayers are the proteins that characterize each of the membranes.

Figure 1.1 is a theoretical representation of these membranes. Note that some of the proteins are embedded in the lipid bilayer so that they extend through it; other proteins are located to face only one side of the membrane. For example, a few of the enzymes of glycolysis are attached to the outer aspect of the outer membrane. Although glycolysis is usually thought of as a cytosolic pathway, some of the enzymes are in fact associated with the mitochondria. Another example is the enzymes of fatty acid oxidation embedded in the inner membrane but oriented so that they protrude into the matrix. The functions of these proteins can be related to their position within and on the membranes that house them.

Diet Effects on Membrane Lipid Composition and Structure

The fatty acid composition of the membrane phospholipids can be influenced by diet (23–30) and hormonal state (31–38). In turn, these differences in composition and fluidity can affect mitochondrial function. Some examples of these influences are shown in Table 1.4 through Table 1.6. Diets rich in

Table 1.4 Diet Effects on Selected Fatty Acids in Mitochondrial Membranes

Treatment	Fatty Acids Mol%					Ref.
	16:0	18:0	18:1	18:2	20:4	
65% sucrose–5% corn oil	20.1 ± 0.7	20.9 ± 0.9	12.7 ± 1.6	20.8 ± 0.7	25.3 ± 0.6	23
65% starch–5% corn oil	19.5 ± 0.5	19.9 ± 0.5	10.3 ± 0.8	24.6 ± 0.4	25.7 ± 0.4	23
65% starch–5% coconut oil	15.4 ± 0.3	22.0 ± 0.4	12.0 ± 0.5	8.0 ± 0.3	18.0 ± 0.7	24
Fat-free diet	16.4	7.0	23.5	4.1	10.5	25
60% sucrose–9% fish oil	30.3 ± 1	9.8 ± 1	27.8 ± 1.5	8.0 ± 0.7	2.3 ± 0.4	Unpub. observ.
60% sucrose–9% beef fat	26.2 ± 1.0	9.8 ± 1.1	42.6 ± 1.0	4.7 ± 0.3	5.7 ± 0.8	Unpub. observ.

Table 1.5 Effects of Hormones on Fatty Acid Profiles
of Isolated Liver Mitochondria[a]

Treatment	Fatty Acids (mol%)						Ref.
	16:0	18:0	18:1	18:2	20:4	22:6	
1 mg GC[a]/day	14.6	22.7	6.6	28.5	19.0	3.5	37
Control	17.0	20.3	7.6	15.1	27.4	1.5	
Thyroidectomy	12.5	21	8.6	19.5	19.0	8.1	33
Control	13.0	20	9.0	17.4	15.7	7.0	
Hypophysectomy	27.1	21.5	13.5	17.0	14.5	2.0	38
Control	27.4	22	12.2	12.1	17.4	2.8	
Diabetes[b]	24.5	23.3	7.0	22.4	15.2	4.2	31
Control	16.9	23.2	9.3	22.7	22.2	3.1	

[a] GC = synthetic glucocorticoid, dexamethasone.
[b] Streptozotocin-induced diabetes.

Table 1.6 Diet Effects on Selected Mitochondrial Functions

Diet	Effect	Ref.
5% coconut oil vs. corn oil	Increased ADP-stimulated malate–aspartate and α-glycerophosphate shuttles; increased state 4 respiration; decreased respiratory control ratio and the ADP:O ratio	24
5% fish oil vs. corn oil	Increased state 3 respiration; decreased transition temperature (more fluid membrane)	26
20% coconut oil vs. safflower oil	Increased desaturase activity	28
50% fish oil vs. vegetable oil	Increased acylcarnitine transferase system and increased fatty acid oxidation; increased membrane fluidity	28

polyunsaturated fatty acids (PUFAs) or saturated fatty acids influence the mitochondrial membranes so that these membranes reflect the fatty acids in the diet (Table 1.4). Diets rich in PUFAs result in membranes with more PUFAs and, as a result, these membranes are more fluid than membranes from animals fed a saturated fat diet or ones from animals fed an essential fatty acid free diet, i.e., a coconut oil diet or a fat free diet (24,25).

Feeding a high-sucrose diet to rats likewise influences these membranes: sucrose is used to support *de novo* fatty acid synthesis and these fatty acids are likely to be saturated ones. They are incorporated into the mitochondrial membranes, thus making these membranes less fluid than those in rats fed a similar amount of starch (23). Animals fed a fish oil diet have fatty acids in their membranes that reflect the fatty acids found in the fish oil (28,29). Beef tallow feeding likewise results in membranes with fatty acids reminiscent of the dietary fat (27,28). Whether the fat in the diet is high or low, the array of the fatty acids in the mitochondrial membranes will reflect the array in the diet, not the level of fat in the diet (27–31). Certain hormones also affect membrane composition and fluidity. Insulin treatment results in an increase in PUFAs (31,32), primarily because this hormone induces the activity of the desaturase enzymes (31).

Desaturation occurs in the endoplasmic reticulum and microsomes. The enzymes that catalyze this desaturation are the $\Delta 4$, $\Delta 5$, or $\Delta 6$ desaturases. Again, desaturation is species specific. Mammals, for example, lack the ability to desaturate fatty acids in the n6 or n3 position. Only plants can do this and even among plants species differences are found. Cold-water plants can desaturate at the n3 position; land plants of warmer regions cannot. The cold-water plants are consumed by cold-water creatures in a food chain that includes fish as well as sea mammals. In turn, these enter the human food supply and become sources of the n3 or omega-3 fatty acids in the marine oils.

In animals, desaturation of *de novo* synthesized fatty acids usually stops with the production of a monounsaturated fatty acid with the double bond in the 9 to 10 positions, counting from the carboxyl end of the molecule. Therefore, palmitic acid (16:0) becomes palmitoleic acid (16:1) and stearic acid (18:0) becomes oleic acid (18:1). In the absence of dietary EFA, most mammals will desaturate eicosenoic acid to produce eicosatrienoic acid. Increases in this fatty acid with unsaturations at $\omega 7$ and $\omega 9$ positions characterize the tissue lipids of EFA-deficient animals (25). They are sometimes called mixed function oxidases because two substrates (fatty acid and NADPH) are oxidized simultaneously. These desaturases prefer substrates with a double bond in the ω6 position, but will also act on fatty acids with a double bond in the ω3 position and on saturated fatty acids.

Desaturation of *de novo* synthesized stearic acid to form oleic acid results in the formation of a double bond at the ω9 position. This is the first committed step of this desaturation/elongation reaction sequence. Oleic acid can also be formed by the desaturation and elongation of palmitic acid. Fatty acid desaturation can be followed by elongation and repeated so that a variety of mono- and polyunsaturated fatty acids can be formed; these contribute fluidity to membranes because of their lower melting point. An increase in the activity of the desaturation pathway is a characteristic response of rats fed a diet high in saturated fatty acids. The body can convert the dietary saturated fatty acids to unsaturated fatty acids, thus maintaining an optimal P:S ratio in the tissues.

Thyroidectomy results in a decrease in mitochondrial levels of linoleic and arachidonic acids (18:2 and 20:4), primarily due to a reduction in fatty acid mobilization (34–36). With a decrease in mobilization, the PUFAs have a longer residence time in the mitochondrial membranes. Treatment with glucocorticoids results in an increase in linoleic and a decrease in arachidonic acids (37). Hypophysectomy, which results in a loss in growth hormone, results in an increase in linoleic acid and a reduction in arachidonic acid (38). Each of these perturbations in membrane lipid composition also results in changes in mitochondrial function — specifically, changes in OXPHOS. Table 1.5 shows the effects of hormones on mitochondrial membrane fatty acid profiles.

Many of the proteins that are part of the OXPHOS system are embedded in the inner mitochondrial membrane and are influenced by the fluidity of that membrane (24,27,28). Their activities can be increased if the membrane is very fluid or decreased if the membrane becomes more rigid. Fluidity can be evaluated by measuring the activity of a given reaction or series of reactions such as oxygen consumption with increases in temperature (39). Careful measurements of oxygen consumption at 3° increments in the temperature of the media surrounding the isolated mitochondria allow assessment of fluidity. By preparing an Arrhenius plot of oxygen consumption vs. temperature, the change in slope at the transition temperature can be visualized — that is, where the membrane changes from a "solid" to a "liquid." Actually, membranes are neither solid nor liquid but exist as a structure with regional differences in consistency based on the fatty acid composition of the local phospholipids and the proteins embedded in these phospholipids.

Fluidity in the outer membrane can be influenced by the ratio of the fatty acid unsaturation index to cholesterol. The fatty acid unsaturation index is defined as the sum of the unsaturated fatty acids (number of unsaturated fatty acids times the number of double bonds in these fatty acids). Because the inner membrane does not contain significant amounts of cholesterol, the fluidity of this membrane is not influenced by its cholesterol content but only by the ratio of unsaturated fatty acids to saturated fatty acids. The preparation of the Arrhenius plot that then describes membrane fluidity uses the equation:

$$d \ln k/dT = Ea/RT^2$$

where k is the rate constant; R is the gas constant (8.312 J/(mol•.K)); Ea, the activation energy; and T the temperature in degrees Kelvin.

Conversion to base-10 logarithms gives the equation:

$$\log[k_2/k_1] = Ea/2.303R \bullet (T_2 - T_1/T_1 T_2)$$

from which it can be seen that the value for Ea can be obtained from the slope of the straight line when the logarithm is plotted against the reciprocal of the absolute temperature.

The transition temperature or breakpoint can be seen as the point at which the two lines intersect. This is a theoretical fluidity value. A typical Arrhenius plot is shown in Figure 1.2 and a comparison of plots using mitochondrial membranes from rats fed diets differing in fat saturation is shown in Figure 1.3. The lines shown in Figure 1.3 are from groups of rats fed a 6% hydrogenated coconut oil or a 6% menhaden oil diet for 4 weeks (25). They were killed at 56 days of age, their

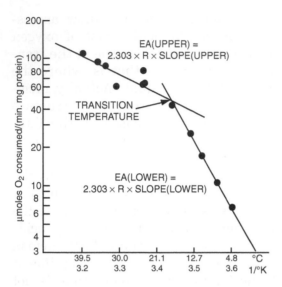

Figure 1.2 A typical Arrhenius plot of the log of the oxygen consumption (micromoles oxygen consumed/milligrams mitochondrial protein) vs. the reciprocal of the temperature in degrees Kelvin.

mitochondria isolated, and the dependence of oxygen consumption on temperature determined.

Note in Figure 1.3 that the transition temperature for the rats fed the menhaden oil diet is lower than that for the rats fed the coconut oil diet. Shown in Table 1.6 are diet-induced effects on several other processes that depend on membrane fluidity. The shuttling of reducing equivalents is one of these, the oxidation of fatty acids is another, and gluconeogenesis is a third.

Age Effects on Membrane Lipids

As animals age, their hormonal status changes, as does the lipid component of their membranes. With age, growth hormone production decreases, the hormones for reproduction increase and then decrease, and, as the animal ages, fat stores become larger. Larger fat cells are resistant to insulin and insulin levels may rise as a result of increased fat cell size

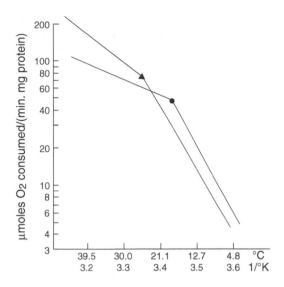

Figure 1.3 Different Arrhenius plots from hepatic mitochondria from rats fed a 6% fish oil (-Δ-) or 6% coconut oil (-•-) diet. (Adapted from M-J.C. Kim and C.D. Berdanier. *FASEB J.* 12: 243–248, 1998.)

(insulin resistance). As mentioned in the preceding section, these hormones can affect the lipid portion of the mitochondrial membranes and thus affect mitochondrial metabolism.

With age, the degree of unsaturation of the membrane fatty acids decreases (40–44) and the number of superoxide radicals increases (45,46). This increase may be responsible for the degradation of the membrane lipids, which, in turn, might explain the age-related changes in membrane function (44). Membranes from aging animals are less fluid and have reduced transport capacities. As animals age, the hepatic mitochondrial respiratory rate declines, the respiratory ratio and the ADP/O ratio (42,47,48) decrease, and the activities of the citric acid cycle enzymes and fatty acid oxidizing enzymes decline (49). In addition, there are reports of an age-related decrease in membrane fatty acid unsaturation coupled with a decrease in membrane fluidity and a decrease in ATP synthesis (44) and a decrease in the exchange of ATP for ADP across the mitochondrial membrane (50–53). A decrease in

ATP synthesis and an amelioration of these age-related decreases in mitochondrial function by restricted feeding (energy intake reduced by 50% over the lifetime of the animals) has also been reported (54,55). Changes in mitochondrial function also take place due to free radical damage, not only to the membranes but also to the mitochondrial DNA (46). Chapter 3 and Chapter 6 in this volume address this issue.

Membrane Function

In the previous section, the importance of diet, age, and hormonal status was described in terms of their influence on the composition of the membrane lipids. Although not emphasized, these compositional differences have important effects on metabolic regulation. This regulation consists of the control of the flux of nutrients, substrates, and/or products into, out of, and between the various compartments of the cell.

The mitochondrial membranes serve as the "gatekeepers" of the organelle. They regulate the influx and efflux of substrates and metabolic products produced or used by the compartment in the course of its metabolic activity. For example, the mitochondrial membrane through its transport of two- and four-carbon intermediates and through its exchange of ADP for ATP regulates the activity of the respiratory chain and ATP synthesis. If too little ADP enters the mitochondria because of decreased ADP transport across the mitochondrial membrane, respiratory chain activity will decrease, less ATP will be synthesized, and other mitochondrial reactions driven by ADP influx or dependent on ATP availability may decrease.

Through its export of citrate from the matrix of the mitochondria it regulates the availability of citrate to the cytosol for cleavage into oxaloacetate and acetyl CoA, the beginning of fatty acid synthesis. If more citrate is exported from the mitochondria than can be split to oxaloacetate and acetyl CoA, this citrate will feed back onto the phosphofructokinase reaction, and glycolysis will be inhibited. Thus, the activity of the mitochondrial membrane tricarboxylate transporter has a role in the control of cytosolic metabolism. Other transporters such as the dicarboxylate transporter or the adenine nucle- otide translocase have similar responsibilities vis-a-vis the control of cytosolic and mitochondrial metabolic activity. Such transporters are necessary because highly charged molecules such as ADP or ATP cannot readily traverse the mitochondrial membranes. Similarly, oxaloacetate, NAD, and NADH cannot traverse the membranes.

Closely packed proteins and lipids are in the mitochon- drial membranes. The membrane-bound proteins have exten- sive hydrophobic regions and usually require their nearby lipids for the maintenance of their activity. Adenylate cyclase, cytochrome b_5, and cytochrome c oxidase have all been shown to have phospholipid affinities. Cytochrome c oxidase has tightly bound aldehyde lipid that cannot be removed without destroying its activity as the enzyme that transfers electrons to molecular oxygen in the final step of respiration.

A number of other membrane proteins have tightly bound fatty acids as part of their structures. These fatty acids are covalently bound to their proteins as a posttranslational event and act to direct, insert, and anchor the proteins in the inner and outer membranes. Some lipids are bound to mem- brane proteins. Some proteins are acylated with fatty acids during their passage from their site of synthesis on the rough endoplasmic reticulum to the membrane, whereas others acquire their lipid component during their placement in the membrane. β-Hydroxybutyrate dehydrogenase, for example, requires the choline head of phosphatidylcholine for its activ- ity. If hepatocytes are caused to increase their synthesis of phosphatidyl-methylethanolamine, which substitutes for phosphatidylcholine in the mitochondrial membrane, β- hydroxybutyrate dehydrogenase activity is reduced (56).

All of these examples illustrate the importance of the mitochondrial membranes in the regulation of metabolism. They illustrate the fact that the gatekeeping property of the membrane is vested in the structure and function of the var- ious transporters and receptors or binding proteins embedded in the bilayer lipid membrane. Whereas the genetic heritage of an individual determines the amino acid sequence of the proteins and thus their function, this function can be modified by the lipid milieu in which they exist. Diet, hormonal state, and genetics in turn control the lipid milieu in terms of the kinds and amounts of the different lipids synthesized within the cell and incorporated into the membrane.

METABOLIC SYSTEMS IN MITOCHONDRIA

Fatty Acid Oxidation

Although fatty acids are important components of membranes, the main function of these molecules in the body is to provide energy to sustain life. This provision is accomplished through oxidation. Regardless of whether the fatty acids come from the diet or are mobilized from the tissue triacylglyceride store, the pathway for oxidation is the same once the triacylglyceride has been hydrolyzed. Fatty acid oxidation occurs primarily in the mitochondrial matrix.

The hydrolysis of stored lipid is catalyzed in a three-step process by one of the lipases specific to mono-, di-, or triacylglycerol. The liberated fatty acids bind to the cytosolic fatty acid binding protein and migrate through the cytoplasm to the outer mitochondrial membrane. At the outer membrane, they are activated by conversion to their CoA thioesters. This activation requires ATP and the enzyme acyl CoA synthase or thiokinase. The several thiokinases differ with respect to their specificity for the different fatty acids. The activation step depends on the release of energy from two high-energy phosphate bonds. ATP is hydrolyzed to AMP and two molecules of inorganic phosphate. Figure 1.4 shows the initial steps in the oxidation of fatty acids.

Once the fatty acid is activated, it is bound to carnitine with the release of CoA. The acyl carnitine is then translocated through both mitochondrial membranes into the mitochondrial matrix via the carnitine acylcarnitine translocase. As one molecule of acylcarnitine is passed into the matrix, one molecule of carnitine is translocated back to the cytosol and the acylcarnitine is converted back to acyl CoA. The acyl CoA can then enter the β-oxidation pathway shown in Figure 1.5.

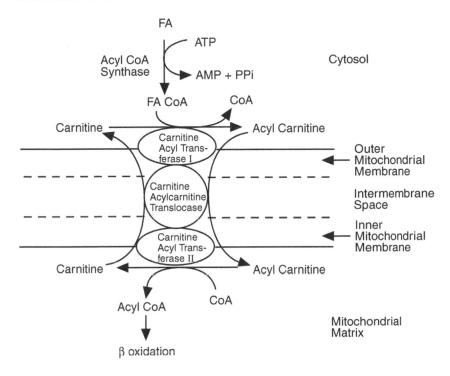

Figure 1.4 Initial steps in the oxidation of fatty acids by the mitochondrial enzymes.

Without carnitine, the oxidation of fatty acids, especially the long chain fatty acids, cannot proceed. Acyl CoA cannot traverse the membrane into the mitochondria by itself; the translocase requires carnitine. Normally, the body synthesizes all the carnitine that it needs for fatty acid oxidation. However, in some instances, endogenous synthesis is inadequate. Premature infants, for example, may require carnitine supplementation and some children with mitochondrial disease may benefit from carnitine supplementation. In the latter case, the benefit is due to the high use of fatty acids as metabolic fuel that, in turn, increases the need for carnitine above that which can be endogenously synthesized.

The oxidation of unsaturated fatty acids follows the same pathway as the saturated fatty acids until the double bonded

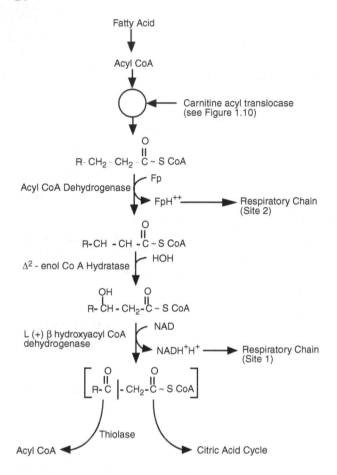

Figure 1.5 The β-oxidation pathway.

carbons are reached. At this point, a few side steps must be taken that involve a few additional enzymes. An example of this pathway using linoleate as the fatty acid being oxidized is shown in Figure 1.6.

Linoleate has two double bonds in the *cis* configuration. β-Oxidation removes three acetyl units, leaving a CoA attached to the terminal carbon just before the first *cis* double bond. At this point, an isomerase enzyme, Δ³ *cis* Δ⁶ *trans* enoyl CoA isomerase, acts to convert the first *cis* bond to a *trans*

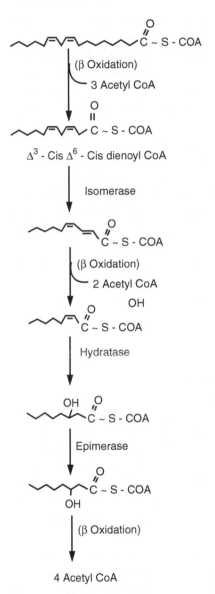

Figure 1.6 Oxidation of linoleic acid showing the opening up of the double bond and the insertion of a hydroxyl group.

bond. The bond is opened up and a hydroxyl group is inserted. Now this part of the molecule can once again enter the β-oxidation sequence and two more acetyl CoA units are released. The second double bond is then opened and, again, a hydroxyl group is inserted. In turn, this hydroxyl group is rotated to the L position and the remaining product then reenters the β-oxidation pathway.

Other unsaturated fatty acids can be similarly oxidized. Each time the double bond is approached, the isomerization and hydroxyl group addition takes place until all of the fatty acid is oxidized. The end products of fatty acid oxidation are acetyl CoAs plus the reducing equivalents released by the dehydrogenase catalyzed reactions in the β-oxidation sequence. The reducing equivalents are sent to the respiratory chain while the acetyl CoAs enter the citric acid cycle. Thus, the fatty acids are oxidized with the production of water by the respiratory chain and the production of carbon dioxide via the citric acid cycle.

Ketogenesis

Some of the acetyl CoA is converted to the ketones, acetoacetate, and β-hydroxybutyrate. The condensation of two molecules of acetyl CoA to acetoacetyl CoA occurs in the mitochondria via the enzyme β-ketothiolase. Acetoacetyl CoA then condenses with another acetyl CoA to form HMG CoA. This HMG CoA is cleaved into acetoacetic acid and acetyl CoA. The acetoacetic acid is reduced to β-hydroxybutyrate; this reduction depends on the ratio of NAD^+ to $NADH^{++}$. The enzyme for this reduction, β-hydroxybutyrate dehydrogenase, is tightly bound to the inner aspect of the mitochondrial membrane. Because of the high activity of the enzyme, the product (β-hydroxybutyrate) and substrate (acetoacetate) are in equilibrium. Measurements of these two compounds can be used to determine the redox state (ratio of oxidized to reduced NAD) of the mitochondrial compartment.

HMG CoA is also synthesized in the cytosol; however, because this compartment lacks the HMG CoA lyase, ketones are not produced here. In the cytosol, HMG CoA is the beginning substrate for cholesterol synthesis. The ketones produced in the mitochondria can ultimately be used as fuel but may appear in the blood, liver, and other tissues at a level of less than 0.2 mM. In starving individuals or people consuming a high-fat diet, blood and tissue ketone levels may rise to ~3 to 5 mM. However, unless ketone levels greatly exceed the body's capacity to use them as fuel (as is the case in uncontrolled diabetes mellitus with levels up to 20 mM), a rise in ketone levels is not a cause for concern. Ketones are choice metabolic fuels for muscle and brain. Although both tissues may prefer to use glucose, the ketones can be used when glucose is in short supply. Ketones are used to spare glucose whenever possible under these conditions.

Although β-oxidation is the main pathway for the oxidation of fatty acids, some fatty acids undergo α-oxidation so as to provide the substrates for the synthesis of sphingolipids. This oxidation occurs in the endoplasmic reticulum and in the mitochondria and involves the mixed function oxidases. The oxidases require molecular oxygen, reduced NAD, and specific cytochromes. The fatty acid oxidation that occurs in organelles other than the mitochondria (i.e., peroxisomes) is an energetically wasteful reaction. It produces heat but no ATP. The peroxisomes do not have the citric acid cycle nor do they have the OXPHOS system. Peroxisomal oxidation in the kidney and liver is an important aspect of drug metabolism.

Gluconeogenesis

The gluconeogenic pathway provides glucose to the body during starvation. It is a pathway that occurs in the cytosol but depends on the mitochondrial compartment for its starting substrate, oxaloacetate. The pathway is shown in Figure 1.7. Gluconeogenesis occurs primarily in the liver and kidney. Most tissues lack the full complement of enzymes needed to run this pathway. In particular, the enzyme phosphoenolpyruvate carboxykinase (PEPCK) is not found outside the liver and kidney. Shown in Figure 1.7 are the enzymes unique to gluconeogenesis. Some of these are found in many cell types, so some of the later steps in glucose release may be in all

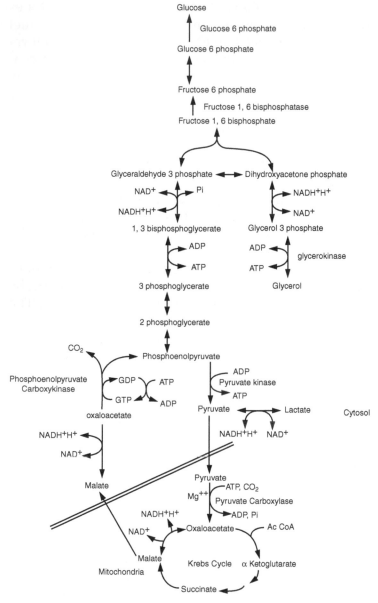

Figure 1.7 Gluconeogenesis. Note the initial step: the malate–aspartate shuttle provides reducing equivalents to the mitochondrial compartment via metabolite exchange across the mitochondrial membrane. The key enzyme in the gluconeogenic pathway is PEPCK. In some species (e.g., the guinea pig), this enzyme is located in the mitochondria.

tissues. The other reactions shown use the same enzymes as glycolysis and do not have control properties with respect to gluconeogenesis. The rate-limiting enzymes of interest are glucose 6-phosphatase, fructose 1,6 biphosphatase, and phosphoenolpyruvate carboxykinase. Pyruvate kinase and pyruvate carboxylase are also of interest because their control is a coordinated one with respect to the regulation of PEPCK.

The control of gluconeogenesis rests in part with the mitochondrial malate–aspartate shuttle (Figure 1.8). This shuttle works to transport reducing equivalents into the mitochondria. It is stimulated by the influx of ADP in exchange for ATP. Malate is transported into the mitochondria whereupon it gives up two reducing equivalents and is transformed into oxaloacetate. Oxaloacetate cannot traverse the mitochondrial membrane, so it is converted to α-ketoglutarate in a coupled reaction that also converts glutamate to aspartate. Aspartate travels out of the mitochondria (along with ATP) and in exchange for glutamate.

Once out in the cytosol, the reactions are reversed. Aspartate is reconverted to glutamate and α-ketoglutarate reconverted to oxaloacetate. In turn, the oxaloacetate can be reduced to malate or decarboxylated to form phosphoenolpyruvate. Measurement of the activity of this shuttle has revealed that, the more active the shuttle is, the more active is gluconeogenesis (58–60) because the shuttle provides a steady supply of oxaloacetate via α-ketoglutarate in the cytosol. This oxaloacetate cannot get there any other way. As mentioned, it is generated by the Krebs cycle in the mitochondria but cannot leave this compartment. The regeneration of oxaloacetate is thus the first step in gluconeogenesis and this involves the mitochondria.

Oxaloacetate is essential to gluconeogenesis because it is the substrate for PEPCK, which catalyzes its conversion to phosphoenolpyruvate (PEP). This is an energy-dependent conversion that overcomes the irreversible final glycolytic reaction catalyzed by pyruvate kinase. The activity of PEPCK is closely coupled with that of pyruvate carboxylase. Whereas the pyruvate kinase reaction produces one ATP, the formation of PEP uses two ATPs: one in the mitochondria for the

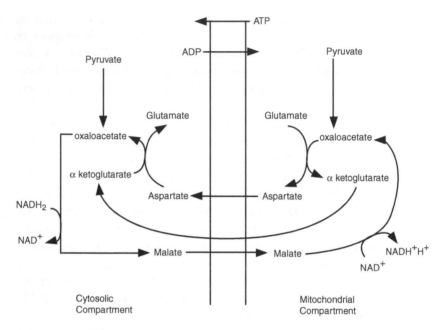

Figure 1.8 The malate–aspartate shuttle. The shuttle is stimulated by an influx of ADP.

pyruvate carboxylase reaction and one in the cytosol for the PEPCK reaction. PEPCK requires GTP provided via the nucleoside diphosphate kinase reaction, which uses ATP. ATP transfers one high-energy bond to GDP to form ADP and GTP.

Urea Cycle

The urea cycle consists of the synthesis of carbamyl phosphate and the synthesis of citrulline, then arginosuccinate, arginine, and finally urea (Figure 1.9). Ornithine and citrulline are shuttled back and forth as the cycle turns to get rid of the excess ammonia via urea release from arginine. The cycle functions to reduce the potentially toxic amounts of ammonia that arise when the ammonia group is removed from amino acids. Most of the ammonia released reflects the coupled action of the transaminases and L-glutamate dehydrogenase.

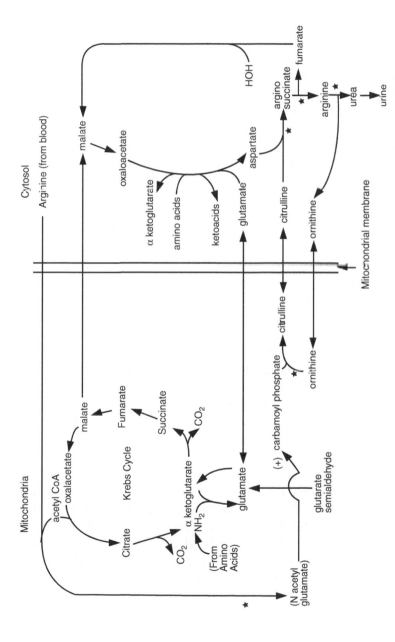

Figure 1.9 The urea cycle. This cycle has two important steps in the mitochondrial matrix. One is the synthesis of carbamyl phosphate and the other is the synthesis of citrulline.

The glutamate dehydrogenase is a bidirectional enzyme that plays a pivotal role in nitrogen metabolism. It is present in kidney, liver, and brain. It uses NAD⁺ or NADP⁺ as a reducing equivalent receiver and operates close to equilibrium, using ATP, GTP, NADH, and ADP depending on the direction of the reaction. In catabolism, it channels NH_3 from glutamate to urea. In anabolism, it channels ammonia to α-ketoglutarate to form glutamate. In the brain, glutamate can be decarboxylated to form γ-aminobutyrate (GABA), an important neurotransmitter. The decarboxylation is catalyzed by the enzyme L-glutamate decarboxylase. Putrescine also can serve as a precursor of GABA, by deamination or via N-acetylated intermediates.

The urea cycle is, energetically speaking, a very expensive process. The synthesis of urea requires 3 mol of ATP for every mole of urea formed. This cycle is very elastic — that is, its enzymes are highly conserved, readily activated, and readily deactivated. Adaptation to a new level of activity is quickly achieved. Urea cycle activity can be high when protein-rich diets are consumed and low when low-protein diets are followed; however, the cycle never shuts down completely. Shown in Figure 1.9, the cycle is fine-tuned by the first reaction, the synthesis of carbamyl phosphate. This reaction, which occurs in the mitochondria, is catalyzed by the enzyme carbamyl phosphate synthetase. The enzyme is inactive in the absence of its allosteric activator, N-acetylglutamate, a compound synthesized from acetyl CoA and glutamate in the liver. As arginine levels increase in the liver, N-acetylglutamate synthetase is activated, which results in an increase in N-acetylglutamate.

The urea cycle is initiated in the hepatic mitochondria and finished in the cytosol. The urea is then liberated from arginine via arginase and released into the circulation whereupon it is excreted from the kidneys in the urine. Ornithine, the other product of the arginase reaction, is recyled back to the mitochondrion only to be joined once again to carbamyl phosphate to make citrulline. Rising levels of arginine turn on mitochondrial N-acetylglutamate synthetase, which provides the N-acetylglutamate that, in turn, activates carbamyl phosphate synthetase, and the cycle goes on.

Shuttle Systems

Changes in the dietary or hormonal status of the individual result in large changes in the cytosolic pathways for glucose and fatty acid use as well as changes in protein turnover. All of these are orchestrated by changes in oxidative phosphorylation, which, in turn, are orchestrated by changes in cytosolic activity. The coordinated decreases and increases have in common a change in compartment redox state (ratio of oxidized to reduced metabolites) and phosphorylation (ratio of ATP to ADP) state. Reducing equivalents generated in the cytosol and carried by NAD+ or NADP+ must be transferred to the mitochondria (using metabolites) for use by the respiratory chain. Through transhydrogenation, reducing equivalents generated in the cytosol through NADP-linked enzymatic reactions are transferred to NAD. In turn, because NAD cannot traverse the mitochondrial membrane, these reducing equivalents must be carried on suitable metabolites into the mitochondrial compartment.

Several shuttle systems exist (57–59); the malate–aspartate shuttle (Figure 1.8) is thought to be the most important. The shuttle requires a stoichiometric influx of malate and glutamate and efflux of aspartate and α-ketoglutarate from the mitochondria. Alterations in the rate of efflux of α-ketoglutarate can significantly alter the shuttle activity in terms of the rate of cytosolic NADH utilization. α-Ketoglutarate efflux depends on mitochondrial ATP/ADP ratios and on the concentration of cytosolic malate. Shuttle activity is controlled by cytosolic ADP levels, malate levels, and availability of NADH.

Malate can also be exchanged for citrate (Figure 1.10) or for phosphate. The malate–citrate exchange involves transport of malate into the mitochondria, where it is converted to citrate via oxaloacetate. The citrate is then transported out of the mitochondria and the reactions reversed. This exchange is thought to be particularly active in lipogenic states because

Mitochondria

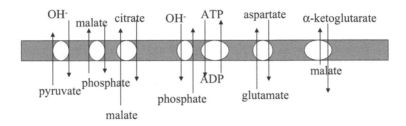

Cytosol

Figure 1.10 Shuttle systems in inner mitochondrial membrane. The monocarboxylate transporter exchanges pyruvate for a hydroxyl group. The dicarboxylate transporter exchanges phosphate for malate. The tricarboxylate transporter exchanges malate for citrate. The phosphate transporter exchanges phosphate for a hydroxyl group. The adenine nucleotide transporter exchanges ADP for ATP. The aspartate–glutamate transporter exchanges glutamate for aspartate and, finally, the malate–α-ketoglutarate transporter exchanges malate for α-ketoglutarate. These shuttles are needed because several metabolic intermediates cannot cross the mitochondrial membrane without a carrier and, if no carrier is available, they must be generated within the compartment. Included in this list are oxaloacetate, NAD, NADH, ATP, and ADP.

it provides citrate to the cytosol for citrate cleavage. Because the cleavage of citrate provides the starting acetate for fatty acid synthesis, this exchange plays a key role in lipogenesis.

As shown in Figure 1.10, other exchanges also take place to varying degrees. One can recognize the components of the malate shuttle among these exchange systems as well as identify the system for the exchange of adenine nucleotides. When the activity of the malate–aspartate is increased, the redox state of the cytosol is decreased as is lipogenesis; however, gluconeogenesis is increased. All of these exchanges of metabolites are related to each other via a coordinated control

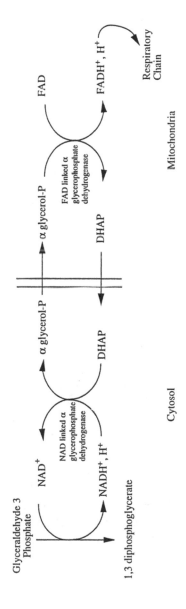

Figure 1.11 The α-glycerophosphate shuttle. This shuttle is located on the outer side of the inner membrane. It is rate controlled by the activity of the FAD-linked mitochondrial α-glycerophosphate dehydrogenase.

exerted by the phosphorylation state, the latter in turn controlled by ADP/ATP exchange.

Of importance to the control of glycolysis is the α-glycerophosphate shuttle (Figure 1.11) located at the outer side of the mitochondrial inner membrane. Transport of α-glycerophosphate and dihydroxyacetone phosphate across the mitochondrial inner membrane is not required. The activity of the shuttle is related to the availability of α-glycerophosphate and the activity of the mitochondrial matrix enzyme, α-glycerophosphate dehydrogenase. It is particularly responsive to thyroid hormone, which is thought to act by increasing the synthesis and activity of the mitochondrial α-glycerophosphate dehydrogenase.

Adenine nucleotide translocation or the exchange of adenine nucleotides across the mitochondrial membrane (Figure 1.10) also plays a role in the regulation of oxidative phosphorylation. Although any one of the adenine nucleotides could theoretically exchange for any other one, this does not happen. AMP movement is very slow (if it occurs at all). Under certain conditions, ADP influx into the mitochondria is many times faster than ATP efflux. This occurs when the need for ATP within the mitochondria exceeds that of the cytosol. An example of this can be seen in starvation, when ATP is needed to initiate urea synthesis and fatty acid oxidation and to support gluconeogenesis.

Under normal dietary conditions, however, the exchange of ADP for ATP is very nearly equal. For every 100 molecules of ADP that enter the mitochondrial compartment, about 87 molecules of ATP exit. The number of molecules of ATP produced by the coupling of the respiratory chain to ATP synthesis depends on the point at which the reducing equivalents enter the chain. If they enter carried by NAD, they are said to enter at site I of the respiratory chain. If they enter carried by FAD, they are said to enter at site II. Site I substrates are those oxidized via NAD-linked dehydrogenases. For example, pyruvate is a site I substrate and succinate is a site II substrate. Reducing equivalents entering at site I will generate the energy for the synthesis of three molecules of ATP.

Those entering at site II will generate enough energy to result in two molecules of ATP.

Knowing how a substrate provides its reducing equivalents to the respiratory chain allows one to estimate how many ATPs will be generated by a particular reaction sequence. Because each ATP has an energy value of about 30.5 kJ or 7.3 kcal/mol, one can also estimate the amount of energy that each sequence will produce or use.

The Citric Acid Cycle

Glucose, fatty acid, and amino acid oxidation result in the production of an activated two-carbon residue, acetyl CoA. Glucose is metabolized to pyruvate, which is converted to acetyl CoA via the pyruvate dehydrogenase complex or converted to oxaloacetate via pyruvate carboxylation (pyruvate carboxylase). Acetyl CoA is joined to oxaloacetate to form citrate and is then processed by the citric acid (also called the Krebs) cycle. This cyclic sequence of reactions is found in every cell type that contains mitochondria. The cycle is catalyzed by a series of enzymes and yields reducing equivalents (H^+) and two molecules of carbon dioxide.

The cycle is illustrated in Figure 1.12. Reducing equivalents and CO_2 are produced when α-ketoglutarate is produced from isocitrate, when ketoglutarate is decarboxylated to produce succinyl CoA. Reducing equivalents are also produced when succinate is converted to fumarate and when malate is converted to oxaloacetate. These reducing equivalents, one pair for each step, are carried to the respiratory chain by way of NAD or FAD. Reducing equivalents from succinate are carried by FAD and those produced by the oxidation of the other substrates are carried by NAD. Once oxaloacetate is formed, it can then pick up another acetate group from acetyl CoA and begin the cycle once again by forming citrate. Thus, for every turn of the cycle, 2 mol of CO_2 and eight pairs of reducing equivalents are produced. As long as oxaloacetate is sufficient to pick up the incoming acetate, the cycle will continue to turn and reducing equivalents will continue to be produced; in turn, these will

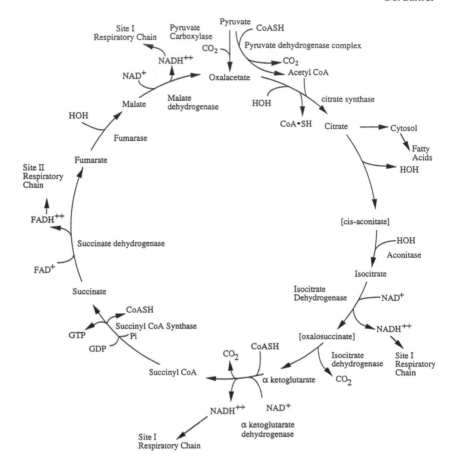

Figure 1.12   The citric acid cycle. This cycle reduces a six-carbon intermediate to a four-carbon one, resulting in the production of two moles of carbon dioxide and reducing equivalents that are sent to the respiratory chain.

be joined with molecular oxygen to produce water, the end product (with carbon dioxide) of the catabolic process.

The preceding simplistic description of the citric acid cycle implies that it is free of controls and proceeds unhindered given adequate supplies of substrates, enzymes, and molecular oxygen. This is not true. Numerous controls regulate the cycle. Among these are the citrate synthase, isocitrate dehydrogenase, and the α-ketoglutarate dehydrogenase reactions. These reactions have large negative ΔG values, meaning that they are essentially one-way reactions. In addition, the concentration of acetyl CoA, ATP, NAD+, and NADH exert regulatory power. Excess NADH, for example, inhibits the reactions catalyzed by pyruvate dehydrogenase, citrate synthase, isocitrate dehydrogenase, and α-ketoglutarate dehydrogenase. The cycle is enhanced by high ADP:ATP and NAD+:NADH ratios. Lastly, excess succinyl CoA can inhibit α-ketoglutarate dehydrogenase and citrate synthase.

In summary, the citric acid cycle produces the reducing equivalents (each with an electron pair) needed by the respiratory chain and the respiratory chain must transfer these hydrogen ions and their associated electrons to the oxygen ion to produce water. In doing so, it generates the electrochemical gradient (the proton gradient) necessary for the formation of the high-energy bonds of ATP. Obviously, then, all of these metabolic processes, the citric acid cycle, the respiratory chain, and ATP synthesis are regulated coordinately.

The Respiratory Chain

The pairs of electrons produced at the four steps described previously in the citric acid cycle as well as the electrons transferred into the mitochondria via other processes are passed down the respiratory chain to the ultimate acceptor, molecular oxygen (60). The respiratory chain enzymes embedded in the mitochondrial inner membrane are particularly complex and have a number of subunits. They catalyze a series of oxidation–reduction reactions. Each reaction is characterized by a redox potential that can be calculated using the Nernst equation:

$$E_h = E_0^1 + \frac{2.303\,RT}{nP} \log \, n\left(\frac{[\text{electron acceptor}]}{[\text{electron donor}]}\right)$$

where
E_h = observed potential

E_0^1 = standard redox potential (pH = 7.0, T = 25°, 1.0-M concentration)

R = gas constant (8.31 J deg^{-1} mol^{-1})

T = temperature (°K)

n = number of electrons transferred

P = Faraday (23,062 cal V^{-1} = 96, 406 J V^{-1})

The more positive the potential is, the greater is the affinity of the negatively charged acceptor for the positively charged electrons. It is this potential that drives the respiratory chain reactions forward towards the formation of water. Each succeeding acceptor donor pair has a higher affinity for the electrons than the preceding pair until the point in the chain at which the product, water, is formed. By comparison to the preceding pairs, water has little tendency to give up its electrons and unless this water is immediately removed, the chain reaction will stop. Of course, water does not accumulate in the mitochondria; it leaves as quickly as it is formed and thus does not feed back to inhibit the chain. If it did accumulate, it would change the concentration of the chain components diluting them. When water accumulates, the mitochondrion swells; if in excess, the mitochondrion would burst and die. Because this does not happen, obviously, water must exit the organelle as soon as it is formed so as to maintain optimal osmotic conditions.

Although the respiratory chain usually proceeds in the forward direction (towards the formation of water) due to the exergonic nature of the reaction cascade, except for the final reaction, all of these steps are fully reversible. In order to be reversed, sufficient energy must be provided to drive the reaction in this direction. For example, the reducing equivalents derived from succinate are usually carried by FAD (as FADH$^+$H$^+$). These can be transferred to NAD (as NADH^{++}H$^+$) with the concomitant hydrolysis of ATP. Electron transport across the other two phosphorylation sites can also be reversed, again, only if sufficient energy is provided.

The respiratory chain consists of four major enzyme complexes (Figure 1.13) located in the inner mitochondrial membrane. The enzymes of the respiratory chain are arranged so as to transport hydrogen ions from the matrix across the inner membrane. When this occurs, a proton gradient develops in close proximity to the F_1F_0ATPase (ATP synthase) complex and provides sufficient energy to drive ATP synthesis by causing a dehydration of ADP and Pi. Reducing equivalents transported into the mitochondrial compartment by the various substrate shuttle systems or generated in the compartment are passed down the respiratory chain in carefully regulated steps. Reducing equivalents enter the chain through the NAD–dehydrogenase complex (complex I) via mitochondrial shuttles or via the FAD–ubiquinone complex (complex II). With respect to the latter complex, reducing equivalents collect via three pathways:

- Succinate contributes its reducing equivalents to a flavoprotein with an iron–sulfur center.
- Glycerol 3-phosphate also uses FAD flavoprotein with an iron–sulfur center, but it is a different protein.
- The products of fatty acid oxidation (a mitochondrial process) are picked up by a FAD-flavoprotein, transferred to an electron-transferring flavoprotein, again with an iron–sulfur center, and then transferred to ubiquinone.

In mammals, complex I consists of about 42 subunits (61). Of these, seven are encoded by the mitochondrial genome and the rest by the nuclear genome synthesized on the ribosomes in the cytoplasm and imported into the mitochondrial compartment. It is the largest of the four respiratory chain complexes. Complex I is known as NADH-coenzyme Q reductase or NADH dehydrogenase. As the name implies, this complex transfers a pair of electrons from NADH to coenzyme Q, a lipid-soluble compound embedded in the inner membrane. Complex I has a molecule of flavin mononucleotide (FMN) and two binuclear iron–sulfur centers and four tetranuclear iron–sulfur centers (62). Because of its FMN, it is called a flavoprotein.

The complex catalyzes the transfer of electrons to complex III via ubiquinone and this transfer is coupled with the vectorial transfer of protons across the mitochondrial membrane. There are two distinct species of tightly bound ubiquinones in complex I that differ in spin relaxation and redox properties. The transfer of electrons leads to the formation of a proton gradient ($\Delta\mu_{H+}$) that, in turn, drives ATP production. The stoichiometry of proton transfer for complex I is 4H+/2e−. This distinguishes this complex from those that follow it in the respiratory chain. The other two H+ translocating complexes (III and IV) have a stoichiometry of 2 H+/2e−.

Complex II, succinate:quinone reductase (succinate dehydrogenase), is the smallest of the respiratory chain complexes (63). None of the subunits of complex II are encoded by the mitochondrial genome. The complex consists of four subunits with several different redox prosthetic groups: a covalently bound FAD, three iron–sulfur clusters, and a cytochrome b. The head of the complex protrudes out into the matrix where its FAD can accept succinate-donated electrons from the citric acid cycle. Actually, this enzyme is a component of the respiratory chain and the citric acid cycle. It too is a flavoprotein because of its FAD content. The FAD is bound to a histidine residue (64). When succinate is converted to fumarate in the citric acid cycle, a concomitant reduction of FAD to $FADH_2$ occurs. This $FADH_2$ transfers its electrons to the iron–sulfur cluster, which, in turn, passes them on to ubiquinone. Because of insufficient energy to elicit a proton gradient, reducing equivalents and associated electrons entering the respiratory chain via complex II yield only two ATPs via OXPHOS rather than the three ATPs generated when entry occurs via complex I.

Once reducing equivalents enter the chain via site 1 or site 2 (the sites correspond to entry via complex I or II), they are passed to complex III, the ubiquinone–cytochrome bc_1 reductase. This complex takes the electrons passed to it from ubiquinone and then passes them to complex IV, cytochrome c oxidase. This passage uses a unique redox pathway called the Q cycle (65). Three different cytochromes (bc, b, c_1) are involved as well as an iron–sulfur protein. The iron is in the middle of a porphorin ring much like that of hemoglobin and oscillates between the reduced and oxidized states (ferrous to ferric).

The Q cycle begins when a molecule of reduced ubiquinone diffuses to the Q_p site on complex III near the outer face of the inner mitochondrial membrane. An electron from the reduced ubiquinone is transferred to a mobile protein called the Rieske protein. The electrons are then transferred to cytochrome c_1. This releases two H+ and leaves UQ^-, a semiquinone anion form of ubiquinone, at the Q_p site. The second electron is then transferred to the b_L heme, converting UQ^- to ubiquinone. The Rieske protein and cytochrome c_1 are similar in structure; each has a globular domain and each is anchored to the inner membrane by a hydophobic segment. The segments differ: the Rieske protein has an N-terminal and the cytochrome c_1 has a C-terminal.

The electron on the b_L heme is passed to the b_H heme against a membrane potential of 0.15 V and is driven by the loss of redox potential as the electron moves from b_L to b_H. The electron is then passed from bH to ubiquinone at the second binding site, converting the ubiquinone to UQ^-. The resulting UQ^- remains firmly bound to the Q_n site. This completes the first half of the Q cycle. The second half is similar in that a second molecule of reduced ubiquinone is oxidized at the Q_p site. One electron is passed to cytochrome c_1 and the other is passed to heme b_H. The b_H electron is transferred to the semiquinone anion UQ^- at the Q_n site. With the addition of two H^+, this produces UQH_2. The UQH_2 is released and returns to the coenzyme Q pool and the Q cycle is complete. Actually, the Q cycle is an unbalanced proton pump. Cytochrome c is a mobile electron carrier, as is ubiquinone. Electrons travel from c to the water-soluble c_1. The c_1 associates loosely with the inner mitochondrial membrane to acquire electrons from the iron–sulfur centers. The c_1 of complex III then migrates along the membrane in a reduced state so as to give these electrons to complex IV.

Complex IV, cytochrome-c oxidase, contains two heme centers (a and a_3) as well as two copper centers. The copper oscillates between the reduced (cuprous) and oxidized (cupric) states. Complexes III and IV elicit a proton gradient and thus

ATP is formed at each of these sites. Complex IV accepts the electrons from cytochrome c and directs them to molecular oxygen to form water. The water thus formed quickly passes out of the compartment into the cytoplasm. As was true of complex I, complexes III and IV have nuclear- and mt-encoded subunits. Complex III has 11 subunits, of which 1 is encoded by the mt genome; complex IV has 13 subunits, of which 3 are mt encoded.

Oxidative Phosphorylation

Oxidative phosphorylation occurs in and on the inner membrane of the mitochondrial compartment and provides ATP not only to this compartment but also to the rest of the cell. The mechanism whereby this ATP synthesis is coupled to the synthesis of water was worked out many years ago. Some of the major players in this field were Mitchell (66), Lehninger (67), Boyer (68,69), Owen and Wilson (70), Brown et al. (71,72), Balaban and Heineman (73), Tager et al. (74), Hinkle and Yu (75), Brand (76), Hackenbrock (77), and Racker (78). In essence, the process links the energy released by the respiratory chain to that needed to synthesize the high-energy bond of the ATP. Not all of this energy is captured. Most of it is released as heat. The efficiency with which the energy is captured as chemical energy vs. that released as heat energy is determined by a number of factors. The composition of the diet, nutritional status, and endocrine status as well as the genetics of the individual can have major effects on this efficiency.

The components of oxidative phosphorylation are divided into five complexes. As described earlier, complexes I through IV plus ubiquinone (UQ) and cytochrome c comprise the respiratory chain. Complex V is the F_1F_0 ATPase shown in Figure 1.14. The electron carriers are the quinoid structures (FMN, FAD, UQ) and the transition metabolic complexes. Complexes I, III, and IV pump protons into the space between the two mitochondrial membranes. The pumping of protons creates a protomotive force that consists of a proton gradient and a membrane potential. This protomotive force is then used by the F_1F_0ATPase (complex V of OXPHOS) to form ATP from ADP and Pi.

The ATPase is a very large complex (mass = 517 to 541 kD) and consists of two parts: F_1 and F_0. The F_1 can dissociate from the complex and act to catalyze the degradation of ATP to ADP and Pi. In this setting, it is called the MgATPase. This is the portion of the complex that sticks out into the matrix. The F_1 consists of five polypeptide chains: α, β, γ, δ, and ε — three units of α and β and one of γ, δ, ε. The latter three subunits make up the stalk that connects the F_1 to the F_0. The α and β subunits alternate around the head of the stalk. In the stalk is a protein called the oligomycin sensitive conferring protein (OSCP).

The F_0 is embedded in the inner membrane and forms the transmembrane pore or channel through which the protons move to drive ATP synthesis. It is composed of three hydrophobic units (a, b, and c) with a stoichiometry of a_1, b_2, and c_{9-12}. Two of these units are mitochondrially encoded (the ATPase 6 and 8 genes, respectively). These two units form the transmembrane pore or proton channel through the inner mitochondrial membrane (79). A mutation in the ATPase 6 gene results in a reduction in ATP synthesis, but respiratory chain activity remains normal.

Because of the difficulty in producing ATP, more energy is released as heat than is trapped in the high-energy bond of the ATP. Leigh's syndrome and NARP (neurogenic muscle weakness, ataxia, and retinitis pigmentosa) have been attributed to base substitutions (T to G or T to C) at position 8993 in the ATPase 6 gene. Other base substitutions in this gene have been described as noted in Chapters 3, 4, 7, 8, and 13. The F_1ATPase has only nuclear-encoded subunits. Mutations in any of the nuclear genes that encode proteins of importance to OXPHOS or the mt genes that encode some of the subunits of OXPHOS can have profound effects on ATP generation and subsequent cell function. In addition, nuclear-encoded factors have control properties with respect to mt gene expression that also affect OXPHOS.

At three stages in the chain, oxidative energy is conserved (Figure 1.13 and Figure 1.14) via coupled vectorial

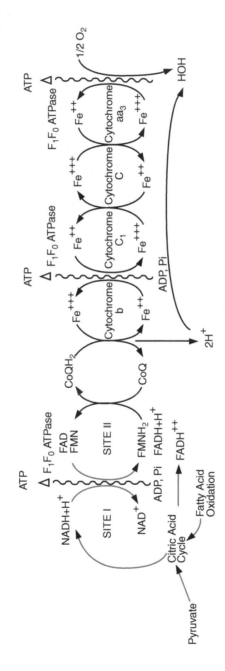

Figure 1.13 The respiratory chain showing each of the sites and where sufficient energy is generated to produce a molecule of ATP.

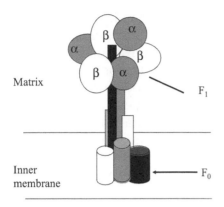

Figure 1.14 Schematic representation of the F_1F_0ATPase.

proton translocations and the creation of a membrane proton gradient ($\Delta \mu H^+$). The thermodynamics of ATP synthesis is discussed in Chapter 2. ATP synthesis depends on the presence of F_1 and F_0 in association and on an intact continuous inner mitochondrial membrane. If the membrane is disrupted, the two parts of the ATPase will be disassociated and ATP synthesis will not occur. In addition, the regulation of ATPase activity is regulated by a number of interrelated factors including (80):

- The previously alluded to thermodynamic poise of the proton gradient across the inner mitochondrial membrane
- An ATPase inhibitor protein that is part of the ATPase complex
- The amount of ADP available for phosphorylation
- Divalent cations such as calcium
- The redox state of the iron–sulfur clusters in the various complexes

The energy of the proton gradient drives the synthesis of ATP. If this energy is dissipated, as happens when agents such as dinitrophenol or one of the uncoupling proteins is present, ATP synthesis does not occur or occurs to a more limited extent. Uncoupling occurs because protons are transported through the membrane from the intramembrane space to the matrix. This "short circuits" the normal energy flow through complex V. In such a circumstance, electrons continue down the respiratory chain, but the protonic energy thus generated is not captured in the high-energy phosphate bond. Instead, this energy is released as heat. Variations or degrees of protonic energy capture *in vivo* are related to the presence or graded absence of naturally occurring agents that serve to inhibit the action of complex V or uncouple the respiratory chain energy generation process from the energy capture process of ATP synthesis.

Uncoupling Proteins

Uncoupling proteins (UCPs) are inner mitochondrial membrane transporters that dissipate the proton gradient. When this gradient is dissipated, heat is released. The uncoupling proteins are similar in size and homologous in amino acid sequence and gene sequence for the majority of their structures. Three distinct UCPs have been identified so far. UCP_1 is uniquely expressed in brown adipose tissue. It uncouples OXPHOS in this tissue — an action that can be stimulated by norepinephrine. Cold exposure and starvation, two events characterized by an initial rise then fall in norepinephrine, stimulate UCP_1 release and heat production by the brown fat. After the initial use in heat production, there is a fall as the animal attempts to conserve energy. UCP homologs have also been found in skeletal muscle and white adipose tissue. However, the control of their release and their function in the heat economy of the whole animal is somewhat different. UCP_2 is found in heart, liver, muscle, and white adipose tissue. It is subject to dietary manipulation and probably functions in body weight regulation. UCP_3 is found in skeletal muscle and is down-regulated during energy restriction or starvation.

Long chain free fatty acids slow state-3 respiration and can appear to act as partial uncoupling agents (81–84). Actually, fatty acids such as arachidonic acid and palmitic acid inhibit adenine nucleotide translocase as well as stimulate the production of the uncoupling proteins. Carboxyatraclylate inhibits the free fatty acid effect on adenine nucleotide trans-locase. When cells are injured, fatty acids from the membrane phospholipids are released; as the levels of these fatty acids rise, the ATP/ADP exchange is correspondingly reduced. The result is a depletion in cellular ATP. Other explanations of the fatty acid effect on OXPHOS include the idea that a large influx of fatty acids serves to increase the volume of the mitochondrion and this in turn serves to separate the respiratory chain and the ATPase spatially.

Some naturally occurring compounds, i.e., catabolic hormones, can serve to reduce coupling efficiency. Some of these stimulate the release of uncoupling proteins and by doing so increase body heat production, which is an important defense reaction to invading pathogens. Increased body heat (fever) serves to reduce the viability of these pathogens. The effects of these compounds on coupling are dose dependent. Other hormones have the reverse effect. Some have direct effects on mitochondrial respiration and coupling; others act through affecting the synthesis and activation of the various protein constituents of the five complexes. Several have multiple modes of action and, in addition, may affect the release of fatty acids from the membrane phospholipids. These effects are listed in Table 1.7.

In addition to naturally occurring uncouplers or decouplers, a number of compounds (drugs, antibiotics, metabolic poisons) can alter the activities of the respiratory complexes and the F_1F_0ATPase. Some of these have been extremely useful to the biochemist seeking to understand how OXPHOS works. These too are shown in Table 1.7. Some drugs act as uncouplers. In this category are dicumarol, 2,4 dinitrophenol, and FCCP (carbonyl cyanide-p-triflouromethoxyphenyl hydrozone). Scientists wishing to study only the ATPase or only the respiratory chain have found these drugs very useful. Complex I inhibitors include rotenone, ptericiden, amytal, demerol, and some mercurials. Amytal and demerol are useful substances to reduce pain. Barbituates likewise are useful for sedation. Scientists wishing to study site 2 or complex II frequently use rotenone to be sure that the electrons passing

Table 1.7 Drugs, Fatty Acids, and Hormone Effects on
Respiration and ATP Synthesis

Hormone	Effect
Thyroxine, triiodothyronine	Dose-dependent increase in respiration and ATP synthesis; increased synthesis of mitochondrial proteins. Excess results in uncoupling.
Epinephrine, norepinephrine	Decreased coupling and increase in heat production
Insulin	Increased mitochondrial protein synthesis, increased coupling efficiency
Glucocorticoid	Dose-dependent increase/decrease
Glucagon	Increased coupling efficiency
Growth hormone	Increased coupling efficiency
Palmitic acid, arachidonic acid	Stimulates uncoupling protein; inhibits ATP/ADP translocase
Dicyclohexyl carbodimide (DCCD)	Inhibits F_1F_0ATPase
Oligomycin	Inhibits F_1F_0ATPase
Rotenone	Blocks electron transport from site 1 to site 3; inhibits complex I activity
2,4-Dinitrophenol	Uncouples respiration by dissipating the proton gradient
Antimycin	Inhibits ubiquinone-cytochrome c reductase
Myxothiazol	Blocks the Q site; inhibits ubiquinone-cytochrome c reductase
2-Thenoyltriflouro acetone	Inhibits complex II
Carboxin	Inhibits complex II
Cyanide, azide, carbon monoxide	Inhibits cytochrome c oxidase
Amytal, barbiturates	Inhibits complex I

down the respiratory chain are contributed only by substrates
entering at site 2 rather than site 1 and 2.

   Carboxin and 2-thenoyltrifluoroacetone specifically
inhibit complex II. The antibiotic, antimycin, inhibits complex
III and oligomycin inhibits the F_1F_0ATPase. Complex IV is
inhibited by cyanide, azide, and carbon monoxide. These
inhibitors are potent inhibitors causing death to the organism
in rather small amounts. Cyanide and azide bind tightly to
the ferric form of cytochrome a_3; carbon monoxide binds to

the ferrous form of the cytochromes and also that of hemo-globin. Because so much more hemoglobin than the cyto-chromes is present, it takes relatively more of the carbon monoxide to kill the animal than cyanide or azide.

Studies of the mechanism of action of the ATP synthase have revealed that this large complex moves within its envi-ronment as it functions to synthesize ATP. Significant move-ment has been found in the αβ pairs, in the γ subunit, and in the ε subunit of *Escherichia coli*. Similarly, evidence suggests that an oligomer of c subunits "rotates" in these microorgan-isms (85). Movement of the ATPsynthase in mammalian mito-chondria is less substantial; nonetheless, some evidence supports the notion that this movement or rotation is an essential feature of ATP synthesis by the F_1F_0ATPase (86). Arguments against this mechanism of action (87) are made as well. If movement is an essential feature of the ATPase, it would explain why mitochondria are less efficient in ATP synthesis when the mitochondrial membrane is less fluid. Less fluid membranes would hinder this movement and thus reduce the amount of ATP synthesized.

Adenine Nucleotide Translocation

As mentioned in the section on mitochondrial shuttles, there is a transporter for the ATP synthesized as the end result of OXPHOS:adenine nucleotide transporter (88–90). The trans-porter is needed because ATP, ADP, and AMP are highly charged molecules. They contribute to the relative pH of the compartment in which they are found. If ATP accumulates in the mitochondrial matrix, for example, the pH of the matrix will fall. The adenine nucleotide transporter protein neutral-izes (somewhat) this tendency for lower pH by binding the newly synthesized ATP and transporting it out of the com-partment in exchange for ADP. The translocator will transport only these two bases and will not transport AMP or Pi. The transporter is bound to the inner aspect of the inner mem-brane and extends through the membrane.

The activity of the translocator is not energy dependent; however, some energy is lost in the process. A 13% loss in energy with translocation is estimated — that is, for every molecule of ADP transported into the matrix, ~0.87 molecule of ATP appears outside the mitochondrion. The translocator is inhibited by atractylate, carboxyatractylate, epiatractylate, bongkrekate, and isobongkrekate. At 18°C it has a translocation activity of 200 μmol/min/g protein and a flux rate of 2.5 nmol/min/cm^2 (88). The activity of the translocator is unaffected by the composition of the membrane lipids or by age, even though rapidly dividing cells seem to accumulate ATP during the growth process (89,90). This accumulation is likely related to the increase in the number of mitochondria in rapid growth.

Mitogenesis occurs at significantly greater rates than does cell replication and this requires significant amounts of ATP. Interestingly, there is a coordinated induction of energy gene expression in rapidly dividing cells, even in those from patients with mitochondrial disease. The transcript levels of the OXPHOS proteins (nuclear and mitochondrially encoded), the pyruvate dehydrogenase complex, creatine phosphokinase, muscle glycogen phosphorylase, the ATP/ADP translocator, hexokinase I muscle phosphofructokinase, and the ubiquinone oxidoreductase were all increased (or decreased) by the same set of stimulants or suppressors (90). These observations suggest that mitochondrial function vis-a-vis energy balance are coordinately regulated at the level of the expression of all the genes encoding critical elements in the OXPHOS system. Not only the individual subunits of the system are coordinately regulated, but also the contributors to the OXPHOS system. This level of control provides a new view of how mitochondrial metabolism is controlled, especially as it relates to cytosolic metabolism.

Metabolic Control

The regulation of metabolism can be viewed as the controls exerted on a single reaction in the cell or as the control of an animal's response to a change in its environment. Neither view is wholly correct because regulation can and does occur at many levels in the body. The simplest level of control is that which is exerted over a single enzymatic reaction. This reaction is controlled by the amount of available substrate, the amount and activity of the enzyme that catalyzes the reaction, the presence of appropriate amounts of required cofactors and/or coenzymes, and the accumulated product.

One can calculate the strength of the control exerted by the enzyme that catalyzes a reaction or by factors that affect that enzyme using measurements of substrate, products, cofactor amounts, coenzyme amounts, and enzyme amounts using the equation for the calculation of the flux control coefficient. This is an expression of the rate at which substrate passes to product through one or more reactions in a series. The equation for the flux control coefficient can be modified to calculate the control properties of a series of reactions within a system such as the respiratory chain or the effect of the respiratory chain on ATP synthesis.

The flux control coefficient for an individual enzyme E_i within a pathway is defined as

$$C \frac{J}{E_i} - (\partial J/J)/(\partial e_i/e_i)$$

where J is the flux of the system and e_i is the activity of any enzyme.

Control strength can vary among enzymes in a given reaction sequence and this can be estimated using metabolic control analysis. The elasticity and responsivity of a reaction or series of reactions to perturbations in reactants or environment can also be calculated. It is expressed as $\varepsilon^v = x\partial v/v\partial x$, where v is the rate of the reaction and x is a variable that modified v. Thus, the flux through a single reaction or a series of reactions can be estimated. Elasticity, response, and control coefficients determine how a steady-state response will be affected by changes in one or more constituents or conditions needed by this reaction or reaction sequence. This provides a quantitative approach to assessing metabolic control.

Overall regulation of respiration or the activity of the intact respiratory chain is vested in the availability of the phosphate acceptor, ADP. A rapid influx of ADP into the mitochondrial compartment ensures a rapid respiratory rate. In the course of metabolism, ATP is hydrolyzed to ADP and Pi. The ADP travels into the mitochondria via the adenine nucleotide translocase and stimulates OXPHOS. Control of OXPHOS is also vested in the [ATP]/[ADP] ratio, the phosphorylation potential, the [NADH]/[NAD$^+$] ratio, [Ca^{++}], and the supply of oxygen.

In muscle, the [NADH]/[NAD$^+$] and the [Ca^{++}] may be more important than ADP influx with respect to the control of OXPHOS. The working muscle uses the energy provided by the hydrolysis of creatine phosphate as well as the hydrolysis of ATP. In muscle, OXPHOS regulation also depends on the activity of creatine kinase, the enzyme that catalyzes the production of creatine phosphate. It has been suggested that the [creatine]/[creatine phosphate] ratio represents a sort of feedback signal for OXPHOS in muscle mitochondria (91). Because mitochondrial creatine kinase is likely to be displaced from equilibrium (a prerequisite for metabolic control), it may be that, in muscle mitochondria, OXPHOS is limited by the creatine kinase reaction.

Ion Flux-Role of Calcium

Calcium flux between the cytoplasm and the mitochondrion regulates mitochondrial activity (92). Ca^{++} uptake is energetically less demanding than its export. Uptake followed by export is an oscillatory process. Ca^{++} flows in until a 200-μM concentration is reached, whereupon it is actively pumped out using the energy of ATP. This active system costs about 63% of the ATP hydrolysis energy. The three separate Ca^{++} transport mechanisms are: a calcium uniporter, a sodium-dependent transporter, and a sodium-independent transporter. The uniporter rapidly sequesters the calcium into the mitochondrial matrix. It also sequesters Fe^{++}, Sr^{++}, Mn^{++}, Ba^{++}, and Pb^{++}. Magnesium^{++} uptake is much slower than calcium uptake so likely it is not sequestered by this uniporter. The other two transporters are more active in facilitating export than import. The sodium-dependent transporter has a velocity that is much greater than the sodium-independent transporter. Together, these transport systems assure a balance of calcium movement in the mitochondrion.

The mitochondrial calcium cycle is designed to regulate intramitochondrial calcium levels that in turn reflect those of the cytoplasm. Surges in cytoplasmic calcium via the cell signaling systems activate a variety of ATP-dependent reactions. In turn, as ATP is used, the ratio of ATP to ADP and AMP changes and this in turn affects mitochondrial activity. More ADP is available to flow into the mitochondria and stimulate the malate–aspartate shuttle and oxidative phosphorylation. Newly formed ATP is then exported along with the calcium ion back to the cytoplasm. Calcium flux is a multifaceted system providing flexibility and responsivity to the balance of anabolic and catabolic pathways via the mitochondria. Just as calcium is important to normal cellular metabolism, it also plays a role in cell death. If the membranes are disrupted via injury or chemical insult, calcium will accumulate in the mitochondria and the organelle will die. If too many mitochondria die, the cell will die. Indeed, abnormal calcium flux may play a role in apoptosis.

SUMMARY

An overview of the structure and function of the mitochondria has been given in the preceding sections. The various metabolic pathways operating in this organelle were briefly summarized so as to provide a platform for understanding the chapters that follow. Clearly, any change in the genes that encode any of the more than 1000 proteins comprising the mitochondrion will result in a compromised mitochondrial state. Following chapters will focus on only the genome that exists in the mitochondrion. Although it encodes only the 13 proteins of OXPHOS, these proteins play a central role in the normal functioning of the system. Any perturbation in the genes that encode these proteins will result in a disturbance in the normal production of ATP. The elasticity of the metabolic system that depends on variation in ATP availability will be lost.

Recognition of the impact of mitochondrial gene mutations has been a recent development in the history of human disease. However, knowing how a shortfall in ATP production can affect metabolism in general and mitochondrial metabolism in particular will help in the understanding of how the disease process develops. In addition, knowledge about the interaction of the nuclear and the mitochondrial genomes will broaden understanding of the dependence on one on the other.

REFERENCES

1. IE Scheffler. A century of mitochondrial research: achievements and perspectives. *Mitochondrion* 1:3–31, 2001.

2. MA Jacobs, SR Paynes, AJ Bendick. Moving pictures and pulsed-field gel electrophoresis show only linear mitochondrial DNA molecules from yeasts with linear mapping and circular-mapping mitochondrial genomes. *Curr Genet* 30:3–11, 1996.

3. C Leblanc, C Boyen, O Richards, G Bonnard, J-M Grienenberger, B Kloarg. Complete sequence of the mitochondrial DNA of rhodophyte, *Chondrus crispus* (gigartinales). Gene content and genome organization. *J Mol Biol* 250:484–495, 1995.

4. G Burger, I Plante, KM Lonergan, MW Gray. The mitochondrial DNA of the amoeboid protozoan, *Acanthamoeba castellanii*: Complete sequence, gene content, and genome organization. *J Mol Biol* 245:522–537, 1995.

5. FN Martin. Linear mitochondrial genome organization *in vivo* in the genus *Pythium*. *Curr Genet* 28:225–234, 1995.

6. BA Roe, D-P Ma, RK Wilson, JF-H Wong. The complete nucleotide sequence of the *Xenopus laevis* mitochondrial genome. *J Biol Chem* 260:9759–9774, 1985.

7. M Sumida. Mitochondrial DNA differentiation in the Japanese brown frog *Rana japonica* as revealed by restriction endonuclease analysis. *Genes Genet Syst* 72:79–90, 1997.

8. M Sumida. Inheritance of mitochondrial DNAs and allozymes in the female hybrid lineage of two Japanese pond frog species. *Zoo Sci* 14:277–286, 1997.

9. MS Meusel, RFA Moritz. Transfer of paternal mitochondrial DNA during fertilization of honeybee (*Apio mellifera*) eggs. *Cur Genet* 24:539–543, 1993.

10. S Heindleder, K Mainz, Y Plante, H Lewalski. Analysis of mitochondrial DNA indicates that domestic sheep are derived from two different ancestral maternal sources: no evidence for contribution from Urial and Argali sheep. *J Heredity* 89:113–120, 1998.

11. R Steinborn, M Muller, G Brem. Genetic variation in functionally important domains of the bovine mitochondrial DNA control region. *Biochem Biophys Acta* 1397:295–304, 1998.

12. RT Loftus, DE MacHugh, LO Ngere, DS Balaian, AM Bachi, DG Bradley, EP Cunningham. Mitochondrial genetic variation in European, African, and Indian cattle populations. *Am Genet* 25:265–271, 1994.

13. S Anderson, HL DeBruijn, AR Coulson, IC Eperon, F Sanger, IG Young. Complete sequence of bovine mitochondrial DNA: conserved features of the mammalian mitochondrial genome. *J Mol Biol* 156:683–717, 1982.

14. 3 Marklund, R Chaudhary, L Marklund, K Sandberg, L Andersson. Extensive mitochondrial DNA diversity in horses revealed by PCR-SSCP analysis. *Am Genet* 26:193–196, 1995.

15. W Wang, A-H Liu, S-Y Lin, H Lan, B Su, D-W Xie, L-M Shi. Multiple genotypes of mitochondrial DNA within a horse population from a small region in Yunnan province of China. *Biochem Genet* 32:371–379, 1994.

16. X Xu, A Gullberg, U Arnason. The complete mitochondrial DNA of donkey and mitochondrial DNA comparisons among four closely related mammalian species-pairs. *J Mol Evol* 43:438–446, 1996.

17. G Gadaleta, G Pepe, G DeCandia, C Quagliariello, E Sbisa, C Saccone. The complete nucleotide sequence of the *Rattus norvegicus* mitochondrial genome: cryptic signals revealed by comparative analysis between vertebrates. *J Mol Evol* 28:497–516, 1989.

18. U Arnason, A Gullberg, X Xu. A complete mitochondrial DNA molecule of the white handed gibbon, *Hylobates lar*, and comparison among individual mitochondrial genes of all hominoid genera. *Hereditas* 124:185–189, 1996.

19. JE Hixson, WM Brown. A comparison of the small ribosomal RNA genes from mitochondrial DNA of the great apes and humans. Sequences, structure, evolution, and phylogenetic implications. *Mol Biol Evol* 3:1–18, 1986.

20. S Anderson, AT Bankier, BG Barrell, MHL Bruijin, AR Coulsen, J Drouin, IC Eperon, DP Nierlich, BA Roe, F Sanger, PH Schreier, AJH Smith, R Staden, IG Young. Sequence and organization of the human mitochondrial genome. *Nature* 290: 457–465, 1981.

21. G Daum, JE Vance. Import of lipids into mitochondria. *Prog Lipid Res* 36:103–130, 1997.

22. CA Mannella, H Tedeschi. The emerging picture of mitochondrial membrane channels. *J Bioenergetics Biomembranes* 24: 3–5, 1992.

23. RC Wander, CD Berdanier. Effects of dietary carbohydrate on mitochondrial composition and function in two strains of rats. *J Nutr* 115:190–199, 1985.

24. OE Deaver, RC Wander, RH McCusker, CD Berdanier. Diet effects on membrane phospholipid fatty acids and membranc function in BHE rats. *J Nutr* 116:1148–1155, 1986.

25. RC Stancliff, MA Williams, K Ittsumi, L Packer. Essential fatty acid deficiency and mitochondrial function. *Arch Biochem Biophys* 131:629–642, 1969.

26. M-JC Kim, CD Berdanier. Nutrient gene interactions determine mitochondrial function: effect of dietary fat. *FASEB J* 12: 243–248, 1998.

27. EJ McMurchie. Dietary lipids and the regulation of membrane fluidity and function. In: *Physiological Regulation of Membrane Fluidity*, (EJ McMurchie, Ed.) 1988: Alan Liss Inc., New York, 189–237.

28. CD Berdanier. Fatty acids and membrane function In: *Fatty Acids in Foods and Their Health Implications*, 2nd ed. (CK Chow, Ed.) 2000: Marcel Dekker, New York, 569–584.

29. RL Wolff, B Entressangles. Compositional changes of fatty acids in the 1 (1″)- and 2(2″)-positions of cardiolipin from liver, heart, and kidney mitochondria of rats fed a low fat diet. *Biochem Biophys Acta* 1082:136–142, 1991.

30. JT Venkatraman, RK Tiwarai, B Cinader, J Flory, T Wiezbicki, MT Clandinin. Influence of genotype on diet-induced changes in membrane phosphatidylcholine and phosphatidylethanolamine composition of splenocytes, liver nuclear envelope and liver mitochondria. *Lipids* 26:198–202, 1991.

31. J-P Poisson, SC Cunnane. Long chain fatty acid metabolism in fasting and diabetes: relation between altered desaturase activity and fatty acid composition. *J Nutr Biochem* 2:60–69, 1991.

32. J-P Poisson. Essential fatty acid metabolism in diabetes. *Nutrition* 5:263–266, 1989.

33. KS Withers, AJ Hulbert. The influence of dietary fatty acids and hypothyroidism on mitochondrial fatty acid composition. *Nutr Res* 7:1139–1150, 1987.

34. D Raederstorff, CA Meier, U Moser, P Walter. Hypothyroidism and throxin substitution affect the n-3 fatty acid composition of rat liver mitochondria. *Lipids* 26:781–787, 1991.

35. FM Ruggiero, C Landriscina, GV Gnoni, E Quagliariello. Lipid composition of liver mitochondria and microsomes in hyperthyroid rats. *Lipids* 19:171–178, 1984.

36. D Raederstorff, CA Meler, U Moser, P Walter. Hypothyroidism and thyroxine substitution affect the n-3 fatty acid composition of rat liver mitochondria. *Lipids* 26:781–787, 1991.

37. Y-S Huang, UN Sad, DF Horrobin. Effect of dexamethasone on the distribution of essential fatty acids in plasma and liver phospholipids. *IRCS Med Sci* 14:180–199, 1986.

38. S Clejan, VT Maddeiah. Growth hormone and liver mitochondria: effects on phospholipid composition and fatty acyl distribution. *Lipids* 21:677–684, 1986.

39. DL Melchior, JM Stein. Thermotropic transitions in biomembranes. *Ann Rev Biophys Bioeng* 5:205–239, 1976.

40. PS Timiras. Biological perspectives. on aging. *Am Sci* 66:605–613, 1978.

41. D Hegner. Age-dependence of molecular and functional changes in biological membrane properties *Mech Aging Develop* 14:101–118, 1980.

42. MT Clandinin, SM Innis. Does mitochondrial ATP synthesis decline as a function of change in the membrane environment with aging? *Mech Aging Develop* 22:205–208, 1983.

43. MB Lewin, PS Timiras. Lipid changes with aging in cardiac mitochondrial membranes. *Mech Aging Develop* 24:343–351, 1984.

44. F Guerrieri, G Vendemaile, N Turturro, A Fratello, A Furio, L Muolo, I Grattagliano, S Papa. Alteration of F_1F_0ATP synthase during aging. Possible involvement of oxygen free radicals. *Ann NY Acad Sci* 786:62–71, 1996.

45. BP Yu, EA Suescun, SY Yang. Effect of age-related lipid peroxidation on membrane fluidity and phospholipase A2 modulation by dietary restriction. *Mech Aging Develop* 65:17–33, 1992.

46. CD Berdanier, HB Everts. Mitochondrial DNA in aging and degenerative disease. *Mutation Res* 475:169–184, 2001.

47. M-E Harper, S Monemdjou, JJ Ramsey, R Weindruch. Age-related increase in mitochondrial proton leak and decrease in ATP turnover reactions in mouse hepatocytes. *Am J Physiol* 275:E197–E206, 1998.

48. CD Berdanier, S McNamara. Aging and mitochondrial activity in BHE and Wistar strains of rats. *Exp Gerontol* 15:519–525, 1980.

49. RG Hansford, F Castro. Age-related changes in the activity of enzymes of the tricarboxylate cycle and lipid oxidation, and of carnitine content in the muscles of rats. *Mech Aging Develop.* 19:191–201, 1982.

50. JH Kim, G Woldgiorgis, CE Elson, E Shrago. Age-related changes in respiration coupled to phosphorylation.1. Hepatic mitochondria. *Mech Aging Develop* 46:263–277, 1988.

51. JH Kim, E Shrago, CE Elson. Age related changes in respiration coupled to phosphorylation. II. Cardiac mitochondria. *Mech Aging Develop* 46:279–290, 1988.

52. H Nohl, R Kramer. Molecular basis of age-dependent changes in the activity of adenine nucleotide translocase. *Mech. Aging Develop* 14:137–144, 1980.

53. G Paradies, FM Ruggiero. Effect of aging on the activity of the phosphate carrier and on the lipid composition in rat liver mitochondria. *Arch Biochem Biophys* 284:332–337, 1991.

54. J Venkatraman, G Fernandes. Modulation of age-related alterations in membrane composition and receptor-associated immune functions by food restriction in Fischer 344 rats. *Mech Aging Develop* 63:27–44, 1992.

55. RH Weindruch, MK Cheung, MA Verity, RL Walford. Modification of mitochondrial respiration by aging and dietary restriction. *Mech Aging Develop* 12:375–392, 1980.

56. RM Clancy, LH McPherson, M Glaser. Effect of changes in the phospholipid composition on the enzymatic activity of D-$-hydroxybutyrate dehydrogenase in rat hepatocytes. *Biochem* 22:2385–2364, 1983.

57. AP Halstrap. The mitochondrial pyruvate carrier. Kinetics and specificity for substrates and inhibitors. *Biochem J* 148:85–96, 1975.

58. B Safer, CM Smith, JR Williamson. Control of the transport of reducing equivalents across the mitochondrial membrane in perfused rat heart. *J Mol Cell Cardiology* 2:111–124, 1974.

59. AJ Meijer, K VanDam. The metabolic significance of anion transport in mitochondria. *Biochem Biophys Acta* 346:213–244, 1974.

60. Y Hatefi. The mitochondrial electron transport chain and oxidative phosphorylation system. *Annu Rev Biochem* 54:1015–1069, 1985.

61. H Weiss, T Friedrich, G Hofhaus, D Preis. The respiratory chain NADH dehydrogenase (complex I) of mitochondria. *Eur J Biochem* 197:563–576, 1991.

62. T Ohnishi. NADH-quinone oxidoreductase, the most complex complex. *J Bioenergetics Biomembranes* 25:325–329, 1993.

63. L Hederstedt, L Ohnishi. Progress in succinate:quinone oxidoreductase research. In: *Molecular Mechanisms in Bioeneregetics* (L Ernster, Ed.) 1992: Elsevier Science Publishers, London, 163–198.

64. HK Paudel, L Yu, C-A Yu. Involvement of a histidine residue in the interaction between membrane anchoring protein (QPs) and succinate dehydrogenase in mitochondrial succinate-ubiquinone reductase. *Biochem Biophys Acta* 1056:159–165, 1991.

65. EC Slater. The Q cycle: a ubiquitous mechanism of electron transfer. *Trends Biochem Sci* 8:239–242, 1983.

66. P Mitchell. Vectorial chemistry and molecular mechanisms of chemiosmotic coupling: power transmission by proticity. *Biochem Soc Trans* 4:399, 1976.

67. A Lehninger. *The Mitochondrion*. 1975: W.A. Benjaman Inc., New York.

68. PD Boyer. A perspective of the binding change mechanism for ATP synthesis. *FASEB J* 3:2164–2178, 1989.

69. PD Boyer. The ATP synthase — a splendid molecular machine. *Annu Rev Biochem* 66:717, 1997.

70. CS Owen, DF Wilson. Control of respiration by mitochondrial phosphorylation state. *Arch Biochem Biophys* 161:581–591, 1974.

71. GC Brown, P Lakin–Thomas, MD Brand. Control of respiration and oxidative phosphorylation in isolated rat liver cells. *Eur J Biochem* 192:355–362, 1990.

72. GC Brown. Control of respiration and ATP synthesis in mammalian mitochondria and cells. *Biochem J* 284:1–13, 1992.

73. RS Balaban, FW Heineman. Control of mitochondrial respiration in the heart *in vivo*. *Mol Cell Biochem* 89:191–197, 1989.

74. JM Tager, RJA Wanders, AK Groen, W Kunz, R Bohnensack, U Kuster, G Letko, G Bohme, J Duszynski, L Wojtczak. Control of mitochondrial respiration. *FEBS Lett* 151:1–9, 1983.

75. PC Hinkle, ML Yu. The phosphorus/oxygen ratio of mitochondrial oxidative phosphorylation. *J Biol Chem* 254:2450–2455, 1979.

76. MD Brand, AL Lehninger. H+/ATP ratio during ATP hydrolysis by mitochondria: modification of the chemiosmotic theory. *Proc Natl Acad Sci USA* 74:1955–1959, 1977.

77. CR Hackenbrook, B Chazotte, SS Gupte. The random collision model and a critical assessment of diffusion and collision in mitochondrial electron transport. *J Bioenergetics Biomembranes* 18:331–368, 1986.

78. E Racker. From Pasteur to Mitchell: a hundred years of bioenergetics. *Fed Proc* 39:210, 1980.

79. T Papakonstantinou, RHP Law, S Manon, RJ Devenish, P Nagley. Relationship of subunit 8 of yeast ATP synthase and the inner mitochondrial membrane. Subunit 8 variants containing multiple lysine residues in the central hydrophobic domain retain function. *Eur J Biochem* 227:745–752, 1995.

80. K Schwerzmann, PL Petersen. Regulation of the mitochondrial ATP synthase/ATPase complex. *Arch Biochem Biophys* 250:1–18, 1986.

81. Y Takeuchi, H Morii, M Tamura, O Hayaishi, Y Watanabe. A possible mechanism of mitochondrial dysfunction during cerebral ischemia: Inhibition of mitochondrial respiration activity by arachidonic acid. *Arch Biochem Biophys* 289:33–38, 1991.

82. AY Andreyev, TO Bondareva, VI Dedukhova, EN Mokhova, VP Skulachev, NI Volkov. Carboxyatractylate inhibits the uncoupling effect of free fatty acids. *FEBS Lett* 226:265–269, 1988.

83. A Vianello, E Petrussa, F Macri. ATP/ADP antiporter is involved in uncoupling of plant mitochondria induced by low concentrations of palmitate. *FEBS Lett* 347:239–242, 1994.

84. KR Chien, A Sen, R Reynolds, A Chang, Y Kim, MD Gunn, LM Buja, JT Willerson. Release of arachidonate from membrane phospholipids in cultured neonatal rat myocardial cells during adenosine triphosphate depletion. Correlation with the progression of cell injury. *J Clin Invest* 75:1770–1780, 1985.

85. PL Petersen. Frontiers in ATP synthase research: understanding the relationship between subunit movements and ATP synthesis. *J Bioenergetics Biomembranes* 28:389–395, 1996.

86. RL Cross, TM Duncan. Subunit rotation in F_1F_0ATP synthases as a means of coupling proton transport through F_0 to the binding changes in F_1. *J Bioenergetics Biomembranes* 28:403–407, 1996.

87. JA Berdan. Rotary movements within the ATP synthase do not constitute an obligatory element of the catalytic mechanism. *IUBMB Life* 55:473–481, 2003.

88. M Klingenberg. The ADP–ATP translocation in mitochondria, a membrane potential controlled transport. *J Membrane Biol* 56:97–105, 1980.

89. J Streicher–Scott, R Lapidus, PM Sokolove. The reconstituted mitochondrial adenine nucleotide translocator: effects of lipid polymorphism. *Arch Biochem Biophys* 315:548–554, 1994.

90. JK Pollaak, R Sutton. The adenine nucleotide translocator in foetal, suckling and adult rat liver mitochondria. *Biochem Biophys Res Commun* 80:193–198, 1978.

91. M Wyss, J Smeitink, RA Wevers, T Wallimann. Mitochondrial creatine kinase: a key enzyme of aerobic energy metabolism. *Biochem Biophys Acta* 1102:119–166, 1992.

92. KK Gunter, TE Gunter. Transport of calcium by mitochondria. *Am J Physiol* 267:C313–C339, 1994.

2

Mitochondrial Bioenergetics in Intact Tissue

RICHARD L. VEECH

CONTENTS

It is impossible to explain honestly the beauties of the
laws of nature in a way that people can feel, without their
having some deep understanding of mathematics. I am
sorry, but this seems to be the case.

Richard P. Feynman,
The Character of Physical Law

A RECENT HISTORY OF THEORIES OF MITOCHONDRIAL ENERGY GENERATION

Peter Mitchell and Classical Biochemists

The modern history of the understanding of mitochondrial
energy generation centers on Peter Mitchell (1,2), a man who
conducted much of his research in his private laboratory in
Bodmin, Cornwall, using family funds. The fundamental
physical chemical laws governing mitochondria generation of
ATP, however, were known for at least a century before. Until
after Mitchell's proposal, it was not known how these laws
were implemented mechanistically.

Mitchell was trained in the laboratory of J.F. Danielli
in the Department of Biochemistry at Cambridge and was

therefore familiar with Danielli and H. Davson's lipid bilayer structure of biological membranes. In 1961, he presaged the theory of mitochondrial chemiosmosis with the postulate that if the lysosomal enzyme glucose 6-phosphatase were located in an isotropic membrane, a pH gradient of three pH units would produce an electrochemical gradient of the H^+ ion of 60 mV and a lowering of the dissociation constant by a factor of 1000. For the lysosome, such a pH gradient of perhaps two pH units is now widely accepted.

Mitchell applied the same thinking to a proposed mitochondrial ATPase, which he postulated would be powered by an electrochemical gradient of the H^+ ion. As a young postdoctoral student, I heard Peter Mitchell give a lecture to the Oxford biochemistry department outlining his chemiosmotic theory. During the course of the lecture, he postulated a pH gradient between mitochondria and cytoplasm of five pH units. Hans Krebs, who scrupulously avoided making critical comments during or after presentations, expostulated, "Now Peter!" He simply could not accept a mitochondrial pH of 11.

Unfortunately, Mitchell's ideas often had errors of detail that irritated the cognoscenti, even those who were not ill disposed. This slowed the acceptance of his ideas, which if wrong in detail, have been proven correct in concept in subsequent years. As I hope to show, Mitchell's groundbreaking theory is actually an expression of very basic physical chemical principles going back to Michael Faraday, Walter Nernst, and J. Willard Gibbs — scientists working before mitochondria were known to exist. It took the work of many others, notably Efriam Racker, Peter Petersen, Brian Chappell and his Bristol group, Paul Boyer, John Walker, and others too numerous to mention, to show how the system proposed by Mitchell really worked.

In this chapter, I have chosen to use the standard nomenclature and symbols of physical chemistry, rather than the more personal nomenclature coined by Mitchell in the hope of increasing clarity. Much of the data cited here have been published previously in the references cited. Most studies on mitochondrial energetics have been done using isolated mitochondria or submitochondrial particles, which can lead to artifacts inherent in any *in vitro* system. However, mitochondria operate in the cytoplasm of the cell, so I have chosen to discuss studies that attempted to define the parameters of mitochondrial energetics in intact tissue. These studies are rare because of the time required to perform them and, probably, because of the quixotic nature required of those that undertake them. I hope that they are informative.

Finally, I am aware, in the words of Paul Srere, that "for each equation you put in a paper, you lose half of your audience." Words can be as slippery as a shower-room floor, and some thoughts are most precisely expressed in equations. I think that precision in thought for the few who want to understand the mechanisms involved in mitochondrial function is worth the loss of the larger numbers of casual readers.

The Unity of Energy: "Chemical" vs. "Chemiosmotic" Theory

The prevalent idea held by most of the best known biochemists in the 1960s was that a high-energy intermediate was involved in ATP synthesis, such as 1,3 diphosphoglycerate in the glycolytic pathway. Opposing this consensus, Mitchell held that no high-energy chemical intermediate was involved in mitochondrial ATP production, but rather only an electrochemical gradient across the mitochondrial membrane that was converted into chemical bond energy. This hypothesis has now been vindicated by a great deal of detailed evidence.

The stoichiometry of two energetic protons per ATP synthesis is, of course, not correct (although he defended this idea until his death). This was based largely on a confusion between the standard free energy of ATP hydrolysis, $\Delta G'^\circ$, and the actual free energy of hydrolysis, $\Delta G'$, which is about double the value of the former. No matter — Mitchell first saw that energy could be captured in biological membranes and that the electrical and concentration energy of a system could be just as real and effective as the energy of a chemical bond. This was a signal achievement.

Another factor in the slow acceptance of Mitchell's ideas was his coining of a whole new vocabulary: the terms "uniporter," "synporter," and "antiporters." Uniporters are now generally called transporters, such as the glucose transporter that carries the uncharged glucose molecule catalytically across a membrane unaccompanied by another molecule. Synporters are now generally called cotransporters in which a molecule is transported across a membrane accompanied by another ion or molecule, e.g., the H^+-Pi cotransporter. Antiporters are now generally called exchangers, in which a compound on one side of the membrane is exchanged for a molecule on the other side, e.g., the HCO_3^-/Cl^- exchanger. Other terms were coined, like "proton motive force," which is in reality the electrochemical energy of the proton between mitochondrial and cytoplasmic phases (see Equation 2.8). These neologisms had the effect of slowing the understanding and acceptance of the theory in wider scientific circles.

Some Fundamental Equations and Definitions from Physical Chemistry

In this chapter, we will try to examine mitochondria in their normal milieu, the cytoplasm of a living cell. Most of the examples are taken from work using the isolated perfused working rat heart because of the relative simplicity of the system and the ability to determine the necessary parameter required to analyze mitochondrial energetics. The terms chosen for a discussion of mitochondrial energy transduction have been taken from those codified by the International Union of Pure and Applied Chemistry (3).

The Free Energy of any Chemical Reaction

The free energy of a chemical reaction — in this case, the hydrolysis of ATP — is

$$\Delta G'_{ATP} = \Delta G^\circ_{ATP} + RT \ln \frac{[\text{ADP}]\,[\text{Pi}]}{[\text{ATP}]} \qquad (2.1)$$

where: $\Delta G'_{ATP}$ is the actual chemical energy released upon the hydrolysis of the pyrophosphate bond of ATP. By convention, water, which is a participant in the hydrolysis or synthesis of the terminal pyrophosphate bond of ATP: ATP^{4-} + $H_2O \rightarrow ADP^{3-}$ + HPO^{2-} + H^+, is given a value of 1 and left out of the equation. The ATP symbol is taken to include all of the ionic species of the molecule present under defined conditions. R is the gas constant whose value is 8.3145×10^{-3} kJ/mol/K and K is degrees Kelvin (temperature in centigrade + 273.15). The term "ln" is the natural logarithm. The actual free energy, $\Delta G'_{ATP}$, thus comprises two terms: the standard free energy, $\Delta G°_{ATP}$, which is a constant under a defined set of conditions, and the natural logarithm of the actual ratio of the concentration of products over reactants times a constant RT.

In most textbooks of biochemistry, tables of $\Delta G°$ are given; this causes confusion in many minds. The actual value of $\Delta G'_{ATP}$ in all living tissues so far examined in detail, including those containing mitochondria like heart muscle and liver and in red cell, which does not contain mitochondria, is remarkably constant between −53 and −60 kJ/mol (4,5). For simplicity, in these discussions, concentration, indicated in square brackets [ ], will be taken to be equivalent to "activity," which is a theoretical concept difficult to measure in physiological situations.

The standard free energy of ATP hydrolysis — that is, the energy of hydrolysis under defined conditions of temperature, ionic strength, pH, and free $[Mg^{2+}]$ where the concentrations of products and reactants are all 1 M — is $\Delta G°_{ATP}$. The concentration of H_2O, the term for which is not written in this abbreviated form, is by convention taken to be 1. The $\Delta G°_{ATP}$ of ATP hydrolysis at temperature T = 311.15 K or 38°C; pH = 7.2; ionic strength = 0.25; free $[Mg^{2+}]$ of 1 mM = −32.6 kJ/mol. Figure 2.1 illustrates the variation of $\Delta G°$ of ATP hydrolysis with changes in pH and $[Mg^{2+}]$ at 38°C and ionic strength 0.25.

Energy Generated from the Redox Reaction of the Electron Transport Chain

The net reaction of the electron transport system is to take the electrons and protons from the free mitochondrial NADH

pH and [Mg^{2+}] Effects on ΔG_{ATP}

ΔG ATP hydrolysis =

$$- RT \ln (K_{\text{ionic ATPhydrolysis}} \times \frac{f\text{ADP} \times f\text{Pi}}{f\text{ATP} \times [\text{H}]}) + RT \ln (\frac{[\Sigma\text{ADP}] \times [\Sigma\text{Pi}]}{[\Sigma\text{ATP}]})$$

$$f\text{ATP} = \left(1 + \frac{[\text{H}^+]}{K_{aATP}} + \frac{[\text{H}^+]^2[\text{ATP}^+]}{K_{aHATP}\ K_{aATP}} + K_{bMgATP}[\text{Mg}^{2+}] + \frac{K_{bMgHATP}[\text{Mg}^{2+}][\text{H}^+]}{K_{aATP}} \right)$$

$$f\text{ADP} = \left(1 + \frac{[\text{H}^+]}{K_{aADP}} + \frac{[\text{H}^+]^2[\text{ADP}^+]}{K_{aHADP}\ K_{aADP}} + K_{bMgADP}[\text{Mg}^{2+}] + \frac{K_{bMgHADP}[\text{Mg}^{2+}][\text{H}^+]}{K_{aADP}} \right)$$

$$f\text{Pi} = \left(1 + \frac{[\text{H}^+]}{K_{aPi}} + K_{bMgPi}[\text{Mg}^{2+}] \right)$$

R: the gas constant, 8.3145×10^{-3} kJ/mol/K^{-1}
T: the absolute temperature, 311.5K

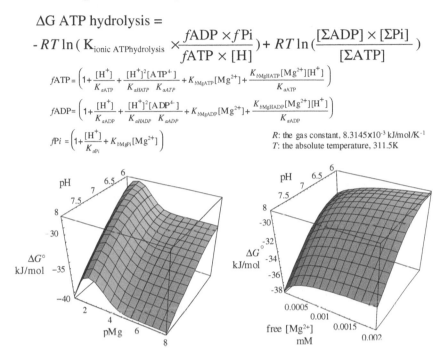

Figure 2.1 The variation of $\Delta G°$ of ATP hydrolysis with changes in pH and free [Mg^{2+}] at 38°C and ionic strength 0.25. As shown in this figure, the actual value of $\Delta G°_{ATP}$ varies significantly with pH and free [Mg^{2+}] within the physiological ranges of free [Mg^{2+}] of 0.2 to 1 mM (38) and a pH range from 6.9 to 7.5. The equations required to correct these constants for variations in free [Mg^{2+}] and pH are available from Veech et al. (39). The joule is the SI unit for energy or heat, and 1 J = 4.184 calories exactly, if other units used commonly in nutritional studies are desired. (From R.L. Veech et al., *Alcohol Clin Exp Res* 18:1040–1056, 1994.)

pool and convert that to H$_2$O, just like a rocket on Cape Canaveral, in the sum reaction:

$$\text{NADH}_{mito} + \text{H}^+_{mito} + \tfrac{1}{2}\,\text{O}_2 \rightarrow \text{NAD}^+_{mito} + \text{H}_2\text{O} \qquad (2.2)$$

The extent and direction of a reaction are determined by the concentrations of the free substrates and free products. The free mitochondrial [NAD$^+$]/[NADH] ratio is determined by measurement of the components of the glutamate dehydrogenase reaction (EC 1.4.1.2), located within the mitochondrial matrix, or the substrates of D-β-hydroxybutyrate reaction (EC 1.1.1.30), located at the inner mitochondrial membrane (6), according to the reactions (7):

$$\frac{[NAD^+]_{mito}}{[NADH]_{mito}} = \frac{[\alpha - ketoglutarate^{2-}]_{total}[NH_4^+]_{mito}[H^+]_{mito}}{[L\text{-glutamate}^-]_{total}} \quad (2.3)$$

$$\times \frac{1}{3.87 \times 10^{-13} M^2}$$

or (6)

$$\frac{[NAD^+]_{mito}}{[NADH]_{mito}} = \frac{[acetoacetate^-]_{total}[H^+]_{mito}}{[D\text{-}\beta - hydroxybutyrate^-]_{total}} \times \frac{1}{4.93 \times 10^{-9} M} \quad (2.4)$$

By analogy with Equation 2.1, the redox potential of the NAD couple at pH 7 in volts may be calculated from the half reaction of NAD (see Clark (8) for details) according to:

$$Eh^7_{NAD/NADH} = E^\circ_{NAD/NADH} + \frac{RT}{nF} \ln \frac{[NAD]_{mito}}{[NADH]_{mito}} \times \frac{10^{-7}}{[H^+]_{mito}} \quad (2.5)$$

where the $E^\circ_{NAD/NADH}$ at pH 7 is –0.32 V (9); n is the number of electrons or 2; F is the Faraday constant or 96.485 kJ/mol/V; and RT is as described earlier.

The redox potential of the mitochondrial NAD couple is about –0.28 to –0.30 V, and the half potential of the O_2/H_2O couple, $Eh^7_{O_2/H_2O}$, is +0.814 V at any concentration of O_2 (8). From knowledge of the free mitochondrial [NAD$^+$]/[NADH], one may calculate the redox energy available from the reduction of O_2 to H_2O by the mitochondrial NADH as described in Equation 2.2 according to:

$$\Delta G'_{Electron\ Transport} = -nF(Eh^7_{O_2/H_2O} - Eh^7_{NAD/NADH}) \quad (2.6)$$

Thus, the chemical energy generated by the redox reactions of the electron transport system is –211 kJ/2 mol of electrons traversing the chain.

These few standard calculations lead to some astounding conclusions. The measured $\Delta G'$ of ATP ranges from –53 to –60 kJ/mol and three ATPs are generated from each substrate producing NADH, requiring a low of –153 to a high of –180 kJ/three ATPs generated. This means that the electron transport system is able to convert an astounding 72 to 85% of its available redox energy into the chemical energy of the ATP bond with only –31 to –52 kJ of heat produced for each one-half O_2 consumed from the –211 kJ generated by the electron transport system. The mitochondria are truly a miraculous chemical transducer.

Another important lesson taught by this examination of mitochondrial function is the importance of distinguishing free [NAD+]/[NADH] from the total measured amounts of NAD and NADH (see Table 2.1). Developed over a generation

Table 2.1 Total[a] Pyridine Nucleotides in Rat Liver

Substance	Fed	Fasted 48 h
NAD total	0.76	0.82
NADH total	0.14	0.16
$NAD_{Total}/NADH_{Total}$	5.4	5.1
Free cytosolic [NAD+]/[NADH] from K_{LDH}	860	477
Free mitochondrial [NAD+]/[NADH] from $K_{\beta HBDH}$	10.1	7.5
NADP total	0.067	0.080
NADPH total	0.31	0.36
$NADP_{Total}/NADPH_{Total}$	0.22	0.22
Free cytoplasmic [NADP+]/[NADPH] from K_{ICDH}	0.014	0.006

[a] Free plus bound.

Notes: Values are in micromoles per gram of fresh weight liver. For method of calculation, see Krebs and Veech.

Source: H.A. Krebs and R.L. Veech, in S. Papa, J.M. Tager, E. Quagliariello, and E.C. Slater, Eds., *The Energy Level and Metabolic Control in Mitochondria*. Adriatica Editrice: Bari, 1969, 329–382.

ago, this knowledge has slipped the attention of modern molecular biologists. The pyridine nucleotides do not pass the mitochondrial membrane. Therefore, two pools of NAD and NADH are within the cell. In addition, large amounts of NADH are bound to protein, particularly the multiple dehydrogenases within the cell. Estimation of the free [NAD+]/[NADH] gives no information as to the redox status of the NAD couple within cytoplasm (10) and nucleus (11) or within the mitochondria.

The ratio of the free [NAD+]/[NADH] in cytoplasm is over two orders of magnitude higher than would be inferred from measurements of the total NAD and NADH. If the free mitochondrial [NAD+]/[NADH] were as oxidized as the cytosol, it would require that electron transport be 93% efficient to achieve the synthesis of three ATPs with a $\Delta G'$ of −60 kJ/mol, the highest energy levels found in living tissue. Recently, a number of transcription factors (12–15) have been found that appear to be controlled by pyridine nucleotides. Knowledge of the free [NAD+]/[NADH] or [NADP+]/[NADPH] ratios and the classical biochemical reasoning underlying these conclusions remains essential in the newer biochemical areas of molecular biology.

LIFE ON THE EDGE

Redox Energy Is Converted to Chemical Bond Energy at a Membrane

The fundamental insight of Peter Mitchell's theory of mitochondrial energy generation was that protons were extruded from the mitochondrial matrix into the cytoplasm during the transfer of electrons from NADH to O_2 to form water. He envisioned that energy was generated at the inner mitochondrial membrane separating two phases. For him, this was energy generation at the edge, and this edge was essential. The key to the proposal was that energy could be conserved by the transfer of a charged particle from one phase to another as long as the phases had differing electrochemical potentials. Assuming a mechanism to capture that energy to be available, no chemical intermediate would be necessary. This is life on the edge.

We now know that protons are extruded at site I; NADH dehydrogenase multienzyme complex (EC 1.6.5.3) catalyzes the reaction:

$$NADH^+ + ubiquinone \leftrightarrow NAD^+ + ubiquinol;$$

at site III, ubiquinone cytochrome C reductase (EC 1.10.2.2) catalyzes the reaction:

$$QH_2 + 2 \text{ ferricytochrome C} \leftrightarrow Q + 2 \text{ ferrocytochrome C;}$$

and complex IV, cytochrome oxidase (EC 1.9.3.1), catalyzes the reaction:

$$2 \text{ ferrocytochrome C} + \tfrac{1}{2} O_2 \to 2 \text{ ferricytochrome C} + H_2O.$$

The mechanism of proton extrusion at each site in the mitochondrial membrane has not been worked out. Furthermore, the exact number of protons extruded has been the subject of a lively controversy for a number of years — particularly in the results obtained from the oxygen pulse experiments used by Mitchell and others, which gave 2 protons extruded at each site. From a thermodynamic point of view, it is unlikely that only 2 mol of protons extruded at sites I, III, and IV could yield the –53 to –60 kJ per mole needed to produce three ATPs.

From the equations cited later, in order for only two protons to provide the energy necessary to synthesize, a molecule of ATP would require an electrical potential between mitochondria and cytoplasm approaching –300 mV — above that which even a black lipid membrane could bear. The inner mitochondrial lipid bilayer, traversed by many transmembrane proteins, could stand much lower electric potentials between phases before it developed "leaks" or was destroyed. An excellent review by Martin Brand summarizing the work of many complex and contradictory experiments done over many years suggests that the classical estimate of three ATPs resulting from the metabolism of NAD-linked substrates is produced from the extrusion of four protons for each ATP generated (16).

The protons translocated out of mitochondria reenter the mitochondria, where they power three separate reactions acting in concert to produce ATP from the energy of the H^+ gradient between cytoplasm and mitochondria: the phosphate–H^+ cotransporter (17):

$$H_2PO_4^-{}_{out} + H^+{}_{out} \rightarrow H_2PO_4^-{}_{in} + H^+{}_{in};$$

the adenine nucleotide translocator (18):

$$ATP^{4-}{}_{in} + ADP^{3-}{}_{out} \rightarrow ATP^{4-}{}_{out} + ADP^{3-}{}_{in};$$

and the FiFo ATP synthase (19–22):

$$3H^+{}_{out} + ADP^{3-}{}_{in} + H_2PO_4^-{}_{in} \rightarrow ATP^{4-}{}_{in} + 3H^+{}_{in};$$

for the net reaction of oxidative phosphorylation:

$$H_2PO_4^-{}_{out} + ADP^{3-}{}_{out} + 4H^+{}_{out} \rightarrow ATP^{4-}{}_{out} + 4H^+{}_{in}.$$

This sequence of reactions shows that three protons traverse the ATP synthase site and one proton enters the mitochondria on the Pi–H^+ cotransporter, yielding a net stoichiometry of four protons from outer to inner phase for each ATP synthesized (16). The entire complex of three separate enzymes has now been isolated from the mitochondrial inner membrane together (23) and comprise the enzymatic mechanism required to transport the precursors and products of the net reaction between cytoplasm and mitochondria and to synthesize ATP.

The first two carrier enzymes are dimeric polypeptides. $H_2PO_4^-$ is transported into the mitochondria powered only by the small energy of the chemical concentration gradient of H^+ between the two phases. In contrast, the exchange of ATP_4^- for ADP_3^- is powered by the much larger negative electrical potential maintained within the mitochondria strongly favoring the extrusion of mitochondria ATP_4^- and importation of cytoplasmic ADP_3^-. This accounts for the observation that total measured ADP is in the millimolar range and the free cytoplasmic ADP is in the micromolar range in all cells containing mitochondria (4).

The ATP synthase is a multienzyme complex of almost unimaginable elegance with over 30 subunits. Remarkably, the γ-subunit spins around inside a triplet of α- and β-subunits synthesizing one ATP at each 120° turn or three ATPs for a complete revolution. The direction of the turning has even been observed and shown to be reversible. The remarkable implication is that the process approaches 100% efficiency and that the ATP synthase complex is a miraculous molecular machine, the details of which could not have been predicted by any scientist prior to the detailed work elucidating its structure and enzymatic mechanism (21).

An Energetic Statement of Mitochondrial Oxidative Phosphorylation

From the preceding discussion, one can write a simple thermodynamic statement that describes the sum reaction of

- Pi–H$^+$ cotransport
- Adenine nucleotide translocation
- ATP synthase reactions

Working together, these comprise mitochondrial ATP synthesis powered by the energy of the electrochemical gradient of protons extruded from mitochondria during electron transport. That statement is:

$$\frac{\Delta G'_{ATP}}{4} = RT\ln\frac{[\text{H}^+]_{mitochondria}}{[\text{H}^+]_{cytoplasm}} + FE_{mito/cyto} \qquad (2.7)$$

where the first term on the right-hand side of the equation is the chemical energy arising from the small difference in H$^+$ ion concentration between the mitochondrial and cytoplasmic phases, and $E_{mito/cyto}$ is the electrical potential of the mitochondrial phase relative to the cytoplasmic phase. The $\Delta G'_{ATP}$ is the chemical energy required to synthesize ATP from ADP and Pi, and 4 is the net number of protons moving from cytoplasm to mitochondria powering the three enzymatic processes required for ATP synthesis. The energy of the proton gradient returning from cytoplasm to mitochondria is:

$$\Delta G'_{\text{H}^+ mito/cyto} = RT\ln\frac{[\text{H}^+]_{mitochondria}}{[\text{H}^+]_{cytoplasm}} + FE_{mito/cyto} \qquad (2.8)$$

We have attempted to determine the factors relevant to Equation 2.7 in intact tissue — namely, the isolated working rat heart perfused with glucose alone, glucose + insulin, glucose + 4 mM D-β-hydroxybutyrate and 1 mM acetoacetate, or a combination of the three (24). The following emerged: the pH of cytosol was about 7 in all conditions. The estimated mitochondrial pH was 7.1 during perfusion with glucose alone and rose to 7.2 on addition of insulin. Addition of ketone bodies elevated mitochondrial pH to 7.5 and the combination to 7.4. This suggests that only a minor portion of the energy of phosphorylation is derived from the chemical energy of the proton gradient.

The majority of the energy was derived from the electrical gradient between the mitochondrial and cytosolic phases. The $\Delta G'_{ATP}$ ranged from −55.5 kJ/mol during perfusion with glucose alone and rose to −58 to −59 kJ/mol under all other conditions. The inferred $E_{mito/cyto}$ was −140 mV during perfusion with glucose or glucose + insulin; this fell to between −120 and −125 mV on addition of ketone bodies. The redox energy between site I (the NADH dehydrogenase multienzyme complex) and site III (ubiquinone cytochrome C reductase) was from −53 kJ/mol during perfusion with glucose alone and rose to −58 kJ/mol on addition of insulin, −59 kJ/mol with ketones, and −60 kJ/mol on addition of the combination. These values agree very well with the independently measured $\Delta G'$ of ATP hydrolysis, which ranged from −55 to −59 kJ/mol (see Figure 2.2).

The picture that emerges from studies of whole tissue is quite compatible with the formulation or generation of protons from the redox energy of the respiratory chain and the net process of ATP synthesis. It is also compatible with the movement of four protons from cytoplasm to the mitochondrial phase in a process powered by an electrical potential between mitochondria and cytoplasm of −120 to −140 mV and the chemical energy of a H^+ ion gradient of between 0.1 and 0.5 pH units. In turn, this is compatible with the original hypothesis put forward by Mitchell, with the exception of the stoichiometry of the protons involved. However, as is the case with all great discoveries, the establishment of Peter Mitchell's original

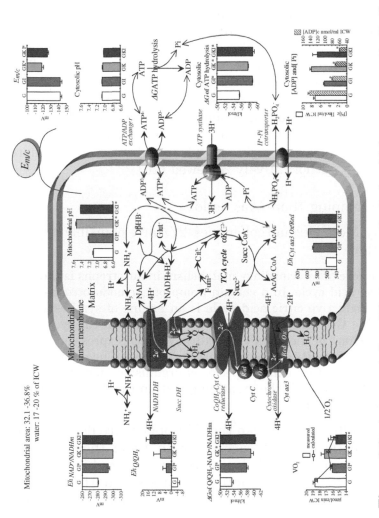

Figure 2.2 A representation of the processes of mitochondrial energy generation in the working rat heart perfused with glucose alone, glucose + insulin, glucose + ketone, or glucose + insulin and ketones. For the measurements required and the method of calculation, see reference 24. (From K. Sato et al, *FASEB J.* 9:651–658, 1995.)

hypothesis took the work of several hundred scientists over many years and was built on energetic foundations going back at least 100 years.

THE ENDOSYMBIOSIS OF ORGANELLES

Mitochondrial Reactions

The Citric Acid Cycle and its Relationship to
        Reducing Power

So far, this discussion has addressed two of the basic metabolic pathways of mitochondria: electron transport and the process of ATP synthesis. The citric acid cycle is another major metabolic pathway that occurs in mitochondria (25). This cyclic pathway accounts for the degradation of most foodstuffs, carbohydrate, most amino acids, and fats to CO_2 and reducing power, in the form of mitochondrial NADH or reduced flavoproteins. The point at which various foodstuffs enter the citric acid cycle —at acetyl CoA, α-ketoglutarate, or elsewhere — has been authoritatively reviewed (26).

In the Beginning There Was Redox

Recently, Morowitz and colleagues (27) have plausibly hypothesized that

> The core of intermediary metabolism in autotrophs is the citric acid cycle. In a certain group of chemoautotrophs, the reductive citric acid cycle is an engine of synthesis, taking CO_2 and synthesizing the molecules of the cycle. The chart of the metabolic pathways is an expression of the universality of intermediary metabolism. The reaction networks of all extant species of organisms map onto a single chart; it is the great unity within the diversity of the living world.

The mitochondria, like the chloroplast, are derived from endosymbiotic bacteria carrying with them their electron transport system (28). Reducing power produced in the citric acid cycle to produce metabolic energy in the form of the pyrophosphate bond of ATP requires that the reducing equivalents produced have O_2 so that the electron transport system can convert the reducing energy into the pyrophosphate bond energy of ATP. However, eukaryotes have existed on Earth for about 1.5 billion years and, until approximately 0.6 billion years ago, the seas were largely anoxic with a high sulfur content (29). What good would it be to have mitochondria with their citric acid cycle if no O_2 was available to convert the reducing power derived from substrates to produce ATP?

Harold Morowitz has postulated that the substrates of the citric acid cycle form the earliest and most fundamental pathway of metabolism, which is central across all life forms (27). He postulates that, in an anoxic world, prior to the generation of O_2 by other endosymbiont chloroplasts, the citric acid cycle took CO_2 and reducing equivalents and converted them into the multiple metabolic intermediates of carbohydrate, fats, and amino acids — instead of converting the ingested carbohydrates, fats, and amino acids to CO_2 and reducing equivalents. Hans Krebs, who was well versed in the absolute constraints of thermodynamics, took great pains to explain how a cyclic, not a linear, chemical process could work. He was highly amused by the young girl who asked him, "Why does the cycle run clockwise, not counterclockwise?" If one accepts Morowitz's reasoning, then the young girl's question was profound. The biochemical point illustrated is that the equilibrium constants of the reactions of the citric acid cycle are such that the direction of the cycle can be determined by the redox state of the cofactors involved (Figure 2.3).

Gibbs and the Heterotrophes Dilemma: the Differing Energy of Substrates

Substrates of Differing Energies Must Traverse the Same Pathway

Willard Gibbs' seminal treatise "On the equilibrium of heterogeneous substances" defines laws that are absolute and must be obeyed (30). They present a problem to heterotrophes that has not, until now, been much considered. However, if all metabolic substrates feed into the same citric acid cycle in the

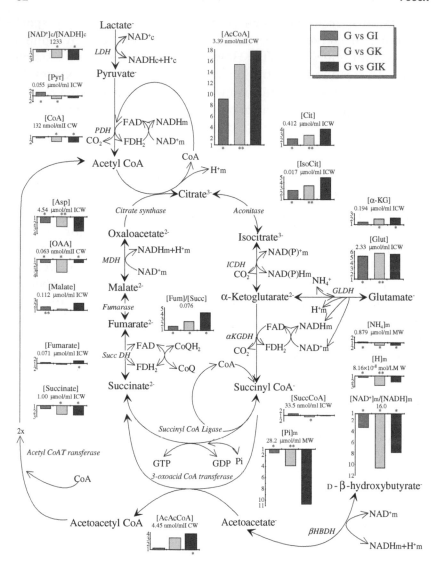

Figure 2.3 Changes in the intermediates of the citric acid cycle in the working rat heart perfused under four different conditions. The bar graphs represent the fold changes in metabolite concentrations and free cytosolic and mitochondrial [NAD⁺]/[NADH] relative to perfusion with glucose alone. For the methods of measurement see reference 24. (From K. Sato et al., *FASEB J.*, 9:651–658, 1995.)

Table 2.2 Heats of Combustion of Substrates in $\Delta H°$

	Formula	C_2 Unit	Kilojoules per C_2 unit
Palmitic	$C_{16}H_{32}O_2$	$C_2H_4O_{1/4}$	−1247
Butyric	$C_4H_8O_2$	C_2H_4O	−1088
D-β-Hydroxybutyric	$C_4H_8O_3$	$C_2H_4O_{3/2}$	−1021
Glucose	$C_6H_{12}O_6$	$C_2H_4O_2$	−933
Acetic	$C_2H_4O_2$	$C_2H_4O_2$	−874
Pyruvic	$C_3H_4O_3$	$C_2H_{8/3}O_2$	−778

present oxidative world, how are various energies inherent in these different compounds accommodated if the $\Delta G'$ of ATP is limited to between −53 and −60 kJ/mol? The fixed stoichiometry of the citric acid cycle and electron transport dictate that all non-nitrogenous foodstuffs are finally converted to CO_2 and H_2O — that is, they are combusted. The differences in the inherent energies contained in the foodstuffs metabolized by these common fixed pathways present a problem for mitochondria (Table 2.2).

The overall reaction of the citric acid cycle is to convert foodstuffs to reducing equivalents and CO_2. The reducing equivalents then enter the electron transport system to combine ultimately with O_2 to form H_2O while releasing their chemical energy. The overall reaction of these two pathways is combustion, but at much lower temperatures. However, even though these substrates enter the citric acid cycle by forming acetyl CoA, they possess different states of oxidation. Put another way, all the C_2 units metabolized by the citric acid cycle were not created equal because their inherent energies released by combustion differ by a factor of almost two. Yet, based on empirical measurements from a selection of freeze-clamped mammalian tissues, the energy of ATP hydrolysis, $\Delta G'_{ATP}$ is kept in a remarkably narrow range of between −53 and −60 kJ/mol (4,5), and varies about 10% — not by a factor of two.

During the mitochondrial metabolism of β-hydroxybutyric or pyruvic acids, four NADHs, one QH_2, and one substrate-level GTP are produced, requiring the formation of 15 ATP equivalents per C_2 unit converted to CO_2. The cost of those 15 terminal pyrophosphate bonds would therefore be between −795 and −900 kJ. One can immediately see that metabolism of pyruvate alone does not have the energy required to make 15 ATPs within the physiological range. Glucose produces one extra NADH per C_2 unit at the GAP dehydrogenase step of glycolysis, which, if transferred into the mitochondria by one or another shuttle for oxidation in the respiratory chain, produces five NADHs, one QH_2, and one GTP, requiring the synthesis of 18 ATP equivalents — not 15.

To produce 18 ATPs with $\Delta G'$ within the physiological range of −53 to −60 kJ/mol therefore requires 954 to −1080 kJ per C_2 unit. Glucose has only −933 kJ/C_2 unit, allowing it to produce ATP with energy at the lower end of the physiological range. This low energy of $\Delta G'_{ATP}$ can be observed in the working perfused rat heart during perfusion with glucose alone (24). In contrast to glucose, D-β-hydroxybutyrate has one-half less O_2 per C_2H_4 unit; that is, it is less oxidized than glucose and therefore possesses more inherent energy when converted to CO_2 and H_2O in a bomb calorimeter or by the reactions of intermediary metabolism. Ketone bodies may therefore be classified as a high-energy metabolite when compared to glucose.

This too can be observed in the working perfused rat heart, where a physiological ratio of ketone bodies was added to the perfusate. The result is a 28% increase in the hydraulic work of the heart — that is, an increase in the joules of hydraulic work observed vs. the joules obtained from converting NADH to H_2O in electron transport (24).

However, what happens when fatty acids, like palmitate, are converted to acetyl CoA in mitochondrial β-oxidation because palmitate is less oxidized than D-β-hydroxybutyrate? The answer to this question resides with the last of the major mitochondrial transport proteins, the uncoupling protein. Martin Klingenberg has argued that the mitochondrial uncoupling protein is simply a H^+–Pi cotransporter that has lost its Pi-binding site (18). However, during the elevation of free fatty acids within the cell, the presence of fatty acids causes this protein to catalyze the transfer of protons from cytoplasm to mitochondria, bypassing the ATP synthase and thus producing heat, not ATP.

Brown fat has been recognized in hibernating animals since the 1500s, and its role in sympathetic nervous controlled thermogenesis was recognized in the early 20th century. Uncoupling protein, specifically UCP1, has been known to be responsible for the generation of heat in brown fat and, by now, five isozymes of UCPs have been identified with variable expression in tissue. UCP2 and UCP3 are found in a variety of tissues, including heart and skeletal muscle. Administration of thyroid hormone increases the expression of UCP2 and UCP3 in heart. When perfused with palmitate, the heart has a marked decrease in cardiac efficiency (31).

Elevation of free fatty acids induces the peroxisomal proliferator-activated nuclear receptors called PPAR α- and, to some extent, β-prostaglandins and leukotrienes that also activate PPAR γ. Activation of these PPARs results in the proliferation of peroxisomes, which oxidize fatty acids without coupling the energy produced, and they increase the transcription of uncoupling proteins in the mitochondria. Their catalysis of the movement of protons into the mitochondrial matrix results in the production of heat rather than ATP. When coupled to uncoupling protein, these controls on protein transcription and action of fatty acid allow for the dispersal the inherently high energy contained in the C_2 units of fatty acids as heat.

The Centrality of the ΔG′ of ATP

It is a truism that ATP is the "energy currency" of the cell because it is used in transport work, mechanical work, and chemical work. A 70-kg man contains about 50 g of ATP; however, consuming and expending 2500 calories per day would result in the synthesis of about 180 kg of ATP (32). By now, this is textbook stuff, as is the statement that the hydrolysis of ATP produces about −30 kJ/mol. I hope that this is recognized to be the $\Delta G°$ of ATP, not the actual $\Delta G'$ of ATP, which is −53 to −60 kJ/mol in all living tissues so far tested. The remarkable constancy of this value is surprising as well as intriguing.

The red cell, which produces all of its ATP by the process of glycolysis, achieves essentially the same $\Delta G'$ of ATP hydrolysis as do heart, skeletal muscle and liver (4,5) producing ATP by entirely different mitochondrial pathways. There is no inherent reason why these two very different processes should produce ATP with the same $\Delta G'$. How could these two very different processes achieve the same energy level in ATP? I have no ready answer to the question. The only speculation is that living mammalian cells must have it that way. The circularity of that statement is obvious, but let me suggest an answer.

Heart uses ATP mainly to do mechanical work, the liver mainly to do chemical work, and the red cell mainly pumps ions. In fact, all of these tissues pump ions. If any tissue is injured by any means and loses cellular energy, they all go through a stereotype response of losing K^+, gaining Na^+ and Ca^{2+}, and swelling. How can this be, if the electrical potential between the inside and outside phase of the cells varies so greatly — from –83 mV in resting excitable tissues of heart, muscle, and nerve to between –26 and –40 mV in visceral organs like liver and to –7 mV in red cell. This occurs despite the wide difference in voltage and the fact that all cells use ATP to power the Na^+ pump (EC 3.6.1.37).

I suggest that the only difference is the degree to which other ions are permeant — that is, the ions free to distribute themselves between inner and outer phases of the cell in accordance with the potential between phases, $E_{out/in}$ differ between the different cell types. In the excitable tissues, K^+ is permeant; in liver Cl^- is permeant; and in red cell, HCO_3^- and Cl^- are permeant. That is, the energy of the chemical concentration gradient is equal to the E between extra- and intracellular phases analogous to Equation 2.8.

In mitochondria, like their bacterial precursors, no ion is permeant, except under catastrophic circumstance when the so-called transition pore opens and the mitochondrial potential and ATP synthetic capacity are destroyed by K^+ movement. Unlike bacterial and plant cells with a rigid wall that can withstand pressure, animal cell walls made of lipoprotein bilayers cannot stand pressure. Furthermore, all cells contain large amounts of charged impermeant anionic materials, like ATP_4^-, PCr_2^-, phosphorylated glycolytic intermediates, and di- and tricarboxcylic anions. Impermeant charged ions form a Gibbs–Donnan equilibrium system, one energy component of which is pressure (33). In order for animal cells to counter this anionic intracellular Donnan, they must create a cationic extracellular Donnan system to make the activity of H_2O equal in both phases, thus avoiding pressure. They use the Na^+ pump to create a cationic extracellular Gibbs–Donnan system equal in magnitude and opposite in sign to the intracellular anionic Gibbs–Donnan system. The reaction catalyzed by the Na pump is:

$$3Na_i^+ + 2K_o^+ + ATP \rightarrow 3Na_o^+ + 2K_i^+ + ADP + Pi$$

in heart, where there is an open K^+ channel and electrical neutrality is achieved by the movement of a K^+ from out to in.

The extent and direction of the Na^+ gradient across and between the intra- and extracellular phase is thus equal to the energy of one-third of the $\Delta G'$ of ATP hydrolysis (33). The basic equations that apply to mitochondrial ion movements between phases with different electrical potential apply to these movements across the plasma membrane.

In tissues with a lower potential, where there is no open K^+ channel, charge neutrality is maintained in the reaction of the pump through the linked movement of other ions to achieve electrical and osmotic neutrality. In these cases, the equation representing the relationship between the energy of the distribution between extracellular and intracellular phases of the nine major inorganic ions and the $\Delta G'$ of ATP hydrolysis may be described by an osmoneutral, electroneutral statement reflecting the sum reaction of the sodium pump and its linked plasma membrane ion transporters (5,33) (see Equation 2.9):

$$\Delta G'_{ATP} + RT \ln$$

$$\frac{[Ca^{2+}]_o[Na^+]_o^2[Mg^{2+}]_i[H^+]_i[K^+]_i^2[Cl^-]_o[HCO_3^-]_i[H_2PO_4^-]_o[HPO_4^{2-}]_i}{[Ca^{2+}]_i[Na^+]_i^2[Mg^{2+}]_o[H^+]_o[K^+]_o^2[Cl^-]_i[HCO_3^-]_o[H_2PO_4^-]_i[HPO_4^{2-}]_o} = 0$$

$$(2.9)$$

The arrangement of the energy of the cations and anions can best be appreciated graphically (Figure 2.4).

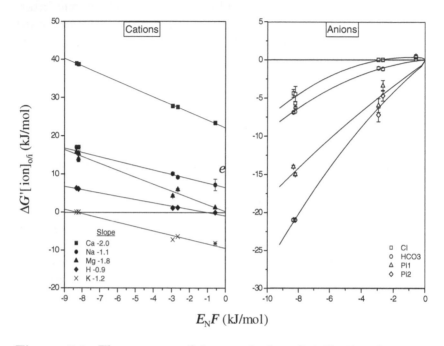

Figure 2.4 The energy of inorganic ion distribution between extra- and intracellular phase of working perfused rat heart, freeze clamped liver, and red cell vs. the measured electric potential between phases × the Faraday. This figure presents the energy of the distribution of the nine major inorganic ions between the intra- and extracellular phases, $\Delta G'$ [ion^z]$_{o/i}$ of mammalian cells of differing electrical potentials, E_N, where z is the valance of the ion and F is the familiar Faraday constant. The measured data are from working perfused rat heart and freeze-clamped rat liver under several conditions and from red cell under one condition (5). The mean $\Delta G'$ [ion^z]$_{o/i}$ ± SEM is plotted on the y-axis for each ion; this was calculated, just as the energy of the proton gradient across mitochondrial membrane was calculated (Equation 2.8), by the formalism:

$$\Delta G'[\text{ion}^z]_{o/i} = RT\ln\frac{[\text{ion}^z]_o}{[\text{ion}^z]_i} - zFE_N$$

Figure 2.4 (continued)

When $\Delta G'$ [ion^z]$_{o/i}$ is zero, this indicates that the concentration energy of the ion gradient was equal in magnitude and opposite in sign to the electrical energy of the ion gradient; in other words, the distribution of the ion acted as if it were traversing an open ion channel. $\Delta G'[K^+]_{o/i}$ was zero in heart and $\Delta G'[Cl^-]_{o/i}$ was zero in liver; $\Delta G'[Cl^-]_{o/i}$ and $\Delta G'[HCO_3^-]_{o/i}$ were zero in red cell. The energy of all the cations, except K^+ in heart, was positive, indicating that all cations were activity "pumped" out of the negative intracellular phase of the cells creating a cationic extracellular Gibbs–Donnan equilibrium system. In contrast, the energy of the anionic gradients was negative, indicating that they were actively "pumped" into the cell contributing to the anionic Gibbs–Donnan system. In contrast to the cations, the slopes of the change of the energy of anionic distributions were not linear, presumably indicating the differing amounts of impermeant intracellular anionic metabolites. The energy of all ions that were not zero contributes to the extra- and intracellular Gibbs–Donnan system; this makes the activity of H_2O on both sides of the membrane zero and thus maintains a constant volume in the steady state. The entire system of balanced ionic gradients depends upon the energy of ATP hydrolysis (Equation 2.9).

Just as the energy of four protons between the mitochondrial and cytosolic phase are equal to the $\Delta G'_{ATP}$, Equation 2.9 says that the energy of the distribution of the major inorganic ions between the extra- and intracellular phases is also related to the energy of hydrolysis of ATP in a Gibbs–Donnan near-equilibrium system — even when the electrical potential between the phases varies from −83 to −7 mV. The same fundamental principles that govern mitochondrial ATP synthesis may be applied to illuminate understanding of the creation of cellular ionic gradients. It explains why, when cellular energy metabolism is disrupted by any means, the stereotype response of ion movement occurs. The converse is also true: to maintain ionic homeostasis requires the maintenance of the $\Delta G'_{ATP}$.

Diseases of Mitochondrial Energy Metabolism and Substrate Selection

Thirty years ago, it was thought that the pathways of mitochondrial energy metabolism were so fundamental that no severe impairment of these pathways was compatible with life. In the last 20 years, diseases specific to mutations within mitochondrial diseases have been recognized (34). In general, these specific diseases of mitochondrial metabolism are rare. Within the last decade it has been proposed that some very common complex diseases result from disordered mitochondrial energy metabolism, including those caused by disordered substrate utilization.

It has recently been reported that the hearts of individuals with type II diabetes, which appear normal in most measurable parameters, have a lowered PCr/ATP ratio indicative of a low $\Delta G'$ of ATP hydrolysis that can be correlated with the levels of circulating free fatty acids (35). Others have presented evidence that common neurodegenerative diseases such as Parkinson's and Alzheimer's diseases result from specific defects in mitochondrial metabolism. It has further been suggested that changes in substrate availability — specifically the provision of the high-energy ketone bodies in the absence of free fatty acids — may provide therapeutic benefit in these disease phenotypes (36,37). Study and thorough understanding of mitochondrial energy metabolism in an age of molecular medicine may be unfashionable, but will probably be very productive.

REFERENCES

1. P Mitchell. Chemiosmotic coupling in oxidative and photosynthetic phosphorylation. *Biol Rev Camb Philos Soc* 41:445–502, 1966.

2. P Mitchell. *Chemiosmotic Coupling and Energy Transduction.* Bodmin: Glynn Research Ltd.; 1968.

3. I Mills, T Cvitas, N Kallay, K Homann, K Kuchitsu. *Quantities, Units and Symbols in Physical Chemistry.* Oxford: Blackwell Scientific; 1988.

4. RL Veech, JWR Lawson, NW Cornell, HA Krebs. Cytosolic phosphorylation potential. *J Biol Chem* 254:6538–6547, 1979.

5. RL Veech, Y Kashiwaya, DN Gates, MT King, K Clarke. The energetics of ion distribution: the origin of the resting electric potential of cells. *IUBMB Life* 54:241–252, 2002.

6. DH Williamson, P Lund, HA Krebs. The redox state of free nicotinamide-adenine dinucleotide in the cytoplasm and mitochondria of rat liver. *Biochem J* 103:514–527, 1967.

7. PC Engle, K Dalziel. The equilibrium constants of the glutamate dehydrogenase systems. *Biochem J* 105(2):691–695, 1967.

8. WM Clark. *Oxidation Reduction Potentials of Organic Systems*. Baltimore: Williams and Wilkins Co.; 1960.

9. K Burton, TH Wilson. The free-energy changes for the reduction of diphosphopyridine nucleotide and the dehydrogenation of l-malate and l-glycerol 1-phosphate. *Biochem J* 54:86–94, 1953.

10. HA Krebs, RL Veech. Pyridine nucleotide interrelations. In: S Papa, JM Tager, E Quagliariello, EC Slater, Eds. *The Energy Level and Metabolic Control in Mitochondria*. Adriatica Editrice: Bari; 1969, 329–382.

11. Q Zhang, DW Piston, RH Goodman. Regulation of corepressor function by nuclear NADH. *Science* 295(5561):1895–1897, 2002.

12. CC Fjeld, WT Birdsong, RH Goodman. Differential binding of NAD+ and NADH allows the transcriptional corepressor carboxyl-terminal binding protein to serve as a metabolic sensor. *Proc Natl Acad Sci USA* 100(16):9202–9207, 2003.

13. L Guarente. SIR2 and aging — the exception that proves the rule. *Trends Genet* 17:391–392, 2001.

14. M Fulco, RL Schiltz, S Iezzi, MT King, P Zhao, Y Kashiwaya. Sir2 regulates skeletal muscle differentiation as a potential sensor of the redox state. *Mol Cell* 12:51–62, 2003.

15. J Rutter, M Reick, LC Wu, SL McKnight. Regulation of clock and NPAS2 DNA binding by the redox state of NAD cofactors. *Science* 293:510–514, 2001.

16. M Brand. The stoichiometry of proton pumping and ATP synthesis in mitochondria. *Biochemist* 16:20–24, 1994.

17. JP Wehrle, PL Pedersen. Phosphate transport processes in eukaryotic cells. *J Membr Biol* 111:199–213, 1989.

18. M Klingenberg. The mitochondrial carrier family involved in energy transduction. In: F Palmieri, E Quagliariello, Eds. *Molecular Basis of Biomembrane Transport*. Amsterdam: Elsevier Science; 1988, 141–153.

19. PD Boyer. A research journey with ATP synthase. *J. Biol. Chem* 277:39045–39061, 2002.

20. C Gibbons, MG Montgomery, AGW Leslie, JE Walker. The structure of the central stalk in bovine F1-ATPase at 2.4-A resolution. *Nat Struc Biol* 7:1055–1061, 2000.

21. K Yasuda, H Noji, K Kinosita, Jr, M Yoshida. F1-ATPase is a highly efficient molecular motor that rotates with discrete 120° steps. *Cell* 93:1117–1124, 1998.

22. PL Pedersen, LM Amzel. ATP synthases. Structure, reaction center, mechanism, and regulation of one of nature's most unique machines. *J Biol Chem* 268:9937–9940, 1993.

23. YH Ko, M Delannoy, J Hullihen, W Chiu, PL Pedersen. Mitochondrial ATP synthasome. Cristae-enriched membranes and a multiwell detergent screening assay yield dispersed single complexes containing the ATP synthase and carriers for Pi and ADP/ATP. *J Biol Chem* 278:12305–12309, 2003.

24. K Sato, Y Kashiwaya, CA Keon, N Tsuchiya, MT King, GK Radda, B Chance, K Clarke, RL Veech. Insulin, ketone bodies, and mitochondrial energy transduction. *FASEB J* 9:651–658, 1995.

25. HA Krebs, WA Johnson. The role of citric acid in intermediate metabolism in animal tissues. *Enzymologia* 4:148–156, 1937.

26. HA Krebs, A Kornberg, K Burton. *Energy Transformations in Living Matter*. Berlin: Springer-Verlag; 1957.

27. HJ Morowitz, JD Kostelnik, J Yang, GD Cody. The origin of intermediary metabolism. *Proc Natl Acad Sci USA* 97:7704–7708, 2000.

28. S Berry. Endosymbiosis and the design of eukaryotic electron transport. *Biochim Biophys Acta* 1606:57–72, 2003.

29. W Martin, C Rotte, M Hoffmeister, U Theissen, G Gelius–Dietrich, S Ahr. Early cell evolution, eukaryotes, anoxia, sulfide, oxygen, fungi first (?), and a tree of genomes revisited. *IUBMB Life* 55:193–204, 2003.

30. JW Gibbs. On the equilibrium of heterogeneous substances. *Trans Conn Acad Arts Sci* 3:108,248,343,524, 1875.

31. EA Boehm, BE Jones, GK Radda, RL Veech, K Clarke. Increased uncoupling proteins and decreased efficiency in palmitate-perfused hyperthyroid rat heart. *Am J Physiol Heart Circ Physiol* 280:H977–H983, 2001.

32. A Kornberg. *For Love of Enzymes.* Cambridge, MA: Harvard University Press; 1989.

33. T Masuda, GP Dobson, RL Veech. The Gibbs–Donnan near-equilibrium system of heart. *J Biol Chem* 265:20321–20334, 1990.

34. DC Wallace. Mitochondrial diseases in man and mouse. *Science* 283:1482–1488, 1999.

35. M Scheuermann–Freestone, PL Madsen, D Manners, AM Blamire, RE Buckingham, P Styles. Abnormal cardiac and skeletal muscle energy metabolism in patients with type 2 diabetes. *Circulation* 107:3040–3046, 2003.

36. RL Veech, B Chance, Y Kashiwaya, HA Lardy, CF Cahill, Jr. Ketone bodies, potential therapeutic uses. *IUBMB Life* 51(4):241–247, 2001.

37. RL Veech. The therapeutic implications of ketone bodies: the effects of ketone bodies in pathological conditions: ketosis, ketogenic diet, redox states, insulin resistance, and mitochondrial metabolism. *Prostaglandins Leukotrienes Essential Fatty Acids* 70:309–319, 2004.

38. D Veloso, GW Guynn, M Oskarsson, RL Veech. The concentrations of free and bound magnesium in rat tissues. Relative constancy of free Mg^{2+} concentrations. *J Biol Chem* 248:4811–4819, 1973.

39. RL Veech, DN Gates, CW Crutchfield, WL Gitomer, Y Kashi-waya, MT King. Metabolic hyperpolarization of liver by ethanol: the importance of Mg2+ and H+ in determining impermeant intracellular anionic charge and energy of metabolic reactions. *Alcohol Clin Exp Res* 18:1040–1056, 1994.

40. RP Feynman. *The Character of Physical Law.* Cambridge, MA: MIT Press; 1967.

3

The Human Mitochondrial Genome: Mechanisms of Expression and Maintenance

JAN-WILLEM TAANMAN AND
SIÔN LLEWELYN WILLIAMS

CONTENTS

INTRODUCTION

Mitochondria are essential eukaryotic organelles whose principal function is to generate ATP through the process of oxidative phosphorylation (1). They also house many additional metabolic pathways and are involved in Ca^{2+} homeostasis, apoptosis, and thermogenesis (2–4). Mitochondria possess two membranes: an outer membrane, which encloses the organelle and separates it from the cytosol, and a larger inner membrane. The latter can be further divided into two contiguous components: the inner boundary membrane that apposes the outer membrane, and the cristae membrane that projects into the matrix compartment (5,6). The five multisubunit complexes (I through V) of the oxidative phosphorylation system required for ATP synthesis are embedded in the cristae membrane (7).

Mitochondria are descendants of α-proteobacteria that formed an endosymbiotic relationship with ancestral eukaryotic organisms (8). The organelle still carries many hallmarks of its proteobacterial progenitor. For instance, chloramphenicol and other inhibitors of prokaryotic protein synthesis also inhibit mitochondrial protein synthesis, which is insensitive to cycloheximide, an inhibitor of cytosolic protein synthesis (9). The first reports that showed the ability of mitochondria to synthesize proteins, published in the late 1950s (10,11), were therefore met with skepticism because many believed that this chloramphenicol-sensitive, protein synthetic activity originated from contaminating bacteria. Nevertheless, during the early 1960s, when it was demonstrated that mammalian mitochondria contained RNA, the notion that mitochondria are able to synthesize proteins won increasing support (12).

In 1963, DNA was discovered within mitochondria (13). This finding suggested that mitochondria not only translate mRNA into protein, but that the very genes for these proteins are also present in the organelles (14). In the following years, it became clear that mitochondrial DNA (mtDNA) consists of multiple copies of a small DNA molecule, and that the organelle encodes its own rRNAs and tRNAs to translate a limited number of protein genes crucial for oxidative phosphorylation. During its evolution into the present-day powerhouse of the eukaryotic cell, the endosymbiont transferred many of its genes to the nucleus and the organelle gained additional eukaryotic functions. Consequently, the vast majority of the mitochondrial proteins are imported nuclear gene products, synthesized on cytosolic ribosomes.

Although mitochondria have become largely dependent on nuclear-encoded factors, some autonomy has remained. Replication of mtDNA is independent of the cell cycle (Figure 3.1) (15,16) and persists for several hours in enucleated cells (16). Replication, transcription, and translation of mtDNA can be maintained in isolated mitochondria for some time (17,18), and replication and transcription can be studied with mitochondrial extracts (19). These experimental opportunities and the fact that mtDNA is a relatively small but abundant DNA species have greatly facilitated the first 25 years of mtDNA research. Since the late 1980s, it has become evident that qualitative and quantitative changes in human mtDNA play a causal role in a large number of clinically heterogeneous diseases (20,21). This recognition has provided a renewed impetus for mtDNA research.

By definition, mitochondria in all organisms are able to carry out two essential functions: the expression of an integral genome and the generation of ATP coupled to electron transport. The structure and function of the mitochondrial ATP-producing system has been reviewed in detail elsewhere (1). This chapter focuses on the mechanisms and factors involved in transcription, translation, and replication of the human mitochondrial genome. When relevant, comparisons with the mitochondrial genome of other organisms are made. We hope that this chapter will provide a useful resource for those involved with diseases of mtDNA.

Throughout the text of this chapter, human gene and protein symbols are used as recommended by the HUGO Gene Nomenclature Committee (http://www.gene.ucl.ac.uk/nomenclature). Human gene symbols consist of italic uppercase Roman letters and Arabic numbers, and human proteins are

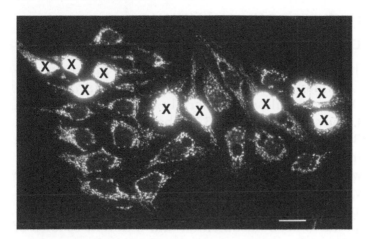

Figure 3.1 Replication of mtDNA is independent of the cell cycle. An asynchronous culture of human HeLa S3 cervical carcinoma cells was grown in the presence of 15 µM 5-bromo-2′-deoxyuridine (Br-dU) for 3 h, followed by fluorescent, immunocytochemical detection of incorporated Br-dU as described (16). Br-dU is a thymidine analog that incorporates into replicating nuclear DNA and mtDNA. Some cells in this image are in the S-phase of the cell cycle and, consequently, incorporate large amounts of Br-dU into their nuclear DNA (cell nuclei indicated with a cross). The cells that do not show nuclear incorporation are in the G_1- or G_2-phase of the cell cycle. All cells show a punctuate staining pattern throughout their cytoplasm. This staining represents mtDNA replicated during the Br-dU labeling period (16). Because all cells show a similar amount of newly replicated mtDNA, it is concluded that mtDNA replication is cell cycle independent. Scale bar: 10 µm.

designated by the same uppercase characters as the gene symbols but in regular font. All mitochondrial genes start with the letters "*mt*." This nomenclature will help the reader to cross reference with human genomic databases and may end some of the confusion caused by alternative names used by different groups over the years; however, common aliases are given in all tables.

GENERAL FEATURES OF HUMAN MTDNA

Structure, Copy Number, and Organization

The mitochondrial genome is located in the matrix of the organelle. Human mtDNA is a closed-circular, double-stranded DNA molecule and generally exists as supercoiled circles (22). The sequence of its 16,569 base pairs was reported in 1981 (23) and the published sequence was reanalyzed and corrected in 1999 (24; for the human mtDNA sequence, see: http://www.ncbi.nlm.nih.gov, file NC_001807 or http://www.mitomap.org/). Human cells may contain thousands of copies of mtDNA per cell, though large variations per cell type have been documented and different studies of the same cell type sometimes show considerable discrepancies (Table 3.1). The copy number of mtDNA is reduced several fold during spermatogenesis (25), but appears to increase dramatically during oogenesis (Table 3.1) (26–28). Experiments with early mouse embryos have revealed that, following fertilization, the total mtDNA copy number of the developing embryo remains constant until gastrulation (29).

In the 1960s and early 1970s, equilibrium CsCl gradient centrifugation experiments of mammalian mtDNA led to a number of important findings. In combination with electron microscopy, the technique revealed the existence of two oligomeric forms of the circular mtDNA molecule, the unicircular dimer (two genomes linked head to tail) and catenated oligomers (two or more monomeric circles interconnected like links in a chain) (22,30–33). The standard monomeric form predominates in normal tissues, which usually have no unicircular dimers and 10% catenated molecules. In malignant tissues, particularly leukemic cells, a substantial fraction of unicircular dimers is present (34,35).

Since 1988, a large number of rearrangements of mtDNA have been identified, in the form of deletions and partial duplications, which are associated with a variety of diseases affecting heart, skeletal muscle, brain, and other organs (36–38). These pathological rearrangements always coexist with wild-type mtDNA in variable proportions, a situation known as heteroplasmy. Very low levels of rearranged

Table 3.1   Reported mtDNA Copy Numbers in Human Tissues and Cultured Cells

Tissue or Cell Type	mtDNA Copy Number (mean ± SD)	Ref.
Cardiac muscle	6970 ± 920/diploid nuclear genome	605
Skeletal muscle	3650 ± 620/diploid nuclear genome	605
	1811 ± 546/diploid nuclear genome	606
	1420 ± 281/diploid nuclear genome	607
Liver	738 ± 121/diploid nuclear genome	607
	1008 ± 198/diploid nuclear genome	607
Platelets	4/cell	608
Erythrocytes	0/cell	608
Spermatozoa	~1000/haploid nuclear genome	602
	~1500/haploid nuclear genome	609
Oocytes	130,000 ± 80,000/haploid nuclear genome	28
	193,000/haploid nuclear genome (range: 20,000–598,000)	27
	795,000 ± 243,000/haploid nuclear genome	26
Cultured primary fibroblasts	231 ± 54/diploid nuclear genome	607
	800 ± 24/diploid nuclear genome	610
	2400–6000/diploid nuclear genome	611
Cultured HeLa cells	7200/aneuploid nuclear genome	611
Cultured HeLa.TK⁻ cells	8800/aneuploid nuclear genome	612
Cultured HeLa S3 cells	7900 ± 1700/aneuploid nuclear genome	613
Cultured 143B.TK⁻ cells	9100 ± 1600/aneuploid nuclear genome	613

mtDNA, which resemble those found in pathological states, have also been detected in normal tissue (28,39). These molecules have been termed sublimons by analogy with plant mtDNA molecules with similar properties (39). Although each sublimon is typically present at a low level (at most, a few copies per cell), sublimon abundance in a given tissue can vary over three orders of magnitude between healthy individuals. They are most prominent in postmitotic tissue and, collectively, can represent a non-negligible percentage of total mtDNA (39). Sublimons accumulate in postmitotic tissues with age (40–43), but to levels at which they do not appear to have a phenotypic effect (43,44).

In addition to low levels of rearranged mtDNA molecules, the existence of mtDNA molecules with single base changes has been demonstrated in healthy individuals (45). It is thought that normal cells may contain hundreds of slightly different mtDNA sequences (45). Intercellular clonal expansions of these sequence variations create distinct spectra of mtDNA polymorphisms in individual cells of human tissues during aging (46,47).

The early CsCl gradient centrifugation experiments also revealed that, under alkaline denaturing conditions, the two strands of the mtDNA duplex have distinct buoyant densities ("heavy" and "light") as result of their different G+T base content (48). Furthermore, CsCl gradient centrifugation combined with electron microscopy showed that a large proportion of vertebrate mtDNA from metabolically active cells contain a short, three-stranded structure, called the displacement or D-loop, in which a short nucleic acid strand, complementary to the light (L) strand, displaces the heavy (H) strand (Figure 3.2) (49,50).

The short nucleic acid strand is traditionally termed the 7S DNA. In humans, the 7S DNA is about 650 nucleotide residues long. The D-loop region is defined to be flanked by the genes for tRNA^{Phe} (*MTTF*) and tRNA^{Pro} (*MTTP*) (51). The region has evolved as the major control site for mtDNA expression, containing the putative origin of replication and the major initiation sites for L- and H-strand transcription (Figure 3.2).

In the 1980s, studies in the budding yeast *Saccharomyces cerevisiae* revealed that the mitochondrial genome is organized as DNA-multiprotein complexes (52–54). These complexes, termed nucleoids, were later also visualized with DNA-specific, fluorescent dyes and probes in cultured mammalian cells as punctate submitochondrial structures (Figure 3.1) (16,55–62). In mammalian cells containing reticular mitochondria, nucleoids are dispersed throughout the mitochondrial network of the cell (55,56,58,61,62), and each organelle appears to contain at least one nucleoid in cells with

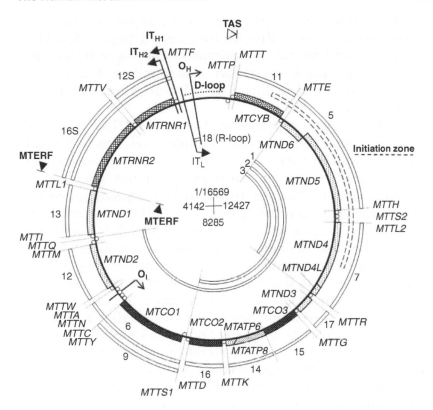

Figure 3.2 Genetic and transcriptional map of the human mitochondrial genome. The continuous circle represents the mtDNA duplex. Shaded boxes at outer side of circle = genes encoded on H-strand; shaded box at inner side of circle = gene encoded on L-strand; 22 tRNA genes = open dots; mapping position of 7S DNA species in control region (D-loop) = dotted line; origins of H-strand (O_H) and L-strand (O_L) replication and direction of DNA synthesis (according to strand-asynchronous replication model) = open arrows; position of TAS-element involved in premature termination of H-strand synthesis = blocked, open arrow head; initiation zone for replication according to strand-synchronous model = dashed open box; initiation of transcription sites for L- (IT_L) and H-strand (IT_{H1}, IT_{H2}), and direction of RNA synthesis = filled arrows; mapping positions of L-strand transcripts = open boxes within mtDNA circle; mapping positions of H-strand transcripts = open boxes outside mtDNA circle; transcription cleavage sites = dotted lines at boundaries of RNA species. Numbering of the nucleotides of the mtDNA duplex (1 to 16569) is indicated in the center. The nomenclature of the genes (in italic capitals) is explained in Table 3.2. The sites where transcription terminates is mediated by the MTERF protein (indicated by blocked, filled arrow heads). Transcripts are numbered according to size (1 to 18), except for the 12S and 16S rRNAs; transcript 18 represents the RNA at the R-loop.

fragmented mitochondria (58). Time-lapse fluorescence microscopy has shown that nucleoids are dynamic structures that divide and follow the movements of the mitochondrial network, presumably to ensure transmission of mtDNA to daughter mitochondria during mitochondrial growth and division (59,61,62).

Early electron microscopy studies of mtDNA, extracted from the human HeLa cervical carcinoma cell line under mild conditions and further purified by CsCl gradient centrifugation, revealed a protein-containing structure and a membrane-like patch attached to the D-loop region, suggesting that mtDNA is bound through protein to the inner mitochondrial membrane at or near the control region (63). More recent work in the toad *Xenopus laevis* showed that mtDNA binds tightly to a limited number of abundant mitochondrial inner membrane proteins that help to anchor it to the inner membrane (64). A combination of genetic and cytochemical experiments with *S. cerevisiae* has demonstrated that a subset of nucleoids lie adjacent to discrete outer membrane structures that contain the outer mitochondrial membrane proteins Mmm1p and/or Mmm2p (65,66). Both proteins have been shown to be required for mtDNA maintenance and also to function in the preservation of mitochondrial morphology, possibly by mediating attachments to extramitochondrial structures such as actin (66–68).

In agreement with these findings, immunocytochemical studies with the human ECV302 epithelial-like cell line have suggested that nucleoids are tethered directly or indirectly through mitochondrial membranes to the kinesin motor implicated in moving mitochondria along microtubules, because a component of that motor, KIF5B, was found to co-localize often with the nucleoids (61). Thus, a picture is emerging in which mtDNA is associated with a discrete proteinaceous structure that spans the outer and inner mitochondrial membranes at contact sites, i.e., regions where the outer membrane is in close apposition with the inner membrane. These proteinaceous structures appear to interact with the dynein/kinesin- and actin/myosin-based motors that allow the organelle to move along the microfilament and microtubule networks in the cytoplasm.

Human mitochondrial nucleoids are thought to contain 2 to 10 mtDNA molecules (55,61,62), but detailed information concerning the number of mtDNA molecules per nucleoid in different cell types and fluctuations in the number during the cell cycle is lacking. Although it is presumed that nucleoids contain many of the proteins necessary for mtDNA transactions, relatively little is known about the protein composition of nucleoids in mammalian cells. Immunocytochemical experiments have revealed that the putative mitochondrial helicase Twinkle (PEO1), the mitochondrial transcription factor TFAM, and the mitochondrial single-stranded DNA-binding protein SSBP1 co-localize with mtDNA (59,60,62), suggesting that these proteins are constituents of the nucleoid. The mtDNA polymerase (DNA polymerase γ), however, appeared to be present throughout the organelles (16,59). Biochemical isolation of nucleoids suggested that TFAM — possibly DNA polymerase γ and various other as yet unidentified proteins, but not SSBP1 — co-purify with mtDNA (59). A better characterization of mammalian nucleoids, using improved purification procedures, will be necessary to understand what controls nucleoid assembly and how they are transmitted to progeny cells.

Gene Content, Codon Usage, and Decoding Mechanism

Human mtDNA contains 37 genes (Table 3.2; Figure 3.2) (23). Most information is encoded on the H-strand, with genes for two rRNAs, 14 tRNAs, and 12 polypeptides. The L-strand codes for eight tRNAs and a single polypeptide. All 13 protein products are essential constituents of the enzyme complexes of the oxidative phosphorylation system (23,69–71). Comparative analysis of mitochondrial genomes from different organisms has indicated that the gene organization is highly conserved among mammals (72,73). The mammalian mitochondrial genome shows exceptional economy of organization. The genes are closely packed and lack introns; some genes

Table 3.2   The Human Mitochondrial Genes

Gene	Strand	Nucleotide Position	Product
MTATP6	8527–9207	Subunit a (ATPase 6) of ATP synthase (complex V)	H
MTATP8	8366–8572	Subunit A6L (ATPase 8) of ATP synthase (complex V)	H
MTCO1	5904–7445	Subunit I of cytochrome c oxidase (complex IV)	H
MTCO2	7586–8269	Subunit II of cytochrome c oxidase (complex IV)	H
MTCO3	9207–9990	Subunit III of cytochrome c oxidase (complex IV)	H
MTCYB	14747–15887	Apocytochrome b subunit of ubiquinol:cytochrome c oxidoreductase (complex III)	H
MTND1	3307–4262	Subunit ND1 of NADH:ubiquinone oxidoreductase (complex I)	H
MTND2	4470–5511	Subunit ND2 of NADH:ubiquinone oxidoreductase (complex I)	H
MTND3	10059–10404	Subunit ND3 of NADH:ubiquinone oxidoreductase (complex I)	H
MTND4	10760–12137	Subunit ND4 of NADH:ubiquinone oxidoreductase (complex I)	H
MTND4L	10470–10766	Subunit ND4L of NADH:ubiquinone oxidoreductase (complex I)	H
MTND5	12337–14148	Subunit ND5 of NADH:ubiquinone oxidoreductase (complex I)	H
MTND6	14149–14673	Subunit ND6 of NADH:ubiquinone oxidoreductase (complex I)	L
MTRNR1	1671–3228	Large ribosomal subunit RNA (16S rRNA)	H
MTRNR2	648–1601	Small ribosomal subunit RNA (12S rRNA)	H
MTTA	5587–5655	tRNA^{Ala}	L
MTTC	5761–5826	tRNA^{Cys}	L
MTTD	7518–7585	tRNA^{Asp}	H
MTTE	14674–14742	tRNA^{Glu}	L
MTTF	577–647	tRNA^{Phe}	H
MTTG	9991–10058	tRNA^{Gly}	H
MTTH	12138–12206	tRNA^{His}	H
MTTI	4263–4331	tRNA^{Ile}	H
MTTK	8295–8364	tRNA^{Lys}	H
MTTL1	3230–3304	tRNA$^{\text{Leu(UUR)a}}$	H

(continued)

Table 3.2 (continued)   The Human Mitochondrial Genes

Gene	Strand	Nucleotide Position	Product
MTTL2	12266–12336	tRNA$^{Leu(CUN)}$a	H
MTTM	4402–4469	tRNA^{Met} and tRNA^{fMet}	H
MTTN	5657–5729	tRNA^{Asn}	L
MTTP	15955–16023	tRNA^{Pro}	L
MTTQ	4329–4400	tRNA^{Gln}	L
MTTR	10405–10469	tRNA^{Arg}	H
MTTS1	7445–7516	tRNA$^{Ser(UCN)}$a	L
MTTS2	12207–12265	tRNA$^{Ser(AGY)}$a	H
MTTT	15888–15953	tRNA^{Thr}	H
MTTV	1602–1670	tRNA^{Val}	H
MTTW	5512–5576	tRNA^{Trp}	H
MTTY	5826–5891	tRNA^{Tyr}	L

[a] The triplets indicted in brackets refer to the sequence of the codons complementary to the tRNA anticodon, with R for purine, Y for pyrimidine, and N for any of the four bases.

even overlap. Except for the approximately 1-kilobase, non-coding D-loop region, intergenetic sequences are absent or limited to a few bases. In most protein genes, the last one or two bases of termination codons are not encoded, but generated post-transcriptionally by polyadenylation of the mRNAs (74).

Soon after the first mtDNA sequences became available in the early 1980s, comparisons with mitochondrial protein sequences revealed deviations from the standard genetic code. Later variations in codon usage were even found between mitochondria from different species (summarized in Osawa et al. [75]). In mammals, TGA is used as a tryptophan codon rather than as a termination codon, AGR (R = A or G) specifies a stop instead of arginine, and AUA codes for methionine instead of isoleucine (Figure 3.3) (76).

Another surprising feature of the mitochondrial genetic system is its use of a simplified decoding mechanism in which a modified tRNA wobble base interaction with mRNA codons allows mitochondria to translate all codons with less than the 32 tRNA species required according to Crick's wobble hypothesis. Mitochondrial tRNAs with unmodified uridine in the first

		Codon Amino acid (Anticodon)	Codon Amino acid (Anticodon)	Codon Amino acid (Anticodon)	Codon Amino acid (Anticodon)	
First position (5' end)	U	UUU UUC Phe (GAA) UUA UUG Leu (U*AA)	UCU UCC UCA UCG Ser (UGA)	UAU UAC Tyr (GUA) UAA UAG stop	UGU UGC Cys (GCA) UGA UGG Trp (U*CA)	U C A G
	C	CUU CUC CUA CUG Leu (UAG)	CCU CCC CCA CCG Pro (UGG)	CAU CAC His (GUG) CAA CAG Gln (U*UG)	CGU CGC CGA CGG Arg (UCG)	U C A G
	A	AUU AUC Ile (GAU) AUA AUG Met (C*AU)	ACU ACC ACA ACG Thr (UGU)	AAU AAC Asn (GUU) AAA AAG Lys (U*UU)	AGU AGC Ser (GCU) AGA AGG stop	U C A G
	G	GUU GUC GUA GUG Val (UAC)	GCU GCC GCA GCG Ala (UGC)	GAU GAC Asp (GUC) GAA GAG Glu (U*UC)	GGU GGC GGA GGG Gly (UCC)	U C A G
		U	C	A	G	

Second position

Figure 3.3 The mammalian mitochondrial genetic code and corresponding anticodons. Codon and anticodon triplets are given 5'→3'. Chemically modified nucleotides in the first (wobble) position of the anticodon are marked with an asterisk.

(wobble) position of the anticodon read all four codons of the four-codon families (76,77), whereas those containing 5-taurinomethyluridine or 5-taurinomethyl-2-thiouridine read both codons of two-codon families with a purine in the third position (Figure 3.3) (78–81). The methionine codons AUG and AUA of mammalian mitochondria, however, are recognized by a single anticodon that has 5-formylcytidine in the wobble position (Figure 3.3) (82). The tRNA carrying this anticodon is charged with *N*-formylmethionine to function as initiator of mitochondrial protein synthesis (83–85) or with methionine for elongation of the polypeptide chain.

Two-codon families with a pyrimidine in the third position are decoded by tRNAs with an unmodified guanosine in the wobble position (Figure 3.3). The altered wobble rules would imply that 23 tRNA species are required to decode mtDNA; however, as mentioned earlier, in mammals AGR codons indicate a stop and the corresponding tRNA is absent. Therefore, the 22 tRNA species encoded by mammalian mtDNA are sufficient to translate all 13 mitochondrial proteins.

Transmission

Transmission of the Mitochondrial Genome

In the mid-1970s, restriction fragment length polymorphism analysis indicated that mtDNA is transmitted through the female germ line in mammals (86–89). The mtDNA copy number of a sperm cell is relatively low and the copy number of an oocyte is exceptionally high (Table 3.1), so the maternal inheritance observed in these early studies could have simply been the consequence of dilution of the paternal contribution beyond the detection limit of the technique on which these studies relied. Nevertheless, later investigations, in which more sensitive, PCR-based techniques were used, confirmed the maternal inheritance of mtDNA.

In 2002, however, the dogma of strict maternal inheritance was challenged by a patient with mitochondrial myopathy due to a pathogenic two-base pair deletion in 90% of his skeletal muscle mtDNA (90). Surprisingly, the mtDNA harboring the mutation was of paternal origin, whereas the remaining 10% of the muscle mtDNA and the mtDNA in all other tissues studied from this patient matched the maternal genotype (90). Although this single case indicates that partial paternal inheritance of mtDNA is possible, inheritance of human paternal mtDNA is considered so rare (91–93) that it should not affect genetic counseling. It may, however, have important implications for forensic science and studies of human evolution and migration.

In the early 1990s, low levels of paternal mtDNA were detected in *inter*specific hybrids of the mice species *Mus musculus* and *Mus spretus* throughout development from pronucleus stage to neonates (94,95). In *intra*specific offspring of *M. musculus* though, paternal mtDNA was only detected in the early pronucleus stage (95). In mammals, including humans, sperm mitochondria are transferred to the egg during fertilization (96), but detailed studies of fertilized eggs from rodents (95,97,98) and cows (99) have indicated that sperm-derived mitochondria are lost in the preimplantation embryo. The leakage of paternal mtDNA in progenies of *inter*-specific mice crosses suggests that this process is species specific. Further studies demonstrated that only sperm or spermatid mitochondria, but not liver mitochondria, contain factors that render them susceptible to elimination from mouse embryos, emphasizing the unique role of the sperm organelles (100).

Elimination of human sperm mtDNA in the egg is supported by the lack of paternal mtDNA in extra-embryonic tissues and offspring after intracytoplasmic sperm injection (101,102). The nature of the molecular surveillance mechanism that selectively destroys sperm-derived mtDNA from the fertilized egg became clear in 1999, when it was discovered that sperm mitochondria inside fertilized cow and rhesus monkey eggs are tagged with the universal proteolytic marker ubiquitin (103). Ubiquitination selectively earmarks sperm mitochondria for degradation by the embryo's proteosomes and lysosomes (104,105). The ubiquitination and destruction were not observed in hybrid embryos of domestic cow eggs and wild bull sperm, substantiating the preceding inferred suggestion of a species-specific recognition apparatus. The evolutionary basis for the destruction may be that paternal mtDNA is compromised by the action of reactive oxygen species encountered by the sperm during spermatogenesis and fertilization (106). This notion is supported by the fact that, in the only reported patient showing paternal mtDNA transmission (90), the paternally derived mtDNA carried a sporadic pathogenic microdeletion.

Transmission of mtDNA Sequence Variants

It has been postulated that the nucleoid is the basic unit of mtDNA transmission (107). This hypothesis is corroborated by observations indicating that each mitochondrion has at least one nucleoid (58), and that nucleoid segregation follows the dynamics of mitochondrial fusion and fission (59,61,62). The transmission of human mtDNA sequence variants is poorly understood (108). Although mtDNA is highly polymorphic in mammalian populations, normally a single sequence variant of mtDNA is transmitted through the female germline. Thus, the overwhelming majority of mtDNA molecules in most individuals are homoplasmic (identical at birth). When a mother does carry heteroplasmic mtDNA, there tends to be a rapid shift in mtDNA variant frequency and a complete switch to one variant can occur within a single generation (109–114). This suggests that, at one or several stages of oogenesis, mtDNA molecules are selectively sampled from a larger population for transmission and amplification.

It is thought that a restriction in the number of mtDNA molecules during oogenesis is behind this process (115) and several studies have been undertaken to determine the number of segregating units (the size of the bottleneck), in order to estimate recurrence risks for mitochondrial disease caused by mtDNA mutations. The group of Shoubridge examined the pattern of segregation in heteroplasmic mice derived from inbred strains, harboring several mtDNA polymorphisms (116). Their study indicated that the pattern of segregation was largely determined by random genetic drift during oogenesis. This conclusion was confirmed in a subsequent investigation, in which the proportion of mutant mtDNA was measured in primary oocytes from a woman who carried the heteroplasmic, pathogenic 3243A→G mtDNA mutation (117); however, oocytes from a woman who carried the heteroplasmic, pathogenic 8993T→G mtDNA mutation showed a skewed distribution of the mutation (118). The latter observation is not compatible with the random genetic drift, but suggests that wild-type or 8993T→G mutated mtDNA is preferentially amplified during oogenesis.

The number of segregating units of mtDNA can be calculated by applying Wright's population genetic theory, provided that the duration of the bottleneck is known (119). Some authors take the 24 synchronous cell divisions needed to produce a full quota of 8.3 million primary oocytes from a single progenitor cell (120) as duration of the bottleneck (116); others assume that a single, restrictive sampling event occurs once during development (115,121) or calculate the number of segregating units for either value (117). Using the 24 synchronous cell divisions as duration of the bottleneck, the estimated size of the bottleneck is about 200 segregating units. When one assumes that the sampling event occurs once during oogenesis, the estimated size of the bottleneck is 1 to 31 segregating units.

To determine when the segregation occurs during oogenesis, Shoubridge's group performed quantitative mtDNA genotyping of cells from their inbred mouse strains through various stages of oogenesis and compared the variance between cells (116). Their results indicated that segregation occurs during the cell divisions between the developmental stages of the primordial germ cell and primary oocytes. Their estimate of the mtDNA copy number of the primordial germ cell is roughly 200 — a number that matches their estimate of the size of the bottleneck. Thus, at least in mice, segregation of mtDNA can be explained by random drift without the need for resorting to any restrictive sampling event. Unfortunately, however, although it appears that mathematical models can be used to predict the range of possible levels of mutant mtDNA in offspring, the resulting range is wide and is of limited value in genetic counseling (115,122).

In addition to variations of the inherited level of heteroplasmy transmitted from a mother to her children, variations of the level of heteroplasmy between cells of an individual arise during development, growth, and aging. Mathematical models have suggested that the continuous mtDNA turnover can lead to clonal expansion of mtDNA mutations through random genetic drift in rapidly dividing cancer cells and postmitotic tissues (123). On the other hand, there is good evidence for nonrandom segregation of mtDNA sequence variants during histogenesis and organ maturation in Shoubridge's heteroplasmic mouse model (124) and in studies of pathogenic mtDNA mutations in patients (125–129).

Furthermore, several cell culture models indicate nonrandom segregation of mutant and wild-type mtDNA (130–134). The segregation pattern varies with the specific mutation, cell type, and nuclear background. This implies that nuclear genes modify the somatic segregation behavior of mtDNA sequence variants and that the modification in function caused by the mtDNA base change influences this process. In further studies of the heteroplasmic mouse model, the laboratory of Shoubridge (135) found that selection of an mtDNA sequence variant in certain tissues was not due to differences in oxidative phosphorylation function or efficiency of mtDNA replication. This suggests that the nuclear genes involved in the tissue-specific and age-dependent selection of mtDNA sequence variants play a more subtle role in mtDNA maintenance.

In an effort to identify these nuclear genes, genome-wide scans of the mouse model have been undertaken (136). These studies showed linkage of the trait to loci on three different chromosomes, but the genes have not yet been identified. Their characterization is eagerly awaited because it should provide insights into basic mechanisms of mtDNA maintenance and may help explain, or even provide approaches to adjust, the segregation of heteroplasmic pathogenic mtDNA mutations in patients.

TRANSCRIPTION OF MTDNA

Initiation of Transcription

cis-Acting Elements

The basic mechanism of mammalian mtDNA transcription was determined in the 1980s, mainly through the efforts of the laboratories of Attardi and Clayton (reviewed in references 137 through 139). Human mtDNA transcription start sites and promoter regions were mapped using a large variety of techniques (140–147). All available data are consistent with the conclusion that there are two major initiations of transcription sites for the H- and L-strand (IT_{H1} and IT_L), situated within 150 base pairs of one another in the D-loop region (Figure 3.4). A promoter element with a pentadecamer consensus sequence motif, 5'-CANACC(G)CC(A)AAAGAYA (N = A, C, G, or T; Y = C or T), encompasses the transcription

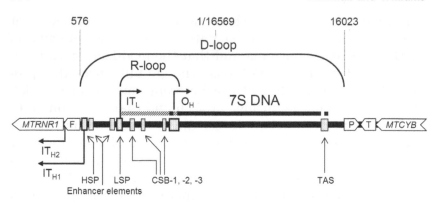

Figure 3.4 Map of the D-loop region of human mtDNA. The thick line with boxes represents part of the mtDNA duplex. The extent of the D-loop and the region putatively displaced by R-loop RNA are marked. Open boxes represent genes. Genes marked with F, P, and T correspond to the tRNA^{Phe}, tRNA^{Pro}, and tRNA^{Thr} genes. The *MTRNR1* and *MTCYB* genes (see Table 3.2) are also marked (both genes are incomplete in the figure). The orientation of the genes is indicated by chevrons. Shaded boxes represent various sequence elements involved in transcription and replication of mtDNA. Transcription initiation sites and direction of RNA synthesis are indicated by bent arrows. In the D-loop region, two major transcription initiation sites are present. Initiation of transcription site IT$_{H1}$, encompassed by the H-strand promoter (HSP), directs the transcription of the H-strand, whereas initiation of transcription site IT$_L$, encompassed by the L-strand promoter (LSP), directs the transcription of the L-strand. A second, minor initiation of transcription site (IT$_{H2}$) for H-strand transcription is located in the gene for tRNA^{Phe}, near the boundary with the *MTRNR1* gene. Enhancer elements upstream of the HSP and LSP that are known to bind the TFAM protein are indicated. The TAS box shows the location of the termination-associated sequence. CBS-1, -2, and -3 mark the conserved sequence blocks 1, 2, and 3. O$_H$ marks the origin of H-strand replication with the box corresponding to the variation in actual start sites. The mapping position of the 7S DNA species is depicted by the line drawn above the line representing mtDNA; the DNA portion is given as a solid line and the RNA portion is hatched. The variable transition from RNA to DNA and the ragged 3'-ends are represented by the stuttered transition. The numbering of the nucleotides is indicated at the top of the figure.

initiation sites (in bold type) and is critical for transcription (144,146).

H-strand transcription starts at nucleotide position 561 (IT$_{H1}$; numbering according to Anderson et al. [23]) located within the H-strand promoter (HSP), whereas L-strand transcription starts at position 407 (IT$_L$) within the L-strand promoter (LSP). Enhancer elements located just upstream of these promoter regions are required for optimal transcription (Figure 3.4). These elements, which were later shown to be binding sites for the transcription factor TFAM (see next section), exhibit sequence similarity, but only if one element is inverted relative to the other (144,146,148). This suggests that these *cis*-acting enhancers are able to function bidirectionally.

Despite the close proximity of the two promoter regions, the early *in vitro* transcription studies demonstrated that these elements are functionally independent (142,144,146,147). This functional autonomy was later confirmed in studies of patients carrying a large-scale, heteroplasmic mtDNA deletion that included the HSP but not the LSP (149,150). *In situ* hybridization experiments of skeletal muscle from these patients revealed focal accumulations of deleted mtDNA and L-strand transcripts with concomitant depletion of H-strand transcripts, thus corroborating the functional independence of the promoters *in vivo*.

Detailed transcription studies have suggested that a second initiation site for H-strand transcription is located around nucleotide position 638 (IT$_{H2}$) in the *MTTF* gene, immediately adjacent to the *MTRNR2* gene (Figure 3.2) (140–142,144). Its promoter region only shows limited similarity with the 15-base pair consensus sequence, and this site is considered to be used less frequently for transcription of the H-strand than IT$_{H1}$.

trans-Acting Factors

Biophysical fractionation studies of human mitochondrial transcription extracts have led to the identification of two nuclear-encoded proteins required for transcription initiation:

Table 3.3 Human Nuclear-Encoded Factors Involved in mtDNA Transcription and Post-Transcriptional RNA Modifications

Factor	Description (Aliases)	M_r (×1000)	Gene Symbol	Cytogenetic Position
MTERF	Mitochondrial transcription termination factor (mTERF, mtTERM, MTTER)	34	*MTERF*	7q21–q22
MTO1	Mitochondrial translation optimalization 1 homolog	67	*MTO1*	6q14.1
POLRMT	Mitochondrial RNA polymerase	138	*POLRMT*	19p13.3
TFAM	Mitochondrial transcription factor A (mtTFA, mtTF1, TCF6)	25	*TFAM*	10q21
TFB1M	Mitochondrial transcription factor B1 (mtTFB1)	38	*TFB1M*	6q25.1– q25.3
TFB2M	Mitochondrial transcription factor B2 (mtTFB2)	43	*TFB2M*	1q44
TRNT1	Mitochondrial ATP(CTP): tRNA nucleotidyl-transferase (mtCCA, CCA1, CGI-47)	47	*TRNT1*	3p25.1

a relatively nonselective core RNA polymerase (151), POL-RMT, and a dissociable transcription factor (148,152), TFAM, which confers promoter selectivity on the polymerase (Table 3.3). The core enzyme is thought to bind to the promoter region. Mitochondrial RNA polymerases have not been purified to homogeneity from any source. Nevertheless, a cDNA predicted to encode the 1230-amino acid residue precursor of POLRMT has been identified by screening of a human expressed sequence tags (ESTs) database with the yeast polymerase sequence (153).

POLRMT is homologous to the single subunit RNA polymerases of T3, T7, and SP6 bacteriophages, though the amino acid similarity is restricted to the C-terminal half of the mitochondrial enzyme (153,154). Structural and mutation studies have indicated that this evolutionarily conserved region is important for promoter selectivity and polymerase activity

(155–158). Mutation experiments in yeast have suggested that the N-terminal half of the polymerase interacts with factors involved in post-transcriptional events and associated with the mitochondrial inner membrane (159–161). Furthermore, immunocytochemical experiments with the human EVC304 cell line have revealed that newly synthesized mitochondrial transcripts remain in a focus that originally lies immediately next to the nucleoid, suggesting that mitochondrial RNAs remain at transcription sites for extended periods after they have been made (61).

In the same study, newly synthesized mitochondrial RNA was located near a subset of the cytosolic ribosomes (61). It also was found to co-localize with cytosolic foci containing nascent peptide-associated protein (NAC) (61), a complex involved in directing nascent cytosolic polypeptides to the appropriate cellular compartment and known to associate with mitochondrial membranes (162). Moreover, newly made mRNA was found to lie near TOM22 (61), a mitochondrial outer membrane component of the system that imports nuclear-encoded proteins into the organelle (163).

These results are consistent with an association of the mitochondrial transcription and translation machinery on one side of the membrane with a cytosolic translation machinery and import system on the other. This arrangement may facilitate coordinate mitochondrial transcription, cytosolic and mitochondrial translation, mitochondrial protein import, and assembly of the mitochondrial and nuclear-encoded protein subunits of the oxidative phosphorylation complexes.

The human transcription factor TFAM, which acts in concert with POLRMT, has been characterized in detail (Table 3.3) (164–169). It is an abundant 25-kDa protein, largely comprising two high-mobility group (HMG) boxes that are separated by a 27-amino acid residue linker and followed by a 25-amino acid residue, C-terminal tail. HMG boxes are DNA-binding motifs found in a diverse family of proteins that include abundant nonhistone components of chromatin, and specific regulators of nuclear transcription and cell differentiation (170).

In organello and *in vitro* DNase I protection studies combined with run-off transcription assays have demonstrated that TFAM binds to the enhancer elements located 12 to 39 base pairs upstream of the respective transcription initiation sites, IT_{H1} and IT_L (Figure 3.4), and that binding at these sites is required for accurate and efficient initiation of transcription (147,148,164,171,172). Mutation analysis of TFAM has indicated that its basic C-terminal tail is important for specific recognition of the transcriptional enhancer elements in the D-loop (Figure 3.4), and the HMG boxes are required for nonspecific DNA-binding properties (173). The HSP and LSP are able to function bidirectionally (174). The asymmetric binding of TFAM relative to the transcription start site may ensure that transcription proceeds primarily in a unidirectional fashion (Figure 3.4). Common with many other members of the HMG family of proteins, TFAM has the capacity to bend and unwind the DNA duplex and to wrap around the distorted DNA strands (175,176). These TFAM-induced conformational changes at the promoter regions may permit access of POLRMT to the template to initiate the transcription process.

The *in vitro* transcription assays with crude POLRMT extracts and purified TFAM carried out during the 1980s revealed that TFAM binding to the upstream enhancer of the HSP is several fold weaker than to the upstream enhancer of the LSP and only moderately stimulates H-strand transcription (148,152,164). More recent *in vitro* assays demonstrated that recombinant POLRMT — on its own or in combination with recombinant TFAM — is incapable of initiating transcription from the natural promoters (177,178).

These observations strongly suggested that at least one additional *trans*-acting factor is necessary for transcription initiation. Because mammalian mitochondrial RNA polymerases have not been purified to homogeneity, accessory factors may be present in the active mitochondrial RNA polymerase fractions. The identification of a 40-kDa protein, obligatory for promoter-directed transcription selectivity of mitochondrial RNA polymerases in yeast (179–182) and *X. laevis* (183,184) prompted searches of human DNA databases for the elusive transcription factor. In 2002, these searches resulted in the discovery of two additional human mitochondrial transcription factors: TFB1M and TFB2M (Table 3.3) (185,186). The two factors show a strong amino acid sequence similarity and TFB1M shows 16% identity with the homolog from the yeast *Schizosaccharomyces pombe*. Northern blot analysis has indicated that the *TFB1M* and *TFB2M* genes are transcribed ubiquitously, suggesting coexpression of the proteins. Each factor alone can support HSP- and LSP-specific transcription in a pure *in vitro* system, containing recombinant POLRMT and TFAM, though TFB2M is at least tenfold more active in promoting transcription than TFB1M (186).

Binding assays of TFB1M or TFB2M with C-terminal deletion mutants of TFAM have indicated that TFB1M and TFB2M interact with the C-terminal tail of TFAM (187). TFB1M and TFB2M also interact directly with POLRMT, but are considered to have weak, sequence-nonspecific DNA-binding activity (186,187); thus, a model has been proposed in which TFB1M or TFB2M act as a bridge between POLRMT bound to the HSP or LSP, and TFAM bound to the upstream enhancer element. The model explains the strict requirement for the natural spacing of ten base pairs between the TFAM-binding site and the initiation of transcription site that was found in an earlier study (188).

Remarkably, primary sequence comparisons and crystallography data have revealed that TFB1M and TFB2M are structurally related to a superfamily of RNA adenosine methyltransferases (185,186,189,190). Consistent with this finding, TFB1M is capable of binding *S*-adenosylmethionine, the requisite methyl-donating substrate of this class of enzymes (185). Further investigations have revealed that TFB1M can functionally complement the *Escherichia coli* KsgA protein, which methylates two adjoining adenosine residues located within a conserved stem-loop structure at the 3′-end of bacterial 16S rRNA (187). The homologous adjoining adenosine residues of the human mitochondrial 12S counterpart are similarly modified (191). Therefore, it is thought that TFB1M and TFB2M are responsible for this modification.

Mutation analysis of the conserved methyltransferase motif of TFB1M has revealed that it stimulates transcription *in vitro* independently of S-adenosylmethionine-binding and rRNA methyltransferase activity (187). Thus, TFB1M and TFB2M appear to be dual-function proteins, acting as transcription factors and rRNA-modification enzymes. However, TFB2M is at least an order of a magnitude more active in promoting transcription than TFB1M, so it is possible that TFB2M is primarily a transcription factor *in vivo*, and TFB1M is primarily an rRNA methyltransferase *in vivo*.

Elongation and Termination of Transcription

After initiation of transcription at the LSP, the L-strand is transcribed as a single polycistronic precursor (192,193). Although the HSP can direct transcription of the entire H-strand in a similar fashion, a more complicated model has been proposed by Attardi and coworkers (141). Proliferating HeLa cells synthesize the two mitochondrial rRNA species at an approximately 50-fold higher rate than the mRNAs encoded on the H-strand (141,194). Attardi and coworkers explain this difference in part by the existence of two functionally distinct transcription events, starting at IT_{H1} or IT_{H2} (Figure 3.4) (140). According to their dual H-strand transcription initiation model, transcription starts relatively frequently at IT_{H1} and then terminates at the *MTRNR2/MTTL1* gene boundary (Figure 3.2). This transcription process is considered to be responsible for the bulk of the mitochondrial rRNAs, tRNA[Phe], and tRNA[Val]. Transcription of the H-strand starts less often at IT_{H2}, but results in a long polycistronic transcript that includes all genetic information of the H-strand.

The existence of two individually controlled, overlapping H-strand transcription units is supported by the observations that ethidium bromide and ATP can modulate the relative H-strand transcription rates of rRNA and mRNA independently (195,196). Consistent with these findings, ethidium bromide- and ATP-dependent modifications in protein–DNA footprints, which correlate with changes in the rate of rRNA synthesis, but not of mRNA synthesis (172), have been demonstrated upstream of IT_{H1}. Moreover, a protein–DNA interaction site has been identified upstream of IT_{H2} that could represent the initiation complex (172). It is, however, not clear how two initiation events occurring less than 100 base pairs apart are able to determine the fate of RNA synthesis at the *MTRNR2/MTTL1* boundary, more than 2500 nucleotide residues downstream.

In contrast to the views of Attardi and coworkers, Clayton and coworkers assume that H-strand transcription exclusively starts at IT_{H1} and that the majority of transcripts terminate at the *MTRNR2/MTTL1* boundary. Thus, Attardi's group explain the higher synthesis rate of rRNA compared to mRNA by a site-specific transcription initiation event and an associated termination event at the *MTRNR2/MTTL1* boundary, and Clayton's group explain the synthesis rate difference solely by an attenuation event at the *MTRNR2/MTTL1* boundary.

The first indication of early termination of H-strand transcription came from single-stranded (S_1) nuclease protection analysis of the 3′-ends of human and mouse 16S rRNA molecules (197,198). These mapping studies revealed that the genomic location of the last template-encoded nucleotide residue of some of the 16S rRNA molecules corresponds to the nucleotide residue immediately adjacent to the 5′-end of the *MTTL1* gene, whereas the last nucleotide residue of other forms maps to any position up to seven nucleotide residues downstream in *MTTL1*. These ragged ends suggest that the process that generates the 3′-end of 16S rRNA is one of imprecise termination of transcription, rather than accurate processing of the primary transcript (197,198). Subsequently, a protein fraction was isolated from mitochondrial HeLa cell lysates, which in DNase I footprinting studies protected the 28-base pair region immediately downstream of the mtDNA sequence corresponding to *in vivo* produced 3′-ends of 16S rRNA molecules and promoted specific termination of transcription (199).

The DNase I footprint comprises an evolutionarily conserved tridecamer sequence motif, 5′-TGGCAGAGCCCGG-3′

(L-strand), located within *MTTL1* (nucleotide positions 3237 to 3249 of human mtDNA; Figure 3.2). *In vitro* mutagenesis and transcription experiments have shown that this sequence element is essential and sufficient for directing termination of transcription (200–202). Nevertheless, the relative transcript levels are unaffected in cell cultures and tissues from patients with mitochondrial myopathy, encephalopathy, lactic acidosis, and stroke-like episodes (MELAS) who carry a heteroplasmic 3243A→G transition in the middle of this element (150,203–206). Thus, it seems doubtful that the *in vitro* observed, defective attenuation of transcription associated with the 3243A→G mutation (201) is of any pathological significance.

The groups of Attardi and Clayton have shown that the mitochondrial protein fraction mediating attenuation of transcription at the *MTRNR2/MTTL1* boundary contains several polypeptides of around 34 kDa (201,207). The human cDNA specifying the major polypeptide from this fraction has been cloned (208). The mitochondrial transcription termination factor encoded by the cDNA has been called MTERF (Table 3.3). The protein contains three leucine zipper motifs bracketed by two basic domains that are all critical for its specific DNA-binding capacity (208).

MTERF exists in two forms in HeLa cell mitochondrial lysates (209). One is a monomeric form exhibiting DNA-binding and transcription termination activities; the other is probably a homotrimeric form lacking these activities. Consequently, it has been proposed that the activity of MTERF is modulated by the transition between an active monomer and an inactive trimer (209). Further studies with the rat counterpart have revealed that MTERF is a phosphoprotein containing four potential phosphorylation sites (210). Although the DNA-binding activity of the rat homolog is unaffected by its phosphorylation state, only the phosphorylated form is able to terminate transcription (210). Thus, phosphorylation of MTERF provides an additional means to regulate H-strand transcription. The apparent complexity of the transcriptional attenuation system is not surprising, given that it should be able to adjust its activity in response to the cellular demand for mitochondrial rRNAs on one hand, and for mitochondrial mRNAs and tRNAs on the other.

Interestingly, *in vitro* transcription assays with the 34-kDa polypeptide fraction have indicated that the MTERF-complex bound to its mtDNA target site functions bidirectionally and shows an even greater efficiency of termination in the reverse orientation relative to the initiation of transcription site (211). Thus, in addition to an attenuating function for H-strand transcription, the protein complex may bring L-strand transcription to an end at a site where no L-strand genes are present downstream (Figure 3.2). The complex induces bending of DNA and is also able to terminate transcription by heterogeneous RNA polymerases (202). It is therefore thought that the MTERF-complex stops the polymerase by constituting a physical barrier, rather than by a specific interaction with the enzyme.

Once the RNA polymerase transcribing the H-strand has passed the *MTRNR2/MTTL1* boundary, the entire strand is transcribed (74,192,193,212,213). The second H-strand transcription termination site of human mtDNA has not been mapped, but it has been suggested that the final H-strand transcription termination site of mouse mtDNA is located beyond the D-loop region, immediately upstream of the *MTTF* gene (214). This region, termed D-TERM, includes the sequence AATAAA. In nuclear mRNAs, AAUAAA serves as a polyadenylation signal and is believed to have a role in 3'-end formation (215). Gel-shift analysis has indicated that the 22-base pair D-TERM sequence forms two major complexes with mouse liver mitochondrial extracts. Protein purification by DNA-affinity chromatography has yielded two major proteins of 45 and 70 kDa. In contrast to the MTERF nucleoprotein complex, the D-TERM complex terminates transcription in a unidirectional manner in an *in vitro* reconstituted system (214).

Processing of Primary Transcripts

Polycistronic mitochondrial transcripts are rapidly processed into mature, adenylated mRNA and rRNA species or tRNA species that carry a 3'-CCA trinucleotide. Human mtDNA transcription products were initially identified and characterized in HeLa cells (74,212,216). They include the 22 tRNAs, two rRNAs (12S and 16S), a precursor comprising both rRNAs, and another 15 poly(A)-containing RNA species. The RNAs in addition to the two rRNAs and 22 tRNAs are conventionally numbered according to their size with RNA 1 the longest and RNA 18 the shortest species (Figure 3.2). Most mature polyadenylated RNAs correspond to a single gene, but two of them, RNA 7 and RNA 14, are dicistronic and contain the messages of the *MTND4* and *MTND4L* genes, and the *MTATP6* and *MTATP8* genes, respectively (Figure 3.2).

In 1981, sequencing of the human mtDNA revealed that the two rRNA- and most mRNA-coding sequences are immediately contiguous to tRNA sequences (Figure 3.2) (23). This extraordinary genetic arrangement led to the proposal that the secondary structure of the tRNA sequences function as punctuation marks in the reading of the mtDNA information (74). Precise endoribonucleolytic excision of the tRNAs from the primary transcript will concomitantly release correctly processed rRNAs and, in most cases, correctly processed mRNAs (74,213). In view of this, processing of the polycistronic H- and L-strand transcripts is considered to be a relatively simple process requiring only a few enzymes. In cases in which an mRNA terminus cannot be accounted for by tRNA excision (e.g., the 3'-end of RNA 14, see Figure 3.2), the endoribonuclease is thought to recognize an RNA secondary structure at the border that shares key features with the cloverleaf configuration of a tRNA.

Excision of mitochondrial tRNAs involves two enzymatic activities: a single endoribonucleolytic cleavage at the 5'-end and a single endoribonucleolytic cleavage at the 3'-end. The hydrolysis of the phosphodiester bond at the 5'-end is catalyzed by mitochondrial ribonuclease (RNase) P. As postulated in 1981 (74), this enzyme is indeed guided by the secondary structure of tRNA sequences in the primary transcript (217). The nuclear and eubacterial counterparts of mitochondrial RNase P are ribonucleoproteins, i.e., they consist of an essential RNA subunit and protein subunits (217,218).

The enzymes in chloroplasts and mitochondria are less well characterized. The most extensively studied chloroplast, RNase P (from spinach), appears to be composed exclusively of protein (217). Yeast mitochondrial RNase P comprises a nuclear-encoded protein and an mtDNA-encoded RNA species (219–221), but preliminary characterization of the mitochondrial enzyme from the protozoan parasite *Trypanosoma brucei* revealed no RNA moiety essential for its activity (222). The composition of mammalian mitochondrial RNase P is still under debate (223,224). Rossmanith and colleagues claim that mitochondrial RNase P is a pure protein enzyme in HeLa cells (225,226), but Attardi and colleagues claim that it is a ribonucleoprotein (227,228). A mitochondrial RNase P with properties similar to those of the enzyme described by Attardi's group has also been reported to occur in rat mitochondria (229). Attardi's group has shown that mitochondrial RNase P from extensively purified HeLa cell mitochondria is a particle with a sedimentation constant of approximately 17S and, surprisingly, that its 340-nucleotide RNA subunit is identical in sequence to the H1 RNA subunit of nuclear RNase P (228).

The protein composition of mammalian mitochondrial RNase P has not yet been determined. Human nuclear RNase P, which has a sedimentation constant similar to that of human mitochondrial RNase P, contains ten protein subunits (Table 3.4) (230). RNase P is evolutionarily related to another RNase, the RNase MRP (mitochondrial RNA processing) complex (231). RNase MRP was originally identified as an RNA-containing endoribonuclease that processes RNA primers for mtDNA replication (see next section); however, most of the RNase MRP is found in the nucleolus (232–235), where it participates in 5.8S rRNA maturation (236–238).

Although RNase P and RNase MRP have a different RNA moiety, both RNAs can fold into a similar, cage-shape secondary structure (239). Moreover, nuclear RNase P and RNase MRP share most, if not all, of their protein subunits (Table 3.4) (218,236). It is possible that mammalian mitochondrial RNase P has not only its RNA moiety in common with nuclear RNase P, but also shares some protein subunits with nuclear RNase P and RNase MRP.

Table 3.4 Protein and RNA Components of Human Nuclear Endoribonuclease P (RNase P) and Mitochondrial RNA Processing Endoribonuclease (RNase MRP)

Component	Description (Aliases)	M_r (×1000)	Gene Symbol	Cytogenetic Position
POP1	Processing of precursor 1; protein subunit shared by RNase P and RNase MRP	113	*POP1*	8q22.11
POP4	Processing of precursor 4; protein subunit shared by RNase P and RNase MRP (RPP29)	25	*POP4*	19q13.11
POP5	Processing of precursor 5; protein subunit shared by RNase P and RNase MRP	19	*POP5*	12q24.31
POP7	Processing of precursor 7; protein subunit RNase P (RPP20)	20	*POP7*	7q22
RMRP	RNA component of mitochondrial RNA processing endoribonuclease (RNase MRP RNA, Th RNA, 7-2 RNA)	—[a]	*RMRP*	9p21–p12
RPP14	14-kDa subunit of ribonuclease P protein	14	*RPP14*	3p21.2
RPP21	21-kDa subunit of ribonuclease P protein	21	*RPP21*	6p21.32
RPP25	25-kDa subunit of ribonuclease P protein	25	*RPP25*	15q23
RPP30	30-kDa subunit of ribonuclease P protein; protein subunit shared by RNase P and RNase MRP	30	*RPP30*	10q23.32–q23.33
RPP38	38-kDa subunit of ribonuclease P protein; protein subunit shared by RNase P and RNase MRP	32	*RPP38*	10p13
RPP40	40-kDa subunit of ribonuclease P protein	40	*RPP40*	6p25.1
RPPH1	Ribonuclease P RNA component H1 (H1 RNA, 8-2 RNA)	—[b]	*RPPH1*	14q

[a] 265 nucleotides.
[b] 340 nucleotides.

In bacteria, maturation of the 3'-ends of precursor tRNAs differs between species: in *E. coli* it is a multistep process involving endo- and exonucleases (240), whereas in *Bacillus subtilis* maturation of the 3'-ends is catalyzed by a single enzyme known as the 3'-tRNA precursor processing endoribonuclease or RNase Z (241). Maturation by a single endoribonucleolytic step is also common in the nuclei of eukaryotes (240). An endoribonucleolytic activity cleaving the precursor tRNA immediately 3' to the discriminator nucleotide of the tRNA (Figure 3.5) has been identified in *in vitro* assays of HeLa cell and rat liver mitochondrial extracts (225,229,242). Therefore, it appears that the 3'-processing of mitochondrial tRNAs resembles the single endoribonuclease step seen in *B. subtilis*.

Silencing of the RNase Z gene (*Jhl-1*) of cultured *Drosophila* S2 cells by RNA interference causes accumulation of nuclear and mitochondrial precursor tRNAs (243). This suggests that *Jhl-1* encodes the nuclear as well as the mitochondrial forms of RNase Z. It is tempting to speculate that a single enzyme functions in both compartments in mammals also. Unfortunately however, the low abundance of the organellar enzyme has hampered its structural characterization. *In vitro* assays with HeLa cell mitochondrial extracts and synthetic precursor tRNA substrates have indicated that sequences downstream of the cleavage site are critical for recognition by the mitochondrial RNase Z (242). The enzyme exhibits no detectable activity with intact precursor tRNAs as substrate, but does convert 5'-processed intermediates to mature tRNAs (225,229,242), suggesting that maturation of the 5'-end of mitochondrial tRNAs precedes maturation of the 3'-end.

Some exceptions exist however. For instance, the human gene for tRNA[Ser(AGY)] (*MTTS2*) is directly flanked by the genes for tRNA[His] (*MTTH*) and tRNA[Leu(CUN)] (*MTTL2*; Figure 3.2). Detailed *in vitro* examination of the processing pathway of tRNA[Ser(AGY)] has suggested that mitochondrial RNase P and RNase Z recognize only the tRNA[His] and tRNA[Leu(CUN)] structures in the primary transcript and excise these two flanking tRNAs, consequently releasing the enclosed tRNA[Ser(AGY)] (244).

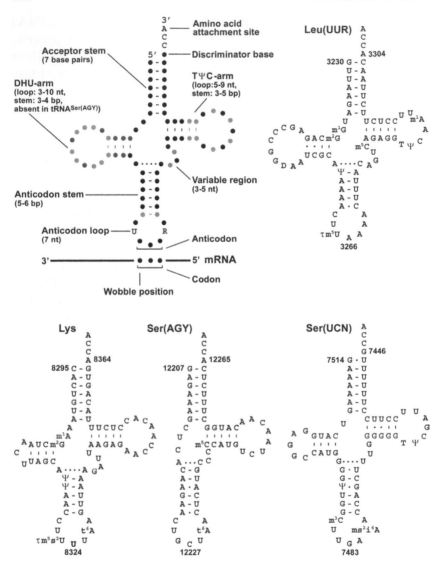

Figure 3.5 Secondary structures and post-transcriptional modifications of certain human mitochondrial tRNAs. The upper left structure indicates the variation in the number of nucleotides that constitute the different secondary regions of the 22 mitochondrial tRNAs. Letters indicate conserved nucleosides, black dots represent conserved nucleoside positions, and gray dots represent nonconserved nucleoside positions. Structural elements are named

Figure 3.5 (continued)   and the interaction with an mRNA is depicted. The remainder of the figure displays the secondary structures of tRNA$^{Leu(UUR)}$, tRNA^{Lys}, tRNA$^{Ser(AGY)}$, and tRNA$^{(UCN)}$. Hyphens indicate Watson and Crick base-pairing; dots indicate possible non-Watson and Crick interactions. Known post-transcriptional modifications have been compiled from references 81, 256, and 603. Abbreviations of residues: A, adenosine; C, cytidine; D, dihydrourdine; G, guanosine; m^{1}A, 1-methyladenosine; m^{3}C, 3-methylcytidine; m^{5}C, 5-methylcytidine; m^{1}G, 1-methylguanosine; m^{2}G, N^2-methylguanosine; ms^{2}i^{6}A, 2-methylthio-N^6-isopentenyladenosine; R, purine (adenosine or guanosine); T, thymidine (5-methyluridine); t^{6}A, N^6-threoninocarbonyladenosine; U, uridine; Ψ, pseudouridine; τm^{5}s^{2}U, 5-taurinomethyl-2-thiouridine; τm^{5}U, 5-taurinomethyluridine.

Further *in vitro* experiments with mitochondrial extracts and synthetic tRNA precursors have suggested that certain disease-associated point mutations in the tRNA$^{Leu(UUR)}$ gene (*MTTL1*) may result in a quantitative effect on the processing (245,246). Human transmitochondrial (cybrid) cell lines carrying 3243A→G-, 3271T→C-, or 3256C→T-mutated *MTTL1* showed a small but consistent increase in the steady-state levels of a novel, partially processed RNA species dubbed RNA 19 (204,247,248). RNA 19 comprises the 16S rRNA, tRNA$^{Leu(UUR)}$, and *MTND1* mRNA sequences, which are contiguous in the primary H-strand transcript (Figure 3.2). Accumulation of RNA 19 has also been seen in tissues from various patients carrying heteroplasmic point mutations in *MTTL1* (206,249,250). The proportion of mutation-carrying RNA 19 was always markedly higher than the mtDNA mutant load, suggesting that mutated RNA 19 is less efficiently processed than the wild-type precursor (250,251). These observations indicate that some mtDNA mutations may result in abnormal mitochondrial RNA processing and, in this way, may contribute to disease.

Steady-state levels of mature 16S rRNA, tRNA$^{Leu(UUR)}$, and *MTND1* mRNA are not affected by the mutations (204,247). Although the steady-state level of RNA 19 is extremely low in the cybrids compared to the level of mature 16S rRNA, the level shows a strong inverse correlation with the rates of oxygen consumption of the cybrids (an indicator of mitochondrial oxidative phosphorylation capacity) (252,253). These findings have led to the hypothesis that RNA 19 may be incorporated into ribosomes, rendering them functionally deficient (204,252). If the incorporation results in stalling of the translation of polyribosomal mRNAs, then a small increase of the immature RNA species could disproportionally interfere with mitochondrial translation and explain the severe oxidative phosphorylation defects observed in patients (252).

Post-Transcriptional Modifications

Modifications of tRNA Molecules

An intriguing aspect of mitochondrial tRNA processing is the occurrence of seemingly overlapping tRNA genes in animal mitochondria. For instance, the human gene for tRNA[Tyr] (*MTTY*) seems to share an adenosine residue with the downstream gene for tRNA[Cys] (*MTTC*), so this nucleotide residue potentially represents not only the first base of tRNA[Cys], but also the discriminator base of tRNA[Tyr] (Table 3.2) (23). Thus, tRNA maturation requires endoribonucleolytic cleavage, as well as the addition of a nucleotide residue by an editing reaction.

In vitro assays with a HeLa cell mitochondrial extract and run-off tRNA[Tyr]/tRNA[Cys] precursors as substrate have revealed that tRNA[Cys] is released in its complete form, but tRNA[Tyr] lacks the nucleotide at the discriminator position (254). Experiments in which tRNA[Cys] was partially deleted or completely replaced have indicated that the cleavage reaction represents the activity of the mitochondrial RNase Z recognizing the upstream tRNA[Tyr] sequence (254). The truncated 3′-end of the tRNA[Tyr] is then completed in an editing reaction that attaches the missing adenosine residue (254,255). It is not clear which enzyme is involved in this editing reaction. Likely candidates are the mitochondrial ATP(CTP):tRNA nucleotidyltransferase (see later section) or a mitochondrial polyadenylation enzyme (255).

Around 17% of the nucleotide residues within eukaryotic cytosolic tRNAs are modified and 11% within eubacterial tRNAs; however, only about 6% are modified in animal mitochondrial tRNAs (Figure 3.5) (256). As discussed at the beginning of this chapter, modifications of the wobble base of some tRNAs are required for decoding accuracy (Figure 3.3). Other base modifications appear to promote cloverleaf folding (257,258) and attachment of the cognate amino acid (259). For instance, methylation of the adenine-9,N^1 in tRNA^{Lys} (Figure 3.5) appears to stabilize the transiently formed cloverleaf structure of the nascent tRNA^{Lys} transcript (260).

Various tRNA modification activities have been detected in protein extracts from highly purified, HeLa cell mitochondrial preparations (260). Although some of these activities were also present in cytosolic protein extract and could potentially represent cytosolic contaminations, other activities, including a tRNA (adenine-9,N^1)-methyltransferase activity, were unique to mitochondrial extracts. So far, only one human gene (*MTO1*) coding for a putative mitochondrial tRNA base-modifying protein has been identified (Table 3.3). The identification is based on homology of the predicted human protein with the *E. coli gidA* gene and the yeast *MTO1* gene products. The bacterial protein is thought to be involved in 5-methylaminomethyl-2-thiouridine modification of tRNAs (261), and the yeast protein was found to play a role in optimizing mitochondrial protein synthesis (262). Therefore, it has been proposed that the human *MTO1* gene may encode a mitochondrial tRNA base-modifying enzyme (263), but functional studies have not been performed.

All tRNAs carry the invariant CCA sequence at their 3'-terminus for amino acid attachment (Figure 3.5) (264). The trinucleotide sequence is not encoded by mitochondrial tRNA genes and its addition is an essential step in the maturation of mitochondrial tRNAs. Point mutations near the 3'-end of tRNAs, as found in some patients, may result in reduced CCA addition (246), leading to decreased expression levels of the mature tRNA. The sequential addition of the three nucleotides is catalyzed by ATP(CTP):tRNA nucleotidyltransferase. Activity of the human mitochondrial CCA-adding enzyme has

been identified in HeLa cell organellar extracts (225). Mitochondrial ATP(CTP):tRNA nucleotidyltransferase has been
partly purified from bovine liver (265). cDNA sequences of the
human and mouse equivalents have been determined by mass
spectrophotometric analysis coupled with EST database
searches (265). The predicted amino acid sequences include a
typical N-terminal region for mitochondrial targeting, which
is thought to be cleaved on import (265).

Interestingly, it appears that the cytosolic CCA-adding
enzyme originates from the same gene, but that its translation
starts at an alternative start site 87 bases downstream (266).
Consequently, this protein product lacks the 29-amino acid
residue cleavable import signal. Also in yeast, one gene codes
for two versions of the enzyme by alternative initiation of
translation (267,268). The human tRNA nucleotidyltransferase has been named TRNT1 (Table 3.3). The crystal structure of recombinantly expressed TRNT1 revealed that the
protein consists of four domains with a cluster of conserved
residues forming a positively charged cleft between the first
two N-terminal domains (269). Based on this structure, a
model has been proposed in which the 3′-end of the tRNA is
placed into the catalytic site, close to a patch of conserved
residues that provide the binding sites for ATP and CTP (269).

The aminoacylation or charging of mitochondrial tRNAs
with their cognate amino acids is catalyzed by nuclear-
encoded aminoacyl-tRNA synthetases that are specific for
each particular tRNA. This reaction represents one of the key
processes of protein synthesis. Several human mitochondrial
aminoacyl-tRNA synthetases have been cloned, mapped, and
biochemically characterized (270–279). These studies have
indicated that, in some cases, two distinct genes encode the
cytosolic and mitochondrial isoenzymes, but in others a single
gene encodes both forms of the protein. Different genetic
mechanisms have evolved that allow a single gene to encode
both forms.

Like the tRNA nucleotidyltransferase gene *TRNT1*, the
glycyl-tRNA synthetase gene (*GARS*) uses alternative transcription start sites, with the furthest upstream start site
resulting in the addition of a mitochondrial targeting

sequence to the N-terminus of the protein (270,280). The gene for lysyl-tRNA synthetase (*KARS*), on the other hand, generates the two isoforms by alternative spicing: the cytosolic isoform is created by splicing of exons 1 and 3, and inclusion of exon 2 between exons 1 and 3 produces the mitochondrial isoform with an N-terminal presequence (275).

Mitochondrial aminoacyl-tRNA synthetases encoded by separate genes tend to show a high degree of homology to the corresponding eubacterial enzyme, but sometimes striking structural differences are present. For instance, prokaryotic as well as (eukaryotic) cytosolic phenylalanyl-tRNA synthetases have an $\alpha_2\beta_2$ tetrameric structure, whereas their mitochondrial counterpart is a monomer containing three sequence motifs that are commonly present in the α subunit and one motif that normally resides in the β subunit (271).

The mammalian mitochondrial translational system utilizes two tRNA^{Ser} species: one specific for codons AGY and the other for UCN. In addition, two tRNA^{Leu} species are used; one is specific for codons UUR and the other for CUN (Figure 3.3). Mitochondrial leucyl-tRNA synthetases have not been studied in detail, but tRNA recognition studies of mammalian mitochondrial seryl-tRNA synthetases have demonstrated that a single enzyme is responsible for serylation of both tRNA^{Ser} species (274,276). This is quite remarkable because the two tRNA^{Ser} iso-acceptors share no common sequence motifs and are topologically quite distinct (Figure 3.5; see later section).

Investigations of the pathogenetic mechanisms of mitochondrial tRNA mutations have suggested that some mutations cause aminoacylation deficiency. Studies of cybrid cell lines carrying near homoplasmic levels of the 8344A→G or 8313G→A mutation in the tRNA^{Lys} gene have revealed a considerable drop in the fraction of aminoacylated tRNA^{Lys} (281,282). Similarly, a number of laboratories have reported a marked decrease in the fraction of aminoacylated tRNA$^{Leu(UUR)}$ in cybrid cell lines carrying the relatively common 3243A→G tRNA$^{Leu(UUR)}$ gene mutation (283–285). A reduction in leucyl-tRNA$^{Leu(UUR)}$ levels was also found *in vivo* in tissue samples from a patient heteroplasmic for the 3243A→G mutation; however, surprisingly, no reduction in lysyl-tRNA^Lys levels was found in muscle tissue from a patient heteroplasmic for the 8344A→G mutation (286).

Functional studies with *in vitro* transcribed or native tRNA^Leu(UUR) have indicated that the 3243A→G mutation results in a 12- to 25-fold decrease in aminoacylation of the tRNA compared to the wild-type (277,287). Interestingly, the less common 3243A→T mutation showed a 300-fold decrease in aminoacylation of the tRNA compared to the wild-type (277). Thus, distinct base changes at the same position can have a dramatically different effect on the efficiency of aminoacylation of the tRNA.

To function as initiator tRNA during translation, part of the mitochondrial methionyl-tRNA pool is converted to *N*-formylmethionyl-tRNA (83–85). This is carried out by mitochondrial methionyl-tRNA transformylase. Bovine methionyl-tRNA transformylase has been cloned (288) and its interaction with methionyl-tRNA^Met has been analyzed (289). These studies revealed that, in contrast to *E. coli* methionyl-tRNA transformylase, which uses a set of nucleotide residues in the acceptor stem of tRNA as identity element, the 40-kDa bovine enzyme employs the aminoacyl moiety of methionyl-tRNA for substrate recognition. This may prevent formylation of other mitochondrial aminoacyl-tRNA species that share structural features with the acceptor stem of methionyl-tRNA^Met (289).

Modifications of rRNA and mRNA Molecules

When the polycistronic transcript is cleaved by mitochondrial RNase P at the 5′-side of each tRNA, the 3′-end of the upstream mRNA or rRNA becomes available for adenylation (74). Mammalian mitochondrial mRNAs carry poly(A) tails of around 55 residues (290,291); the 16S rRNA is polyadenylated by up to ten residues and the 12S rRNA is monoadenylated (197,198). In the cytosol of eukaryotes, polyadenylation increases the stability of mRNAs, whereas in bacteria and chloroplasts it promotes rapid decay (292,293). Poly(A) tails have been shown to stimulate the degradation of mRNAs in mitochondria of *T. brucei* (294). Their role in modulating rates of RNA turnover in mammalian mitochondria is not known, but subsequent deadenylation may possibly cause enhanced mRNA decay (295).

A poly(A) polymerizing activity has been identified in bovine and rat mitochondrial extracts (265,296,297). Purification of the rat enzyme yielded a protein with an apparent molecular weight of 60 kDa (298). Purification of the bovine enzyme has failed so far, due to instability of the enzyme (265). Searches of human genome databases have not yielded any candidate. Unlike nuclear mRNAs, mitochondrial mRNAs do not have upstream AAUAAA polyadenylation signals, suggesting that recognition of the messenger by mitochondrial poly(A) polymerase is fundamentally different from the recognition process in the nucleus.

Similar to mitochondrial tRNAs, mitochondrial rRNAs contain a much smaller number of modified nucleotide residues than their bacterial or cytosolic counterparts. The few methylated nucleotide residues present in mammalian mitochondrial rRNAs appear to be concentrated in domains corresponding to the subunit interfaces (299). Human 16S mitochondrial rRNA contains only a single pseudouridine (Ψ) compared to 4 to 9 in the bacterial homologs and 55 in the human cytosolic homolog (300). The single pseudouridine of human 16S mitochondrial rRNA is located at an evolutionarily conserved site within the peptidyl transferase domain — arguing strongly for the functional importance of this modification — but its precise role is unknown. Human 12S mitochondrial rRNA contains two adjacent, methylated adenosine residues within a conserved stem-loop structure at the 3′-end. As inferred earlier, these bases may be modified by TFB1M (Table 3.3) (191). Other mitochondrial enzymes that catalyze modifications of rRNA nucleotide residues have not been identified.

MITOCHONDRIAL PROTEIN SYNTHESIS

Mitochondrial tRNAs

Essential elements of protein synthesis, tRNAs interact with numerous factors, such as aminoacyl tRNA synthetases (see previous discussion), elongation factors, mRNA, and different sites on the ribosome (see next section). All known tRNAs from archaea, bacteria, and the eukaryotic cytosol, as well as those from chloroplasts and the mitochondria of fungi and plants, possess a four-armed, cloverleaf-like secondary structure and an "L"-shaped tertiary structure (301). Studies of bacterial and cytosolic tRNAs have indicated that this tertiary structure is critical for the function of tRNAs. As inferred from the mtDNA sequences of the early 1980s, it was originally thought that many animal mitochondrial tRNAs were structurally quite different. The most extreme examples are tRNA$^{Ser(AGY)}$, which lacks the entire arm with the dihydrouridine (DHU) loop (302), and the iso-acceptor tRNA$^{Ser(UGA)}$, which possesses an extended anticodon stem of six base pairs instead of the usual five (Figure 3.5) (303).

Another striking feature of mammalian mitochondrial tRNAs is that their "variable region" is restricted to 3 to 5 nucleotide residues (Figure 3.5); in tRNAs of other origin, it may have up to 23 nucleotide residues (256). Nevertheless, more recent alignment studies of the tRNA genes of 31 fully sequenced, mammalian mitochondrial genomes revealed that, with the exception of serine-specific tRNAs, all mammalian mitochondrial tRNAs fold into canonical cloverleaf structures (304). However, as result of their decreased G+C content and increased frequency of non-Watson–Crick base pairs in stem regions (Figure 3.5), mitochondrial tRNAs have lower intrinsic thermodynamic stabilities than tRNAs of other origin. Thermodynamic instability might predispose mitochondrial tRNAs to deactivation by point mutations because a nucleotide substitution may disrupt these fragile tRNAs more easily than those holding more robust secondary interactions (305,306).

Mammalian mitochondrial tRNAs show considerable deviations in the size of the DHU and TΨC loops (Figure 3.5). In addition, the conserved UUC sequence in the TΨC loop of bacterial and cytosolic tRNAs (converted into TΨC by posttranscriptional modification) is absent in most mammalian mitochondrial tRNA genes. Furthermore, the GG doublet normally present in the dihydrouridine loop of bacterial and cytosolic tRNAs is only present in a minority of mammalian mitochondrial tRNAs. The frequent absence of conserved nucleotide residues in both loops suggests that the conventional tertiary interactions between the loops do not occur (304).

The structures of three mammalian mitochondrial tRNAs — tRNA[Ser(AGY)], tRNA[Ser(UCN)], and tRNA[Phe] — have been experimentally investigated (307–311). These studies confirmed the unusual tertiary folding rules for mammalian mitochondrial tRNAs; however, further experimental work is needed to define fully the intramolecular interactions that govern the tertiary structures of mitochondrial tRNAs. These studies are of great importance for understanding of the impact of tRNA mutations on tRNA conformation and possible subsequent recognition failures between mutated tRNAs and other factors of the mitochondrial translational machinery.

Mitochondrial Ribosomes

Mammalian mitochondrial ribosomes were originally isolated in the second half of the 1960s (312,313). The particles reside in the matrix of the organelle and have a diameter of about 320 Å in the direction of the longest axis (314). A large body of evidence derived from studies in yeast has indicated that mitochondrial translation is primarily, if not exclusively, an inner membrane-associated process (315–319). This concept is supported by fractionation studies of bovine mitochondria, which showed that about half of the ribosomes are associated with the inner membrane (320). This number probably represents an underestimate of the actual degree of association because the time needed for the large-scale preparation of the mitochondrial fractions is likely to result in the release of ribosomes from the membrane.

All 13 mammalian mtDNA-encoded proteins are mitochondrial inner membrane proteins. A physical association of mitochondrial ribosomes with the inner membrane may facilitate cotranslational membrane insertion of the nascent polypeptide. As mentioned in earlier paragraphs, studies in yeast have also indicated that mitochondrial RNA polymerase binds to mitochondrial inner membrane-associated factors and that at least a subset of the mitochondrial nucleoids is associated with proteinaceous structures that span the outer and inner membranes. Together, these observations suggest that transcription, translation, and membrane insertion are coordinated at the inner membrane, in the vicinity of the nucleoids.

Mitochondrial translation is considered to be prokaryotic in nature because, as indicated at the beginning of this chapter, the spectrum of antibiotics inhibiting mitochondrial protein synthesis resembles that of bacteria. In addition, like bacterial ribosomes, mitochondrial ribosomes use N-formyl-methionyl-tRNA for polypeptide initiation (83–85,321). Moreover, mitochondrial translational initiation and elongation factors are also functional on bacterial ribosomes *in vitro* (322–325). Even so, the physical and biochemical properties of mitochondrial ribosomes differ considerably from their cytosolic and bacterial counterparts (326).

Mammalian mitochondrial ribosomes have a remarkably low RNA content, resulting in a low sedimentation coefficient (~55S) compared to bacterial (~70S) and cytosolic (~80S) ribosomes (313,327–330). Sedimentation studies of the early 1970s showed that the ~39S and ~28S mitochondrial ribosomal subunits respectively contain the 16S and 12S rRNA species encoded by mtDNA (327,328). Human 16S rRNA is 1558 nucleotide residues long, and human 12S rRNA is 954 nucleotide residues long (excluding their adenyl tails). This is 40 to 50% shorter than the bacterial homologs. Although mammalian mitochondrial ribosomes lack several of the major RNA stem structures of bacterial ribosomes, they do have a relatively high protein content (326). As a result, the total mass of mitochondrial ribosomes is higher than that of bacterial ribosomes (331). It has been suggested that the additional and/or larger proteins of the mitochondrial ribosome compensate for the shortened rRNAs (326); however, the three-dimensional cryo-electron microscopic map of the bovine mitochondrial ribosome, published in 2003, demonstrated that many of the proteins occupy new positions within the ribosome and that most missing RNA segments are not replaced by any protein mass (314).

The 13.5-Å resolution structure further revealed that although characteristic structural features for both ribosomal subunits are instantly recognizable in the cryo-electron microscopic map, the overall structural organization in both subunits is markedly divergent from that of other ribosomes. Mitochondrial ribosomal subunits are held together by five RNA–RNA bridges, two RNA–protein bridges, seven protein–protein bridges, and one bridge that involves RNA and protein components from both subunits (314). In contrast, cytosolic ribosomes are held together predominantly by RNA–RNA bridges (332,333). These findings indicate that during the evolution of the mitochondrial ribosome, proteins have replaced certain rRNA functions, including much of their participation in the intersubunit communication.

In the sedimentation studies of the early 1970s, 5S rRNA was not detected in fungal and animal mitochondrial ribosomes (327,328,334), even though this approximately 120-nucleotide RNA species is present in all bacterial, chloroplast, and cytosolic ribosomes, as well as mitochondrial ribosomes of higher plants and some algae, where it is encoded by mtDNA (335). Notwithstanding the absence of 5S rRNA in mammalian mitochondrial ribosomes, nuclear-encoded 5S rRNA was later found to be tightly associated with highly purified mitochondrial fractions of mammalian cells (336,337); intriguingly, the number of mitochondrially imported 5S rRNA molecules appeared to be compatible with the predicted average number of ribosomes per organelle (338). The mitochondrial ribosome cryo-electron microscopic map of 2003 confirmed, however, the absence of 5S rRNA and revealed that about half of the 5S rRNA volume is replaced by proteins (314). This observation has put an end to all speculation of a functional role for nuclear-encoded 5S rRNA in mammalian mitochondrial ribosomes. The genuine role of 5S rRNA in mitochondria still remains to be clarified.

Rapid progress in genome sequencing and proteomics technology led to the identification of the full complement of human mitochondrial ribosomal proteins (MRPs) in 2001 (for

nomenclature, see http://www.gene.ucl.ac.uk/nomencla-
ture/genefamily/MRPs.html) (339–342). The chromosomal
locations of essentially all human MRP genes have been
mapped (343,344); some genes have been characterized in
detail (345,346) and 120 pseudogenes have been identified
(347). There are 29 different proteins in the small (28S) ribosomal subunit, 14 of which are homologous to proteins present
in the *E. coli* small ribosomal subunit. The remaining 15
proteins have no apparent homologs in prokaryotic, chloroplast, or cytosolic ribosomes, but are unique to mitochondrial
ribosomes. The human large (39S) mitochondrial ribosomal
subunit contains 48 distinct proteins, of which 28 are homologous to proteins present in the *E. coli* large ribosomal subunit. The remaining 20 are "new" proteins specific to
mitochondrial ribosomes.

Surprisingly, three sequence variants have been found
for one of the "prokaryotic" proteins of the small ribosomal
subunit MRPS18 (339). The three isoforms, MRPS18A,
MRPS18B, and MRPS18C, differ remarkably in size, ranging
from 11.7 to 27 kDa. In analogy with bacterial ribosomes,
each mitochondrial ribosome probably contains a single copy
of MRPS18. Therefore, the expression of three MRPS18 isoforms suggests the existence of a heterogeneous population
of mitochondrial ribosomes, which may have different kinetic
or decoding properties. Another surprising finding is that two
of the "new" proteins in the small mitochondrial ribosomal
subunit, MRPS29 and MRPS30, were earlier identified as the
proapoptotic proteins DAP3 and PDCD, respectively
(341,348). Thus, apparently, MRPS29/DAP3 and
MRPS30/PDCD have dual functions. Whether their role in
apoptosis is exerted from their position in the mitochondrial
ribosome remains an open question.

Mammalian mitochondrial ribosomes contain a high-
affinity binding site for GTP (349). Photoaffinity labeling
experiments have mapped this site to MRPS29 (326). Perhaps
the ribosome-based, GTP-binding function of MRPS29/DAP3
is entirely separate from its role in apoptosis. On the other
hand, MRPS29/DAP3 and MRPS30/PDCD may cooperate in

triggering a mitochondrial ribosome-induced, GTP-dependent apoptotic cascade.

Initiation of Translation

Although isolated mitochondria faithfully carry out protein synthesis, an *in vitro* mitochondrial translational system, using only mitochondrial extracts, is not available. Isolated mitochondrial ribosomes show poly(U)-directed, phenylalanine polymerizing activity (331), but polymerizing activity directed by natural mRNAs has never been demonstrated. Due to this persistent lack, some aspects of mitochondrial protein synthesis are poorly understood. Notwithstanding the absence of a true *in vitro* model of mitochondrial translation, screening of human EST databases with protein sequences of bacterial initiation and elongation factors has led to the identification of several mitochondrial translational factors that subsequently have been characterized in bacterial *in vitro* translational systems.

Initiation of mitochondrial translation is intriguing because mammalian mitochondrial mRNAs are devoid of significant upstream untranslated regions (23,213) and lack a 7-methylguanylate cap structure at their 5'-end (350). Thus, unlike prokaryotic messengers, mitochondrial messengers have no leader sequences to facilitate ribosome binding, and the cap recognition and scanning mechanism for directing the ribosome to the initiator codon, as used in the cytosolic compartment, can also be ruled out. The low translational efficiency of mitochondrial messengers (351) may in fact be the result of the absence of a distinct ribosome recognition site at the 5'-end of mitochondrial mRNAs and may necessitate the observed abundance of mitochondrial messengers compared to nuclear messengers (352,353) to ensure a sufficient level of translation. In contrast to mammalian mitochondrial mRNAs, yeast mitochondrial mRNAs have long 5'-untranslated regions. The fungal organelles require nuclear-encoded, mRNA-specific translational activator proteins for initiation of most of the major mRNAs (354). Homologs of the yeast *trans*-activators have not been identified in animal EST databases (355). This suggests that quite different strategies have developed in yeast and mammalian mitochondria for initiation of translation.

In vitro experiments with bovine mitochondrial ribosomes have demonstrated that, in contrast to their bacterial and cytosolic equivalents, the small (28S) mitochondrial ribosomal subunit has the ability to bind mRNA tightly in a sequence-independent fashion and in the apparent absence of auxiliary initiation factors or initiating N-formylmethionyl-tRNA (356). As judged from the size of the mRNA fragment protected from RNase T_1 digestion, the principal interaction between the 28S subunit and the mRNA strand occurs over a 30- to 80-nucleotide stretch (357,358), but a minimum length of about 400 nucleotide residues is required for efficient binding (358). This may explain why the two shortest open reading frames of mammalian mtDNA, *MTATP8* and *MTND4L* (<300 nucleotides), are both part of overlapping genes, *MTATP8/MTATP6* and *MTND4L/MTND4* (Figure 3.2), which are transcribed as dicistronic messengers (23,216). Monocistronic messengers of *MTATP8* and *MTND4L* are possibly too short to interact effectively with the 28S ribosomal subunit.

Natural mitochondrial mRNAs have an extensive secondary structure at their 5′-terminal regions with the initiation codon sequestered in a stem structure (357). However, this secondary structure does not appear to be a critical factor in 28S ribosomal subunit–mRNA interaction because binding studies of synthetic mRNAs with reduced secondary structure (prepared by substituting inosine 5′-triphosphate (ITP) for GTP during *in vitro* transcription reactions) behaved similarly to normal mRNAs (358). Moreover, the 28S subunit–mRNA complex forms as readily on circular mRNAs as on linear mRNAs (359), indicating that a free 5′-terminus on the mRNA is not required for the initial interaction.

Therefore, it is thought that the interaction between the 28S subunit and the messenger occurs in a sequence-independent manner. Initial binding is believed to be followed by movement of the 28S subunit to the initiation codon and melting of the 5′-terminal secondary structure of the mRNA.

This process is probably mediated by (as yet unspecified) auxiliary factors. The cryo-electron microscopic map of bovine mitochondrial ribosomes revealed that mitochondrial ribosomes have a unique, gate-like protein structure at their mRNA entry site (314). Perhaps this structure is involved in recruitment of the unusual mitochondrial mRNAs and regulation of translational initiation.

Mitochondrial protein synthesis is considered to follow the classical model of protein synthesis, as described for *E. coli*. Accordingly, during the first step of mitochondrial protein synthesis, *N*-formylmethionyl-tRNA will bind to the peptidyl (P) site on the ribosome, while the other two sites for tRNA molecules, the aminoacyl (A) site and the exit (E) site, remain empty. The first identified and most thoroughly studied initiation factor of mammalian mitochondria is mitochondrial translational initiation factor 2 (MTIF2; Table 3.5) (323,360–363). It is a monomeric protein of about 78 kDa homologous to the bacterial initiation factor IF-2. MTIF2 belongs to the family of GTPases that are molecular switches capable of alternating between an active (MTIF2·GTP) and an inactive (MTIF2·GDP) conformation. Reminiscent of its bacterial equivalent, MTIF2 promotes binding of *N*-formylmethionyl-tRNA to the 26S ribosomal subunit in a GTP- and mRNA-dependent reaction.

Detailed *in vitro* characterization of bovine MTIF2 has suggested that the factor may bind to the 26S subunit prior to its interaction with GTP, but that GTP enhances the binding affinity between MTIF2 and the 26S subunit and allows *N*-formylmethionyl-tRNA to join the complex (325,360). Hydrolysis of GTP is thought to facilitate the release of MTIF2 and the concomitant association of the 39S ribosomal subunit to form the 55S initiation complex; however, GTP hydrolysis appears not to be crucial for subunit association because nonhydrolysable analogs of GTP can still promote formation of the 55S initiation complex (360).

In *E. coli*, formylation of methionyl-tRNA is necessary for the interaction of the tRNA with IF-2. Formylation also eliminates any significant interaction with elongation factor EF-Tu. Studies with purified bovine MTIF2, isolated bovine

Table 3.5 Human Nuclear-Encoded Mitochondrial Translation
Factors

Factor	Description (Aliases)	M_r (×1000)	Gene Symbol	Cytogenetic Position
EFG1	Mitochondrial elongation factor G1 (mtEF-G1, EF-G1$_{mt}$, GFM)	80	*EFG1*	3q25.1–q26.2
EFG2	Mitochondrial elongation factor G2 (mtEF-G2, EF-G2$_{mt}$)	80	*EFG2*	5q13
MTIF2	Mitochondrial translation initiation factor 2 (mtIF2, IF-2$_{mt}$)	78	*MTIF2*	2p14–p16
MTIF3	Mitochondrial translation initiation factor 3 (mtIF3, IF-3$_{mt}$)	27	*MTIF3*	13p12.2
MTRF1	Mitochondrial translation release factor 1 (mtRF-1, RF1$_{mt}$)	52	*MTRF1*	13q14.1–q14.3
MRRF	Mitochondrial ribosome recycling factor (mtRRF1, RRF1$_{mt}$)	29	*MRRF*	9q32–q34.1
TSFM	Mitochondrial elongation factor Ts (mtEF-Ts, EF-Ts$_{mt}$)	31	*TSFM*	12q13–q14
TUFM	Mitochondrial elongation factor Tu (mtEF-Tu, EF-Tu$_{mt}$)	45	*TUFM*	16p11.2

mitochondria, and *N*-formylmethionyl-tRNA or methionyl-tRNA from *E. coli* or yeast have demonstrated that the binding of nonformylated methionyl-tRNA to mitochondrial ribosomes is about 20- (yeast methionyl–tRNA) to 50-fold (*E. coli* methionyl–tRNA) lower than observed with their formylated counterparts (360). These results suggest that also in mitochondria the added formyl group reinforces the binding of the initiator tRNA to MTIF2.

Unexpectedly, however, two different yeast mutant strains, deficient in formylation of mitochondrial methionyl-tRNAs, exhibited normal mitochondrial function (364). This implies that, at least in yeast, formylation of initiator methionyl-tRNA is not required for mitochondrial protein synthesis per se. Furthermore, it was shown that expression of bovine

MTIF2 is able to support mitochondrial translation of a yeast double mutant deficient for the yeast MTIF2 equivalent and methionyl-tRNA formyltransferase (365). This suggests that formylation of the initiator methionyl-tRNA may not be strictly required for mammalian mitochondrial protein synthesis either.

A second mitochondrial translational initiation factor, MTIF3, has been identified during human EST database searches (Table 3.5) (366). The approximately 27-kDa protein is homologous to the prokaryotic and chloroplast initiation factor IF3, but has diverged considerably and does not appear to be well conserved throughout the animal kingdom. Prokaryotic IF3 plays a role in the discrimination of the initiation codon (AUG or, occasionally, GUG or UUG) from other codons. Interestingly, the IF3 amino acid residues implicated in this function are not conserved in human MTIF3 (366). AUG and AUA serve as initiation codons in human mitochondria. Consequently, the proofreading properties of MTIF3 may be quite different from those of bacterial IF3s.

Bacterial IF3 also acts as a ribosome dissociation factor. IF3 binds to the small ribosomal subunit and inhibits its association with the large ribosomal subunit, thus ensuring a supply of small ribosomal subunits for initiation of protein synthesis. Sucrose gradient centrifugation experiments with mitochondrial ribosomes in the presence or absence of MTIF3 have suggested that the mitochondrial factor has a similar function (366). In line with its ability to increase the availability of free 28S subunits required for the activity of MTIF2, purified MTIF3 has been found to promote *N*-formylmethionyl-tRNA binding on mitochondrial ribosomes in the presence of MTIF2, and either poly(A,U,G) or an *in vitro* transcript of the mitochondrial *MTCO2* gene as mRNA (366).

Despite extensive EST database searches, no mitochondrial counterpart has been found for the bacterial and chloroplast initiation factor IF1 (366). IF1 is a small (~70 amino acids) and not well conserved protein that may, therefore, be easily overlooked in *in silico* searches. Bacterial IF1 binds to the small ribosomal subunit in a region that will become the A site (367). By binding to this site, IF1 is postulated to prevent accidental initiation from the A site and to promote the correct positioning of *N*-formylmethionyl-tRNA in the P site (368,369). To assess the possible requirement for a factor equivalent to IF1 in the mitochondrial system, the effect of *E. coli* IF1 on initiation complex formation was examined using *E. coli* and (bovine) mitochondrial ribosomes as well as *E. coli* IF2 and IF3, and MTIF2 and MTIF3 (366). The presence of IF1 had essentially no effect on initiation complex formation on mitochondrial ribosomes in the presence of MTIF2 and MTIF3. This observation suggests that the mitochondrial system may not need a factor directly corresponding to IF1.

Elongation of Translation

The mitochondrial elongation factors (Table 3.5) are homologous to the three elongation factors found in *E. coli*: EF-Tu, EF-Ts, and EF-G. These factors have been purified from bovine liver mitochondria (322,324); their cDNAs have been cloned from various mammalian sources (370–376) and the human genes have been partly characterized (374–378). These studies revealed that *E. coli* EF-G has two homologs in humans: EFG1 and EFG2 (Table 3.5) (376). Both proteins are phylogenetically conserved through evolution (376). The overall amino acid identity between EFG1 and EFG2 is only 33%, but some domains are highly conserved. Comparison of northern blot hybridization results (375,376) suggests that EFG1 and EFG2 are coexpressed in all tissues, pointing to complementary roles for EFG1 and EFG2 during elongation of the polypeptide chain.

In the *E. coli* elongation cycle, EF-Tu promotes the codon-dependent placement of aminoacyl-tRNA at the A-site of the ribosome. This process requires the formation of a ternary complex comprising aminoacyl-tRNA, EF-Tu, and GTP. After delivery of the correct aminoacyl-tRNA to the A site, GTP is hydrolyzed and EF-Tu·GDP dissociates from the ribosome. The inactive EF-Tu·GDP is recycled to the active EF-Tu·GTP by exchange of GDP by GTP. This step is mediated by the nucleotide exchange factor EF-Ts that displaces GDP, which in turn is replaced by GTP to reform EF-Tu·GTP.

Following peptide bond formation catalyzed by the large ribosomal subunit, EF-G drives the translocation of the tRNAs at the A and P sites of the ribosome to the P and E sites, respectively. Concomitantly, the mRNA is moved to expose the next codon to the A site. During the last step of the elongation cycle, the deacylated tRNA at the E site is released from the ribosome. Similar to EF-Tu, EF-G is a GTPase that undergoes a cycle in which it alternates between an active state when bound to GTP and an inactive state when bound to GDP (379).

The mammalian mitochondrial elongation factors homologous to *E. coli* EF-Tu and EF-Ts are TUFM and TSFM, respectively (Table 3.5). Elongation of the nascent mitochondrial polypeptide chain is assumed to proceed in a similar fashion to that of the elongation process in *E. coli*. There are, however, subtle differences in the manner in which TUFM interacts with other components of the translational machinery compared to the bacterial EF-Tu.

A binary complex of TUFM with its nucleotide exchange factor has been isolated from bovine liver mitochondria (322). This TUFM·TSFM complex is very stable and, unlike its bacterial equivalent, cannot easily be dissociated by guanosine nucleotides (322). Under certain experimental conditions, however, the TUFM·TSFM complex will dissociate in the presence of GTP and phenylalanyl-tRNA, resulting in a ternary complex similar to that observed in *E. coli* (380).

Careful evaluation of the equilibrium dissociation constants (K) of bovine TUFM with its ligands has indicated that the K_{GDP} as well as the K_{GTP} for TUFM are about two orders of magnitude higher than the dissociation constants of the corresponding complexes formed by *E. coli* EF-Tu (381). The $K_{aminoacyl-tRNA}$ and K_{TSFM} for the aminoacyl-tRNA·TUFM·TSFM complex are around 16- and 3-fold higher, respectively, than those of the bacterial ternary complex (382). Even though some dissociation constants governing the elongation cycle are strikingly different in mammalian mitochondria and *E. coli*, when the concentrations of the various components under *in vivo* conditions are taken into account, calculations indicate that the ternary complex will be the major form of TUFM/EF-Tu in both systems (382).

Resolution of the crystal structure of bovine TUFM·GDP at 1.94 Å (383) has revealed that the residues of TUFM that directly interact with GDP are nearly all identical to those of *E. coli* EF-Tu and are therefore unlikely to be responsible for the large difference in affinity for the ligand. The decrease in affinity may, however, be due to an increased mobility of parts of TUFM around the binding site. The x-ray diffraction data have further indicated that the C-terminal extension of TUFM, which is not present in bacterial EF-Tu, shows structural similarities with DNA-binding zinc fingers (383), suggesting that this extension may be involved in RNA binding.

Termination of Translation

On bacterial as well as cytosolic ribosomes, termination of protein synthesis takes place as response to an mRNA stop rather than a sense codon in the A site. Because all mitochondrial mRNA species normally contain a stop codon at their 3'-end, a similar process is thought to be responsible for termination of protein synthesis in mitochondria. Surprisingly, however, a patient cell culture has been reported that lacked a functional *MTATP6* mRNA stop codon but was still capable of efficient MTATP6 protein synthesis (384). This observation indicates that the presence of a stop codon may not be an absolute requirement for termination of translation in human mitochondria.

Termination of protein synthesis involves the action of several auxiliary proteins, termed release factors. In *E. coli*, release factor RF1 recognizes the stop codons UAA and UAG, and RF2 recognizes the stop codons UAA and UGA. A third factor, RF3, stimulates the activities of RF1 and RF2. Binding of RF1 and RF2 to a stop codon at the A site results in hydrolysis of the bond between the nascent polypeptide chain and the tRNA at the P site. The detached polypeptide then leaves the ribosome. Subsequent dissociation of the tRNA, mRNA, and ribosome is mediated by a fourth factor, the ribosome recycling factor RRF (also called RF4). The two subunits of the ribosome disassemble and are set to start a new round of protein synthesis (385–387).

An approximately 39-kDa release factor has been partly purified from rat liver mitochondria (388). Like bacterial RF1, the rat mitochondrial factor recognizes codons UAA and UAG in an *in vitro* assay, but not UGA (which serves as a codon for tryptophan in mammalian mitochondria; Figure 3.3). For that reason, the factor has been named MTRF1. Rat MTRF1 does not recognize the codons AGG or AGA, which are used as alternative terminators in human mitochondria (23); however, rat mitochondria only employ UAA as stop codon (389), so the single factor is probably sufficient for the recognition of stop codons in this species.

A putative human MTRF1 cDNA and its corresponding gene (Table 3.5) have been identified by screening of human EST with the *E. coli* RF-1 amino acid sequence (390,391). The cDNA is predicted to code for a 52-kDa protein that has significant homology to its counterparts in prokaryotes and mitochondria of lower eukaryotes. EST searches have also revealed human cDNAs that encode a protein with homology to *E. coli* RRF (390). The hypothetical, 29-kDa protein has been named MRRF (Table 3.5) (391). MRRF shows 25 to 30% sequence identity to prokaryotic RRFs, but only 19% sequence identity with the mitochondrial ribosome recycling factor Fil1p from *S. cerevisiae* (390,392,393). So far, mitochondrial homologs of *E. coli* RF-2 and RF-3 have not been identified. Therefore, it would be particularly interesting to know whether human MTRF1 is able to recognize all four different mitochondrial stop codons used by human mitochondria, or that additional factors are necessary.

REPLICATION OF MTDNA

General Features of Replication

There appears to be a constant turnover of mtDNA in mammalian tissues (394). The apparent half-life of rat mtDNA was found to be 6.7 days in heart, 9.4 days in liver, 10.4 days in kidney, and about 31 days in brain (395). In the 1970s, labeling experiments with the thymidine analog 5-bromo-2′-deoxyuridine indicated that mtDNA molecules of exponentially growing HeLa cells may be replicated more than once or not at all within a cell generation time (396). Furthermore, using radioactivity profiles of [³H]thymidine and bromodeoxyuridine double-labeled mtDNA resolved on buoyant CsCl gradients, it was calculated that mtDNA molecules are randomly selected for replication and that replication occurs at a constant rate throughout the cell cycle of mouse L-cells (15).

More recent *in situ* localization experiments of replicating mtDNA in various human cell cultures are consistent with the view that replication is a continuous process (Figure 3.1), even in quiescent cells (16). In addition, these studies revealed that actively replicating nucleoids are present throughout the mitochondrial network (16) and that each mtDNA molecule replicates independently of others in the nucleoid (61). Despite this seemingly relaxed nature of mtDNA replication, mtDNA copy number is maintained at a remarkably constant level in proliferating and quiescent cell cultures (397). In an exponentially growing, human renal cell line transfected with a plasmid encoding inactive mitochondrial DNA polymerase, mtDNA depleted with an apparent half-life of 2 to 3 days when the expression of the transgene was induced (398). Depletion of mtDNA was reversible in this cell line, as demonstrated by restoration of mtDNA copy number to normal within 10 days when the expression of mutant polymerase was suppressed following a 3-day induction period (398). These observations indicate a tight control of mtDNA copy numbers.

Studies with yeast and rat liver mitochondria have suggested that mtDNA is associated with the mitochondrial membrane during replication (399,400). Replication of mtDNA is an inherently slow process and has been estimated to occur at a rate 200 times slower than that of *E. coli* DNA replication (401). The synthesis of full-length daughter strands requires approximately 1 h and the entire replication cycle is completed in approximately 2 h (399,402). For many years, the prevailing model for replication of mtDNA involved a strand-asynchronous mechanism of duplication. This model is primarily advocated by Clayton (401). In 2000, Holt and Jacobs proposed a strand-synchronous mechanism (403). The validity of either model has been hotly debated (404–406).

The Strand-Asynchronous Model of Replication

Basic Mechanism

As discussed at the beginning of this chapter, early molecular studies of mammalian mtDNA relied for an important part upon equilibrium CsCl gradient centrifugation techniques. In these experiments, putative replicative intermediates of mtDNA were isolated from exponentially growing cell cultures, facilitated by the fortuitously slow rate of replication (407). Electron micrographs of the putative replicative intermediates suggested that mtDNA molecules are replicated unidirectionally from two spatially and temporally distinct, strand-specific origins (407).

The strand-asynchronous (or strand-displacement) model of mammalian mtDNA replication is directly based on these electron microscopic observations. According to this model (Figure 3.6) (401), replication begins with unidirectional synthesis of a daughter H-strand and consequent displacement of the parental H-strand. The origin of H-strand replication (O_H) is collectively formed by closely spaced, defined sites within the D-loop region, downstream of the LSP (Figure 3.4). The origin of L-strand replication (O_L) is collectively formed by closely spaced, defined sites at two-thirds of the genomic distance away from O_H, in a cluster of five tRNA genes (Figure 3.2). Synthesis of the daughter L-strand is initiated when the replication fork of the leading (daughter H) strand has passed O_L and thus exposes O_L on the displaced H-strand in single-stranded form. Synthesis of the lagging (daughter L) strand proceeds in the opposite direction to that of H-strand replication.

A key aspect of this model is that both strands are replicated continuously from temporally and physically distinct, strand-specific origins, without frequent priming of short,

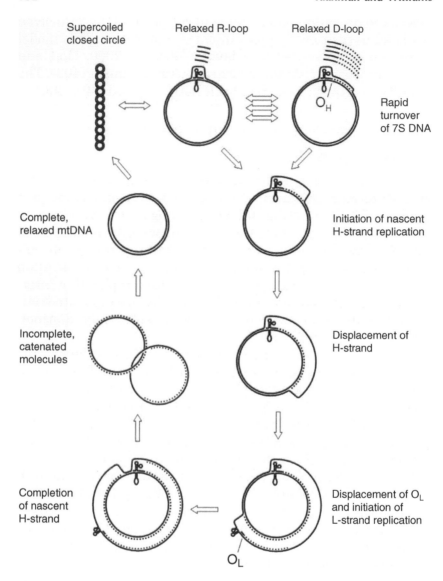

Figure 3.6 The strand-asynchronous or strand-displacement model of human mtDNA replication. The solid, circular lines represent the parental H- and L-strands. Nascent H- and L-strands are shown as stippled lines. Both strands are synthesized continuously from physically and temporally distinct sites termed O_H and O_L. Primer RNAs for initiation of H- and L-strand synthesis

Figure 3.6 (continued) are shown as short, solid lines. The highly structured R-loop is depicted with a hypothetical Holliday junction and includes looping of the L-strand that would be required for junction formation. The putative stem-loop secondary structure of the displaced O_L is also shown. 7S DNA is depicted as nascent H-strand in the D-loop triplex. The additional R- and D-loop RNA and DNA species, shown above the mtDNA molecule, point to their rapid turnover. Open arrows indicate the progression of the replication cycle. Double arrows reflect the metabolic instability of the R-loop and D-loop strands and consequent equilibrium between supercoiled, closed circular mtDNA and relaxed R-loop and D-loop mtDNA. For further details see text.

lagging-strand (Okazaki) fragments that would otherwise be required for polymerization in the 5′→3′ direction. The delay in L-strand synthesis results in a replication intermediate with a long, single-stranded parental H-strand (Figure 3.6). When the replication process is near completion, two distinct daughter molecules are formed: a duplex circle with a newly synthesized leading strand and a gapped circle with a partial, newly synthesized lagging strand. In each case, the final steps of synthesis and ligation result in closed circular mtDNA products (Figure 3.6).

Support for the strand-asynchronous model comes from S_1 nuclease protection studies and ligation-mediated PCR studies, which confirmed the presence of initiation sites at O_L in addition to initiation sites at O_H (408,409). Furthermore, the overwhelming majority of naturally occurring, deleted forms of human tissue mtDNA contain O_H and O_L (for a compilation of deletion junctions, see http://www.mitomap.org). This suggests that molecules lacking O_H or O_L are unable to replicate. An alternative explanation for the conservation of these two sequence elements in deleted mtDNA species is, however, possible (see next section).

Initiation of H-Strand Synthesis

According to the strand-asynchronous model, mtDNA replication starts with H-strand synthesis in the D-loop (Figure

3.4). For initiation of DNA synthesis, all known DNA polymerases require the presence of a preexisting oligonucleotide primer hybridized to the template strand. *De novo* synthesis of an oligoribonucleotide by an RNA polymerase or a DNA primase is the major mechanism of primer production.

Fine mapping of RNA and DNA species in the D-loop region of human and mouse mtDNA has suggested that short mitochondrial transcripts, originating at IT_L, serve as primers for synthesis of the 7S DNA strand, which displaces the H-strand in the D-loop region (410,411). Although a direct precursor–product relationship between the D-loop 7S DNA and productive H-strand replication has never been demonstrated, the strand-asynchronous replication model assumes that the 7S DNA species is the precursor of the nascent H-strand (Figure 3.6). This assumption is supported by the observation that the 5′-termini of 7S DNA are identical to the 5′-termini of nascent H-strands (408,409). No differences between the initiation of L-strand transcription and initiation of RNA primer formation for 7S DNA synthesis are known and it is not clear which mechanism decides between transcript elongation or 7S DNA synthesis (137). Transitions from RNA to DNA synthesis take place at several distinct sites that jointly make up O_H in a region of three short, evolutionarily conserved sequence blocks, termed CSB-1, -2, and -3 (Figure 3.4) (51). In human mtDNA, the most prominent origin of replication site has been mapped to nucleotide position 191 (412).

Only a few initiation events at O_H are thought actually to result in the synthesis of a full-length mtDNA molecule; most are terminated at specific sites, depending on the vertebrate species, 500 to 1000 bases downstream to yield the 7S DNA species (413). Arrested nascent H-strands remain annealed to their template L-strand and create the triplex D-loop structure (401). The 3′-ends of prematurely terminated H-strands map downstream of a short, evolutionarily conserved element, termed the termination-associated sequence (TAS) element (414).

This has led to the proposal that TAS elements are involved in regulation of the premature stalling of H-strand synthesis (414). The numbers of TAS elements vary per vertebrate species. The human mitochondrial genome contains a single TAS element (TAS-D) and a single trinucleotide stop point, 51 to 53 nucleotides downstream from the template (L-strand) TAS element (3′-TAACCCAAAAATACA; nucleotide positions 16158 to 16172) (415). The mechanism that determines whether a nascent H-strand terminates downstream of the TAS element or elongates over the entire length of the genome is not known, but is considered to be a key regulation mechanism of cellular mtDNA copy number.

Initiation of L-Strand Synthesis

Mapping studies of the 5′-termini of *in vivo* nascent L-strands have suggested that replication of the human L-strand starts at two distinct points, separated by 37 nucleotides (408). These two points collectively form O_L. The origins are located in a small noncoding region, flanked by five tRNA genes (Figure 3.2). According to the strand-asynchronous replication model, O_L is only activated when the parental H-strand is displaced by the growing daughter H-strand (Figure 3.6). After being exposed as a single-stranded entity, both origins are thought to adopt a distinctive stem-loop structure (408).

In vitro run-off replication studies have suggested that this configuration serves as recognition structure for a mitochondrial DNA primase that provides an RNA primer for L-strand synthesis (416,417). The location of O_L in a tight cluster of tRNA genes and the absence of the potential stem-loop structure in some vertebrate species (414) suggest that additional secondary structures may also contribute to DNA primase recognition. RNA priming of the major human L-strand origin starts at a T-rich stretch of the predicted loop. Transition from RNA to DNA synthesis occurs at a unique site near a critical GC-rich stretch, at the base of the hairpin (416,417).

The human mitochondrial DNA primase believed to be involved in priming of L-strand replication has only been partly purified (418). The enzyme has many characteristics distinguishing it from other prokaryotic and eukaryotic primases and shows an unusual sedimentation behavior as a result of its association with RNA species. These RNAs seem to play a critical role in the replicative function of the enzyme because degradation of the RNAs with nucleases leads to rapid inactivation of primase activity. The predominant RNA moiety cofractioning with DNA primase activity appears to be the nuclear gene product 5.8S rRNA (418); however, because only crude fractions have been analyzed, the cofractionation of 5.8S rRNA might have been the result of an adventitious contamination.

The Strand-Synchronous Model of Replication

The strand-asynchronous model was seriously challenged by Holt and Jacobs in 2000, when their application of Brewer and Fangman, two-dimensional, agarose gel electrophoresis to replicating mtDNA (419) revealed two classes of replication intermediates in mammalian cells (403). One was resistant to S_1 nuclease digestion and displayed the mobility properties of coupled leading- and lagging-strand replication products, whereas the other one was sensitive to S_1 nuclease digestion and was presumed to derive from the traditional, strand-asynchronous mode of mtDNA replication.

The strand-synchronous (or strand-coupled) replication model (Figure 3.7) corresponds to the replication mode of chromosomal DNA. A fundamental aspect of this mode of replication is that the lagging-strand is replicated discontinuously, with frequent RNA priming of short, lagging-strand Okazaki fragments required for polymerization in the $5' \rightarrow 3'$ direction. Using the same two-dimensional electrophoresis technique, strand-synchronous replication had previously been described for mtDNA from sea urchin (420) and *S. pombe* (421).

More recently, Holt and Jacobs reported that bona fide replication intermediates from highly purified mitochondria are essentially duplex throughout their length, but contain widespread regions of RNA–DNA hybrid as a result of the incorporation of ribonucleotides on the L-strand that subsequently are converted to DNA (422). The authors suggested that excision of ribonucleotides during DNA extraction may

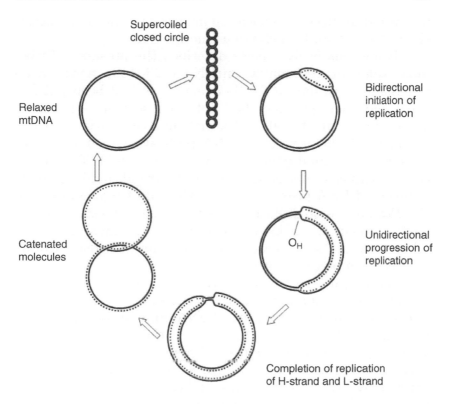

Figure 3.7   The strand-synchronous or strand-coupled model of human mtDNA replication. The solid, circular lines represent the parental H- and L-strands. Nascent H- and L-strands are shown as stippled lines. Both strands are initially synthesized bidirectionally from replication origins within a broad initiation zone (see Figure 3.2). Subsequently, arrest of one of the forks occurs at O_H. Replication at the other fork continues, resulting in unidirectional advance of replication. Open arrows indicate the progression of the replication cycle. For further details see text.

modify the properties of replication intermediates and yield partly single-stranded DNA molecules. Thus, the biased incorporation of ribonucleotides on the L-strand accounts for the seemingly strand-asynchronous replication mode of mammalian mtDNA. Holt and Jacobs interpreted their results to indicate that mtDNA replication proceeds mainly, if not exclusively, by a strand-synchronous mechanism.

Numerous early reports described the presence of ribonucleotides in mammalian mtDNA (423–428); however, no functional explanation for their existence had been put forward and their presence had not been seen as a potential source for artifacts (401). Nevertheless, as early as 1974, evidence of synchronous mtDNA replication in mammalian cells was presented using electron microscopy (429). Incorporated ribonucleotides on the nascent L-strand could represent leftovers of RNA primers used for Okazaki fragment synthesis. Proteins with RNase H1 activity have been shown to participate in Okazaki fragment primer removal, although other enzymes are necessary to eliminate all ribonucleotide residues completely (430).

Interestingly, a fraction of RNase H1 is targeted to mitochondria in mammalian cells (431). Moreover, *Rnaseh1* knockout mice fail to generate mtDNA and show embryonic lethality (431). Although this observation does not provide direct support for the strand-synchronous model of replication because the enzyme might have a role in removal of RNA primers in either model, it does indicate an essential role for RNase H1 in mtDNA maintenance.

In a detailed mapping study of replication origins, Holt's group has further refined the strand-synchronous model (432). Based on their results with human, mouse, and rat mtDNA, the authors have proposed that initiation of strand-synchronous replication does not occur near to O_H, as previously suggested (422), but instead initiates from multiple origins scattered throughout a broad zone, stretching from within the *MTCYB* gene to within the *MTND4* gene (Figure 3.2) (432). They suggest that origins are not necessarily evenly dispersed across this initiation zone, nor do all origins need to be activated at equal frequency.

According to their revised strand-synchronous model, mtDNA replicates initially bidirectionally from the initiation zone but, after fork arrest near O_H, replication is restricted to one direction only (Figure 3.7). Thus, in the revised model, O_H acts not as the origin of strand-synchronous replication but as the terminus for mtDNA replication. Short regions of triplex DNA are known to inhibit transcription. Similarly, the function of the D-loop may be to mediate replication fork arrest by forming a structural barrier (432).

As mentioned earlier, naturally occurring deletions of mtDNA spare O_H and the D-loop, but the region that contains the initiation zone is frequently deleted. Holt and coworkers (432) suggest that perhaps any DNA located downstream from the D-loop could function as an origin zone because of its proximity to the D-loop and membrane components (63). In earlier work, the authors located a possible replication fork barrier close to O_L (403). The fork barriers near O_H and O_L may offer a partial explanation for the location of pathological mtDNA mutations: molecules in which replication has stalled at the fork barriers near O_H and O_L may be favored substrates for illicit recombination (432).

It should be noted that the strand-synchronous model of mtDNA replication proposed by Holt and Jacobs is based exclusively on the Brewer and Fangman gel electrophoresis procedure. Although this is a well-established technique for revealing DNA replication intermediates, it has been argued by Bogenhagen and Clayton that the complicated two-dimensional gel patterns can be interpreted in more than one way (404). The strand-synchronous model of mtDNA replication is attractive because it does not require the unusually large, single-stranded loops that form the basis of the asynchronous model; however, it will only become fully accepted once it is supported by other experimental approaches. For example, fully duplex replicative intermediates should be visualized by electron microscopy and Okazaki fragments produced by coupled lagging-strand synthesis should be documented.

Maturation of the Daughter mtDNA Molecules

Regardless of the mode of mtDNA replication, after synthesis of the two nascent mtDNA strands is completed, the daughter mtDNA molecules must be decatenated, the RNA primers removed, and the corresponding gaps filled in and ligated; finally, the closed circular mtDNA must adopt its tertiary structure through the introduction of superhelical turns (Figure 3.6 and Figure 3.7). Although significant progress has been made concerning possible *trans*-acting factors involved in these events, knowledge of the true nature of these processes is still in its infancy.

The mtDNA is negatively supercoiled *in vivo* (433). Studies of mouse L-cells have suggested that the segregated daughter molecules are initially converted to closed circles with few, if any, superhelical turns (402,433). These forms appear to have a half-life of less than 1 h while being subjected to a process that introduces superhelical turns into the molecules. The end product has a superhelical density consistent with that of many other mature, closed circular DNAs ($\sigma \approx -0.06$), which translates into around 100 superhelical turns per mtDNA molecule. The supercoiled molecule serves as a substrate for synthesis of a new D-loop. D-loop formation relaxes the mtDNA molecule. In mouse L-cells, the half-life of the supercoiled molecule is less than 1 h and the majority of the mtDNA molecules carries a D-loop (434), suggesting that D-loop formation can be an aggressive process in metabolically active tumor cells. The kinetics may be very different in tissues.

trans-Acting Factors Involved in Replication

Mitochondrial RNases

According to the strand-asynchronous model, short mitochondrial transcripts originating at IT_L serve as primers for synthesis of the 7S DNA and the daughter H-strand. As the precursor RNA primer extends beyond the transition sites of RNA to DNA synthesis, the primary transcript is thought to be enzymatically processed to yield the mature primer RNA 3′-ends. Because of their location, it has been speculated that CSB-1, -2, and -3 direct the precise cleavage of primary transcripts to provide the appropriate primer species (137).

In vitro transcription studies of the human O_H region with mitochondrial RNA polymerase fractions and purified TFAM have indicated that the precursor RNA primer exists as a stable and persistent RNA–DNA hybrid, also known as an R-loop (Figure 3.4) (435). Hybrid formation requires the

GC-rich sequence block CSB-2 and is also affected by mutations in CSB-3 (435). The R-loop of mammalian mtDNA is thought to be similar to the highly structured R-loop at the bacterial ColE1-type replication origin (436).

The search by Clayton's group for a catalytic activity capable of processing L-strand transcripts containing O_H sequences has led to the identification of RNase MRP (Table 3.4). In their original studies, single-stranded, O_H-containing RNA species were used as substrate; the *in vitro* RNase MRP cleavage sites did not correspond with the *in vivo* 5'-termini of the nascent human and mouse H-strands (410,411,437, 438). In more recent studies, however, it was demonstrated that human and mouse RNase MRP cut the precursor RNA in the context of a triple-stranded R-loop configuration *in vitro* at virtually all of the major 3'-priming sites found *in vivo* (439,440).

As explained earlier in this chapter, RNase MRP is a ribonucleoprotein that is evolutionarily related to RNase P (231). Although the two enzymes have a distinct RNA moiety, they do have several, if not all, of their protein subunits in common (Table 3.4) (218,236). The predominant nucleolar location of RNase MRP has led to controversy as to its mitochondrial function (233,441). Ultrastructural *in situ* hybridization experiments, however, have validated the presence of RNase MRP RNA in nucleoli as well as mitochondria (234) — consistent with its dual role in maturation of nuclear rRNAs and mitochondrial RNA primers.

Genetic evidence for a functional role of RNase MRP RNA in mitochondria has come from studies in yeast. Although mutation analysis of the RNase MRP RNA gene in *S. cerevisiae* could only confirm the nuclear function of RNase MRP RNA (442), an *S. pombe* strain with a dominant mutation in its RNase MRP RNA gene was shown to require the mitochondrially associated, nuclear mutation *ptp-1* for viability (443), thus linking RNase MRP RNA to mitochondrial biogenesis. Notwithstanding this evidence, a direct involvement of RNase MRP RNA in mtDNA replication *in vivo* has not been demonstrated.

Human RNase MRP RNA is 265 nucleotides long (444). Mutations in the human RNase MRP RNA gene (*RMRP*; Table 3.4) are responsible for the recessively inherited developmental disorder, cartilage-hair hypoplasia (445). The clinical features of this disease are attributed to reduced levels of RNase MRP RNA in nucleoli rather than in mitochondria because the reduced levels of RNase MRP RNA in patients are considered still greatly in excess of those needed for mitochondria to function properly.

Human RNase MRP contains at least six protein subunits, all of which are also part of the nuclear RNase P ribonucleoprotein complex (Table 3.4). The specific functions of these protein subunits remain to be established. Immunological studies indicated a prevalent nucleolar localization of the subunits; evidence of an additional mitochondrial location has not been presented (446–449). Therefore, it is at present unclear whether the protein composition of mitochondrially located RNase MRP matches that of nucleolar RNase MRP.

Originally, endonuclease G, a protein belonging to the large family of DNA/RNA nonspecific, $\beta\beta\alpha$-Me-finger nucleases (450), was also implicated in processing of precursor RNA primers (451). The mitochondrial location of endonuclease G is undisputed (451). This protein has a rather wide spectrum of nucleolytic activities: it cleaves GC-rich, double-stranded, and single-stranded DNA tracts; RNA; and an RNA–DNA heteroduplex containing mouse O_H (451). However, the *in vitro* RNA cleavage sites of the heteroduplex do not align with all *in vivo* priming sites. Deletion of the homologous gene in yeast does not affect mtDNA metabolism (452). In addition, more recent studies have demonstrated that, during the early stages of programmed cell death (apoptosis), endonuclease G is released from mitochondria and translocates to the nucleus, where it facilitates oligonucleosomal DNA fragmentation (453–455).

Further studies have also indicated that endonuclease G, like the proapoptotic protein cytochrome-*c*, resides inside the mitochondrial intermembrane space and not inside the matrix where R-loop processing occurs (454,456,457). In addition, involvement of endonuclease G in mtDNA replication has been ruled out in a transgenic mouse study (458).

DNA Polymerase γ

Synthesis of mtDNA is performed by DNA polymerase γ (459–461). The enzyme belongs to the family A class of DNA polymerases, which includes the two-subunit bacteriophage T7 DNA polymerase (462). DNA polymerase γ activity represents less than 1% of the total cellular DNA polymerase activity. In contrast to other eukaryotic DNA polymerases, DNA polymerase γ is resistant to the antibiotic aphidicolin, but is highly sensitive to dideoxynucleoside triphosphates (463). As a consequence, long-term treatment with antiviral nucleoside analog drugs, such as 3′-azido-3′-deoxythymidine (zidovudine, AZT) or 2′,3′-dideoxycytidine (zalcitabine, ddC), can give rise to severe oxidative phosphorylation defects triggered by mtDNA depletion and mutations resulting from inhibition of DNA polymerase γ (464).

Although DNA polymerase γ is thought to function primarily as a DNA-directed DNA polymerase *in vivo*, it has a potent RNA-directed DNA polymerase (reverse transcriptase) activity *in vitro* (465), suggesting that this latter activity may be physiologically relevant. As indicated earlier, ribonucleotides are found in mtDNA at scattered positions. Because the DNA polymerase γ holo-enzyme is highly processive, it is unlikely that it falls off during mtDNA synthesis and an RNA polymerase subsequently incorporates one ribonucleotide. Instead, DNA polymerase γ probably incorporates a ribonucleotide occasionally (465).

Alternatively, the incorporated ribonucleotides may be remnants of RNA primers used to prime synthesis of the Okazaki fragments of the lagging strand during strand-synchronous replication of mtDNA (422). The reverse transcriptase activity of DNA polymerase γ may allow the enzyme to continue DNA synthesis when a ribonucleotide is encountered on the template strand. The reverse transcriptase activity of DNA polymerase γ is generally used to measure the enzyme's activity in biochemical assays (466,467).

Combined structural and biochemical data suggest that the mammalian DNA polymerase γ holo-enzyme is an αβ$_2$ heterotrimer comprising a 140-kDa POLG subunit and two 54-kDa POLG2 subunits (Table 3.6) (468–472). Studies of recombinantly expressed, purified human POLG have demonstrated that the 5′→3′ DNA polymerase as well as a 3′→5′ exonuclease activity reside in this subunit (473,474). The exonuclease proofreading activity is highly mismatch-specific and ensures faithful replication of mtDNA (475).

Table 3.6 Human Nuclear-Encoded Factors Directly Involved in mtDNA Synthesis

Factor	Description (Aliases)	M$_r$ (×1000)	Gene Symbol	Cytogenetic Position
LIG3	DNA ligase III	96 or 103	*LIG3*	17q11.2–q12
POLG	Catalytic subunit of DNA polymerase (POLG1, POLGA, PolγA, POLGα, p140)	140	*POLG*	15q25
POLG2	Accessory subunit of DNA polymerase (POLGB, PolγB, POLGβ, p55)	54	*POLG2*	17q21
PEO1	Progressive external ophthalmoplegia 1 (bacteriophage T7 primase/helicase gene 4-like protein, Twinkle)	77	*PEO1* (*C10orf2*)	10q24
SSBP1	Single-stranded DNA-binding protein 1 (mtSSB, SSBP, SSB)	15	*SSBP1*	7q34
TOP1MT	Mitochondrial DNA topoisomerase I	66	*TOP1MT*	8q24.3
TOP2B	DNA topoisomerase IIβ (TOPO IIβ)	150	*TOP2B*	3p24
TOP3A	DNA topoisomerase IIIα (TOP3; TOPO IIIα)	106	*TOP3A*	17p12–p11.2

In contrast to the holo-enzyme, the 5′→3′ polymerization rate of POLG alone is relatively poor and the protein shows a processivity of only 100 to 300 nucleotides elongated per binding event (473,476). Disruption experiments of the orthologous gene in yeast have indicated that POLG has no basic function outside mitochondria (477).

Sequence comparisons of POLG from different species have revealed three conserved exonuclease (*exo*) motifs (I, II, and III) at the N-terminal half of the protein and three conserved polymerase (*pol*) motifs (A, B, and C) at the C-terminal half of the protein (469,478). The functional importance of these motifs has been confirmed by site-directed mutagenesis (398,479–481). Naturally occurring missense mutations in POLG are a frequent cause of progressive external ophthalmoplegia (482–484), a disease characterized by accumulations of multiple, large-scale deletions of mtDNA. Autosomal dominant mutations cluster in and around the *pol* B motif; no specific hotspots have been identified for autosomal recessive mutations.

Analysis of the deletion breakpoints of skeletal muscle mtDNA from patients with autosomal progressive external ophthalmoplegia has suggested that stalling of DNA polymerase γ at regions of difficulty for the enzyme, such as homopolymeric nucleotide runs, microsatellite-type repeats, and the TAS region, is the primary cause of mtDNA deletion formation (485). Patients with *POLG* mutations also display an increased incidence of apparently novel mtDNA point mutations in the D-loop region (485,486). Predictably, the highest frequency of mtDNA point mutations is seen in patients harboring *POLG* mutations affecting one of the *exo* motifs, but D-loop point mutations have also been found in patients harboring *POLG* mutations in one of the *pol* motifs.

A homozygous *POLG* nonsense mutation has been identified in three Alpers' syndrome patients from two unrelated families (487). Children presenting with this rare genetic disorder are normal at birth and progress normally over the first few weeks to years of life; however, they subsequently develop neurodegenerative symptoms and die of liver failure (488). Previous research had indicated that the disease was associated with depletion of mtDNA and deficiency of DNA polymerase γ reverse transcriptase activity in one of the patients (489). The mutation is predicted to result in a large, C-terminal truncation of POLG, which includes all three *pol* motifs (487). It is difficult to comprehend how such a harsh mutation affecting the polymerase activity of the enzyme is compatible with the initial normal development of the patients. A possible explanation suggested by the authors is that the context of the nonsense mutation is one that permits a proportion of initiated ribosomes to read through the mutant stop codon and that this is controlled in an age-dependent and tissue-specific manner (487). However, no evidence is presented to support the postulated stop codon suppression.

POLG contains an N-terminal polyglutamine tract encoded by a CAG microsatellite repeat (469). Analysis of *POLG* genotypes in different European populations identified an association between absence of the common, ten-repeat allele and male subfertility (490,491). The molecular mechanism leading to the impairment of fertility in patients with this polymorphism is unknown. Spermatozoa depend heavily on oxidative phosphorylation for motility, so decreased energy metabolism has long been hypothesized to contribute to infertility (492). Possibly, a suboptimal POLG affects the integrity of mtDNA during spermatogenesis, consequently leading to reduced oxidative phosphorylation and motility defects.

In vitro reconstitution experiments have indicated that the processivity of POLG is dramatically improved by binding of the POLG2 subunit (472,476). In addition, POLG2 binding stimulates the polymerization rate of POLG (472,476). Consistent with these important roles of POLG2 in mtDNA synthesis, mutations in the equivalent subunit of *Drosophila melanogaster* cause lethality during early pupation, concomitant with loss of mtDNA and mitochondrial mass, and reduced cellular proliferation in the central nervous system (493). Resolution of the crystal structure of recombinantly expressed, purified mouse POLG2 has indicated that POLG2 forms a dimer (471).

POLG2 can be divided into three structural domains. Domains 1 and 3 are homologous to regions of class IIa prokaryotic aminoacyl-tRNA synthetases; domain 2 is unique. It is, however, improbable that POLG2 is a functional tRNA synthetase because amino acid substitutions in the regions of POLG2 corresponding to conserved sites of aminoacyl-tRNA synthetases strongly suggest that the protein cannot operate as a tRNA synthetase (471). Deletion analysis of POLG2 has indicated that blocks of sequence in common with type IIa aminoacyl-tRNA synthetases are essential for stimulation of POLG activity and for binding of POLG2 in complexes with POLG and primer–template oligonucleotides (470).

It has also been shown that POLG2 binds double-stranded DNA longer than 40 base pairs, with little apparent sequence specificity (471,494). Critical basic amino acid residues involved in DNA binding have been mapped to surface loops on opposite sides of the POLG2 dimer (494). Both DNA-binding sites are required for high-affinity binding, suggesting that an individual DNA molecule must wrap around the dimer to interact simultaneously with both sites (494). The ability to bind double-stranded DNA is not necessary for POLG2 stimulation of POLG activity *in vitro*, but may play a role in DNA replication or repair *in vivo* (471,494).

Mitochondrial Single-Stranded Binding Protein

In primer extension assays, the rate of initiation of DNA synthesis by human DNA polymerase γ is dramatically enhanced by addition of the 15-kDa mitochondrial single-stranded DNA-binding protein, SSBP1 (Table 3.6) (495). The rat protein has been visualized on the displaced, single-strand portions of D-loops and expanding D-loops by electron microscopy (496). In HeLa cells, SSBP1 is 3000-fold more abundant than mtDNA and appears to stabilize the D-loop structure (497). These observations suggest that the biological role of SSBP1 is to stabilize single-stranded regions of mtDNA in D-loop structures and other replicative intermediates, thus preventing the formation of secondary, single-stranded DNA structures that could hinder the progress of DNA polymerase γ.

Genetic studies in yeast and *Drosophila* have indicated that SSBP1 is crucial for mtDNA maintenance *in vivo* (498–500). SSBP1 is distinct from nuclear single-stranded DNA-binding proteins, but resembles the single-stranded DNA-binding protein from *E. coli* in overall structure as well as in DNA-binding properties (498,501–503). Like its bacterial counterpart, SSBP1 acts as a homotetramer (502,504,505). The crystal structure of SSBP1 suggests that single-stranded DNA wraps around the tetrameric complex through electropositive channels guided by flexible loops (505).

Mitochondrial DNA Helicase and Possible Primase

The unwinding of duplex DNA is a prerequisite for DNA replication and repair, providing the single-stranded DNA template for DNA polymerase to copy. The disruption of the hydrogen bonds that hold the two strands together is catalyzed by DNA helicases. Unwinding is accomplished in a reaction that is coupled to the hydrolysis of a nucleoside or deoxynucleoside 5′-triphosphate (NTP or dNTP, respectively) (506).

In 2001, during a search for mutations associated with chromosome 10q24-linked, autosomal dominant progressive external ophthalmoplegia, Spelbrink and colleagues identified *PEO1* (60). The gene encodes the 77-kDa PEO1 protein (also called Twinkle; Table 3.6) that co-localizes with mtDNA in mitochondrial nucleoids. PEO1 shows primary sequence similarity to the bacteriophage T7 gene 4 primase/helicase and other helicases that form hexamers *in vivo* when bound to DNA in the presence of Mg^{2+} and NTP or dNTP (60). PEO1 shows the greatest similarity with the domain responsible for helicase activity, at the C-terminal half of the bacteriophage T7 primase/helicase. Primase domain motifs found in the N-terminal half of the bacteriophage T7 primase/helicase are not present in PEO1; therefore, a role for the human protein as a mitochondrial primase is unlikely.

Remarkably, many of the identified mutations associated with progressive external ophthalmoplegia cluster in a small region of about 30 amino acid residues corresponding to the linker region of the bacteriophage protein between the primase and helicase domains (60,507). The exact function of this region is unknown, but it has been suggested that mutations in this region might have a subtle effect on subunit interactions (60) because the corresponding segment of the bacteriophage protein is required for hexamer formation (508). Examination of the mtDNA deletion breakpoints in skeletal muscle from patients carrying mutated *PEO1* revealed that the mechanism by which mtDNA deletions are introduced is probably similar to the frequent replication stalling mechanism by which *POLG* missense mutations cause mtDNA deletions (see earlier discussion) (485).

The enzymatic properties of PEO1, with and without its putative partners SSBP1 and holo-DNA polymerase γ, have been characterized in a series of *in vitro* experiments with purified recombinant proteins (509,510). These experiments have shown that PEO1 possesses a 5′→3′ DNA helicase activity. The enzyme requires the presence of a stretch of ten nucleotides of single-stranded DNA on the 5′-side of the duplex and also a short, single-stranded 3′-tail of the duplex to unwind duplex DNA. In addition, PEO1 has an absolute requirement for hydrolysis of an NTP. SSBP1 has a stimulatory effect on the rate of DNA unwinding by PEO1, and this effect is specific because the related *E. coli* single-stranded DNA-binding protein cannot substitute for SSBP1 (509).

The *in vitro* experiments have further demonstrated that, on its own, holo-DNA polymerase γ cannot use double-stranded DNA as a template (510). Similarly, PEO1 alone was found unable to unwind long stretches of double-stranded DNA (510). In combination, however, holo-DNA polymerase γ and PEO1 formed a processive replication machinery capable of using double-stranded DNA to synthesize single-stranded DNA of about 2000 bases. Addition of SSBP1 stimulated the reaction further, generating DNA products of approximately the size of an mtDNA molecule (510). These results indicate that holo-DNA polymerase γ, PEO1, and SSBP1 act together at the DNA replication fork to form a macromolecular replisome.

Strand-synchronous replication of mtDNA necessitates a continuous DNA primase activity at the replication fork for coordinated leading and lagging DNA strand synthesis, but strand-asynchronous replication only requires primase activity for initiation of the L-strand DNA synthesis at O_L. As mentioned earlier, a human mitochondrial DNA primase implicated in priming of L-strand synthesis at O_L has only partly been purified (418). Although primary sequence comparisons suggest that PEO1 has lost its primase activity, such an activity cannot be ruled out without thorough biochemical analysis. Other mitochondrial enzyme activities needed at the replication fork, such as a primer removal activity, also remain uncharacterized.

Mitochondrial DNA Topoisomerases

DNA topoisomerases catalyze the breaking and rejoining of the DNA phosphodiester backbone in a way that allows DNA strands to pass through one another, thus altering the topology of DNA (511,512). Mitochondrial DNA topoisomerase activities are necessary for decatenation of the closed circular mtDNA molecules after replication, and for supercoiling and relaxation (Figure 3.6 and Figure 3.7). Type I topoisomerases transiently break a single DNA strand, whereas the type II enzymes transiently break two strands of duplex DNA.

Type I topoisomerases have been subdivided further into two groups: type IA and type IB. Type IA enzymes break the DNA by forming a temporary covalent bond to the 5'-end of the broken DNA, and type IB topoisomerases link covalently to the 3'-end of the break. In the catalytic intermediate of type II topoisomerases, one enzyme monomer binds to each of the 5'-ends of the double strand break through a conserved tyrosine residue (512,513). Type I and type II DNA topoisomerases are sensitive to a different spectrum of enzyme inhibitors. Type II enzymes require ATP hydrolysis for activity, but type I enzymes function independently of ATP.

A mitochondrial inner membrane-associated, type I topoisomerase has been characterized in various mammalian tissues, including rat and bovine liver (514–516), calf thymus

(517–519), human leukemia cells (520,521), and human platelets (522,523). The enzyme catalyzes the relaxation of negative as well as positive supercoils in an ATP-independent reaction. It is distinguishable from the main, nuclear type I topoisomerase (TOP1) by molecular weight, pH profile, thermal stability, chromatographic properties, and sensitivity to dimethylsulfoxide, ethidium bromide, and the trypanocidal drug berenil. A human cDNA sequence that probably encodes this mitochondrial-specific, type I topoisomerase has been identified (524). The 66-kDa protein specified by this cDNA has been called TOP1MT (Table 3.6). The protein is highly homologous to TOP1, except for its N-terminal domain, which is much shorter than that of TOP1 and contains a mitochondrial targeting sequence.

Biochemically, TOP1MT is a type IB topoisomerase that requires a divalent cation (Ca^{2+} and Mg^{2+} work equally well *in vitro*) and alkaline pH for optimal activity (524). A possible role of the enzyme is to relieve the torsional stress introduced into mtDNA when complementary strands are separated by the replicative mitochondrial DNA helicase PEO1 during progression of the replication fork. TOP1MT may also relieve torsional stress in mtDNA produced during transcription. The strong conservation between the nuclear and mitochondrial enzymes suggests that their genes evolved from a common ancestor gene because prokaryotic type IB topoisomerases are structurally very different from the vertebrate enzymes (512,525).

Initially, the human nuclear type IA topoisomerase, topoisomerase IIIα (TOP3A), was thought to be exclusively a nuclear enzyme acting in concert with a RecQ-type DNA helicase (526,527); however, more recent biochemical investigations have indicated a dual localization of human TOP3A to the nucleus and mitochondria (528). The *TOP3A* gene has two potential start codons within the same exon for the synthesis of proteins of 1001 and 976 amino acids in length (529). The N-terminal sequence of the 1001-residue isoform contains a mitochondrial targeting signal (528). TOP3A can relax negatively supercoiled DNA (529), but the precise role of TOP3A in mitochondria has yet to be elucidated.

In *E. coli*, the concerted activity of RecQ DNA helicase and topoisomerase III is able to catalyze the catenation and decatenation of covalently closed, circular double-stranded DNA molecules, and it has been suggested that such an activity could act to decatenate newly replicated daughter DNA molecules or disrupt early recombination intermediates between inappropriately paired DNA molecules (530). Type IB DNA topoisomerases are known to act on a double-stranded DNA segment (513), so TOP1MT is unlikely to permit the removal of the last few parental DNA strand intertwines during mtDNA replication. It is thus plausible that the mitochondrial isoform of TOP3A participates in the resolution of mtDNA rings at the end of the replication cycle. More detailed biochemical and genetic studies are needed to verify this conjecture.

A putative, mammalian mitochondrial type II DNA topoisomerase activity was initially identified in rat liver (531), human leukemia cells (520), and calf thymus (532). More recently, a type II DNA topoisomerase activity was found among mtDNA replicative proteins recovered from complexes of mtDNA and protein isolated from bovine heart mitochondria (533). The enzyme is able to relax a negatively supercoiled DNA template *in vitro*, in a reaction that exhibits a dependency for Mg^{2+} and ATP. The relaxation activity is inhibited by the anticancer drug etoposide (VP-16) and other inhibitors of eukaryotic type II DNA topoisomerases (533).

The purified activity has been assigned to a 150-kDa truncated version of DNA topoisomerase IIβ (TOP2B; Table 3.6) (533) — one of the two DNA topoisomerase type II enzymes known to exist in mammalian nuclei (534,535). Mass spectrophotometric and immunological data suggest that the mitochondrial isoform lacks the 30-kDa C-terminal region of the 180-kDa nuclear enzyme (533). This deleted C-terminal region constitutes one of the three structural and functional domains present within all eukaryotic, nuclear type II enzymes (512). The N-terminal and central domains contain the motif involved in ATP hydrolysis and the tyrosine residue required for DNA cleavage, respectively. In contrast to the C-terminal domain, these other two domains are highly conserved among all eukaryotic type II enzymes and crucial for activity.

The C-terminal region of the human 180-kDa TOP2B isoform has been shown to contain the signal for nuclear localization (534,535). Consequently, truncation of the C-terminal domain may be essential for the 150-kDa isoform to be targeted to the mitochondrion and to assume a role in mtDNA maintenance. Truncation may be achieved by alternative splicing or proteolytic processing, but the mechanism is at this point speculative. The specific role of TOP2B in the nucleus is not yet clear (535). It has been suggested that the mitochondrial isoform may serve to decatenate newly replicated mtDNA circles from one another at the final stage of replication (533), but further studies are necessary to confirm this.

Mitochondrial DNA Ligase

DNA ligases catalyze the joining of double- and single-strand breaks (nicks) in the phosphodiester backbone of duplex DNA during DNA replication, repair, and recombination (536). Four distinct ATP-dependent DNA ligases have been identified in the nuclei of human cells: DNA ligase I (LIG1), III (LIG3), IV (LIG4), and V (LIG5). A fifth ligase, originally named DNA ligase II, is now known to be a 70-kDa proteolytic derivative of LIG3 (537). All DNA ligases appear to have descended from a common ancestral nucleotidyltransferase enzyme (538). Although the biological role of LIG5 has not yet been defined (539), functions of the other three nuclear DNA ligases have been studied in detail.

LIG1 functions in the joining of Okazaki fragments (540) and has been implicated in long patch excision repair (541). LIG4 appears to be involved in double strand break repair (542) and nonhomologous end-joining (543). LIG3 is expressed as two different isoforms: LIG3α and LIG3β, which are produced by alternative splicing and differ in their C-terminal sequences (544). Their common N-terminus carries a zinc finger that appears to function in the recognition of DNA secondary structures that resemble intermediates of DNA metabolism. The 103-kDa LIG3α isoform has a C-terminal BRCT (**BR**CA1 **C**-terminal-related) domain. This is an autonomously folding protein module that was first identified in the C-terminus of the BRCA1 tumor suppressor protein, but has since been found in a range of proteins implicated in DNA replication and repair. The BRCT domain of LIG3α interacts with the XRCC1 protein known to be involved in base excision repair, thus implicating LIG3α in this process (545). The 96-kDa LIG3β isoform, which lacks the BRCT domain, is only expressed in germ-line tissues (544,546). The function of this isoform is not known.

Although considerable information is available about nuclear DNA ligases, far less is known about mitochondrial DNA ligases. In 1976, a report describing a ligase activity of rat liver mitochondria was published (547). More recently, it was discovered that human LIG3 (Table 3.6) is directed to nuclear and mitochondrial subcellular compartments. Like the *TOP3A* gene, the *LIG3* gene uses an alternative transcription initiation site to incorporate an N-terminal mitochondrial targeting sequence (548). It remains to be established whether mitochondrial LIG3 is a product of the *LIG3α* or *LIG3β* mRNA.

LIG3 antisense mRNA expression in the human, tumor-derived HT1080 cell line has been shown to lead to reduced mitochondrial LIG3 levels and a decrease in cellular mtDNA content (549). The mtDNA in these cells contained numerous single-strand breaks not present in control cells, suggesting that LIG3 plays an essential role in mtDNA maintenance. *In vitro* reconstitution experiments with highly purified enzymes from *X. laevis* mitochondria have indicated that, together with DNA polymerase γ, LIG3 is probably involved in mitochondrial base excision repair (550); however, there is no evidence that the nuclear-binding partner of LIG3α, XRCC1, is present in mitochondria.

Hamster EM9 cells, which are devoid of XRCC1, were found to be far more sensitive to the cytotoxic effects of ionizing radiation than wild-type cells due to a defect in nuclear DNA repair; however, there was no apparent difference in the ability of EM9 and wild-type cells to restore their mtDNA to preirradiation levels (551). This implies that XRCC1 is not required for the function of mitochondrial LIG3. Mitochondrial lysates from *X. laevis* oocytes were demonstrated to contain the LIG3α and LIG3β homologs (546). Gel filtration, sedimentation, and native gel electrophoresis experiments indicated that *X. laevis* mitochondrial LIG3α exists as a high molecular weight complex; *X. laevis* mitochondrial LIG3β behaved as a monomer (546). It is tempting to speculate that vertebrate mitochondria may have an XRCC1 homolog that stabilizes mitochondrial LIG3α.

Although it was originally proposed that the *LIG4* gene could also encode a mitochondrial form of LIG4 in vertebrates (550), later experiments have indicated that this is improbable (546,549). Interestingly, in yeast, it is the LIG1 homolog, Cdc9p, that functions as DNA ligase in the nucleus as well as in mitochondria (552). Cdc9p appears to be required for mtDNA replication and for repair of damaged mtDNA, including double-strand breaks (553). It is, however, unlikely that LIG1 is dual-targeted in mammalian cells because adenylation experiments performed on human mitochondrial extracts detected only a single DNA ligase protein that corresponded to LIG3 (549). Based on the apparent absence of additional forms of DNA ligases in vertebrate mitochondria, it is expected that mitochondrial LIG3 not only is involved in base excision repair, but also seals nicks formed during replication of the mitochondrial genome (Figure 3.6 and Figure 3.7).

CONTROL OF MTDNA COPY NUMBER

Variations in mtDNA Copy Numbers

Copy numbers of mtDNA vary widely in different tissues (Table 3.1; Figure 3.8) and are strictly controlled during embryonic development of mammals (29). Copy numbers of mtDNA have been shown to increase with increased metabolic activity, e.g., during continual muscular exercise (554). It is often assumed that a simple linear relationship exists between the mitochondrial gene dosage and the oxidative phosphorylation capacity of a cell, but this relationship may

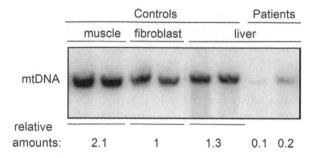

Figure 3.8 Variable levels of mtDNA in tissues of controls and patients with a defect in mitochondrial nucleotide synthesis. Autoradiogram of a Southern blot loaded with 3 μg of *Pvu*II-digested, total genomic DNA per lane and hybridized with a ^{32}P-labeled probe for mtDNA. Even loading of the samples was confirmed by subsequent hybridization with a nuclear DNA probe (not shown). Both patients were deficient for deoxyguanosine kinase and presented with hepatocerebral mtDNA depletion (561). Relative amounts of mtDNA present in the tissues are indicated below the autoradiogram.

not be that straightforward (555,556). One study has shown that the expression of mitochondrial genes and the activity of cytochrome-*c* oxidase increase in parallel with the mtDNA copy number during chronic stimulation of striated rabbit muscle (557). This finding supports the view that the mtDNA copy number is indeed an important feature underlying the oxidative phosphorylation capacity.

In another study, however, no direct correlation was found between mtDNA content and cytochrome-*c* oxidase content during the transition of human fibroblasts from proliferating to resting cells (397). Furthermore, in a study in which the relationship between mtDNA copy number and cytochrome-*c* oxidase activity was compared in a number of human tissues, tissue-specific variations in mtDNA copy number were not proportional to tissue-specific variations in cytochrome-*c* oxidase activity (352). Therefore, it appears that, although modifications in mtDNA copy number may be paralleled by changes in oxidative phosphorylation capacity within certain tissues, the mtDNA copy number plays a less significant role in determining differences in oxidative phosphorylation capacity between different tissues.

Interestingly, Tang and colleagues have found that the mtDNA copy number in human osteosarcoma cells containing normal-sized, duplicated, partly deleted dimer, or deleted mtDNA is inversely proportional to the size of the respective mtDNA molecules (558), suggesting that these cells maintain a constant mass of mtDNA rather than a constant number of mtDNA molecules. Based on their observations, these researchers proposed that the regulation of mtDNA copy number be provided by the organellar nucleotide pool size.

The hypothesis of substrate limitation as regulator of mtDNA copy number is in part supported by the findings that defects in mitochondrial nucleotide synthesis are associated with depletion of mtDNA (Figure 3.8) (559–563) and that this depletion can be prevented by nucleotide supplementation (397). The mitochondrial nucleotide pool size is likely to fluctuate through the cell cycle due to import of nucleotides from the cytosol during the S-phase when *de novo* nucleotide synthesis is up-regulated (564). As discussed at the beginning of this chapter, however, replication of mtDNA is constant throughout the cell cycle. Moreover, the mtDNA copy number remains constant during the transition of cultured fibroblasts from proliferating to resting cells (397). Thus, although a low organellar nucleotide pool size may limit mtDNA copy numbers, above a certain threshold the pool size does not appear to affect mtDNA copy numbers.

Transcriptional Control of Nuclear Genes Encoding Mitochondrial Factors

Physiological changes are thought to influence mtDNA copy number through transcriptional activation of nuclear genes coding for mitochondrial transcription or replication factors. Two classes of nuclear transcriptional mediators have been implicated in mitochondrial biogenesis. The first includes DNA-binding transcription factors typified by the nuclear respiratory factor (NRF)-1 and NRF-2 that act on many nuclear genes involved in mitochondrial function and biogenesis (565). Notably, NRF1 and -2 responsive elements are found in the genes *TFAM*, *TFB1M*, *TFB2M*, and *RMRP*, which are directly involved in mtDNA transcription and maintenance (Table 3.3 and Table 3.4).

The second class includes nuclear coactivators typified by the peroxisome proliferator-activated receptor-γ coactivator (PGC)-1α, PGC-1β (also termed PGC-1-related estrogen receptor coactivator or PERC) and the PGC-1-related coactivator (PRC) (566–568). These proteins do not bind DNA but rather work through interactions with DNA-bound transcription factors to regulate gene expression. An important feature of these coactivators is that their expression is responsive to physiological signals mediating thermogenesis, cell proliferation, and glucogenesis. For instance, PGC-1α is cold inducible in brown fat and interacts with multiple transcription factors to orchestrate a program of adaptive thermogenesis (569). PGC-1α can induce nuclear genes that are required for mitochondrial biogenesis through stimulation of NRF1 and -2 expression and, in addition, through direct binding to the NRF1 protein and coactivating its transcriptional function (570).

PGC-1β is structurally closely related to PGC-1α and expressed in a comparable but not identical tissue-specific manner; it also interacts with NRF1 (571,572). The overall homology between PGC-1α and PRC is lower, although they do share some structural domains, including a domain for interaction with nuclear receptors and RNA recognition motifs (573). Unlike PGC-1α, PRC is not induced significantly during thermogenesis, but is cell-cycle regulated in cultured cells under conditions in which PGC-1α is not expressed (573). PRC has a transcriptional specificity very similar to that of PGC-1α, especially in its interaction with NRF1 and in the activation of NRF1 target genes (573). Differences in regulation and tissue-specific expression patterns of the PGC-1α family members suggest that each confers distinct physiological responses. Thus, these coactivators allow organisms to react to complex physiological changes by coordinating programs of gene expression essential to cellular energetics, including mtDNA transcription and maintenance.

Control of mtDNA Copy Number by *cis*- and *trans*-Acting Factors

In their study of human osteosarcoma cells containing normal or rearranged mtDNA (558), Tang and coworkers found no evidence that the number of replication origins affected the copy number of mtDNA. Cell lines with partly deleted, dimeric mtDNA contained twice the number of replication origins compared to similarly sized, wild-type mtDNA, yet the copy number was almost identical between cells harboring these two mtDNA types. On the other hand, cells with the corresponding duplicated mtDNA containing twice the number of replication origins had a copy number markedly lower than the copy number found in wild-type cells (558). If the number of replication origins had influenced the cellular mtDNA copy number, then the number of mtDNAs per cell should have corresponded to the number of origins.

As pointed out earlier, the majority of nascent H-strands synthesized from the O_H are prematurely terminated downstream of the TAS element, defining the 3' boundary of the D-loop (Figure 3.4). Extension of nascent H-strands beyond the TAS element is an evident requirement for strand-asynchronous mtDNA replication. Therefore, the mechanism determining the extension could represent an important regulator of mtDNA copy number. To investigate this potential mechanism, a quantitative, ligation-mediated PCR assay was used to determine the level of total and extended nascent H-strands during proliferation of human T lymphocytes (574) and after pharmacologically mediated mtDNA depletion of mouse LA9 cells (575).

Both studies revealed that the ratio of (extended nascent strands):(total nascent strands) increased, suggesting that cell proliferation and mtDNA repletion lead to a partial release of premature termination. Thus, the regulation of H-strand synthesis at the TAS element may indeed play a significant role in copy number control in these systems. *In vivo* and *in organello* footprinting studies have indicated a protein-binding site encompassing the TAS element of human and rat mtDNA (576). Furthermore, a 48-kDa protein has been isolated from bovine mitochondria with a DNA-binding activity specific for the bovine TAS-like A element (577). These findings suggest that a nuclear-encoded, *trans*-acting protein interacts with the *cis*-acting TAS elements and regulates the equilibrium between D-loop formation and H-strand replication. Further details of this system remain to be established.

In principle, any critical mtDNA transcription or replication factor could be a limiting factor in determining maximum mtDNA copy number (578). Although a large number of housekeeping transcription and replication factors have been identified (Table 3.3 and Table 3.6), only the impacts of a few of these have been investigated. In rabbits, mRNA encoding the POLG homolog was found to be present at similar levels in different tissues and did not seem to be regulated by physiological changes that modified mtDNA copy numbers (579). Moreover, overexpression of POLG in human cells did not alter mtDNA levels (480). Thus, it appears that mtDNA copy numbers are not regulated by limiting concentrations of the protein directly responsible for mtDNA synthesis. In the same rabbit study, the levels of the mRNA encoding the SSBP1 homolog did appear to be regulated coordinately with variations in abundance of mtDNA among tissues and increased in association with enhanced mitochondrial biogenesis (579).

Unfortunately, no information has been published concerning the relationship between mtDNA copy number and SSBP1 protein levels. Because post-transcriptional regulation of SSBP1 protein levels cannot be excluded, the role of SSBP1 in regulation of mtDNA copy number remains to be confirmed. Some evidence suggests that TFB1M and TFB2M may be limiting factors in determining mtDNA copy number. In *Drosophila* Schneider cell cultures, the mtDNA transcription level and the mtDNA copy number correlate with the expression level of the fly's TFB2M equivalent (580). This suggests that TFB2M may increase mtDNA copy number by increasing the availability of RNA primers for initiation of mtDNA synthesis. Furthermore, solid evidence indicates that, in addition to its role in mitochondrial transcription, TFAM has a direct function in regulation of mtDNA copy number.

Multiple Roles for TFAM

Binding of TFAM to mtDNA is not restricted to the enhancer elements upstream of the HSP and LSP (Figure 3.4). The protein is inherently flexible in its recognition of mtDNA sequences (171,175) and various lines of evidence have suggested that it is involved in packaging of the entire mtDNA molecule. The TFAM homolog of *S. cerevisiae*, Abf2p, is an abundant yeast mitochondrial protein (581). Unlike TFAM, Abf2p is not required for transcriptional initiation of mtDNA (176). *S. cerevisiae* strains with a disrupted *ABF2* gene lose their mtDNA when cultured in the presence of fermentable carbon sources (581); however, this loss can be rescued by the bacterial histone-like protein HU (582), suggesting that Abf2p maintains mtDNA as an architectural factor.

Because human TFAM can also substitute for Abf2p (583), TFAM is likely to share common properties with Abf2p and HU. Originally, TFAM was considered to be just a mitochondrial transcription factor because the TFAM:mtDNA ratio was estimated to be only 15: to 30:1 (164,584); however, a more recent study indicated that this ratio is 1700:1 in HeLa cells and 940:1 in human placenta (497). The abundance of TFAM suggests a role for this protein in addition to its function as transcriptional activator because TFAM fully activates HSP- and LSP-specific transcription in the presence of TFB1M or TFB2M *in vitro* at a TFAM:DNA template ratio of 10: to 100:1 (186).

Coimmunoprecipitation experiments have indicated that TFAM and mtDNA are closely associated with each other (585). Based on the fact that TFAM is sufficiently abundant to wrap mtDNA completely, these data suggest that human mtDNA is packaged with TFAM. Atomic force microscopy images have shown that Abf2p binding to DNA induces drastic bends in the DNA backbone for linear as well as circular DNA (586). At a high concentration of Abf2p, DNA collapses into a tight nucleoprotein complex. The interaction forces between Abf2p and DNA were observed to be relatively weak as evidenced by the fast Abf2p off-rate (k_{off} = 0.014 ± 0.001 s^{-1}) (587). These observations suggest that, similar to Abf2p, TFAM may compact mtDNA through a relatively simple mechanism that involves bending of the backbone. The weak binding forces may leave mtDNA accessible for transcription and replication.

Numerous studies have shown that TFAM protein levels vary concomitantly with levels of mtDNA. Low levels of TFAM are seen in human cells depleted transiently or permanently of mtDNA (Figure 3.9), and patients with mitochondrial myopathies have increased levels of TFAM in "ragged-red" muscle fibers with elevated levels of mtDNA (588–591). Knockdown of the *Drosophila* TFAM counterpart by RNA interference led to a striking reduction in mtDNA content of *Drosophila* Kc167 cell cultures without marked inhibition of mitochondrial transcription (592). In *Drosophila* Schneider and chicken DT40 cell cultures, overexpression of the TFAM counterparts resulted in a dose-dependent increase of the mtDNA copy number, but mitochondrial transcripts increased only marginally (580,593).

Heterozygous *Tfam* knock-out mice exhibited decreased mtDNA levels, but homozygous knock-out embryos were devoid of mtDNA and died (594). Despite its strong DNA-binding activity, human TFAM is a poor activator of murine mitochondrial transcription (595). Transgenic mice overexpressing human TFAM showed an increase in mtDNA copy numbers in a variety of tissues; this was directly proportional to the TFAM protein levels, but was without an increase of oxidative phosphorylation capacity or mitochondrial mass (595). Taken together, these findings implicate TFAM as key regulator of mtDNA copy number. TFAM appears to stabilize mtDNA copy number by being the limiting factor of the mtDNA-stabilizing protein scaffold constituting the mitochondrial nucleoid.

Although relatively low levels of TFAM may be sufficient to activate mitochondrial transcription, including the possible synthesis of primers for strand-asymmetric mtDNA replica-

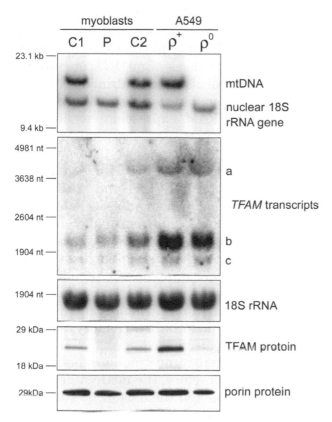

Figure 3.9 TFAM protein levels vary concomitantly with levels of mtDNA. Comparative Southern, northern, and western blot analyses of control myoblast cultures (C1 and C2); a myoblast culture from a patient almost completely devoid of mtDNA (P); the lung carcinoma cell line A549 (ρ^+); and its derivative (ρ^0), which was pharmacologically depleted of mtDNA (604). Blots with *Pvu*II-digested, total genomic DNA (2.5 µg/lane), total cellular RNA (4 µg/lane), or whole-cell protein extracts (5 µg/lane) were developed with DNA probes and antibodies as indicated. The probe for the nuclear 18S rRNA gene and transcript, and the antibody against porin were used to verify equal loading of the samples. Migration of molecular size standards is indicated. Note that TFAM protein levels are markedly decreased in samples lacking mtDNA, but *TFAM* mRNA levels are not affected. Note also the three different *TFAM* transcripts marked "a" to "c." In addition to the major, 2100-nucleotide *TFAM* transcript (b), a smaller *TFAM* transcript (c) is present that lacks the exon 5 sequence. The 4000-nucleotide *TFAM* transcript (a) has not been characterized further.

tion, relatively high levels of TFAM, found *in vivo*, allow the mtDNA to adopt its stable, higher order structure. As mentioned previously, Tang and colleagues have proposed that the tight maintenance of a constant mass of mtDNA in cultured cells may be regulated by control of mitochondrial dNTP pools (558). An alternative explanation for their observations might be that tight regulation of TFAM expression results in a constant level of TFAM, which protects a constant length of mtDNA.

Transcription of the *TFAM* gene is highly dependent on activation of the *TFAM* promoter by NRF1 and -2 (596). In one study, skeletal muscle of aged human subjects showed an up-regulation of NRF1 and TFAM, and had higher mtDNA levels compared to that of younger subjects (597). These correlations suggest that the nuclear transcriptional activator NRF1 may control mtDNA copy number via regulation of *TFAM* gene expression. This view is further supported by a mouse knock-out model of NRF1 (598). Homozygosity of the *null* allele resulted in embryonic lethality. Homozygous *null* blastocysts exhibited severely decreased levels of mtDNA, whereas mature oocytes of heterozygous mothers had a normal complement of mtDNA. Therefore, the mtDNA depletion must occur between fertilization and the blastocyst stage and probably results from the loss of an NRF1-dependent pathway of mtDNA maintenance.

The human *TFAM* gene is composed of seven exons (167,169). Transcriptional analysis has revealed a number of *TFAM* mRNA isoforms (Figure 3.9) (599). In addition to the major mRNA species, a smaller isoform that lacks the exon 5 sequence is present in most somatic tissues (167,169). Exon 5 encodes the second HMG box. Either the minor mRNA isoform does not appear to be translated into protein or the protein is rapidly degraded because only a single protein band of 25 kDa corresponding to full-length TFAM is detected on protein blots of cell lysates probed with a TFAM antiserum (Figure 3.9). In addition to the ubiquitously expressed mitochondrial TFAM protein, the mouse *Tfam* gene encodes a protein isoform that lacks the mitochondrial targeting sequence and is present in the nucleus of spermatocytes and elongated spermatids (600). The mRNA species encoding the testes-specific isoform uses an alternate first exon (600).

A similar testes-specific mRNA isoform has been detected in humans and rats but, unlike the murine testes-specific transcript, this mRNA appears not to be translated (599,601). Mitochondrial TFAM protein levels decrease during human spermatogenesis (599). It has been speculated that the presence of a testes-specific *TFAM* promoter down-regulates TFAM protein levels in male germ cell mitochondria by transcriptional interference between the testes-specific and ubiquitous *TFAM* promoters (599,601). This down-regulation of TFAM protein levels could serve as a mechanism for the documented reduction of mtDNA copy number during spermatogenesis (602) and could thus be part of the system that ensures maternal inheritance of mtDNA.

CONCLUDING REMARKS

Since the discovery of mtDNA in 1963, our knowledge of the structure, function, and cellular mechanisms involved in its expression has advanced spectacularly. The expansion of genomics and proteomics in the late 1990s has greatly facilitated the identification of mitochondrial proteins and it appears that we may now be approaching the point at which the full complement of nuclear-encoded factors required for mtDNA expression is known. This will no doubt have an important impact on the development of diagnostic tools for diseases that involve mtDNA.

Despite this dramatic progress, many exciting challenges still exist. Of key importance is to develop a better understanding of the functional aspects of many of the recently identified mitochondrial proteins involved in mtDNA expression. Although the fundamental principles of mitochondrial transcription and translation are now known, many of the molecular mechanisms of both systems remain poorly understood. Likewise, our understanding of mtDNA replication has recently been shaken by the emergence of a model that challenges the complex mechanisms accepted since the early 1980s. The paucity of *in vivo* data relating to all aspects of mtDNA biology also needs to be addressed if we are to comprehend fully the complex relationship in mtDNA diseases between a patient's genotype and clinical phenotype. As the number of diseases associated with mtDNA continues to increase, the impetus for research remains as strong as ever, thus ensuring that the future of this dynamic field will be as stimulating as the last 40 years have been.

REFERENCES

1. J-W Taanman, SL Williams. Structure and function of the mitochondrial oxidative phosphorylation system. In: AHV Schapira, S DiMauro, Eds. *Mitochondrial Disorders in Neurology* 2. Boston: Butterworth Heinemann, 2002, pp 1–34.

2. IM Scheffler. *Mitochondria*. New York: Wiley–Liss, 1999, pp 246–272.

3. OH Petersen. Calcium signal compartmentalization. *Biol Res* 35:177–182, 2002.

4. G Kroemer, JC Reed. Mitochondrial control of cell death. *Nat Med* 6:513–519, 2000.

5. TG Frey, CA Mannella. The internal structure of mitochondria. *Trends Biochem Sci* 25:319–324, 2000.

6. L Griparic, AM van der Bliek. The many shapes of mitochondrial membranes. *Traffic* 2:235–244, 2001.

7. RW Gilkerson, JM Selker, RA Capaldi. The cristal membrane of mitochondria is the principal site of oxidative phosphorylation. *FEBS Lett* 546:355–358, 2003.

8. BF Lang, MW Gray, G Burger. Mitochondrial genome evolution and the origin of eukaryotes. *Annu Rev Genet* 33:351–397, 1999.

9. P Borst, LA Grivell. Mitochondrial ribosomes. *FEBS Lett* 13:73–88, 1971.

10. MV Simpson, JR McLean, GI Cohen, IK Brandt. *In vitro* incorporation of leucine-1-C^{14} into the protein of liver mitochondria. *Federation Proc* 16:249–250, 1957.

11. JR McLean, GL Cohn, IK Brandt, MV Simpson. Incorporation of labeled amino acids into protein of muscle and liver mitochondria. *J Biol Chem* 233:657–663, 1958.

12. JM Tager, S Papa, E Quagliariello, EC Slater. *Regulation of Metabolic Processes in Mitochondria*. Amsterdam: Elsevier, 1966, pp 5–289.

13. MMK Nass, S Nass. Intramitochondrial fibers with DNA characteristics. *J Cell Biol* 19:593–629, 1963.

14. AM Kroon. Protein synthesis in mitochondria III: on the effects of inhibitors on the incorporation of amino acids into protein by intact mitochondria and digitonin fractions. *Biochim Biophys Acta* 108:275–284, 1965.

15. D Bogenhagen, DA Clayton. Mouse L cell mitochondrial DNA molecules are selected randomly for replication throughout the cell cycle. *Cell* 11:719–727, 1977.

16. J Magnusson, M Orth, P Lestienne, JW Taanman. Replication of mitochondrial DNA occurs throughout the mitochondria of cultured human cells. *Exp Cell Res* 289:133–142, 2003.

17. JA Enríquez, J Ramos, A Pérez–Martos, MJ López–Pérez, J Montoya. Highly efficient DNA synthesis in isolated mitochondria from rat liver. *Nucleic Acids Res* 22:1861–1865, 1994.

18. JA Enríquez, P Fernández–Silva, A Pérez–Martos, MJ López–Pérez, J Montoya. The synthesis of mRNA in isolated mitochondria can be maintained for several hours and is inhibited by high levels of ATP. *Eur J Biochem* 237:601–610, 1996.

19. GM Attardi, A Chomyn, Eds. *Methods in Enzymology*, vol 264. *Mitochondrial Biogenesis and Genetics Part B*. San Diego: Academic Press, 1996, pp 3–261.

20. JV Leonard, AHV Schapira. Mitochondrial respiratory chain disorders I: mitochondrial DNA defects. *Lancet* 355:299–304, 2000.

21. S DiMauro, EA Schon. Mitochondrial respiratory-chain diseases. *N Engl J Med* 348:2656–2668, 2003.

22. R Radloff, W Bauer, J Vinograd. A dye-buoyant-density method for the detection and isolation of closed circular duplex DNA: the closed circular DNA in HeLa cells. *Proc Natl Acad Sci USA* 57:1514–1521, 1967.

23. S Anderson, AT Bankier, BG Barrell, MH de Bruijn, AR Coulson, J Drouin, IC Eperon, DP Nierlich, BA Roe, F Sanger, PH Schreier, AJ Smith, R Staden, IG Young. Sequence and organization of the human mitochondrial genome. *Nature* 290:457–465, 1981.

24. RM Andrews, I Kubacka, PF Chinnery, RN Lightowlers, DM Turnbull, N Howell. Reanalysis and revision of the Cambridge reference sequence for human mitochondrial DNA. *Nat Genet* 23:147, 1999.

25. A Rantanen, NG Larsson. Regulation of mitochondrial DNA copy number during spermatogenesis. *Hum Reprod* 15 Suppl 2:86–91, 2000.

26. JA Barritt, M Kokot, J Cohen, N Steuerwald, CA Brenner. Quantification of human ooplasmic mitochondria. *Reprod Biomed Online* 4:243–247, 2002.

27. P Reynier, P May–Panloup, MF Chretien, CJ Morgan, M Jean, F Savagner, P Barriere, Y Malthiery. Mitochondrial DNA content affects the fertilizability of human oocytes. *Mol Hum Reprod* 7:425–429, 2001.

28. X Chen, R Prosser, S Simonetti, J Sadlock, G Jagiello, EA Schon. Rearranged mitochondrial genomes are present in human oocytes. *Am J Hum Genet* 57:239–247, 1995.

29. L Pikó, KD Taylor. Amounts of mitochondrial DNA and abundance of some mitochondrial gene transcripts in early mouse embryos. *Dev Biol* 123:364–374, 1987.

30. B Hudson, J Vinograd. Catenated circular DNA molecules in HeLa cell mitochondria. *Nature* 216:647–652, 1967.

31. DA Clayton, J Vinograd. Circular dimer and catenate forms of mitochondrial DNA in human leukaemic leucocytes. *J Pers* 35:652–657, 1967.

32. DA Clayton, CA Smith, JM Jordan, M Teplitz, J Vinograd. Occurrence of complex mitochondrial DNA in normal tissues. *Nature* 220:976–979, 1968.

33. DR Wolstenholme, JD McLaren, K Koike, EL Jacobson. Catenated oligomeric circular DNA molecules from mitochondria of malignant and normal mouse and rat tissues. *J Cell Biol* 56:247–255, 1973.

34. DA Clayton, J Vinograd. Complex mitochondrial DNA in leukemic and normal human myeloid cells. *Proc Natl Acad Sci USA* 62:1077–1084, 1969.

35. DA Clayton, RW Davis, J Vinograd. Homology and structural relationships between the dimeric and monomeric circular forms of mitochondrial DNA from human leukemic leukocytes. *J Mol Biol* 47:137–153, 1970.

36. IJ Holt, AE Harding, JA Morgan–Hughes. Deletions of muscle mitochondrial DNA in patients with mitochondrial myopathies. *Nature* 331:717–719, 1988.

37. P Lestienne, G Ponsot. Kearns–Sayre syndrome with muscle mitochondrial DNA deletion. *Lancet* 1:885, 1988.

38. J Poulton, ME Deadman, RM Gardiner. Duplications of mitochondrial DNA in mitochondrial myopathy. *Lancet* 1:236–240, 1989.

39. OA Kajander, AT Rovio, K Majamaa, J Poulton, JN Spelbrink, IJ Holt, PJ Karhunen, HT Jacobs. Human mtDNA sublimons resemble rearranged mitochondrial genomes found in pathological states. *Hum Mol Genet* 9:2821–2835, 2000.

40. GA Cortopassi, D Shibata, NW Soong, N Arnheim. A pattern of accumulation of a somatic deletion of mitochondrial DNA in aging human tissues. *Proc Natl Acad Sci USA* 89:7370–7374, 1992.

41. M Corral–Debrinski, T Horton, MT Lott, JM Shoffner, MF Beal, DC Wallace. Mitochondrial DNA deletions in human brain: regional variability and increase with advanced age. *Nat Genet* 2:324–329, 1992.

42. S Melov, JM Shoffner, A Kaufman, DC Wallace. Marked increase in the number and variety of mitochondrial DNA rearrangements in aging human skeletal muscle. *Nucleic Acids Res* 23:4122–4126, 1995.

43. OA Kajander, PJ Karhunen, HT Jacobs. The relationship between somatic mtDNA rearrangements, human heart disease and aging. *Hum Mol Genet* 11:317–324, 2002.

44. EJ Brierley, MA Johnson, OF James, DM Turnbull. Mitochondrial involvement in the ageing process. Facts and controversies. *Mol Cell Biochem* 174:325–328, 1997.

45. E Nekhaeva, ND Bodyak, Y Kraytsberg, SB McGrath, NJ Van Orsouw, A Pluzhnikov, JY Wei, J Vijg, K Khrapko. Clonally expanded mtDNA point mutations are abundant in individual cells of human tissues. *Proc Natl Acad Sci USA* 99:5521–5526, 2002.

46. K Khrapko, E Nekhaeva, Y Kraytsberg, W Kunz. Clonal expansions of mitochondrial genomes: implications for *in vivo* mutational spectra. *Mutat Res* 522:13–19, 2003.

47. Y Kraytsberg, E Nekhaeva, NB Bodyak, K Khrapko. Mutation and intracellular clonal expansion of mitochondrial genomes: two synergistic components of the aging process? *Mech Ageing Dev* 124:49–53, 2003.

48. H Kasamatsu, J Vinograd. Replication of circular DNA in eukaryotic cells. *Annu Rev Biochem* 43:695–719, 1974.

49. A Arnberg, EF van Bruggen, P Borst. The presence of DNA molecules with a displacement loop in standard mitochondrial DNA preparations. *Biochim Biophys Acta* 246:353–357, 1971.

50. H Kasamatsu, DL Robberson, J Vinograd. A novel closed-circular mitochondrial DNA with properties of a replicating intermediate. *Proc Natl Acad Sci USA* 68:2252–2257, 1971.

51. MW Walberg, DA Clayton. Sequence and properties of the human KB cell and mouse L cell D-loop regions of mitochondrial DNA. *Nucleic Acids Res* 9:5411–5421, 1981.

52. I Miyakawa, N Sando, S Kawano, S Nakamura, T Kuroiwa. Isolation of morphologically intact mitochondrial nucleoids from the yeast, *Saccharomyces cerevisiae. J Cell Sci* 88:431–439, 1987.

53. I Miyakawa, H Aoi, N Sando, T Kuroiwa. Fluorescence microscopic studies of mitochondrial nucleoids during meiosis and sporulation in the yeast, *Saccharomyces cerevisiae. J Cell Sci* 66:21–38, 1984.

54. BA Kaufman, SM Newman, RL Hallberg, CA Slaughter, PS Perlman, RA Butow. In organello formaldehyde crosslinking of proteins to mtDNA: identification of bifunctional proteins. *Proc Natl Acad Sci USA* 97:7772–7777, 2000.

55. M Satoh, T Kuroiwa. Organization of multiple nucleoids and DNA molecules in mitochondria of a human cell. *Exp Cell Res* 196:137–140, 1991.

56. M Dellinger, M Gèze. Detection of mitochondrial DNA in living animal cells with fluorescence microscopy. *J Microsc* 204:196–202, 2001.

57. AF Davis, DA Clayton. *In situ* localization of mitochondrial DNA replication in intact mammalian cells. *J Cell Biol* 135:883–893, 1996.

58. DH Margineantu, WG Cox, L Sundell, SW Sherwood, JM Beechem, RA Capaldi. Cell cycle dependent morphology changes and associated mitochondrial DNA redistribution in mitochondria of human cell lines. *Mitochondrion* 1:425–435, 2002.

59. N Garrido, L Griparic, E Jokitalo, J Wartiovaara, AM van der Bliek, JN Spelbrink. Composition and dynamics of human mitochondrial nucleoids. *Mol Biol Cell* 14:1583–1596, 2003.

60. JN Spelbrink, F-Y Li, V Tiranti, K Nikali, Q-P Yuan, M Tariq, S Wanrooij, N Garrido, G Comi, L Morandi, L Santoro, A Toscano, GM Fabrizi, H Somer, R Croxen, D Beeson, J Poulton, A Suomalainen, HT Jacobs, M Zeviani, C Larsson. Human mitochondrial DNA deletions associated with mutations in the gene encoding Twinkle, a phage T7 gene 4-like protein localized in mitochondria. *Nat Genet* 28:223–231, 2001.

61. FJ Iborra, H Kimura, PR Cook. The functional organization of mitochondrial genomes in human cells. *BMC Biol* 2:9, 2004.

62. F Legros, F Malka, P Frachon, A Lombès, M Rojo. Organization and dynamics of human mitochondrial DNA. *J Cell Sci* 117:2653–2662, 2004.

63. M Albring, J Griffith, G Attardi. Association of a protein structure of probable membrane derivation with HeLa cell mitochondrial DNA near its origin of replication. *Proc Natl Acad Sci USA* 74:1348–1352, 1977.

64. DF Bogenhagen, Y Wang, EL Shen, R Kobayashi. Protein components of mitochondrial DNA nucleoids in higher eukaryotes. *Mol Cell Proteomics* 2:1205–1216, 2003.

65. AEA Hobbs, M Srinivasan, JM McCaffery, RE Jensen. Mmm1p, a mitochondrial outer membrane protein, is connected to mitochondrial DNA (mtDNA) nucleoids and required for mtDNA stability. *J Cell Biol* 152:401–410, 2001.

66. MJ Youngman, AEA Hobbs, SM Burgess, M Srinivasan, RE Jensen. Mmm2p, a mitochondrial outer membrane protein required for yeast mitochondrial shape and maintenance of mtDNA nucleoids. *J Cell Biol* 164:677–688, 2004.

67. SM Burgess, M Delannoy, RE Jensen. *MMM1* encodes a mitochondrial outer membrane protein essential for establishing and maintaining the structure of yeast mitochondria. *J Cell Biol* 126:1375–1391, 1994.

68. I Boldogh, N Vojtov, S Karmon, LA Pon. Interaction between mitochondria and the actin cytoskeleton in budding yeast requires two integral mitochondrial outer membrane proteins, Mmm1p and Mdm10p. *J Cell Biol* 141:1371–1381, 1998.

69. IG Macreadie, CE Novitski, RJ Maxwell, U John, BG Ooi, GL McMullen, HB Lukins, AW Linnane, P Nagley. Biogenesis of mitochondria: the mitochondrial gene (*aap1*) coding for mitochondrial ATPase subunit 8 in *Saccharomyces cerevisiae*. *Nucleic Acids Res* 11:4435–4451, 1983.

70. A Chomyn, P Mariottini, MW Cleeter, CI Ragan, A Matsuno–Yagi, Y Hatefi, RF Doolittle, G Attardi. Six unidentified reading frames of human mitochondrial DNA encode components of the respiratory-chain NADH dehydrogenase. *Nature* 314:592–597, 1985.

71. A Chomyn, MW Cleeter, CI Ragan, M Riley, RF Doolittle, G Attardi. URF6, last unidentified reading frame of human mtDNA, codes for an NADH dehydrogenase subunit. *Science* 234:614–618, 1986.

72. DR Wolstenholme. Animal mitochondrial DNA: structure and evolution. *Int Rev Cytol* 141:173–216, 1992.

73. D Jameson, AP Gibson, C Hudelot, PG Higgs. OGRe: a relational database for comparative analysis of mitochondrial genomes. *Nucleic Acids Res* 31:202–206, 2003.

74. D Ojala, J Montoya, G Attardi. tRNA punctuation model of RNA processing in human mitochondria. *Nature* 290:470–474, 1981.

75. S Osawa, TH Jukes, K Watanabe, A Muto. Recent evidence for evolution of the genetic code. *Microbiol Rev* 56:229–264, 1992.

76. BG Barrell, S Anderson, AT Bankier, MH de Bruijn, E Chen, AR Coulson, J Drouin, IC Eperon, DP Nierlich, BA Roe, F Sanger, PH Schreier, AJ Smith, R Staden, IG Young. Different pattern of codon recognition by mammalian mitochondrial tRNAs. *Proc Natl Acad Sci USA* 77:3164–3166, 1980.

77. SG Bonitz, R Berlani, G Coruzzi, M Li, G Macino, FG Nobrega, MP Nobrega, BE Thalenfeld, A Tzagoloff. Codon recognition rules in yeast mitochondria. *Proc Natl Acad Sci USA* 77:3167–3170, 1980.

78. JE Heckman, J Sarnoff, B Alzner–DeWeerd, S Yin, UL RajBhandary. Novel features in the genetic code and codon reading patterns in *Neurospora crassa* mitochondria based on sequences of six mitochondrial tRNAs. *Proc Natl Acad Sci USA* 77:3159–3163, 1980.

79. RP Martin, AP Sibler, CW Gehrke, K Kuo, CG Edmonds, JA McCloskey, G Dirheimer. 5-[[(carboxymethyl)amino] methyl]uridine is found in the anticodon of yeast mitochondrial tRNAs recognizing two-codon families ending in a purine. *Biochemistry* 29:956–959, 1990.

80. T Yasukawa, T Suzuki, N Ishii, S Ohta, K Watanabe. Wobble modification defect in tRNA disturbs codon–anticodon interaction in a mitochondrial disease. *EMBO J* 20:4794–4802, 2001.

81. T Suzuki, T Suzuki, T Wada, K Saigo, K Watanabe. Taurine as a constituent of mitochondrial tRNAs: new insights into the functions of taurine and human mitochondrial diseases. *EMBO J* 21:6581–6589, 2002.

82. J Moriya, T Yokogawa, K Wakita, T Ueda, K Nishikawa, PF Crain, T Hashizume, SC Pomerantz, JA McCloskey, G Kawai, N Hayashi, S Yokoyama, K Watanabe. A novel modified nucleoside found at the first position of the anticodon of methionine tRNA from bovine liver mitochondria. *Biochemistry* 33:2234–2239, 1994.

83. JB Galper, JE Darnell. The presence of *N*-formyl-methionyl-tRNA in HeLa cell mitochondria. *Biochem Biophys Res Commun* 34:205–214, 1969.

84. JL Epler, LR Shugart, WE Barnett. *N*-formylmethionyl transfer ribonucleic acid in mitochondria from *Neurospora*. *Biochemistry* 9:3575–3579, 1970.

85. A Halbreich, M Rabinowitz. Isolation of *Saccharomyces cerevisiae* mitochondrial formyltetrahydrofolic acid:methionyl-tRNA transformylase and the hybridization of mitochondrial fMet-tRNA with mitochondrial DNA. *Proc Natl Acad Sci USA* 68:294–298, 1971.

86. CA Hutchison, III, JE Newbold, SS Potter, MH Edgell. Maternal inheritance of mammalian mitochondrial DNA. *Nature* 251:536–538, 1974.

87. AM Kroon, WM de Vos, H Bakker. The heterogeneity of rat-liver mitochondrial DNA. *Biochim Biophys Acta* 519:269–273, 1978.

88. JI Hayashi, H Yonekawa, O Gotoh, J Watanabe, Y Tagashira. Strictly maternal inheritance of rat mitochondrial DNA. *Biochem Biophys Res Commun* 83:1032–1038, 1978.

89. RE Giles, H Blanc, HM Cann, DC Wallace. Maternal inheritance of human mitochondrial DNA. *Proc Natl Acad Sci USA* 77:6715–6719, 1980.

90. M Schwartz, J Vissing. Paternal inheritance of mitochondrial DNA. *N Engl J Med* 347:576–580, 2002.

91. RW Taylor, MT McDonnell, EL Blakely, PF Chinnery, GA Taylor, N Howell, M Zeviani, E Briem, F Carrara, DM Turnbull. Genotypes from patients indicate no paternal mitochondrial DNA contribution. *Ann Neurol* 54:521–524, 2003.

92. M Filosto, M Mancuso, C Vives–Bauza, MR Vila, S Shanske, M Hirano, AL Andreu, S Dimauro. Lack of paternal inheritance of muscle mitochondrial DNA in sporadic mitochondrial myopathies. *Ann Neurol* 54:524–526, 2003.

93. M Schwartz, J Vissing. No evidence for paternal inheritance of mtDNA in patients with sporadic mtDNA mutations. *J Neurol Sci* 218:99–101, 2004.

94. U Gyllensten, D Wharton, A Josefsson, AC Wilson. Paternal inheritance of mitochondrial DNA in mice. *Nature* 352:255–257, 1991.

95. H Kaneda, J Hayashi, S Takahama, C Taya, K Fisher Lindahl, H Yonekawa. Elimination of paternal mitochondrial DNA in intraspecific crosses during early mouse embryogenesis. *Proc Natl Acad Sci USA* 92:4542–4546, 1995.

96. F Ankel–Simons, JM Cummins. Misconceptions about mitochondria and mammalian fertilization: implications for theories on human evolution. *Proc Natl Acad Sci USA* 93:13859–13863, 1996.

97. J Hiraoka, Y Hirao. Fate of sperm tail components after incorporation into the hamster egg. *Gamete Res* 19:369–380, 1988.

98. R Shalgi, A Magnus, R Jones, DM Phillips. Fate of sperm organelles during early embryogenesis in the rat. *Mol Reprod Dev* 37:264–271, 1994.

99. P Sutovsky, CS Navara, G Schatten. Fate of the sperm mitochondria, and the incorporation, conversion, and disassembly of the sperm tail structures during bovine fertilization. *Biol Reprod* 55:1195–1205, 1996.

100. H Shitara, H Kaneda, A Sato, K Inoue, A Ogura, H Yonekawa, JI Hayashi. Selective and continuous elimination of mitochondria microinjected into mouse eggs from spermatids, but not from liver cells, occurs throughout embryogenesis. *Genetics* 156:1277–1284, 2000.

101. C Danan, D Sternberg, A Van Steirteghem, C Cazeneuve, P Duquesnoy, C Besmond, M Goossens, W Lissens, S Amselem. Evaluation of parental mitochondrial inheritance in neonates born after intracytoplasmic sperm injection. *Am J Hum Genet* 65:463–473, 1999.

102. DR Marchington, MS Scott Brown, VK Lamb, RJ van Golde, JA Kremer, JH Tuerlings, EC Mariman, AH Balen, J Poulton. No evidence for paternal mtDNA transmission to offspring or extra-embryonic tissues after ICSI. *Mol Hum Reprod* 8:1046–1049, 2002.

103. P Sutovsky, RD Moreno, J Ramalho–Santos, T Dominko, C Simerly, G Schatten. Ubiquitin tag for sperm mitochondria. *Nature* 402:371–372, 1999.

104. P Sutovsky, RD Moreno, J Ramalho–Santos, T Dominko, C Simerly, G Schatten. Ubiquitinated sperm mitochondria, selective proteolysis, and the regulation of mitochondrial inheritance in mammalian embryos. *Biol Reprod* 63:582–590, 2000.

105. P Sutovsky, TC McCauley, M Sutovsky, BN Day. Early degradation of paternal mitochondria in domestic pig (*Sus scrofa*) is prevented by selective proteasomal inhibitors lactacystin and MG132. *Biol Reprod* 68:1793–1800, 2003.

106. RJ Aitken. Free radicals, lipid peroxidation and sperm function. *Reprod Fertil Dev* 7:659–668, 1995.

107. HT Jacobs, SK Lehtinen, JN Spelbrink. No sex please, we're mitochondria: a hypothesis on the somatic unit of inheritance of mammalian mtDNA. *Bioessays* 22:564–572, 2000.

108. RN Lightowlers, PF Chinnery, DM Turnbull, N Howell. Mammalian mitochondrial genetics: heredity, heteroplasmy and disease. *Trends Genet* 13:450–455, 1997.

109. WW Hauswirth, PJ Laipis. Mitochondrial DNA polymorphism in a maternal lineage of Holstein cows. *Proc Natl Acad Sci USA* 79:4686–4690, 1982.

110. CM Koehler, GL Lindberg, DR Brown, DC Beitz, AE Freeman, JE Mayfield, AM Myers. Replacement of bovine mitochondrial DNA by a sequence variant within one generation. *Genetics* 129:247–255, 1991.

111. PA Bolhuis, EM Bleeker–Wagemakers, NJ Ponne, MJ van Schooneveld, A Westerveld, BC Van den, HF Tabak. Rapid shift in genotype of human mitochondrial DNA in a family with Leber's hereditary optic neuropathy. *Biochem Biophys Res Commun* 170:994–997, 1990.

112. MT Lott, AS Voljavec, DC Wallace. Variable genotype of Leber's hereditary optic neuropathy patients. *Am J Ophthalmol* 109:625–631, 1990.

113. MH Tulinius, M Houshmand, NG Larsson, E Holme, A Oldfors, E Holmberg, J Wahlstrom. *De novo* mutation in the mitochondrial ATP synthase subunit 6 gene (T8993G) with rapid segregation resulting in Leigh syndrome in the offspring. *Hum Genet* 96:290–294, 1995.

114. IFM de Coo, HJM Smeets, FJM Gabreëls, N Arts, BA Van Oost. Isolated case of mental retardation and ataxia due to a de novo mitochondrial T8993G mutation. *Am J Hum Genet* 58:636–638, 1996.

115. J Poulton, V Macaulay, DR Marchington. Mitochondrial genetics '98. Is the bottleneck cracked? *Am J Hum Genet* 62:752–757, 1998.

116. JP Jenuth, AC Peterson, K Fu, EA Shoubridge. Random genetic drift in the female germline explains the rapid segregation of mammalian mitochondrial DNA. *Nat Genet* 14:146–151, 1996.

117. DT Brown, DC Samuels, EM Michael, DM Turnbull, PF Chinnery. Random genetic drift determines the level of mutant mtDNA in human primary oocytes. *Am J Hum Genet* 68:533–536, 2001.

118. RB Blok, DA Gook, DR Thorburn, H-HM Dahl. Skewed segregation of the mtDNA nt 8993 (T→G) mutation in human oocytes. *Am J Hum Genet* 60:1495–1501, 1997.

119. M Solignac, J Génermont, M Monnerot, J C Mounolou. Genetics of mitochondria in *Drosophila*: mtDNA inheritance in heteroplasmic strains of *D. mauritana*. *Mol Gen Genet* 197:183–188, 1984.

120. TG Bakker. A quantitative and cytological study of germ cells in human ovaries. *Proc R Soc Lond B Biol Sci* 158:417–423, 1963.

121. DR Marchington, V Macaulay, GM Hartshorne, D Barlow, J Poulton. Evidence from human oocytes for a genetic bottleneck in an mtDNA disease. *Am J Hum Genet* 63:769–775, 1998.

122. PF Chinnery, N Howell, RN Lightowlers, DM Turnbull. Genetic counseling and prenatal diagnosis for mtDNA disease. *Am J Hum Genet* 63:1908–1911, 1998.

123. PF Chinnery, DC Samuels, J Elson, DM Turnbull. Accumulation of mitochondrial DNA mutations in ageing, cancer, and mitochondrial disease: is there a common mechanism? *Lancet* 360:1323–1325, 2002.

124. JP Jenuth, AC Peterson, EA Shoubridge. Tissue-specific selection for different mtDNA genotypes in heteroplasmic mice. *Nat Genet* 16:93–95, 1997.

125. NG Larsson, MH Tulinius, E Holme, A Oldfors, O Andersen, J Wahlstrom, J Aasly. Segregation and manifestations of the mtDNA tRNA(Lys) A→G(8344) mutation of myoclonus epilepsy and ragged-red fibers (MERRF) syndrome. *Am J Hum Genet* 51:1201–1212, 1992.

126. J Poulton, S O'Rahilly, KJ Morten, A Clark. Mitochondrial DNA, diabetes and pancreatic pathology in Kearns–Sayre syndrome. *Diabetologia* 38:868–871, 1995.

127. K Weber, JN Wilson, L Taylor, E Brierley, MA Johnson, DM Turnbull, LA Bindoff. A new mtDNA mutation showing accumulation with time and restriction to skeletal muscle. *Am J Hum Genet* 60:373–380, 1997.

128. S Rahman, J Poulton, D Marchington, A Suomalainen. Decrease of 3243 A→G mtDNA mutation from blood in MELAS syndrome: a longitudinal study. *Am J Hum Genet* 68:238–240, 2001.

129. PF Chinnery, PJ Zwijnenburg, M Walker, N Howell, RW Taylor, RN Lightowlers, L Bindoff, DM Turnbull. Nonrandom tissue distribution of mutant mtDNA. *Am J Med Genet* 85:498–501, 1999.

130. J Hayashi, S Ohta, A Kikuchi, M Takemitsu, Y Goto, I Nonaka. Introduction of disease-related mitochondrial DNA deletions into HeLa cells lacking mitochondrial DNA results in mitochondrial dysfunction. *Proc Natl Acad Sci USA* 88:10614–10618, 1991.

131. M Yoneda, A Chomyn, A Martinuzzi, O Hurko, G Attardi. Marked replicative advantage of human mtDNA carrying a point mutation that causes the MELAS encephalomyopathy. *Proc Natl Acad Sci USA* 89:11164–11168, 1992.

132. DR Dunbar, PA Moonie, HT Jacobs, IJ Holt. Different cellular backgrounds confer a marked advantage to either mutant or wild-type mitochondrial genomes. *Proc Natl Acad Sci USA* 92:6562–6566, 1995.

133. IJ Holt, DR Dunbar, HT Jacobs. Behaviour of a population of partially duplicated mitochondrial DNA molecules in cell culture: segregation, maintenance and recombination dependent upon nuclear background. *Hum Mol Genet* 6:1251–1260, 1997.

134. Y Tang, G Manfredi, M Hirano, EA Schon. Maintenance of human rearranged mitochondrial DNAs in long-term cultured transmitochondrial cell lines. *Mol Biol Cell* 11:2349–2358, 2000.

135. BJ Battersby, EA Shoubridge. Selection of an mtDNA sequence variant in hepatocytes of heteroplasmic mice is not due to differences in respiratory chain function or efficiency of replication. *Hum Mol Genet* 10:2469–2479, 2001.

136. BJ Battersby, JC Loredo–Osti, EA Shoubridge. Nuclear genetic control of mitochondrial DNA segregation. *Nat Genet* 33:183–186, 2003.

137. DA Clayton. Replication and transcription of vertebrate mitochondrial DNA. *Annu Rev Cell Biol* 7:453–478, 1991.

138. GS Shadel, DA Clayton. Mitochondrial transcription initiation. Variation and conservation. *J Biol Chem* 268:16083–16086, 1993.

139. P Fernández–Silva, JA Enriquez, J Montoya. Replication and transcription of mammalian mitochondrial DNA. *Exp Physiol* 88:41–56, 2003.

140. J Montoya, T Christianson, D Levens, M Rabinowitz, G Attardi. Identification of initiation sites for heavy-strand and light-strand transcription in human mitochondrial DNA. *Proc Natl Acad Sci USA* 79:7195–7199, 1982.

141. J Montoya, GL Gaines, G Attardi. The pattern of transcription of the human mitochondrial rRNA genes reveals two overlapping transcription units. *Cell* 34:151–159, 1983.

142. MW Walberg, DA Clayton. *In vitro* transcription of human mitochondrial DNA. Identification of specific light strand transcripts from the displacement loop region. *J Biol Chem* 258:1268–1275, 1983.

143. BK Yoza, DF Bogenhagen. Identification and *in vitro* capping of a primary transcript of human mitochondrial DNA. *J Biol Chem* 259:3909–3915, 1984.

144. DD Chang, DA Clayton. Precise identification of individual promoters for transcription of each strand of human mitochondrial DNA. *Cell* 36:635–643, 1984.

145. DF Bogenhagen, EF Applegate, BK Yoza. Identification of a promoter for transcription of the heavy strand of human mtDNA: *in vitro* transcription and deletion mutagenesis. *Cell* 36:1105–1113, 1984.

146. JE Hixson, DA Clayton. Initiation of transcription from each of the two human mitochondrial promoters requires unique nucleotides at the transcriptional start sites. *Proc Natl Acad Sci USA* 82:2660–2664, 1985.

147. JN Topper, DA Clayton. Identification of transcriptional regulatory elements in human mitochondrial DNA by linker substitution analysis. *Mol Cell Biol* 9:1200–1211, 1989.

148. RP Fisher, JN Topper, DA Clayton. Promoter selection in human mitochondria involves binding of a transcription factor to orientation-independent upstream regulatory elements. *Cell* 50:247–258, 1987.

149. CT Moraes, F Andreetta, E Bonilla, S Shanske, S DiMauro, EA Schon. Replication-competent human mitochondrial DNA lacking the heavy-strand promoter region. *Mol Cell Biol* 11:1631–1637, 1991.

150. SR Hammans, MG Sweeney, DA Wicks, JA Morgan–Hughes, AE Harding. A molecular genetic study of focal histochemical defects in mitochondrial encephalomyopathies. *Brain* 115:343–365, 1992.

151. DJ Shuey, G Attardi. Characterization of an RNA polymerase activity from HeLa cell mitochondria, which initiates transcription at the heavy strand rRNA promoter and the light strand promoter in human mitochondrial DNA. *J Biol Chem* 260:1952–1958, 1985.

152. RP Fisher, DA Clayton. A transcription factor required for promoter recognition by human mitochondrial RNA polymerase. Accurate initiation at the heavy- and light-strand promoters dissected and reconstituted *in vitro*. *J Biol Chem* 260:11330–11338, 1985.

153. V Tiranti, A Savoia, F Forti, M-F D'Apolito, M Centra, M Rocchi, M Zeviani. Identification of the gene encoding the human mitochondrial RNA polymerase (h-mtRPOL) by cyber-screening of the Expressed Sequence Tags database. *Hum Mol Genet* 6:615–625, 1997.

154. BS Masters, LL Stohl, DA Clayton. Yeast mitochondrial RNA polymerase is homologous to those encoded by bacteriophages T3 and T7. *Cell* 51:89–99, 1987.

155. D Patra, EM Lafer, R Sousa. Isolation and characterization of mutant bacteriophage T7 RNA polymerases. *J Mol Biol* 224:307–318, 1992.

156. G Bonner, D Patra, EM Lafer, R Sousa. Mutations in T7 RNA polymerase that support the proposal for a common polymerase active site structure. *EMBO J* 11:3767–3775, 1992.

157. L Gross, WJ Chen, WT McAllister. Characterization of bacteriophage T7 RNA polymerase by linker insertion mutagenesis. *J Mol Biol* 228:488–505, 1992.

158. LP Gardner, KA Mookhtiar, JE Coleman. Initiation, elongation, and processivity of carboxyl-terminal mutants of T7 RNA polymerase. *Biochemistry* 36:2908–2918, 1997.

159. MS Rodcheffer, BE Boone, AC Bryan, GS Shadel. Nam1p, a protein involved in RNA processing and translation, is coupled to transcription through an interaction with yeast mitochondrial RNA polymerase. *J Biol Chem* 276:8616–8622, 2001.

160. AC Bryan, MS Rodeheffer, CM Wearn, GS Shadel. Sls1p is a membrane-bound regulator of transcription-coupled processes involved in *Saccharomyces cerevisiae* mitochondrial gene expression. *Genetics* 160:75–82, 2002.

161. MS Rodeheffer, GS Shadel. Multiple interactions involving the amino-terminal domain of yeast mtRNA polymerase determine the efficiency of mitochondrial protein synthesis. *J Biol Chem* 278:18695–18701, 2003.

162. R George, P Walsh, T Beddoe, T Lithgow. The nascent polypeptide-associated complex (NAC) promotes interaction of ribosomes with the mitochondrial surface in vivo. *FEBS Lett* 516:213–216, 2002.

163. N Pfanner, N Wiedemann. Mitochondrial protein import: two membranes, three translocases. *Curr Opin Cell Biol* 14:400–411, 2002.

164. RP Fisher, DA Clayton. Purification and characterization of human mitochondrial transcription factor 1. *Mol Cell Biol* 8:3496–3509, 1988.

165. MA Parisi, DA Clayton. Similarity of human mitochondrial transcription factor 1 to high mobility group proteins. *Science* 252:965–969, 1991.

166. K Tominaga, S Akiyama, Y Kagawa, S Ohta. Upstream region of a genomic gene for human mitochondrial transcription factor 1. *Biochim Biophys Acta* 1131:217–219, 1992.

167. K Tominaga, J Hayashi, Y Kagawa, S Ohta. Smaller isoform of human mitochondrial transcription factor 1: its wide distribution and production by alternative splicing. *Biochem Biophys Res Commun* 194:544–551, 1993.

168. V Tiranti, E Rossi, A Ruiz–Carrillo, G Rossi, M Rocchi, S DiDonato, O Zuffardi, M Zeviani. Chromosomal localization of mitochondrial transcription factor A (TCF6), single-stranded DNA-binding protein (SSBP), and endonuclease G (ENDOG), three human housekeeping genes involved in mitochondrial biogenesis. *Genomics* 25:559–564, 1995.

169. A Reyes, M Mezzina, G Gadaleta. Human mitochondrial transcription factor A (mtTFA): gene structure and characterization of related pseudogenes. *Gene* 291:223–232, 2002.

170. R Grosschedl, K Giese, J Pagel. HMG domain proteins: architectural elements in the assembly of nucleoprotein structures. *Trends Genet* 10:94–100, 1994.

171. SC Ghivizzani, CS Madsen, MR Nelen, CV Ammini, WW Hauswirth. In organello footprint analysis of human mitochondrial DNA: human mitochondrial transcription factor A interactions at the origin of replication. *Mol Cell Biol* 14:7717–7730, 1994.

172. V Micol, P Fernández–Silva, G Attardi. Functional analysis of *in vivo* and *in organello* footprinting of HeLa cell mitochondrial DNA in relationship to ATP and ethidium bromide effects on transcription. *J Biol Chem* 272:18896–18904, 1997.

173. DJ Dairaghi, GS Shadel, DA Clayton. Addition of a 29 residue carboxyl-terminal tail converts a simple HMG box-containing protein into a transcriptional activator. *J Mol Biol* 249:11–28, 1995.

174. DD Chang, JE Hixson, DA Clayton. Minor transcription initiation events indicate that both human mitochondrial promoters function bidirectionally. *Mol Cell Biol* 6:294–301, 1986.

175. RP Fisher, T Lisowsky, MA Parisi, DA Clayton. DNA wrapping and bending by a mitochondrial high mobility group-like transcriptional activator protein. *J Biol Chem* 267:3358–3367, 1992.

176. JFX Diffley, B Stillman. DNA binding properties of an HMG1-related protein from yeast mitochondria. *J Biol Chem* 267:3368–3374, 1992.

177. S-C Nam, C Kang. Expression of cloned cDNA for the human mitochondrial RNA polymerase in *Escherichia coli* and purification. *Protein Expr Purif* 21:485–491, 2001.

178. A Prieto–Martín, J Montoya, F Martinez–Azorín. A study on the human mitochondrial RNA polymerase activity points to existence of a transcription factor B-like protein. *FEBS Lett* 503:51–55, 2001.

179. T Lisowsky, G Michaelis. A nuclear gene essential for mitochondrial replication suppresses a defect of mitochondrial transcription in *Saccharomyces cerevisiae*. *Mol Gen Genet* 214:218–223, 1988.

180. SH Jang, JA Jaehning. The yeast mitochondrial RNA polymerase specificity factor, MTF1, is similar to bacterial σ factors. *J Biol Chem* 266:22671–22677, 1991.

181. JA Carrodeguas, S Yun, GS Shadel, DA Clayton, DF Bogenhagen. Functional conservation of yeast mtTFB despite extensive sequence divergence. *Gene Expr* 6:219–230, 1996.

182. B Xu, DA Clayton. Assignment of a yeast protein necessary for mitochondrial transcription initiation. *Nucleic Acids Res* 20:1053–1059, 1992.

183. I Antoshechkin, DF Bogenhagen. Distinct roles for two purified factors in transcription of *Xenopus* mitochondrial DNA. *Mol Cell Biol* 15:7032–7042, 1995.

184. DF Bogenhagen. Interaction of mtTFB and mtRNA polymerase at core promoters for transcription of *Xenopus laevis* mtDNA. *J Biol Chem* 271:12036–12041, 1996.

185. V McCulloch, BL Seidel–Rogol, GS Shadel. A human mitochondrial transcription factor is related to RNA adenine methyltransferases and binds *S*-adenosylmethionine. *Mol Cell Biol* 22:1116–1125, 2002.

186. M Falkenberg, M Gaspari, A Rantanen, A Trifunovic, N-G Larsson, CM Gustafsson. Mitochondrial transcription factors B1 and B2 activate transcription of human mtDNA. *Nat Genet* 31:289–294, 2002.

187. V McCulloch, GS Shadel. Human mitochondrial transcription factor B1 interacts with the C-terminal activation region of h-mtTFA and stimulates transcription independently of its RNA methyltransferase activity. *Mol Cell Biol* 23:5816–5824, 2003.

188. DJ Dairaghi, GS Shadel, DA Clayton. Human mitochondrial transcription factor A and promoter spacing integrity are required for transcription initiation. *Biochim Biophys Acta* 1271:127–134, 1995.

189. FD Schubot, CJ Chen, JP Rose, TA Dailey, HA Dailey, BC Wang. Crystal structure of the transcription factor sc-mtTFB offers insights into mitochondrial transcription. *Protein Sci* 10:1980–1988, 2001.

190. GS Shadel. A dual-function mitochondrial transcription factor tunes out deafness. *Mol Genet Metab* 82:1–3, 2004.

191. BL Seidel–Rogol, V McCulloch, GS Shadel. Human mitochondrial transcription factor B1 methylates ribosomal RNA at a conserved stem-loop. *Nat Genet* 33:23–24, 2003.

192. WI Murphy, B Attardi, C Tu, G Attardi. Evidence for complete symmetrical transcription *in vivo* of mitochondrial DNA in HeLa cells. *J Mol Biol* 99:809–814, 1975.

193. Y Aloni, G Attardi. Symmetrical *in vivo* transcription of mitochondrial DNA in HeLa cells. *Proc Natl Acad Sci USA* 68:1757–1761, 1971.

194. R Gelfand, G Attardi. Synthesis and turnover of mitochondrial ribonucleic acid in HeLa cells: the mature ribosomal and messenger ribonucleic acid species are metabolically unstable. *Mol Cell Biol* 1:497–511, 1981.

195. G Gaines, G Attardi. Intercalating drugs and low temperatures inhibit synthesis and processing of ribosomal RNA in isolated human mitochondria. *J Mol Biol* 172:451–466, 1984.

196. G Gaines, C Rossi, G Attardi. Markedly different ATP requirements for rRNA synthesis and mtDNA light strand transcription vs. mRNA synthesis in isolated human mitochondria. *J Biol Chem* 262:1907–1915, 1987.

197. DT Dubin, J Montoya, KD Timko, G Attardi. Sequence analysis and precise mapping of the 3′ ends of HeLa cell mitochondrial ribosomal RNAs. *J Mol Biol* 157:1–19, 1982.

198. RA Van Etten, JW Bird, DA Clayton. Identification of the 3′-ends of the two mouse mitochondrial ribosomal RNAs. The 3′-end of 16 S ribosomal RNA contains nucleotides encoded by the gene for transfer RNA[LeuUUR]. *J Biol Chem* 258:10104–10110, 1983.

199. B Kruse, N Narasimhan, G Attardi. Termination of transcription in human mitochondria: identification and purification of a DNA binding protein factor that promotes termination. *Cell* 58:391–397, 1989.

200. TW Christianson, DA Clayton. A tridecamer DNA sequence supports human mitochondrial RNA 3′-end formation *in vitro*. *Mol Cell Biol* 8:4502–4509, 1988.

201. JF Hess, MA Parisi, JL Bennett, DA Clayton. Impairment of mitochondrial transcription termination by a point mutation associated with the MELAS subgroup of mitochondrial encephalomyopathies. *Nature* 351:236–239, 1991.

202. J Shang, DA Clayton. Human mitochondrial transcription termination exhibits RNA polymerase independence and biased bipolarity *in vitro*. *J Biol Chem* 269:29112–29120, 1994.

203. A Chomyn, A Martinuzzi, M Yoneda, A Daga, O Hurko, D Johns, ST Lai, I Nonaka, C Angelini, G Attardi. MELAS mutation in mtDNA binding site for transcription termination factor causes defects in protein synthesis and in respiration but no change in levels of upstream and downstream mature transcripts. *Proc Natl Acad Sci USA* 89:4221–4225, 1992.

204. MP King, Y Koga, M Davidson, EA Schon. Defects in mitochondrial protein synthesis and respiratory chain activity segregate with the tRNA$^{Leu(UUR)}$ mutation associated with mitochondrial myopathy, encephalopathy, lactic acidosis, and strokelike episodes. *Mol Cell Biol* 12:480–490, 1992.

205. CT Moraes, E Ricci, E Bonilla, S DiMauro, EA Schon. The mitochondrial tRNA$^{Leu(UUR)}$ mutation in mitochondrial encephalomyopathy, lactic acidosis, and strokelike episodes (MELAS): genetic, biochemical, and morphological correlations in skeletal muscle. *Am J Hum Genet* 50:934–949, 1992.

206. P Kaufmann, Y Koga, S Shanske, M Hirano, S DiMauro, MP King, EA Schon. Mitochondrial DNA and RNA processing in MELAS. *Ann Neurol* 40:172–180, 1996.

207. A Daga, V Micol, D Hess, R Aebersold, G Attardi. Molecular characterization of the transcription termination factor from human mitochondria. *J Biol Chem* 268:8123–8130, 1993.

208. P Fernandez–Silva, F Martinez–Azorin, V Micol, G Attardi. The human mitochondrial transcription termination factor (mTERF) is a multizipper protein but binds to DNA as a monomer, with evidence pointing to intramolecular leucine zipper interactions. *EMBO J* 16:1066–1079, 1997.

209. J Asin–Cayuela, M Helm, G Attardi. A monomer-to-trimer transition of the human mitochondrial transcription termination factor (mTERF) is associated with a loss of *in vitro* activity. *J Biol Chem* 279:15670–15677, 2004.

210. A Prieto–Martín, J Montoya, F Martínez–Azorín. Phosphorylation of rat mitochondrial transcription termination factor (mTERF) is required for transcription termination but not for binding to DNA. *Nucleic Acids Res* 32:2059–2068, 2004.

211. TW Christianson, DA Clayton. *In vitro* transcription of human mitochondrial DNA: accurate termination requires a region of DNA sequence that can function bidirectionally. *Proc Natl Acad Sci USA* 83:6277–6281, 1986.

212. D Ojala, G Attardi. Fine mapping of the ribosomal RNA genes of HeLa cell mitochondrial DNA. *J Mol Biol* 138:411–420, 1980.

213. J Montoya, D Ojala, G Attardi. Distinctive features of the 5′-terminal sequences of the human mitochondrial mRNAs. *Nature* 290:465–470, 1981.

214. V Camasamudram, J-K Fang, NG Avadhani. Transcription termination at the mouse mitochondrial H-strand promoter distal site requires an A/T rich sequence motif and sequence specific DNA binding proteins. *Eur J Biochem* 270:1128–1140, 2003.

215. NJ Proudfoot, A Furger, MJ Dye. Integrating mRNA processing with transcription. *Cell* 108:501–512, 2002.

216. F Amalric, C Merkel, R Gelfand, G Attardi. Fractionation of mitochondrial RNA from HeLa cells by high-resolution electrophoresis under strongly denaturing conditions. *J Mol Biol* 118:1–25, 1978.

217. DN Frank, NR Pace. Ribonuclease P: unity and diversity in a tRNA processing ribozyme. *Annu Rev Biochem* 67:153–180, 1998.

218. S Xiao, F Scott, CA Fierke, DR Engelke. Eukaryotic ribonuclease P: a plurality of ribonucleoprotein enzymes. *Annu Rev Biochem* 71:165–189, 2002.

219. DL Miller, NC Martin. Characterization of the yeast mitochondrial locus necessary for tRNA biosynthesis: DNA sequence analysis and identification of a new transcript. *Cell* 34:911–917, 1983.

220. YL Dang, NC Martin. Yeast mitochondrial RNase P. Sequence of the *RPM2* gene and demonstration that its product is a protein subunit of the enzyme. *J Biol Chem* 268:19791–19796, 1993.

221. ER Seif, L Forget, NC Martin, BF Lang. Mitochondrial RNase P RNAs in ascomycete fungi: lineage-specific variations in RNA secondary structure. *RNA* 9:1073–1083, 2003.

222. R Salavati, AK Panigrahi, KD Stuart. Mitochondrial ribonu-clease P activity of *Trypanosoma brucei*. *Mol Biochem Parasitol* 115:109–117, 2001.

223. W Rossmanith, T Potuschak. Difference between mitochondrial RNase P and nuclear RNase P. *Mol Cell Biol* 21:8236–8237, 2001.

224. RS Puranam, G Attardi. Difference between mitochondrial RNase P and nuclear RNase P — authors' reply. *Mol Cell Biol* 21:8236–8237, 2001.

225. W Rossmanith, A Tullo, T Potuschak, R Karwan, E Sbisà. Human mitochondrial tRNA processing. *J Biol Chem* 270:12885–12891, 1995.

226. W Rossmanith, RM Karwan. Characterization of human mito-chondrial RNase P: novel aspects in tRNA processing. *Biochem Biophys Res Commun* 247:234–241, 1998.

227. C-J Doersen, C Guerrier–Takada, S Altman, G Attardi. Char-acterization of an RNase P activity from HeLa cell mitochon-dria. Comparison with the cytosol RNase P activity. *J Biol Chem* 260:5942–5949, 1985.

228. RS Puranam, G Attardi. The RNase P associated with HeLa cell mitochondria contains an essential RNA component iden-tical in sequence to that of the nuclear RNase P. *Mol Cell Biol* 21:548–561, 2001.

229. S Manam, GC Van Tuyle. Separation and characterization of 5′- and 3′-tRNA processing nucleases from rat liver mitochon-dria. *J Biol Chem* 262:10272–10279, 1987.

230. N Jarrous. Human ribonuclease P: subunits, function, and intranuclear localization. *RNA* 8:1–7, 2002.

231. E Hartmann, RK Hartmann. The enigma of ribonuclease P evolution. *Trends Genet* 19:561–569, 2003.

232. G Reimer, I Raska, U Scheer, EM Tan. Immunolocalization of 7-2-ribonucleoprotein in the granular component of the nucle-olus. *Exp Cell Res* 176:117–128, 1988.

233. T Kiss, W Filipowicz. Evidence against a mitochondrial location of the 7-2/MRP RNA in mammalian cells. *Cell* 70:11–16, 1992.

234. K Li, CS Smagula, WJ Parsons, JA Richardson, M Gonzalez, HK Hagler, RS Williams. Subcellular partitioning of MRP RNA assessed by ultrastructural and biochemical analysis. *J Cell Biol* 124:871–882, 1994.

235. MR Jacobson, LG Cao, YL Wang, T Pederson. Dynamic localization of RNase MRP RNA in the nucleolus observed by fluorescent RNA cytochemistry in living cells. *J Cell Biol* 131:1649–1658, 1995.

236. H Van Eenennaam, N Jarrous, WJ Van Venrooij, GJM Pruijn. Architecture and function of the human endonucleases RNase P and RNase MRP. *IUBMB Life* 49:265–272, 2000.

237. S Chu, RH Archer, JM Zengel, L Lindahl. The RNA of RNase MRP is required for normal processing of ribosomal RNA. *Proc Natl Acad Sci USA* 91:659–663, 1994.

238. Z Lygerou, C Allmang, D Tollervey, B Séraphin. Accurate processing of a eukaryotic precursor ribosomal RNA by ribonuclease MRP *in vitro*. *Science* 272:268–270, 1996.

239. AC Forster, S Altman. Similar cage-shaped structures for the RNA components of all ribonuclease P and ribonuclease MRP enzymes. *Cell* 62:407–409, 1990.

240. M Mörl, A Marchfelder. The final cut. The importance of tRNA 3′-processing. *EMBO Rep* 2:17–20, 2001.

241. O Pellegrini, J Nezzar, A Marchfelder, H Putzer, C Condon. Endonucleolytic processing of CCA-less tRNA precursors by RNase Z in *Bacillus subtilis*. *EMBO J* 22:4534–4543, 2003.

242. L Levinger, O Jacobs, M James. *In vitro* 3′-end endonucleolytic processing defect in a human mitochondrial tRNA$^{Ser(UCN)}$ precursor with the U7445C substitution, which causes nonsyndromic deafness. *Nucleic Acids Res* 29:4334–4340, 2001.

243. EB Dubrovsky, VA Dubrovskaya, L Levinger, S Schiffer, A Marchfelder. *Drosophila* RNase Z processes mitochondrial and nuclear pre-tRNA 3′ ends *in vivo*. *Nucleic Acids Res* 32:255–262, 2004.

244. W Rossmanith. Processing of human mitochondrial tRNA$^{Ser(AGY)}$: a novel pathway in tRNA biosynthesis. *J Mol Biol* 265:365–371, 1997.

245. W Rossmanith, RM Karwan. Impairment of tRNA processing by point mutations in mitochondrial tRNA^Leu(UUR) associated with mitochondrial diseases. *FEBS Lett* 433:269–274, 1998.

246. L Levinger, I Oestreich, C Florentz, M Mörl. A pathogenesis-associated mutation in human mitochondrial tRNA^Leu(UUR) leads to reduced 3'-end processing and CCA addition. *J Mol Biol* 337:535–544, 2004.

247. Y Koga, M Davidson, EA Schon, MP King. Fine mapping of mitochondrial RNAs derived from the mtDNA region containing a point mutation associated with MELAS. *Nucleic Acids Res* 21:657–662, 1993.

248. H Hao, CT Moraes. Functional and molecular mitochondrial abnormalities associated with a C→T transition at position 3256 of the human mitochondrial genome. The effects of a pathogenic mitochondrial tRNA point mutation in organelle translation and RNA processing. *J Biol Chem* 271:2347–2352, 1996.

249. LA Bindoff, N Howell, J Poulton, DA McCullough, KJ Morten, RN Lightowlers, DM Turnbull, K Weber. Abnormal RNA processing associated with a novel tRNA mutation in mitochondrial DNA. A potential disease mechanism. *J Biol Chem* 268:19559–19564, 1993.

250. A Koga, Y Koga, Y Akita, R Fukiyama, I Ueki, S Yatsuga, T Matsuishi. Increased mitochondrial processing intermediates associated with three tRNA^Leu(UUR) gene mutations. *Neuromuscul Disord* 13:259–262, 2003.

251. Y Koga, M Yoshino, H Kato. MELAS exhibits dominant negative effects on mitochondrial RNA processing. *Ann Neurol* 43:835, 1998.

252. EA Schon, Y Koga, M Davidson, CT Moraes, MP King. The mitochondrial tRNA^Leu(UUR) mutation in MELAS: a model for pathogenesis. *Biochim Biophys Acta* 1101:206–209, 1992.

253. Y Koga, M Davidson, EA Schon, MP King. Analysis of cybrids harboring MELAS mutations in the mitochondrial tRNA^Leu(UUR) gene. *Muscle Nerve* 3:S119–S123, 1995.

254. A Reichert, U Rothbauer, M Mörl. Processing and editing of overlapping tRNAs in human mitochondria. *J Biol Chem* 273:31977–31984, 1998.

255. AS Reichert, M Mörl. Repair of tRNAs in metazoan mitochondria. *Nucleic Acids Res* 28:2043–2048, 2000.

256. M Sprinzl, C Horn, M Brown, A Ioudovitch, S Steinberg. Compilation of tRNA sequences and sequences of tRNA genes. *Nucleic Acids Res* 26:148–153, 1998.

257. M Helm, H Brulé, F Degoul, C Cepanec, JP Leroux, R Giegé, C Florentz. The presence of modified nucleotides is required for cloverleaf folding of a human mitochondrial tRNA. *Nucleic Acids Res* 26:1636–1643, 1998.

258. M Helm, R Giegé, C Florentz. A Watson–Crick base-pair-disrupting methyl group (m^{1}A9) is sufficient for cloverleaf folding of human mitochondrial tRNA^{Lys}. *Biochemistry* 38:13338–13346, 1999.

259. F Degoul, H Brulé, C Cepanec, M Helm, C Marsac, J Leroux, R Giegé, C Florentz. Isoleucylation properties of native human mitochondrial tRNA^{Ile} and tRNA^{Ile} transcripts. Implications for cardiomyopathy-related point mutations (4269, 4317) in the tRNA^{Ile} gene. *Hum Mol Genet* 7:347–354, 1998.

260. M Helm, G Attardi. Nuclear control of cloverleaf structure of human mitochondrial tRNA^{Lys}. *J Mol Biol* 337:545–560, 2004.

261. D Brégeon, V Colot, M Radman, F Taddei. Translational misreading: a tRNA modification counteracts a +2 ribosomal frameshift. *Genes Dev* 15:2295–2306, 2001.

262. G Colby, M Wu, A Tzagoloff. *MTO1* codes for a mitochondrial protein required for respiration in paromomycin-resistant mutants of *Saccharomyces cerevisiae*. *J Biol Chem* 273:27945–27952, 1998.

263. X Li, R Li, X Lin, MX Guan. Isolation and characterization of the putative nuclear modifier gene *MTO1* involved in the pathogenesis of deafness-associated mitochondrial 12 S rRNA A1555G mutation. *J Biol Chem* 277:27256–27264, 2002.

264. K Tamura, N Nameki, T Hasegawa, M Shimizu, H Himeno. Role of the CCA terminal sequence of tRNA(Val) in aminoacylation with valyl-tRNA synthetase. *J Biol Chem* 269:22173–22177, 1994.

265. T Nagaike, T Suzuki, Y Tomari, C Takemoto–Hori, F Negayama, K Watanabe, T Ueda. Identification and characterization of mammalian mitochondrial tRNA nucleotidyltransferases. *J Biol Chem* 276:40041–40049, 2001.

266. AS Reichert, DL Thurlow, M Mörl. A eubacterial origin for the human tRNA nucleotidyltransferase? *Biol Chem* 382:1431–1438, 2001.

267. JY Chen, PB Joyce, CL Wolfe, MC Steffen, NC Martin. Cytoplasmic and mitochondrial tRNA nucleotidyltransferase activities are derived from the same gene in the yeast *Saccharomyces cerevisiae. J Biol Chem* 267:14879–14883, 1992.

268. CL Wolfe, AK Hopper, NC Martin. Mechanisms leading to and the consequences of altering the normal distribution of ATP(CTP):tRNA nucleotidyltransferase in yeast. *J Biol Chem* 271:4679–4686, 1996.

269. MA Augustin, AS Reichert, H Betat, R Huber, M Mörl, C Steegborn. Crystal structure of the human CCA-adding enzyme: insights into template-independent polymerization. *J Mol Biol* 328:985–994, 2003.

270. SJ Mudge, JH Williams, HJ Eyre, GR Sutherland, PJ Cowan, DA Power. Complex organization of the 5′-end of the human glycine tRNA synthetase gene. *Gene* 209:45–50, 1998.

271. JM Bullard, YC Cai, B Demeler, LL Spremulli. Expression and characterization of a human mitochondrial phenylalanyl-tRNA synthetase. *J Mol Biol* 288:567–577, 1999.

272. JM Bullard, YC Cai, LL Spremulli. Expression and characterization of the human mitochondrial leucyl-tRNA synthetase. *Biochim Biophys Acta* 1490:245–258, 2000.

273. R Jørgensen, TM Søgaard, AB Rossing, PM Martensen, J Justesen. Identification and characterization of human mitochondrial tryptophanyl-tRNA synthetase. *J Biol Chem* 275:16820–16826, 2000.

274. T Yokogawa, N Shimada, N Takeuchi, L Benkowski, T Suzuki, A Omori, T Ueda, K Nishikawa, LL Spremulli, K Watanabe. Characterization and tRNA recognition of mammalian mitochondrial seryl-tRNA synthetase. *J Biol Chem* 275:19913–19920, 2000.

275. E Tolkunova, H Park, J Xia, MP King, E Davidson. The human lysyl-tRNA synthetase gene encodes both the cytoplasmic and mitochondrial enzymes by means of an unusual alternative splicing of the primary transcript. *J Biol Chem* 275:35063–35069, 2000.

276. N Shimada, T Suzuki, K Watanabe. Dual mode recognition of two isoacceptor tRNAs by mammalian mitochondrial seryl-tRNA synthetase. *J Biol Chem* 276:46770–46778, 2001.

277. B Sohm, M Frugier, H Brulé, K Olszak, A Przykorska, C Florentz. Towards understanding human mitochondrial leucine aminoacylation identity. *J Mol Biol* 328:995–1010, 2003.

278. ZH Shah, M Toompuu, T Hakkinen, AT Rovio, C van Ravenswaay, EMR De Leenheer, RJH Smith, FPM Cremers, CWRJ Cremers, HT Jacobs. Novel coding-region polymorphisms in mitochondrial seryl-tRNA synthetase (*SARSM*) and mitoribosomal protein S12 (*RPMS12*) genes in *DFNA4* autosomal dominant deafness families. *Hum Mutat* 17:433–434, 2001.

279. AC Spencer, A Heck, N. Watanabe, LL Spremulli. Characterization of the human mitochondrial methionyl-tRNA synthetase. *Biochem.* 43:9743–9754, 2004.

280. K Shiba, P Schimmel, H Motegi, T Noda. Human glycyl-tRNA synthetase. Wide divergence of primary structure from bacterial counterpart and species-specific aminoacylation. *J Biol Chem* 269:30049–30055, 1994.

281. JA Enriquez, A Chomyn, G Attardi. MtDNA mutation in MERRF syndrome causes defective aminoacylation of tRNA[Lys] and premature translation termination. *Nat Genet* 10:47–55, 1995.

282. SR Bacman, DP Atencio, CT Moraes. Decreased mitochondrial tRNA[Lys] steady-state levels and aminoacylation are associated with the pathogenic G8313A mitochondrial DNA mutation. *Biochem J* 374:131–136, 2003.

283. A El Meziane, SK Lehtinen, IJ Holt, HT Jacobs. Mitochondrial tRNA[Leu] isoforms in lung carcinoma cybrid cells containing the np 3243 mtDNA mutation. *Hum Mol Genet* 7:2141–2147, 1998.

284. GMC Janssen, JA Maassen, JMW van den Ouweland. The diabetes-associated 3243 mutation in the mitochondrial tRNA^Leu(UUR) gene causes severe mitochondrial dysfunction without a strong decrease in protein synthesis rate. *J Biol Chem* 274:29744–29748, 1999.

285. A Chomyn, JA Enriquez, V Micol, P Fernandez–Silva, G Attardi. The mitochondrial myopathy, encephalopathy, lactic acidosis, and stroke-like episode syndrome-associated human mitochondrial tRNA^Leu(UUR) mutation causes aminoacylation deficiency and concomitant reduced association of mRNA with ribosomes. *J Biol Chem* 275:19198–19209, 2000.

286. GV Börner, M Zeviani, V Tiranti, F Carrara, S Hoffmann, KD Gerbitz, H Lochmüller, D Pongratz, T Klopstock, A Melberg, E Holme, S Pääbo. Decreased aminoacylation of mutant tRNAs in MELAS but not in MERRF patients. *Hum Mol Genet* 9:467–475, 2000.

287. H Park, E Davidson, MP King. The pathogenic A3243G mutation in human mitochondrial tRNA^Leu(UUR) decreases the efficiency of aminoacylation. *Biochemistry* 42:958–964, 2003.

288. N Takeuchi, M Kawakami, A Omori, T Ueda, LL Spremulli, K Watanabe. Mammalian mitochondrial methionyl-tRNA transformylase from bovine liver. Purification, characterization, and gene structure. *J Biol Chem* 273:15085–15090, 1998.

289. N Takeuchi, L Vial, M Panvert, E Schmitt, K Watanabe, Y Mechulam, S Blanquet. Recognition of tRNAs by Methionyl-tRNA transformylase from mammalian mitochondria. *J Biol Chem* 276:20064–20068, 2001.

290. D Ojala, G Attardi. Identification of discrete polyadenylate-containing RNA components transcribed from HeLa cell mitochondrial DNA. *Proc Natl Acad Sci USA* 71:563–567, 1974.

291. M Hirsch, S Penman. Post-transcriptional addition of polyadenylic acid to mitochondrial RNA by a cordycepin-insensitive process. *J Mol Biol* 83:131–142, 1974.

292. M Dreyfus, P Régnier. The poly(A) tail of mRNAs: bodyguard in eukaryotes, scavenger in bacteria. *Cell* 111:611–613, 2002.

293. R Hayes, J Kudla, W Gruissem. Degrading chloroplast mRNA: the role of polyadenylation. *Trends Biochem Sci* 24:199–202, 1999.

294. CM Ryan, KT Militello, LK Read. Polyadenylation regulates the stability of *Trypanosoma brucei* mitochondrial RNAs. *J Biol Chem* 278:32753–32762, 2003.

295. RJ Temperley, SH Seneca, K Tonska, E Bartnik, LA Bindoff, RN Lightowlers, ZM Chrzanowska–Lightowlers. Investigation of a pathogenic mtDNA microdeletion reveals a translation-dependent deadenylation decay pathway in human mitochondria. *Hum Mol Genet* 12:2341–2348, 2003.

296. KM Rose, ST Jacob. Poly(adenylic acid) synthesis in isolated rat liver mitochondria. *Biochemistry* 15:5046–5052, 1976.

297. ST Jacob, KM Rose, HP Morris. Expression of purified mitochondrial poly(A)polymerase of hepatomas by an endogenous primer from liver. *Biochim Biophys Acta* 361:312–320, 1974.

298. KM Rose, HP Morris, ST Jacob. Mitochondrial poly(A) polymerase from a poorly differentiated hepatoma: purification and characteristics. *Biochemistry* 14:1025–1032, 1975.

299. RJ Baer, DT Dubin. Methylated regions of hamster mitochondrial ribosomal RNA: structural and functional correlates. *Nucleic Acids Res* 9:323–337, 1981.

300. J Ofengand, A Bakin. Mapping to nucleotide resolution of pseudouridine residues in large subunit ribosomal RNAs from representative eukaryotes, prokaryotes, archaebacteria, mitochondria and chloroplasts. *J Mol Biol* 266:246–268, 1997.

301. D Söll, UL RajBhandary, TLE RajBhandary. *tRNA: Structure, Biosynthesis and Function.* Washington, DC: American Society for Microbiology, 1995, pp 1–575.

302. MH de Bruijn, PH Schreier, IC Eperon, BG Barrell, EY Chen, PW Armstrong, JF Wong, BA Roe. A mammalian mitochondrial serine transfer RNA lacking the "dihydrouridine" loop and stem. *Nucleic Acids Res* 8:5213–5222, 1980.

303. T Yokogawa, Y Watanabe, Y Kumazawa, T Ueda, I Hirao, K Miura, K Watanabe. A novel cloverleaf structure found in mammalian mitochondrial tRNA[Ser] (UCN). *Nucleic Acids Res* 19:6101–6105, 1991.

304. M Helm, H Brulé, D Friede, R Giegé, D Pütz, C Florentz. Search for characteristic structural features of mammalian mitochondrial tRNAs. *RNA* 6:1356–1379, 2000.

305. LM Wittenhagen, MD Roy, SO Kelley. The pathogenic U3271C human mitochondrial tRNA$^{Leu(UUR)}$ mutation disrupts a fragile anticodon stem. *Nucleic Acids Res* 31:596–601, 2003.

306. LM Wittenhagen, SO Kelley. Impact of disease-related mitochondrial mutations on tRNA structure and function. *Trends Biochem Sci* 28:605–611, 2003.

307. MHL de Bruijn, A Klug. A model for the tertiary structure of mammalian mitochondrial transfer RNAs lacking the entire "dihydrouridine" loop and stem. *EMBO J* 2:1309–1321, 1983.

308. K Wakita, Y Watanabe, T Yokogawa, Y Kumazawa, S Nakamura, T Ueda, K Watanabe, K Nishikawa. Higher-order structure of bovine mitochondrial tRNA^{Phe} lacking the "conserved" GG and TΨCG sequences as inferred by enzymatic and chemical probing. *Nucleic Acids Res* 22:347–353, 1994.

309. Y Watanabe, G Kawai, T Yokogawa, N Hayashi, Y Kumazawa, T Ueda, K Nishikawa, I Hirao, K Miura, K Watanabe. Higher-order structure of bovine mitochondrial tRNA^{Ser}UGA: chemical modification and computer modeling. *Nucleic Acids Res* 22:5378–5384, 1994.

310. I Hayashi, T Yokogawa, G Kawai, T Ueda, K Nishikawa, K Watanabe. Assignment of imino proton signals of G-C base pairs and magnesium ion binding: an NMR study of bovine mitochondrial tRNA$^{Ser\ GCU}$ lacking the entire D arm. *J Biochem (Tokyo)* 121:1115–1122, 1997.

311. I Hayashi, G Kawai, K Watanabe. Higher-order structure and thermal instability of bovine mitochondrial tRNA^{Ser}UGA investigated by proton NMR spectroscopy. *J Mol Biol* 284:57–69, 1998.

312. TW O'Brien, GF Kalf. Ribosomes from rat liver mitochondria. I. Isolation procedure and contamination studies. *J Biol Chem* 242:2172–2179, 1967.

313. TW O'Brien, GF Kalf. Ribosomes from rat liver mitochondria. II. Partial characterization. *J Biol Chem* 242:2180–2185, 1967.

314. MR Sharma, EC Koc, PP Datta, TM Booth, LL Spremulli, RK Agrawal. Structure of the mammalian mitochondrial ribosome reveals an expanded functional role for its component proteins. *Cell* 115:97–108, 2003.

315. TW Spithill, MK Trembath, HB Lukins, AW Linnane. Mutations of the mitochondrial DNA of *Saccharomyces cerevisiae* which affect the interaction between mitochondrial ribosomes and the inner mitochondrial membrane. *Mol Gen Genet* 164:155–162, 1978.

316. S Marzuki, AR Hibbs. Are all mitochondrial translation products synthesized on membrane-bound ribosomes? *Biochim Biophys Acta* 866:120–124, 1986.

317. ME Sanchirico, TD Fox, TL Mason. Accumulation of mitochondrially synthesized *Saccharomyces cerevisiae* Cox2p and Cox3p depends on targeting information in untranslated portions of their mRNAs. *EMBO J* 17:5796–5804, 1998.

318. L Jia, M Dienhart, M Schramp, M McCauley, K Hell, RA Stuart. Yeast Oxa1 interacts with mitochondrial ribosomes: the importance of the C-terminal region of Oxa1. *EMBO J* 22:6438–6447, 2003.

319. G Szyrach, M Ott, N Bonnefoy, W Neupert, JM Herrmann. Ribosome binding to the Oxa1 complex facilitates co-translational protein insertion in mitochondria. *EMBO J* 22:6448–6457, 2003.

320. M Liu, L Spremulli. Interaction of mammalian mitochondrial ribosomes with the inner membrane. *J Biol Chem* 275:29400–29406, 2000.

321. F Feldman, HR Mahler. Mitochondrial biogenesis. Retention of terminal formylmethionine in membrane proteins and regulation of their synthesis. *J Biol Chem* 249:3702–3709, 1974.

322. CJ Schwartzbach, LL Spremulli. Bovine mitochondrial protein synthesis elongation factors. Identification and initial characterization of an elongation factor Tu-elongation factor Ts complex. *J Biol Chem* 264:19125–19131, 1989.

323. HX Liao, LL Spremulli. Identification and initial characterization of translational initiation factor 2 from bovine mitochondria. *J Biol Chem* 265:13618–13622, 1990.

324. HK Chung, LL Spremulli. Purification and characterization of elongation factor G from bovine liver mitochondria. *J Biol Chem* 265:21000–21004, 1990.

325. J Ma, LL Spremulli. Expression, purification, and mechanistic studies of bovine mitochondrial translational initiation factor 2. *J Biol Chem* 271:5805–5811, 1996.

326. TW O'Brien. Properties of human mitochondrial ribosomes. *IUBMB Life* 55:505–513, 2003.

327. G Attardi, D Ojala. Mitochondrial ribosome in HeLa cells. *Nat New Biol* 229:133–136, 1971.

328. A Brega, C Vesco. Ribonucleoprotein particles involved in HeLa mitochondrial protein synthesis. *Nat New Biol* 229:136–139, 1971.

329. MG Hamilton, TW O'Brien. Ultracentrifugal characterization of the mitochondrial ribosome and subribosomal particles of bovine liver: molecular size and composition. *Biochemistry* 13:5400–5403, 1974.

330. VB Patel, CC Cunningham, RR Hantgan. Physiochemical properties of rat liver mitochondrial ribosomes. *J Biol Chem* 276:6739–6746, 2001.

331. A Cahill, DL Baio, CC Cunningham. Isolation and characterization of rat liver mitochondrial ribosomes. *Anal Biochem* 232:47–55, 1995.

332. CMT Spahn, R Beckmann, N Eswar, PA Penczek, A Sali, G Blobel, J Frank. Structure of the 80S ribosome from *Saccharomyces cerevisiae*: tRNA–ribosome and subunit–subunit interactions. *Cell* 107:373–386, 2001.

333. MM Yusupov, GZ Yusupova, A Baucom, K Lieberman, TN Earnest, JH Cate, HF Noller. Crystal structure of the ribosome at 5.5 Å resolution. *Science* 292:883–896, 2001.

334. PM Lizardi, DJ Luck. Absence of a 5S RNA component in the mitochondrial ribosomes of *Neurospora crassa*. *Nat New Biol* 229:140–142, 1971.

335. BF Lang, LJ Goff, MW Gray. A 5 S rRNA gene is present in the mitochondrial genome of the protist *Reclinomonas americana* but is absent from red algal mitochondrial DNA. *J Mol Biol* 261:407–413, 1996.

336. S Yoshionari, T Koike, T Yokogawa, K Nishikawa, T Ueda, K Miura, K Watanabe. Existence of nuclear-encoded 5S-rRNA in bovine mitochondria. *FEBS Lett* 338:137–142, 1994.

337. PJ Magalhães, AL Andreu, EA Schon. Evidence for the presence of 5S rRNA in mammalian mitochondria. *Mol Biol Cell* 9:2375–2382, 1998.

338. NS Entelis, OA Kolesnikova, S Dogan, RP Martin, IA Tarassov. 5 S rRNA and tRNA import into human mitochondria. Comparison of *in vitro* requirements. *J Biol Chem* 276:45642–45653, 2001.

339. EC Koc, W Burkhart, K Blackburn, A Moseley, LL Spremulli. The small subunit of the mammalian mitochondrial ribosome. Identification of the full complement of ribosomal proteins present. *J Biol Chem* 276:19363–19374, 2001.

340. T Suzuki, M Terasaki, C Takemoto–Hori, T Hanada, T Ueda, A Wada, K Watanabe. Structural compensation for the deficit of rRNA with proteins in the mammalian mitochondrial ribosome. Systematic analysis of protein components of the large ribosomal subunit from mammalian mitochondria. *J Biol Chem* 276:21724–21736, 2001.

341. T Suzuki, M Terasaki, C Takemoto–Hori, T Hanada, T Ueda, A Wada, K Watanabe. Proteomic analysis of the mammalian mitochondrial ribosome. Identification of protein components in the 28 S small subunit. *J Biol Chem* 276:33181–33195, 2001.

342. EC Koc, W Burkhart, K Blackburn, MB Moyer, DM Schlatzer, A Moseley, LL Spremulli. The large subunit of the mammalian mitochondrial ribosome. Analysis of the complement of ribosomal proteins present. *J Biol Chem* 276:43958–43969, 2001.

343. N Kenmochi, T Suzuki, T Uechi, M Magoori, M Kuniba, S Higa, K Watanabe, T Tanaka. The human mitochondrial ribosomal protein genes: mapping of 54 genes to the chromosomes and implications for human disorders. *Genomics* 77:65–70, 2001.

344. JE Sylvester, N Fischel–Ghodsian, EB Mougey, TW O'Brien. Mitochondrial ribosomal proteins: candidate genes for mitochondrial disease. *Genet Med* 6:73–80, 2004.

345. DF Johnson, M Hamon, N Fischel–Ghodsian. Characterization of the human mitochondrial ribosomal S12 gene. *Genomics* 52:363–368, 1998.

346. O Spirina, Y Bykhovskaya, AV Kajava, TW O'Brien, DP Nierlich, EB Mougey, JE Sylvester, H-R Graack, B Wittmann–Liebold, N Fischel–Ghodsian. Heart-specific splice-variant of a human mitochondrial ribosomal protein (mRNA processing; tissue specific splicing). *Gene* 261:229–234, 2000.

347. Z Zhang, M Gerstein. Identification and characterization of over 100 mitochondrial ribosomal protein pseudogenes in the human genome. *Genomics* 81:468–480, 2003.

348. EC Koc, A Ranasinghe, W Burkhart, K Blackburn, H Koc, A Moseley, LL Spremulli. A new face on apoptosis: death-associated protein 3 and PDCD9 are mitochondrial ribosomal proteins. *FEBS Lett* 492:166–170, 2001.

349. ND Denslow, JC Anders, TW O'Brien. Bovine mitochondrial ribosomes possess a high affinity binding site for guanine nucleotides. *J Biol Chem* 266:9586–9590, 1991.

350. K Grohmann, F Amairic, S Crews, G Attardi. Failure to detect "cap" structures in mitochondrial DNA-coded poly(A)-containing RNA from HeLa cells. *Nucleic Acids Res* 5:637–651, 1978.

351. P Cantatore, Z Flagella, F Fracasso, AM Lezza, MN Gadaleta, A de Montalvo. Synthesis and turnover rates of four rat liver mitochondrial RNA species. *FEBS Lett* 213:144–148, 1987.

352. C Van den Bogert, H De Vries, M Holtrop, P Muus, HL Dekker, MJ Van Galen, PA Bolhuis, J-W Taanman. Regulation of the expression of mitochondrial proteins: relationship between mtDNA copy number and cytochrome-c oxidase activity in human cells and tissues. *Biochim Biophys Acta* 1144:177–183, 1993.

353. HL Garstka, M Fäcke, JR Escribano, RJ Wiesner. Stoichiometry of mitochondrial transcripts and regulation of gene expression by mitochondrial transcription factor A. *Biochem Biophys Res Commun* 200:619–626, 1994.

354. S Naithani, SA Saracco, CA Butler, TD Fox. Interactions among *COX1*, *COX2*, and *COX3* mRNA-specific translational activator proteins on the inner surface of the mitochondrial inner membrane of *Saccharomyces cerevisiae*. *Mol Biol Cell* 14:324–333, 2003.

355. EC Koc, LL Spremulli. RNA-binding proteins of mammalian mitochondria. *Mitochondrion* 2:277–291, 2003.

356. HX Liao, LL Spremulli. Interaction of bovine mitochondrial ribosomes with messenger RNA. *J Biol Chem* 264:7518–7522, 1989.

357. ND Denslow, GS Michaels, J Montoya, G Attardi, TW O'Brien. Mechanism of mRNA binding to bovine mitochondrial ribosomes. *J Biol Chem* 264:8328–8338, 1989.

358. HX Liao, LL Spremulli. Effects of length and mRNA secondary structure on the interaction of bovine mitochondrial ribosomes with messenger RNA. *J Biol Chem* 265:11761–11765, 1990.

359. MA Farwell, J Schirawski, PW Hager, LL Spremulli. Analysis of the interaction between bovine mitochondrial 28 S ribosomal subunits and mRNA. *Biochim Biophys Acta* 1309:122–130, 1996.

360. HX Liao, LL Spremulli. Initiation of protein synthesis in animal mitochondria. Purification and characterization of translational initiation factor 2. *J Biol Chem* 266:20714–20719, 1991.

361. L Ma, LL Spremulli. Cloning and sequence analysis of the human mitochondrial translational initiation factor 2 cDNA. *J Biol Chem* 270:1859–1865, 1995.

362. DS Bonner, JE Wiley, MA Farwell. Assignment1 of the mitochondrial translational initiation factor 2 gene (*MTIF2*) to human chromosome 2 bands p16→p14 by *in situ* hybridization and with somatic cell hybrids. *Cytogenet Cell Genet* 83:80–81, 1998.

363. RG Overman Jr, PJ Enderle, JM Farrow, III, JE Wiley, MA Farwell. The human mitochondrial translation initiation factor 2 gene (*MTIF2*): transcriptional analysis and identification of a pseudogene. *Biochim Biophys Acta* 1628:195–205, 2003.

364. Y Li, WB Holmes, DR Appling, UL RajBhandary. Initiation of protein synthesis in *Saccharomyces cerevisiae* mitochondria without formylation of the initiator tRNA. *J Bacteriol* 182:2886–2892, 2000.

365. AS Tibbetts, L Oesterlin, SY Chan, G Kramer, B Hardesty, DR Appling. Mammalian mitochondrial initiation factor 2 supports yeast mitochondrial translation without formylated initiator tRNA. *J Biol Chem* 278:31774–31780, 2003.

366. EC Koc, LL Spremulli. Identification of mammalian mitochondrial translational initiation factor 3 and examination of its role in initiation complex formation with natural mRNAs. *J Biol Chem* 277:35541–35549, 2002.

367. AP Carter, WM Clemons, Jr., DE Brodersen, RJ Morgan–Warren, T Hartsch, BT Wimberly, V Ramakrishnan. Crystal structure of an initiation factor bound to the 30S ribosomal subunit. *Science* 291:498–501, 2001.

368. D Moazed, RR Samaha, C Gualerzi, HF Noller. Specific protection of 16 S rRNA by translational initiation factors. *J Mol Biol* 248:207–210, 1995.

369. S Brock, K Szkaradkiewicz, M Sprinzl. Initiation factors of protein biosynthesis in bacteria and their structural relationship to elongation and termination factors. *Mol Microbiol* 29:409–417, 1998.

370. C Barker, A Makris, C Patriotis, SE Bear, PN Tsichlis. Identification of the gene encoding the mitochondrial elongation factor G in mammals. *Nucleic Acids Res* 21:2641–2647, 1993.

371. J Wells, F Henkler, M Leversha, R Koshy. A mitochondrial elongation factor-like protein is over-expressed in tumours and differentially expressed in normal tissues. *FEBS Lett* 358:119–125, 1995.

372. H Xin, V Woriax, W Burkhart, LL Spremulli. Cloning and expression of mitochondrial translational elongation factor Ts from bovine and human liver. *J Biol Chem* 270:17243–17249, 1995.

373. VL Woriax, W Burkhart, LL Spremulli. Cloning, sequence analysis and expression of mammalian mitochondrial protein synthesis elongation factor Tu. *Biochim Biophys Acta* 1264:347–356, 1995.

374. M Ling, F Merante, HS Chen, C Duff, AM Duncan, BH Robinson. The human mitochondrial elongation factor tu (EF-Tu) gene: cDNA sequence, genomic localization, genomic structure, and identification of a pseudogene. *Gene* 197:325–336, 1997.

375. J Gao, L Yu, P Zhang, J Jiang, J Chen, J Peng, Y Wei, S Zhao. Cloning and characterization of human and mouse mitochondrial elongation factor G, *GFM* and *Gfm*, and mapping of *GFM* to human chromosome 3q25.1-q26.2. *Genomics* 74:109–114, 2001.

376. M Hammarsund, W Wilson, M Corcoran, M Merup, S Einhorn, D Grandér, O Sangfelt. Identification and characterization of two novel human mitochondrial elongation factor genes, *hEFG2* and *hEFG1*, phylogenetically conserved through evolution. *Hum Genet* 109:542–550, 2001.

377. JL Vernon, PC Burr, JE Wiley, MA Farwell. Assignment of the mitochondrial translation elongation factor Ts gene (*TSFM*) to human chromosome 12 bands q13→q14 by *in situ* hybridization and with somatic cell hybrids. *Cytogenet Cell Genet* 89:145–146, 2000.

378. ZH Shah, V Migliosi, SCM Miller, A Wang, TB Friedman, HT Jacobs. Chromosomal locations of three human nuclear genes (RPSM12, TUFM, and AFG3L1) specifying putative components of the mitochondrial gene expression apparatus. *Genomics* 48:384–388, 1998.

379. KH Nierhaus. Protein synthesis. An elongation factor turn on. *Nature* 379:491–492, 1996.

380. VL Woriax, JM Bullard, L Ma, T Yokogawa, LL Spremulli. Mechanistic studies of the translational elongation cycle in mammalian mitochondria. *Biochim Biophys Acta* 1352:91–101, 1997.

381. Y-C Cai, JM Bullard, NL Thompson, LL Spremulli. Interaction of mammalian mitochondrial elongation factor EF-Tu with guanine nucleotides. *Protein Sci* 9:1791–1800, 2000.

382. Y-C Cai, JM Bullard, NL Thompson, LL Spremulli. Interaction of mitochondrial elongation factor Tu with aminoacyl-tRNA and elongation factor Ts. *J Biol Chem* 275:20308–20314, 2000.

383. GR Andersen, S Thirup, LL Spremulli, J Nyborg. High resolution crystal structure of bovine mitochondrial EF-Tu in complex with GDP. *J Mol Biol* 297:421–436, 2000.

384. ZM Chrzanowska–Lightowlers, RJ Temperley, PM Smith, SH Seneca, RN Lightowlers. Functional polypeptides can be synthesized from human mitochondrial transcripts lacking termination codons. *Biochem J* 377:725–731, 2004.

385. Y Nakamura, K Ito, LA Isaksson. Emerging understanding of translation termination. *Cell* 87:147–150, 1996.

386. L Janosi, S Mottagui–Tabar, LA Isaksson, Y Sekine, E Ohtsubo, S Zhang, S Goon, S Nelken, M Shuda, A Kaji. Evidence for *in vivo* ribosome recycling, the fourth step in protein biosynthesis. *EMBO J* 17:1141–1151, 1998.

387. L Kisselev, M Ehrenberg, L Frolova. Termination of translation: interplay of mRNA, rRNAs and release factors? *EMBO J* 22:175–182, 2003.

388. CC Lee, KM Timms, CN Trotman, WP Tate. Isolation of a rat mitochondrial release factor. Accommodation of the changed genetic code for termination. *J Biol Chem* 262:3548–3552, 1987.

389. G Gadaleta, G Pepe, G De Candia, C Quagliariello, E Sbisà, C Saccone. The complete nucleotide sequence of the *Rattus norvegicus* mitochondrial genome: cryptic signals revealed by comparative analysis between vertebrates. *J Mol Evol* 28:497–516, 1989.

390. Y Zhang, LL Spremulli. Identification and cloning of human mitochondrial translational release factor 1 and the ribosome recycling factor. *Biochim Biophys Acta* 1443:245–250, 1998.

391. LL Hansen, R Jorgensen, J Justesen. Assignment of the human mitochondrial translational release factor 1 (*MTRF1*) to chromosome 13q14.1→q14.3 and of the human mitochondrial ribosome recycling factor (*MRRF*) to chromosome 9q32→q34.1 with radiation hybrid mapping. *Cytogenet Cell Genet* 88:91–92, 2000.

392. T Kanai, S Takeshita, H Atomi, K Umemura, M Ueda, A Tanaka. A regulatory factor, Fil1p, involved in derepression of the isocitrate lyase gene in *Saccharomyces cerevisiae*. A possible mitochondrial protein necessary for protein synthesis in mitochondria. *Eur J Biochem* 256:212–220, 1998.

393. E Teyssier, G Hirokawa, A Tretiakova, B Jameson, A Kaji, H Kaji. Temperature-sensitive mutation in yeast mitochondrial ribosome recycling factor (RRF). *Nucleic Acids Res* 31:4218–4226, 2003.

394. ML Collins, S Eng, R Hoh, MK Hellerstein. Measurement of mitochondrial DNA synthesis *in vivo* using a stable isotope-mass spectrometric technique. *J Appl Physiol* 94:2203–2211, 2003.

395. NJ Gross, GS Getz, M Rabinowitz. Apparent turnover of mitochondrial deoxyribonucleic acid and mitochondrial phospholipids in the tissues of the rat. *J Biol Chem* 244:1552–1562, 1969.

396. PJ Flory, Jr., J Vinograd. 5-bromodeoxyuridine labeling of monomeric and catenated circular mitochondrial DNA in HeLa cells. *J Mol Biol* 74:81–94, 1973.

397. JW Taanman, JR Muddle, AC Muntau. Mitochondrial DNA depletion can be prevented by dGMP and dAMP supplementation in a resting culture of deoxyguanosine kinase-deficient fibroblasts. *Hum Mol Genet* 12:1839–1845, 2003.

398. M Jazayeri, A Andreyev, Y Will, M Ward, CM Anderson, W Clevenger. Inducible expression of a dominant negative DNA polymerase-γ depletes mitochondrial DNA and produces a ρ^0 phenotype. *J Biol Chem* 278:9823–9830, 2003.

399. CW Shearman, GF Kalf. DNA replication by a membrane–DNA complex from rat liver mitochondria. *Arch Biochem Biophys* 182:573–586, 1977.

400. S Meeusen, J Nunnari. Evidence for a two membrane-spanning autonomous mitochondrial DNA replisome. *J Cell Biol* 163:503–510, 2003.

401. DA Clayton. Replication of animal mitochondrial DNA. *Cell* 28:693–705, 1982.

402. AJ Berk, DA Clayton. Mechanism of mitochondrial DNA replication in mouse L-cells: asynchronous replication of strands, segregation of circular daughter molecules, aspects of topology and turnover of an initiation sequence. *J Mol Biol* 86:801–824, 1974.

403. IJ Holt, HE Lorimer, HT Jacobs. Coupled leading- and lagging-strand synthesis of mammalian mitochondrial DNA. *Cell* 100:515–524, 2000.

404. DF Bogenhagen, DA Clayton. The mitochondrial DNA replication bubble has not burst. *Trends Biochem Sci* 28:357–360, 2003.

405. IJ Holt, HT Jacobs. Response: the mitochondrial DNA replication bubble has not burst. *Trends Biochem Sci* 28:355–356, 2003.

406. DF Bogenhagen, DA Clayton. Concluding remarks: the mito-chondrial DNA replication bubble has not burst. *Trends Biochem Sci* 28:404–405, 2003.

407. DL Robberson, H Kasamatsu, J Vinograd. Replication of mito-chondrial DNA. Circular replicative intermediates in mouse L-cells. *Proc Natl Acad Sci USA* 69:737–741, 1972.

408. DP Tapper, DA Clayton. Mechanism of replication of human mitochondrial DNA. Localization of the 5′ ends of nascent daughter strands. *J Biol Chem* 256:5109–5115, 1981.

409. D Kang, K Miyako, Y Kai, T Irie, K Takeshige. *In vivo* determination of replication origins of human mitochondrial DNA by ligation-mediated polymerase chain reaction. *J Biol Chem* 272:15275–15279, 1997.

410. DD Chang, DA Clayton. Priming of human mitochondrial DNA replication occurs at the light-strand promoter. *Proc Natl Acad Sci USA* 82:351–355, 1985.

411. DD Chang, WW Hauswirth, DA Clayton. Replication priming and transcription initiate from precisely the same site in mouse mitochondrial DNA. *EMBO J* 4:1559–1567, 1985.

412. S Crews, D Ojala, J Posakony, J Nishiguchi, G Attardi. Nucleotide sequence of a region of human mitochondrial DNA containing the precisely identified origin of replication. *Nature* 277:192–198, 1979.

413. D Bogenhagen, DA Clayton. Mechanism of mitochondrial DNA replication in mouse L-cells: kinetics of synthesis and turnover of the initiation sequence. *J Mol Biol* 119:49–68, 1978.

414. GS Shadel, DA Clayton. Mitochondrial DNA maintenance in vertebrates. *Annu Rev Biochem* 66:409–435, 1997.

415. JN Doda, CT Wright, DA Clayton. Elongation of displacement-loop strands in human and mouse mitochondrial DNA is arrested near specific template sequences. *Proc Natl Acad Sci USA* 78:6116–6120, 1981.

416. TW Wong, DA Clayton. In vitro replication of human mitochondrial DNA: accurate initiation at the origin of light-strand synthesis. *Cell* 42:951–958, 1985.

417. JE Hixson, TW Wong, DA Clayton. Both the conserved stem-loop and divergent 5'-flanking sequences are required for initiation at the human mitochondrial origin of light-strand DNA replication. *J Biol Chem* 261:2384–2390, 1986.

418. TW Wong, DA Clayton. DNA primase of human mitochondria is associated with structural RNA that is essential for enzymatic activity. *Cell* 45:817–825, 1986.

419. BJ Brewer, WL Fangman. The localization of replication origins on ARS plasmids in *S. cerevisiae*. *Cell* 51:463–471, 1987.

420. AG Mayhook, AM Rinaldi, HT Jacobs. Replication origins and pause sites in sea urchin mitochondrial DNA. *Proc R Soc Lond B Biol Sci* 248:85–94, 1992.

421. Z Han, C Stachow. Analysis of *Schizosaccharomyces pombe* mitochondrial DNA replication by two dimensional gel electrophoresis. *Chromosoma* 103:162–170, 1994.

422. MY Yang, M Bowmaker, A Reyes, L Vergani, P Angeli, E Gringeri, HT Jacobs, IJ Holt. Biased incorporation of ribonucleotides on the mitochondrial L-strand accounts for apparent strand-asymmetric DNA replication. *Cell* 111:495–505, 2002.

423. MMK Nass. Mitochondrial DNA. II. Structure and physicochemical properties of isolated DNA. *J Mol Biol* 42:529–545, 1969.

424. M Miyaki, K Koide, T Ono. RNase and alkali sensitivity of closed circular mitochondrial DNA of rat ascites hepatoma cells. *Biochem Biophys Res Commun* 50:252–258, 1973.

425. F Wong–Staal, J Mendelsohn, M Goulian. Ribonucleotides in closed circular mitochondrial DNA from HeLa cells. *Biochem Biophys Res Commun* 53:140–148, 1973.

426. LI Grossman, R Watson, J Vinograd. The presence of ribonucleotides in mature closed-circular mitochondrial DNA. *Proc Natl Acad Sci USA* 70:3339–3343, 1973.

427. DM Lonsdale, IG Jones. Ribonuclease-sensitivity of covalently closed rat liver mitochondrial deoxyribonucleic acid. *Biochem J* 141:155–158, 1974.

428. A Brennicke, DA Clayton. Nucleotide assignment of alkali-sensitive sites in mouse mitochondrial DNA. *J Biol Chem* 256:10613–10617, 1981.

429. K Koike, DR Wolstenholme. Evidence for discontinuous replication of circular mitochondrial DNA molecules from Novikoff rat ascites hepatoma cells. *J Cell Biol* 61:14–25, 1974.

430. JA Rumbaugh, LA Henricksen, MS DeMott, RA Bambara. Cleavage of substrates with mismatched nucleotides by Flap endonuclease-1. Implications for mammalian Okazaki fragment processing. *J Biol Chem* 274:14602–14608, 1999.

431. SM Cerritelli, EG Frolova, C Feng, A Grinberg, PE Love, RJ Crouch. Failure to produce mitochondrial DNA results in embryonic lethality in *Rnaseh1* null mice. *Mol Cell* 11:807–815, 2003.

432. M Bowmaker, MY Yang, T Yasukawa, A Reyes, HT Jacobs, JA Huberman, IJ Holt. Mammalian mitochondrial DNA replicates bidirectionally from an initiation zone. *J Biol Chem* 278:50961–50969, 2003.

433. D Bogenhagen, DA Clayton. Mechanism of mitochondrial DNA replication in mouse L-cells: introduction of superhelical turns into newly replicated molecules. *J Mol Biol* 119:69–81, 1978.

434. AJ Berk, DA Clayton. Mechanism of mitochondrial DNA replication in mouse L-cells: topology of circular daughter molecules and dynamics of catenated oligomer formation. *J Mol Biol* 100:85–92, 1976.

435. B Xu, DA Clayton. RNA–DNA hybrid formation at the human mitochondrial heavy-strand origin ceases at replication start sites: an implication for RNA–DNA hybrids serving as primers. *EMBO J* 15:3135–3143, 1996.

436. T Ohsato, T Muta, A Fukuoh, H Shinagawa, N Hamasaki, D Kang. R-loop in the replication origin of human mitochondrial DNA is resolved by RecG, a Holliday junction-specific helicase. *Biochem Biophys Res Commun* 255:1–5, 1999.

437. DD Chang, DA Clayton. A novel endoribonuclease cleaves at a priming site of mouse mitochondrial DNA replication. *EMBO J* 6:409–417, 1987.

438. DD Chang, DA Clayton. A mammalian mitochondrial RNA processing activity contains nucleus-encoded RNA. *Science* 235:1178–1184, 1987.

439. DY Lee, DA Clayton. RNase mitochondrial RNA processing correctly cleaves a novel R-loop at the mitochondrial DNA leading-strand origin of replication. *Genes Dev* 11:582–592, 1997.

440. DY Lee, DA Clayton. Initiation of mitochondrial DNA replication by transcription and R-loop processing. *J Biol Chem* 273:30614–30621, 1998.

441. JN Topper, JL Bennett, DA Clayton. A role for RNAase MRP in mitochondrial RNA processing. *Cell* 70:16–20, 1992.

442. GS Shadel, GA Buckenmeyer, DA Clayton, ME Schmitt. Mutational analysis of the RNA component of *Saccharomyces cerevisiae* RNase MRP reveals distinct nuclear phenotypes. *Gene* 245:175–184, 2000.

443. JL Paluh, DA Clayton. A functional dominant mutation in *Schizosaccharomyces pombe* RNase MRP RNA affects nuclear RNA processing and requires the mitochondrial-associated nuclear mutation *ptp1-1* for viability. *EMBO J* 15:4723–4733, 1996.

444. JN Topper, DA Clayton. Characterization of human MRP/Th RNA and its nuclear gene: full length MRP/Th RNA is an active endoribonuclease when assembled as an RNP. *Nucleic Acids Res* 18:793–799, 1990.

445. M Ridanpää, H Van Ecnennaam, K Pelin, R Chadwick, C Johnson, B Yuan, W vanVenrooij, G Pruijn, R Salmela, S Rockas, O Mäkitie, I Kaitila, A de la Chapelle. Mutations in the RNA component of RNase MRP cause a pleiotropic human disease, cartilage-hair hypoplasia. *Cell* 104:195–203, 2001.

446. N Jarrous, JS Wolenski, D Wesolowski, C Lee, S Altman. Localization in the nucleolus and coiled bodies of protein subunits of the ribonucleoprotein ribonuclease P. *J Cell Biol* 146:559–572, 1999.

447. H Van Eenennaam, GJM Pruijn, WJ Van Venrooij. hPop4: a new protein subunit of the human RNase MRP and RNase P ribonucleoprotein complexes. *Nucleic Acids Res* 27:2465–2472, 1999.

448. H Van Eenennaam, A van der Heijden, RJRJ Janssen, WJ Van Venrooij, GJM Pruijn. Basic domains target protein subunits of the RNase MRP complex to the nucleolus independently of complex association. *Mol Biol Cell* 12:3680–3689, 2001.

449. H Van Eenennaam, D Lugtenberg, JHP Vogelzangs, WJ Van Venrooij, GJM Pruijn. hPop5, a protein subunit of the human RNase MRP and RNase P endoribonucleases. *J Biol Chem* 276:31635–31641, 2001.

450. P Schafer, SR Scholz, O Gimadutdinow, IA Cymerman, JM Bujnicki, A Ruiz–Carrillo, A Pingoud, G Meiss. Structural and functional characterization of mitochondrial EndoG, a sugar non-specific nuclease which plays an important role during apoptosis. *J Mol Biol* 338:217–228, 2004.

451. J Côté, A Ruiz–Carrillo. Primers for mitochondrial DNA replication generated by endonuclease G. *Science* 261:765–769, 1993.

452. HP Zassenhaus, TJ Hofmann, R Uthayashanker, RD Vincent, M Zona. Construction of a yeast mutant lacking the mitochondrial nuclease. *Nucleic Acids Res* 16:3283–3296, 1988.

453. J Parrish, L Li, K Klotz, D Ledwich, X Wang, D Xue. Mitochondrial endonuclease G is important for apoptosis in *C. elegans*. *Nature* 412:90–94, 2001.

454. LY Li, X Luo, X Wang. Endonuclease G is an apoptotic DNase when released from mitochondria. *Nature* 412:95–99, 2001.

455. G van Loo, P Schotte, M van Gurp, H Demol, B Hoorelbeke, K Gevaert, I Rodriguez, A Ruiz–Carrillo, J Vandekerckhove, W Declercq, R Beyaert, P Vandenabeele. Endonuclease G: a mitochondrial protein released in apoptosis and involved in caspase-independent DNA degradation. *Cell Death Differ* 8:1136–1142, 2001.

456. T Ohsato, N Ishihara, T Muta, S Umeda, S Ikeda, K Mihara, N Hamasaki, D Kang. Mammalian mitochondrial endonuclease G. Digestion of R-loops and localization in intermembrane space. *Eur J Biochem* 269:5765–5770, 2002.

457. AM Davies, S Hershman, GJ Stabley, JB Hoek, J Peterson, A Cahill. A Ca^{2+}-induced mitochondrial permeability transition causes complete release of rat liver endonuclease G activity from its exclusive location within the mitochondrial intermembrane space. Identification of a novel endo-exonuclease activity residing within the mitochondrial matrix. *Nucleic Acids Res* 31:1364–1373, 2003.

458. J Zhang, M Dong, L Li, Y Fan, P Pathre, J Dong, D Lou, JM Wells, D Olivares–Villagomez, L Van Kaer, X Wang, M Xu. Endonuclease G is required for early embryogenesis and normal apoptosis in mice. *Proc Natl Acad Sci USA* 100:15782–15787, 2003.

459. A Bolden, GP Noy, A Weissbach. DNA polymerase of mitochondria is a γ-polymerase. *J Biol Chem* 252:3351–3356, 1977.

460. U Hubscher, CC Kuenzle, S Spadari. Functional roles of DNA polymerases β and γ. *Proc Natl Acad Sci USA* 76:2316–2320, 1979.

461. P Lestienne. Evidence for a direct role of the DNA polymerase gamma in the replication of the human mitochondrial DNA *in vitro. Biochem Biophys Res Commun* 146:1146–1153, 1987.

462. DK Braithwaite, J Ito. Compilation, alignment, and phylogenetic relationships of DNA polymerases. *Nucleic Acids Res* 21:787–802, 1993.

463. TSF Wang. Eukaryotic DNA polymerases. *Annu Rev Biochem* 60:513–552, 1991.

464. W Lewis, BJ Day, WC Copeland. Mitochondrial toxicity of NRTI antiviral drugs: an integrated cellular perspective. *Nat Rev Drug Discov* 2:812–822, 2003.

465. E Murakami, JY Feng, H Lee, J Hanes, KA Johnson, KS Anderson. Characterization of novel reverse transcriptase and other RNA-associated catalytic activities by human DNA polymerase γ: importance in mitochondrial DNA replication. *J Biol Chem* 278:36403–36409, 2003.

466. RMI Kapsa, AF Quigley, TF Han, MJB Jean–Francois, P Vaughan, E Byrne. mtDNA replicative potential remains constant during ageing: polymerase γ activity does not correlate with age-related cytochrome oxidase activity decline in platelets. *Nucleic Acids Res* 26:4365–4373, 1998.

467. RK Naviaux, D Markusic, BA Barshop, WL Nyhan, RH Haas. Sensitive assay for mitochondrial DNA polymerase γ. *Clin Chem* 45:1725–1733, 1999.

468. H Gray, TW Wong. Purification and identification of subunit structure of the human mitochondrial DNA polymerase. *J Biol Chem* 267:5835–5841, 1992.

469. PA Ropp, WC Copeland. Cloning and characterization of the human mitochondrial DNA polymerase, DNA polymerase γ. *Genomics* 36:449–458, 1996.

470. JA Carrodeguas, DF Bogenhagen. Protein sequences conserved in prokaryotic aminoacyl-tRNA synthetases are important for the activity of the processivity factor of human mitochondrial DNA polymerase. *Nucleic Acids Res* 28:1237–1244, 2000.

471. JA Carrodeguas, K Theis, DF Bogenhagen, C Kisker. Crystal structure and deletion analysis show that the accessory subunit of mammalian DNA polymerase gamma, PolγB, functions as a homodimer. *Mol Cell* 7:43–54, 2001.

472. SE Lim, MJ Longley, WC Copeland. The mitochondrial p55 accessory subunit of human DNA polymerase γ enhances DNA binding, promotes processive DNA synthesis, and confers *N*-ethylmaleimide resistance. *J Biol Chem* 274:38197–38203, 1999.

473. SW Graves, AA Johnson, KA Johnson. Expression, purification, and initial kinetic characterization of the large subunit of the human mitochondrial DNA polymerase. *Biochemistry* 37:6050–6058, 1998.

474. MJ Longley, PA Ropp, SE Lim, WC Copeland. Characterization of the native and recombinant catalytic subunit of human DNA polymerase γ: identification of residues critical for exonuclease activity and dideoxynucleotide sensitivity. *Biochemistry* 37:10529–10539, 1998.

475. AA Johnson, KA Johnson. Exonuclease proofreading by human mitochondrial DNA polymerase. *J Biol Chem* 276:38097–38107, 2001.

476. AA Johnson, Y Tsai, SW Graves, KA Johnson. Human mitochondrial DNA polymerase holoenzyme: reconstitution and characterization. *Biochemistry* 39:1702–1708, 2000.

477. F Foury. Cloning and sequencing of the nuclear gene *MIP1* encoding the catalytic subunit of the yeast mitochondrial DNA polymerase. *J Biol Chem* 264:20552–20560, 1989.

478. N Lecrenier, P Van Der Bruggen, F Foury. Mitochondrial DNA polymerases from yeast to man: a new family of polymerases. *Gene* 185:147–152, 1997.

479. S Vanderstraeten, S Van den Brûle, J Hu, F Foury. The role of 3'-5' exonucleolytic proofreading and mismatch repair in yeast mitochondrial DNA error avoidance. *J Biol Chem* 273:23690–23697, 1998.

480. JN Spelbrink, JM Toivonen, GAJ Hakkaart, JM Kurkela, HM Cooper, SK Lehtinen, N Lecrenier, JW Back, D Speijer, F Foury, HT Jacobs. *In vivo* functional analysis of the human mitochondrial DNA polymerase POLG expressed in cultured human cells. *J Biol Chem* 275:24818–24828, 2000.

481. MV Ponamarev, MJ Longley, D Nguyen, TA Kunkel, WC Copeland. Active site mutation in DNA polymerase γ associated with progressive external ophthalmoplegia causes error-prone DNA synthesis. *J Biol Chem* 277:15225–15228, 2002.

482. G Van Goethem, B Dermaut, A Löfgren, J-J Martin, C Van Broeckhoven. Mutation of *POLG* is associated with progressive external ophthalmoplegia characterized by mtDNA deletions. *Nat Genet* 28:211–212, 2001.

483. E Lamantea, V Tiranti, A Bordoni, A Toscano, F Bono, S Servidei, A Papadimitriou, H Spelbrink, L Silvestri, G Casari, GP Comi, M Zeviani. Mutations of mitochondrial DNA polymerase γA are a frequent cause of autosomal dominant or recessive progressive external ophthalmoplegia.*Ann Neurol* 52:211–219, 2002.

484. A Agostino, L Valletta, PF Chinnery, G Ferrari, F Carrara, RW Taylor, AM Schaefer, DM Turnbull, V Tiranti, M Zeviani. Mutations of *ANT1*, *Twinkle*, and *POLG1* in sporadic progressive external ophthalmoplegia (PEO). *Neurology* 60:1354–1356, 2003.

485. S Wanrooij, P Luoma, G Van Goethem, C Van Broeckhoven, A Suomalainen, JN Spelbrink. Twinkle and POLG defects enhance age-dependent accumulation of mutations in the control region of mtDNA. *Nucleic Acids Res* 32:3053–3064, 2004.

486. R Del Bo, A Bordoni, M Sciacco, A Di Fonzo, S Galbiati, M Crimi, N Bresolin, GP Comi. Remarkable infidelity of polymerase γA associated with mutations in *POLG1* exonuclease domain. *Neurology* 61:903–908, 2003.

487. RK Naviaux, KV Nguyen. *POLG* mutations associated with Alpers' syndrome and mitochondrial DNA depletion. *Ann Neurol* 55:706–712, 2004.

488. BN Harding. Progressive neuronal degeneration of childhood with liver disease (Alpers–Huttenlocher syndrome): a personal review. *J Child Neurol* 5:273–287, 1990.

489. RK Naviaux, WL Nyhan, BA Barshop, J Poulton, D Markusic, NC Karpinski, RH Haas. Mitochondrial DNA polymerase γ deficiency and mtDNA depletion in a child with Alpers' syndrome. *Ann Neurol* 45:54–58, 1999.

490. AT Rovio, DR Marchington, S Donat, H-C Schuppe, J Abel, E Fritsche, DJ Elliott, P Laippala, AL Ahola, D McNay, RF Harrison, B Hughes, T Barrett, DMD Bailey, D Mehmet, AM Jequier, TB Hargreave, S-H Kao, JM Cummins, DE Barton, HJ Cooke, Y-H Wei, L Wichmann, J Poulton, HT Jacobs. Mutations at the mitochondrial DNA polymerase (*POLG*) locus associated with male infertility. *Nat Genet* 29:261–262, 2001.

491. M Jensen, H Leffers, JH Petersen, AA Nyboe, N Jørgensen, E Carlsen, T Kold Jensen, NE Skakkebæk, E Rajpert–De Meyts. Frequent polymorphism of the mitochondrial DNA polymerase gamma gene (*POLG*) in patients with normal spermiograms and unexplained subfertility. *Hum Reprod* 19:65–70, 2004.

492. JM Cummins, AM Jequier, R Kan. Molecular biology of human male infertility: links with aging, mitochondrial genetics, and oxidative stress? *Mol Reprod Dev* 37:345–362, 1994.

493. B Iyengar, N Luo, CL Farr, LS Kaguni, AR Campos. The accessory subunit of DNA polymerase γ is essential for mitochondrial DNA maintenance and development in *Drosophila melanogaster*. *Proc Natl Acad Sci USA* 99:4483–4488, 2002.

494. JA Carrodeguas, KG Pinz, DF Bogenhagen. DNA binding properties of human pol γB. *J Biol Chem* 277:50008–50014, 2002.

495. R Genuario, TW Wong. Stimulation of DNA polymerase γ by a mitochondrial single-strand DNA binding protein. *Cell Mol Biol Res* 39:625–634, 1993.

496. GC Van Tuyle, PA Pavco. The rat liver mitochondrial DNA–protein complex: displaced single strands of replicative intermediates are protein coated. *J Cell Biol* 100:251–257, 1985.

497. C Takamatsu, S Umeda, T Ohsato, T Ohno, Y Abe, A Fukuoh, H Shinagawa, N Hamasaki, D Kang. Regulation of mitochondrial D-loops by transcription factor A and single-stranded DNA-binding protein. *EMBO Rep* 3:451–456, 2002.

498. E Van Dyck, F Foury, B Stillman, SJ Brill. A single-stranded DNA binding protein required for mitochondrial DNA replication in *S. cerevisiae* is homologous to *E. coli* SSB. *EMBO J* 11:3421–3430, 1992.

499. D Maier, CL Farr, B Poeck, A Alahari, M Vogel, S Fischer, LS Kaguni, S Schneuwly. Mitochondrial single-stranded DNA-binding protein is required for mitochondrial DNA replication and development in *Drosophila melanogaster. Mol Biol Cell* 12:821–830, 2001.

500. CL Farr, Y Matsushima, AT Lagina III, N Luo, LS Kaguni. Physiological and biochemical defects in functional interactions of mitochondrial DNA polymerase and DNA-binding mutants of single-stranded DNA-binding protein. *J Biol Chem* 279:17047–17053, 2004.

501. V Tiranti, M Rocchi, S DiDonato, M Zeviani. Cloning of human and rat cDNAs encoding the mitochondrial single-stranded DNA-binding protein (SSB). *Gene* 126:219–225, 1993.

502. U Curth, C Urbanke, J Greipel, H Gerberding, V Tiranti, M Zeviani. Single-stranded-DNA-binding proteins from human mitochondria and *Escherichia coli* have analogous physico-chemical properties. *Eur J Biochem* 221:435–443, 1994.

503. G Webster, J Genschel, U Curth, C Urbanke, C Kang, R Hilgenfeld. A common core for binding single-stranded DNA: structural comparison of the single-stranded DNA-binding proteins (SSB) from *E. coli* and human mitochondria. *FEBS Lett* 411:313–316, 1997.

504. VS Mikhailov, DF Bogenhagen. Effects of *Xenopus laevis* mitochondrial single-stranded DNA-binding protein on primer-template binding and $3' \rightarrow 5'$ exonuclease activity of DNA polymerase γ. *J Biol Chem* 271:18939–18946, 1996.

505. C Yang, U Curth, C Urbanke, C Kang. Crystal structure of human mitochondrial single-stranded DNA binding protein at 2.4 Å resolution. *Nat Struct Biol* 4:153–157, 1997.

506. SW Matson, KA Kaiser–Rogers. DNA helicases. *Annu Rev Biochem* 59:289–329, 1990.

507. S Lewis, W Hutchison, D Thyagarajan, H-HM Dahl. Clinical and molecular features of adPEO due to mutations in the Twinkle gene. *J Neurol Sci* 201:39–44, 2002.

508. S Guo, S Tabor, CC Richardson. The linker region between the helicase and primase domains of the bacteriophage T7 gene 4 protein is critical for hexamer formation. *J Biol Chem* 274:30303–30309, 1999.

509. JA Korhonen, M Gaspari, M Falkenberg. TWINKLE has 5' → 3' DNA helicase activity and is specifically stimulated by mitochondrial single-stranded DNA-binding protein. *J Biol Chem* 278:48627–48632, 2003.

510. JA Korhonen, XH Pham, M Pellegrini, M Falkenberg. Reconstitution of a minimal mtDNA replisome *in vitro*. *EMBO J* 23:2423–2429, 2004.

511. JC Wang. DNA topoisomerases. *Annu Rev Biochem* 65:635–692, 1996.

512. KD Corbett, JM Berger. Structure, molecular mechanisms, and evolutionary relationships in DNA topoisomerases. *Annu Rev Biophys Biomol Struc* 33:95–118, 2004.

513. JJ Champoux. DNA topoisomerases: structure, function, and mechanism. *Annu Rev Biochem* 70:369–413, 2001.

514. FR Fairfield, WR Bauer, MV Simpson. Mitochondria contain a distinct DNA topoisomerase. *J Biol Chem* 254:9352–9354, 1979.

515. FR Fairfield, WR Bauer, MV Simpson. Studies on mitochondrial type I topoisomerase and on its function. *Biochim Biophys Acta* 824:45–57, 1985.

516. JH Lin, FJ Castora. Response of purified mitochondrial DNA topoisomerase I from bovine liver to camptothecin and m-AMSA. *Arch Biochem Biophys* 324:293–299, 1995.

517. GM Lazarus, JP Henrich, WG Kelly, SA Schmitz, FJ Castora. Purification and characterization of a type I DNA topoisomerase from calf thymus mitochondria. *Biochemistry* 26:6195–6203, 1987.

518. JH Lin, GM Lazarus, FJ Castora. DNA topoisomerase I from calf thymus mitochondria is associated with a DNA binding, inner membrane protein. *Arch Biochem Biophys* 293:201–207, 1992.

519. Z Topcu, FJ Castora. Mammalian mitochondrial DNA topoisomerase I preferentially relaxes supercoils in plasmids containing specific mitochondrial DNA sequences. *Biochim Biophys Acta* 1264:377–387, 1995.

520. FJ Castora, GM Lazarus, D Kunes. The presence of two mitochondrial DNA topoisomerases in human acute leukemia cells. *Biochem Biophys Res Commun* 130:854–866, 1985.

521. FJ Castora, WG Kelly. ATP inhibits nuclear and mitochondrial type I topoisomerases from human leukemia cells. *Proc Natl Acad Sci USA* 83:1680–1684, 1986.

522. MJ Kosovsky, G Soslau. Mitochondrial DNA topoisomerase I from human platelets. *Biochim Biophys Acta* 1078:56–62, 1991.

523. MJ Kosovsky, G Soslau. Immunological identification of human platelet mitochondrial DNA topoisomerase I. *Biochim Biophys Acta* 1164:101–107, 1993.

524. H Zhang, JM Barceló, B Lee, G Kohlhagen, DB Zimonjic, NC Popescu, Y Pommier. Human mitochondrial topoisomerase I. *Proc Natl Acad Sci USA* 98:10608–10613, 2001.

525. H Zhang, L-H Meng, DB Zimonjic, NC Popescu, Y Pommier. Thirteen-exon-motif signature for vertebrate nuclear and mitochondrial type IB topoisomerases. *Nucleic Acids Res* 32:2087–2092, 2004.

526. L Wu, SL Davies, PS North, H Goulaouic, J-F Riou, H Turley, KC Gatter, ID Hickson. The Bloom's syndrome gene product interacts with topoisomerase III. *J Biol Chem* 275:9636–9644, 2000.

527. FB Johnson, DB Lombard, NF Neff, MA Mastrangelo, W Dewolf, NA Ellis, RA Marciniak, Y Yin, R Jaenisch, L Guarente. Association of the Bloom syndrome protein with topoisomerase IIIα in somatic and meiotic cells. *Cancer Res* 60:1162–1167, 2000.

528. Y Wang, YL Lyu, JC Wang. Dual localization of human DNA topoisomerase IIIα to mitochondria and nucleus. *Proc Natl Acad Sci USA* 99:12114–12119, 2002.

529. R Hanai, PR Caron, JC Wang. Human *TOP3*: a single-copy gene encoding DNA topoisomerase III. *Proc Natl Acad Sci USA* 93:3653–3657, 1996.

530. FG Harmon, RJ DiGate, SC Kowalczykowski. RecQ helicase and topoisomerase III comprise a novel DNA strand passage function: a conserved mechanism for control of DNA recombination. *Mol Cell* 3:611–620, 1999.

531. FJ Castora, MV Simpson. Search for a DNA gyrase in mammalian mitochondria. *J Biol Chem* 254:11193–11195, 1979.

532. JH Lin, FJ Castora. DNA topoisomerase II from mammalian mitochondria is inhibited by the antitumor drugs, m-AMSA and VM-26. *Biochem Biophys Res Commun* 176:690–697, 1991.

533. RL Low, S Orton, DB Friedman. A truncated form of DNA topoisomerase IIβ associates with the mtDNA genome in mammalian mitochondria. *Eur J Biochem* 270:4173–4186, 2003.

534. PM Watt, ID Hickson. Structure and function of type II DNA topoisomerases. *Biochem J* 303:681–695, 1994.

535. CA Austin, KL Marsh. Eukaryotic DNA topoisomerase IIβ. *Bioessays* 20:215–226, 1998.

536. IV Martin, SA MacNeill. ATP-dependent DNA ligases. *Genome Biol 3*: reviews 3005.1–3005.7, 2002.

537. AE Tomkinson, DS Levin. Mammalian DNA ligases. *Bioessays* 19:893–901, 1997.

538. S Shuman, B Schwer. RNA capping enzyme and DNA ligase: a superfamily of covalent nucleotidyl transferases. *Mol Microbiol* 17:405–410, 1995.

539. AP Johnson, MP Fairman. The identification and purification of a novel mammalian DNA ligase. *Mutat Res* 383:205–212, 1997.

540. JH Petrini, Y Xiao, DT Weaver. DNA ligase I mediates essential functions in mammalian cells. *Mol Cell Biol* 15:4303–4308, 1995.

541. C Prigent, MS Satoh, G Daly, DE Barnes, T Lindahl. Aberrant DNA repair and DNA replication due to an inherited enzymatic defect in human DNA ligase I. *Mol Cell Biol* 14:310–317, 1994.

542. U Grawunder, M Wilm, X Wu, P Kulesza, TE Wilson, M Mann, MR Lieber. Activity of DNA ligase IV stimulated by complex formation with XRCC4 protein in mammalian cells. *Nature* 388:492–495, 1997.

543. TE Wilson, U Grawunder, MR Lieber. Yeast DNA ligase IV mediates nonhomologous DNA end joining. *Nature* 388:495–498, 1997.

544. ZB Mackey, W Ramos, DS Levin, CA Walter, JR McCarrey, AE Tomkinson. An alternative splicing event which occurs in mouse pachytene spermatocytes generates a form of DNA ligase III with distinct biochemical properties that may function in meiotic recombination. *Mol Cell Biol* 17:989–998, 1997.

545. RA Nash, KW Caldecott, DE Barnes, T Lindahl. XRCC1 protein interacts with one of two distinct forms of DNA ligase III. *Biochemistry* 36:5207–5211, 1997.

546. RM Perez–Jannotti, SM Klein, DF Bogenhagen. Two forms of mitochondrial DNA ligase III are produced in *Xenopus laevis* oocytes. *J Biol Chem* 276:48978–48987, 2001.

547. CJ Levin, SB Zimmerman. A DNA ligase from mitochondria of rat liver. *Biochem Biophys Res Commun* 69:514–520, 1976.

548. U Lakshmipathy, C Campbell. The human DNA ligase III gene encodes nuclear and mitochondrial proteins. *Mol Cell Biol* 19:3869–3876, 1999.

549. U Lakshmipathy, C Campbell. Antisense-mediated decrease in DNA ligase III expression results in reduced mitochondrial DNA integrity. *Nucleic Acids Res* 29:668–676, 2001.

550. KG Pinz, DF Bogenhagen. Efficient repair of abasic sites in DNA by mitochondrial enzymes. *Mol Cell Biol* 18:1257–1265, 1998.

551. U Lakshmipathy, C Campbell. Mitochondrial DNA ligase III function is independent of Xrcc1. *Nucleic Acids Res* 28:3880–3886, 2000.

552. M Willer, M Rainey, T Pullen, CJ Stirling. The yeast *CDC9* gene encodes both a nuclear and a mitochondrial form of DNA ligase I. *Curr Biol* 9:1085–1094, 1999.

553. SL Donahue, BE Corner, L Bordone, C Campbell. Mitochondrial DNA ligase function in *Saccharomyces cerevisiae. Nucleic Acids Res* 29:1582–1589, 2001.

554. RS Williams, S Salmons, EA Newsholme, RE Kaufman, J Mellor. Regulation of nuclear and mitochondrial gene expression by contractile activity in skeletal muscle. *J Biol Chem* 261:376–380, 1986.

555. P Nagley. Coordination of gene expression in the formation of mammalian mitochondria. *Trends Genet* 7:1–4, 1991.

556. RJ Wiesner. Regulation of mitochondrial gene expression: transcription vs. replication. *Trends Genet* 8:264–265, 1992.

557. RS Williams. Mitochondrial gene expression in mammalian striated muscle. Evidence that variation in gene dosage is the major regulatory event. *J Biol Chem* 261:12390–12394, 1986.

558. Y Tang, EA Schon, E Wilichowski, ME Vazquez–Memije, E Davidson, MP King. Rearrangements of human mitochondrial DNA (mtDNA): new insights into the regulation of mtDNA copy number and gene expression. *Mol Biol Cell* 11:1471–1485, 2000.

559. H Mandel, R Szargel, V Labay, O Elpeleg, A Saada, A Shalata, Y Anbinder, D Berkowitz, C Hartman, M Barak, S Eriksson, N Cohen. The deoxyguanosine kinase gene is mutated in individuals with depleted hepatocerebral mitochondrial DNA. *Nat Genet* 29:337–341, 2001.

560. A Saada, A Shaag, H Mandel, Y Nevo, S Eriksson, O Elpeleg. Mutant mitochondrial thymidine kinase in mitochondrial DNA depletion myopathy. *Nat Genet* 29:342–344, 2001.

561. J-W Taanman, I Kateeb, AC Muntau, M Jaksch, N Cohen, H Mandel. A novel mutation in the deoxyguanosine kinase gene causing depletion of mitochondrial DNA. *Ann Neurol* 52:237–239, 2002.

562. L Salviati, S Sacconi, M Mancuso, D Otaegui, P Camano, A Marina, S Rabinowitz, R Shiffman, K Thompson, CM Wilson, A Feigenbaum, AB Naini, M Hirano, E Bonilla, S DiMauro, TH Vu. Mitochondrial DNA depletion and dGK gene mutations. *Ann Neurol* 52:311–317, 2002.

563. M Mancuso, L Salviati, S Sacconi, D Otaegui, P Camano, A Marina, S Bacman, CT Moraes, JR Carlo, M Garcia, M Garcia–Alvarez, L Monzon, AB Naini, M Hirano, E Bonilla, AL Taratuto, S DiMauro, TH Vu. Mitochondrial DNA depletion: mutations in thymidine kinase gene with myopathy and SMA. *Neurology* 59:1197–1202, 2002.

564. G Pontarin, L Gallinaro, P Ferraro, P Reichard, V Bianchi. Origins of mitochondrial thymidine triphosphate: dynamic relations to cytosolic pools. *Proc Natl Acad Sci USA* 100:12159–12164, 2003.

565. RC Scarpulla. Nuclear activators and coactivators in mammalian mitochondrial biogenesis. *Biochim Biophys Acta* 1576:1–14, 2002.

566. H Esterbauer, H Oberkofler, F Krempler, W Patsch. Human peroxisome proliferator activated receptor gamma coactivator 1 (*PPARGC1*) gene: cDNA sequence, genomic organization, chromosomal localization, and tissue expression. *Genomics* 62:98–102, 1999.

567. A Meirhaeghe, V Crowley, C Lenaghan, C Lelliott, K Green, A Stewart, K Hart, S Schinner, J Sethi, G Yco, M Brand, R Cortright, S O'Rahilly, C Montague, A Vidal–Puig. Characterization of the human, mouse and rat PGC1β (peroxisome-proliferator-activated receptor-γ co-activator 1β) gene *in vitro* and *in vivo*. *Biochem J* 373:155–165, 2003.

568. DP Kelly, RC Scarpulla. Transcriptional regulatory circuits controlling mitochondrial biogenesis and function. *Genes Dev* 18:357–368, 2004.

569. P Puigserver, Z Wu, CW Park, R Graves, M Wright, BM Spiegelman. A cold-inducible coactivator of nuclear receptors linked to adaptive thermogenesis. *Cell* 92:829–839, 1998.

570. Z Wu, P Puigserver, U Andersson, C Zhang, G Adelmant, V Mootha, A Troy, S Cinti, B Lowell, RC Scarpulla, BM Spiegelman. Mechanisms controlling mitochondrial biogenesis and respiration through the thermogenic coactivator PGC-1. *Cell* 98:115–124, 1999.

571. J Lin, P Puigserver, J Donovan, P Tarr, BM Spiegelman. Peroxisome proliferator-activated receptor γ coactivator 1β (PGC-1β), a novel PGC-1-related transcription coactivator associated with host cell factor. *J Biol Chem* 277:1645–1648, 2002.

572. D Kressler, SN Schreiber, D Knutti, A Kralli. The PGC-1-related protein PERC is a selective coactivator of estrogen receptor α. *J Biol Chem* 277:13918–13925, 2002.

573. U Andersson, RC Scarpulla. PGC-1-related coactivator, a novel, serum-inducible coactivator of nuclear respiratory factor 1-dependent transcription in mammalian cells. *Mol Cell Biol* 21:3738–3749, 2001.

574. Y Kai, K Miyako, T Muta, S Umeda, T Irie, N Hamasaki, K Takeshige, D Kang. Mitochondrial DNA replication in human T lymphocytes is regulated primarily at the H-strand termination site. *Biochim Biophys Acta* 1446:126–134, 1999.

575. TA Brown, DA Clayton. Release of replication termination controls mitochondrial DNA copy number after depletion with 2′,3′-dideoxycytidine. *Nucleic Acids Res* 30:2004–2010, 2002.

576. M Roberti, C Musicco, PL Polosa, F Milella, MN Gadaleta, P Cantatore. Multiple protein-binding sites in the TAS-region of human and rat mitochondrial DNA. *Biochem Biophys Res Commun* 243:36–40, 1998.

577. CS Madsen, SC Ghivizzani, WW Hauswirth. Protein binding to a single termination-associated sequence in the mitochondrial DNA D-loop region. *Mol Cell Biol* 13:2162–2171, 1993.

578. CT Moraes. What regulates mitochondrial DNA copy number in animal cells? *Trends Genet* 17:199–205, 2001.

579. RA Schultz, SJ Swoap, LD McDaniel, B Zhang, EC Koon, DJ Garry, K Li, RS Williams. Differential expression of mitochondrial DNA replication factors in mammalian tissues. *J Biol Chem* 273:3447–3451, 1998.

580. Y Matsushima, R Garesse, LS Kaguni. *Drosophila* mitochondrial transcription factor B2 regulates mitochondrial DNA copy number and transcription in Schneider cells. *J Biol Chem* 279:26900–26905, 2004.

581. JFX Diffley, B Stillman. A close relative of the nuclear, chromosomal high-mobility group protein HMG1 in yeast mitochondria. *Proc Natl Acad Sci USA* 88:7864–7868, 1991.

582. TL Megraw, CB Chae. Functional complementarity between the HMG1-like yeast mitochondrial histone HM and the bacterial histone-like protein HU. *J Biol Chem* 268:12758–12763, 1993.

583. MA Parisi, B Xu, DA Clayton. A human mitochondrial transcriptional activator can functionally replace a yeast mitochondrial HMG-box protein both *in vivo* and *in vitro*. *Mol Cell Biol* 13:1951–1961, 1993.

584. RP Fisher, T Lisowsky, GA Breen, DA Clayton. A rapid, efficient method for purifying DNA-binding proteins. Denaturation–renaturation chromatography of human and yeast mitochondrial extracts. *J Biol Chem* 266:9153–9160, 1991.

585. TI Alam, T Kanki, T Muta, K Ukaji, Y Abe, H Nakayama, K Takio, N Hamasaki, D Kang. Human mitochondrial DNA is packaged with TFAM. *Nucleic Acids Res* 31:1640–1645, 2003.

586. LR Brewer, R Friddle, A Noy, E Baldwin, SS Martin, M Corzett, R Balhorn, RJ Baskin. Packaging of single DNA molecules by the yeast mitochondrial protein Abf2p. *Biophys J* 85:2519–2524, 2003.

587. RW Friddle, JE Klare, SS Martin, M Corzett, R Balhorn, EP Baldwin, RJ Baskin, A Noy. Mechanism of DNA compaction by yeast mitochondrial protein Abf2p. *Biophys J* 86:1632–1639, 2004.

588. N-G Larsson, A Oldfors, E Holme, DA Clayton. Low levels of mitochondrial transcription factor A in mitochondrial DNA depletion. *Biochem Biophys Res Commun* 200:1374–1381, 1994.

589. J Poulton, K Morten, C Freeman–Emmerson, C Potter, C Sewry, V Dubowitz, H Kidd, J Stephenson, W Whitehouse, FJ Hansen, M Parisi, G Brown. Deficiency of the human mitochondrial transcription factor h-mtTFA in infantile mitochondrial myopathy is associated with mtDNA depletion. *Hum Mol Genet* 3:1763–1769, 1994.

590. AF Davis, PA Ropp, DA Clayton, WC Copeland. Mitochondrial DNA polymerase γ is expressed and translated in the absence of mitochondrial DNA maintenance and replication. *Nucleic Acids Res* 24:2753–2759, 1996.

591. BL Seidel–Rogol, GS Shadel. Modulation of mitochondrial transcription in response to mtDNA depletion and repletion in HeLa cells. *Nucleic Acids Res* 30:1929–1934, 2002.

592. A Goto, Y Matsushima, T Kadowaki, Y Kitagawa. *Drosophila* mitochondrial transcription factor A (*d*-TFAM) is dispensable for the transcription of mitochondrial DNA in Kc167 cells. *Biochem J* 354:243–248, 2001.

593. Y Matsushima, K Matsumura, S Ishii, H Inagaki, T Suzuki, Y Matsuda, K Beck, Y Kitagawa. Functional domains of chicken mitochondrial transcription factor A for the maintenance of mitochondrial DNA copy number in lymphoma cell line DT40. *J Biol Chem* 278:31149–31158, 2003.

594. N-G Larsson, J Wang, H Wilhelmsson, A Oldfors, P Rustin, M Lewandoski, GS Barsh, DA Clayton. Mitochondrial transcription factor A is necessary for mtDNA maintenance and embryogenesis in mice. *Nat Genet* 18:231–236, 1998.

595. MI Ekstrand, M Falkenberg, A Rantanen, CB Park, M Gaspari, K Hultenby, P Rustin, CM Gustafsson, N-G Larsson. Mitochondrial transcription factor A regulates mtDNA copy number in mammals. *Hum Mol Genet* 13:935–944, 2004.

596. JV Virbasius, RC Scarpulla. Activation of the human mitochondrial transcription factor A gene by nuclear respiratory factors: a potential regulatory link between nuclear and mitochondrial gene expression in organelle biogenesis. *Proc Natl Acad Sci USA* 91:1309–1313, 1994.

597. AMS Lezza, V Pesce, A Cormio, F Fracasso, J Vecchiet, G Felzani, P Cantatore, MN Gadaleta. Increased expression of mitochondrial transcription factor A and nuclear respiratory factor-1 in skeletal muscle from aged human subjects. *FEBS Lett* 501:74–78, 2001.

598. L Huo, RC Scarpulla. Mitochondrial DNA instability and peri-implantation lethality associated with targeted disruption of nuclear respiratory factor 1 in mice. *Mol Cell Biol* 21:644–654, 2001.

599. N-G Larsson, A Oldfors, JD Garman, GS Barsh, DA Clayton. Down-regulation of mitochondrial transcription factor A during spermatogenesis in humans. *Hum Mol Genet* 6:185–191, 1997.

600. N-G Larsson, JD Garman, A Oldfors, GS Barsh, DA Clayton. A single mouse gene encodes the mitochondrial transcription factor A and a testis-specific nuclear HMG-box protein. *Nat Genet* 13:296–302, 1996.

601. A Rantanen, M Jansson, A Oldfors, N-G Larsson. Downregulation of Tfam and mtDNA copy number during mammalian spermatogenesis. *Mamm Genome* 12:787–792, 2001.

602. C Díez–Sánchez, E Ruiz–Pesini, AC Lapeña, J Montoya, A Pérez–Martos, JA Enríquez, MJ López–Pérez. Mitochondrial DNA content of human spermatozoa. *Biol Reprod* 68:180–185, 2003.

603. C Florentz, B Sohm, P Tryoen–Tóth, J Pütz, M Sissler. Human mitochondrial tRNAs in health and disease. *Cell Mol Life Sci* 60:1356–1375, 2003.

604. J W Taanman, AG Rodnar, JM Cooper, AAM Morris, PT Clayton, JV Leonard, RHV Schapira. Molecular mechanisms in mitochondrial DNA depletion syndrome. *Hum Mol Genet* 6:935–942, 1997.

605. FJ Miller, FL Rosenfeldt, C Zhang, AW Linnane, P Nagley. Precise determination of mitochondrial DNA copy number in human skeletal and cardiac muscle by a PCR-based assay: lack of change of copy number with age. *Nucleic Acids Res* 31:e61, 2003.

606. C Barthélémy, H Ogier de Baulny, J Diaz, MA Cheval, P Frachon, N Romero, F Goutieres, M Fardeau, A Lombès. Late-onset mitochondrial DNA depletion: DNA copy number, multiple deletions, and compensation. *Ann Neurol* 49:607–617, 2001.

607. B Chabi, B Mousson de Camaret, H Duborjal, J-P Issartel, G Stepien. Quantification of mitochondrial DNA deletion, depletion, and overreplication: application to diagnosis. *Clin Chem* 49:1309–1317, 2003.

608. RC Shuster, AJ Rubenstein, DC Wallace. Mitochondrial DNA in anucleate human blood cells. *Biochem Biophys Res Commun* 155:1360–1365, 1988.

609. G Manfredi, D Thyagarajan, LC Papadopoulou, F Pallotti, EA Schon. The fate of human sperm-derived mtDNA in somatic cells. *Am J Hum Genet* 61:953–960, 1997.

610. ED Robin, R Wong. Mitochondrial DNA molecules and virtual number of mitochondria per cell in mammalian cells. *J Cell Physiol* 136:507–513, 1988.

611. RJ Shmookler Reis, S Goldstein. Mitochondrial DNA in mortal and immortal human cells. Genome number, integrity, and methylation. *J Biol Chem* 258:9078–9085, 1983.

612. D Bogenhagen, DA Clayton. The number of mitochondrial deoxyribonucleic acid genomes in mouse L- and human HeLa cells. Quantitative isolation of mitochondrial deoxyribonucleic acid. *J Biol Chem* 249:7991–7995, 1974.

613. MP King, G Attardi. Human cells lacking mtDNA: repopulation with exogenous mitochondria by complementation. *Science* 246:500–503, 1989.

4

Oxidative Phosphorylation Disease: Diagnosis and Pathogenesis

JOHN M. SHOFFNER

CONTENTS

Normal ATP generation by oxidative phosphorylation (OXPHOS) is a complex process requiring the coordinate expression of two genomes: the nuclear DNA (nDNA) and the mitochondrial DNA (mtDNA). Much of our knowledge as well as many recent questions concerning how the nDNA and mtDNA interact comes from the detailed clinical, biochemical, and genetic analysis of OXPHOS diseases. The first pathogenetic mutations in the mtDNA (1–3) were discovered approximately 7 years after the complete human mtDNA sequence was published (4). This relatively long lag time reflected, in part, some of the controversies involved in recognizing OXPHOS disease phenotypes and the complexities of mtDNA analysis. Since that time, the number of pathogenic mtDNA mutations and nuclear DNA mutations has increased dramatically, resulting in a deeper understanding of how OXPHOS genetics apply to human disease.

The diagnosis of OXPHOS diseases is a technically challenging and time-consuming process. Over the past several years, the term "mitochondrial medicine" emerged to encompass the complex synthesis of clinical, biochemical, pathological, and genetic information required for patient diagnosis. The multiorgan involvement and clinical heterogeneity of these diseases can make diagnosis difficult, thus emphasizing the importance of a well-organized approach to these patients.

OXIDATIVE PHOSPHORYLATION BIOCHEMISTRY AND GENETICS

Mitochondria are cytoplasmic structures of about 0.1 to 0.5 µm in diameter, with an inner and outer membrane separated by an intermembrane space. The outer membrane is permeable to most small molecules and ions and it contains a variety of proteins such as monoamine oxidase, long-chain acyl-CoA synthetase, carnitine palmitoyl transferase 1 (CPT1), and mitochondrial protein import proteins. The inner mitochondrial membrane is impermeable to most metabolites. It has a convoluted structure with multiple folds called cristae. The inner membrane has a high content of protein and cardiolipin. It contains the enzymes of oxidative phosphorylation as well as multiple classes of translocases. The space surrounded by the inner mitochondrial membrane, called the mitochondrial matrix, contains, in addition to mtDNA, an array of enzymes including those for:

- The Kreb cycle (tricarboxylic acid cycle)
- The pyruvate dehydrogenase complex (PDC)
- β-Oxidation of fatty acids
- Ketone metabolism
- Amino acid metabolism
- Heme metabolism
- Nucleotide metabolism
- The peptidases plus chaperonins necessary for mitochondrial protein import and OXPHOS enzyme assembly and maintenance

Oxidative phosphorylation (OXPHOS) is an oxygen-dependent biochemical process localized to the mitochondrial inner membrane that produces most of the adenosine triphosphate (ATP) required by cells for normal function. A complex array of nuclear DNA and mitochondrial DNA (mtDNA) genes operates coordinately to produce functional OXPHOS enzymes. The mtDNA is a 16,569-nucleotide pair, double-stranded, circular molecule that codes for two ribosomal RNAs (rRNA), 22 transfer RNAs (tRNA), and 13 of the 82 structural proteins of the mitochondrial electron transport chain.

Four highly complex enzymes, referred to as the respiratory chain, receive electrons from the catabolism of carbohydrates, fats, and proteins in order to generate a proton gradient across the inner mitochondrial membrane (Figure 4.1). Complex I and complex II (succinate dehydrogenase) collect electrons from the catabolism of fats, proteins, and carbohydrates and transfer them sequentially to coenzyme Q_{10}, complex III, and complex IV (cytochrome c oxidase). Complex II is also part of the tricarboxylic acid cycle (Kreb cycle).

These four enzyme complexes contain flavins, coenzyme Q_{10} (ubiquinone), iron–sulfur clusters, hemes, and protein-bound copper. Complexes I, III, and IV utilize the energy in electron transfer to pump protons across the inner mitochondrial membrane, producing a proton gradient used by complex V (ATP synthase) to condense ADP and inorganic phosphate into ATP. The adenine nucleotide translocase (ANT) delivers ATP to the cytoplasm in exchange for ADP.

As discussed later, the determination of OXPHOS enzyme activities is central to the diagnosis of patients with OXPHOS diseases. Due to the complexity of these enzymes, as well as their biochemical reactions, the clinical application of diagnostic assays makes patient diagnosis extremely difficult and requires extensive experience for accurate interpretation of assay results. Complex I is the largest enzyme in the respiratory chain and is composed of 42 or 43 protein subunits of which 7 are encoded by the mtDNA. One flavin mononucleotide, seven or eight iron–sulfur (FeS) centers, covalently bound lipid, and at least three bound coenzyme Q_{10} molecules are also essential elements of this enzyme. Complex I is an enormous enzyme complex that is over 900 kD — approximately the size of a ribosome.

Complex II (succinate dehydrogenase) contains four subunits coded by the nuclear DNA plus one flavin mononucleotide, and several FeS centers. It is attached to the inner

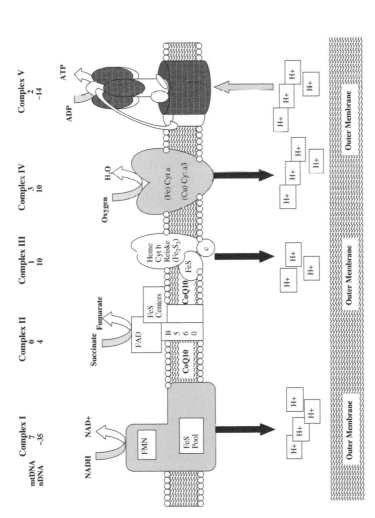

Figure 4.1 Oxidative phosphorylation. Electrons are collected from carbohydrate, fatty acid, and protein oxidation by oxidative phosphorylation. Complexes I, III, and IV pump protons into the space between the inner and outer mitochondrial membranes. This stored energy is used by complex V to generate ATP. Abbreviations: mtDNA, mitochondrial DNA-coded subunits; nDNA, nuclear DNA-coded subunits; iron–sulfur groups, FeS and Reiske (Fe$_2$S$_2$); reduced and oxidized nicotinamide adenine nucleotide, NADH and NAD+. Cytochromes are B560, Cyt a, Cyt a3, Cyt b, and c; coenzyme Q10, CoQ10; flavin mononucleotide, FMN; flavin adenine dinucleotide, FAD; adenosine diphosphate, ADP; adenosine triphosphate, ATP; protons, H+.

membrane by a b-type cytochrome. Complex III is composed of 11 subunits, only 1 of which is coded by the mtDNA (cytochrome b). The metalloprotein core of complex III is essential to the proton-pumping function of this enzyme and is composed of cytochrome b, two hemes, a membrane-attached FeS protein with a Rieske type center (Fe_2S_2), and a membrane-attached cytochrome c_1. An interesting feature of complex III is that this enzyme may participate in mitochondrial protein import via the two core proteins that face the mitochondrial matrix. These core proteins are homologous to mitochondrial processing peptidases.

Complex IV, which is more commonly known as cytochrome c oxidase, contains 13 subunits. Three of the subunits (subunits I, II, and III) are coded by the mtDNA and form the functional core of complex IV. Cytochrome c is a heme protein that donates electrons to complex IV at the cytoplasmic side of the inner mitochondrial membrane. The active site of complex IV contains a heme iron (cytochrome a) and a copper (cytochrome a_3) that are associated with two of the mtDNA-coded subunits (subunits I and II). The function of subunit III is unclear. Complex V (ATP synthase) in mammals contains as many as 16 different subunits; 2 are coded by the mtDNA (ATP6 and ATP8). These polypeptide subunits form three main components: a membrane sector (F_0) that contains the proton channel, a catalytic component (F_1) that can synthesize ATP or hydrolyze ATP, and a stalk structure around which the catalytic component (F_1) rotates in a fashion analogous to a small motor. The fundamental details about how ATP is synthesized by complex V are still unclear.

Oxidative Phosphorylation Disease Genetics

Because genes for OXPHOS are located in two distinct genomes, the inheritance of OXPHOS diseases may occur by maternal or Mendelian (autosomal dominant, autosomal recessive, X-linked) patterns. Sporadic mutations in the mtDNA or the nDNA (5) may also produce OXPHOS disease.

The location of the mtDNA within the mitochondria is associated with a unique inheritance pattern called maternal inheritance, which refers to the *nearly* exclusive transmission of mtDNAs from a mother to her children. Sperm mtDNA disappear in early embryogenesis by selective destruction, inactivation, or dilution by the approximately 150,000-oocyte mtDNAs (6). Rarely, mtDNA can be passed paternally as was observed in a patient with a mitochondrial myopathy associated with a two-base-pair deletion in the ND2 polypeptide of complex I. This mutation occurred on the paternal mtDNA background and was present in the patient as a mixture of maternal and paternal mtDNAs (7,8). In addition to these cases, paternal mtDNA has been identified in abnormal embryos (9).

When a pathogenic mtDNA mutation is present, it is transmitted in a *homoplasmic* or *heteroplasmic* fashion. The mtDNAs within a cell or tissue are referred to as homoplasmic when all the mtDNAs share the same sequence and are referred to as heteroplasmic when mtDNAs with different sequences coexist. Normal and mutant mtDNA sequences differ only at the nucleotide or nucleotides that have been mutated. As a general rule, pathogenic mtDNA mutations can be homoplasmic or heteroplasmic, whereas neutral polymorphisms are almost always homoplasmic. Segregation of the normal and mutant mtDNAs with cell division is a complex process influenced by a number of variables, including the specific mtDNA mutation present, the nuclear DNA background, and the specific cell type. Consequently, the precise ratio of mutant to normal mtDNA does not correlate well with the clinical phenotype or the severity of the biochemical defect. Three loci in the nuclear DNA on chromosomes 2, 5, and 6 of a mouse model appear to account for approximately 12 to 36% of the variance in mtDNA segregation (10).

OXIDATIVE PHOSPHORYLATION DISEASE
DIAGNOSIS

OXPHOS diseases are characterized by several hundred phenotypes. Due to the extreme clinical heterogeneity inherent in these diseases, classification is based on the disease mechanism. The major categories are produced by mutations in nDNA genes or in mtDNA genes. Mutations in mtDNA produce defects in OXPHOS enzyme polypeptides or in mitochondrial protein synthesis. Nuclear gene mutations produce defects in

- OXPHOS enzyme subunits
- Proteins responsible for OXPHOS enzyme assembly
- Proteins that incorporate metals into OXPHOS enzymes
- Proteins that regulate intergenomic communication
- Cofactors that transfer electrons between OXPHOS enzymes
- Proteins responsible for movement of mitochondria within cells
- Proteins that deliver ATP to the cytoplasm

Figure 4.2 summarizes the major sites of cellular dysfunction that produce OXPHOS diseases.

The basic steps for patient diagnosis involve phenotype recognition, metabolic testing, muscle histology, and immunohistochemistry, OXPHOS enzyme analysis, and, in some patients, mtDNA or nDNA analysis. OXPHOS disease diagnosis is difficult and requires a careful evaluation to exclude other classes of disease that can produce secondary OXPHOS defects. A mitochondrial disease specialist can be helpful in assessing the complex clinical, biochemical, and genetic data for appropriate diagnosis as well as for construction of a management plan.

Phenotype Recognition

Phenotype recognition is complex and may require assessment by an individual who specializes in mitochondrial disease. (Shoffner (11) offers a detailed discussion of OXPHOS disease phenotypes.) Symptoms can be confined to a single organ (monosymptomatic), be present in a few organs (oligosymptomatic), or be systemic. The course may be degenerative or relatively static over time.

Central nervous system dysfunction is very common in pediatric OXPHOS diseases. Cognitive abnormalities are

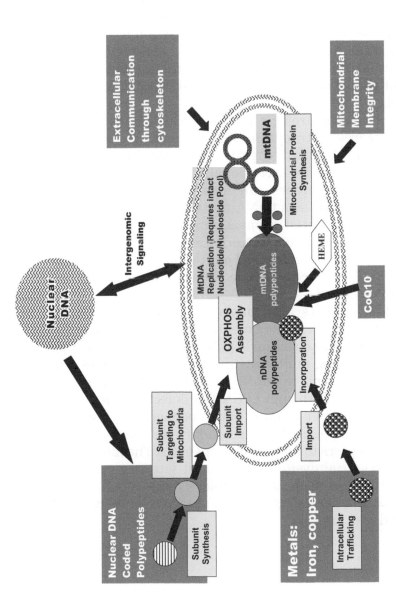

Figure 4.2 Mechanisms for oxidative phosphorylation disease pathogenesis. OXPHOS disease is produced by a broad array of mechanisms that impair various intracellular processes. Pathways known to be impaired are represented in the diagram. Refer to the text for a detailed description of each.

common in patients with OXPHOS diseases and are most likely related to the high degree of reliance of the brain on OXPHOS for normal functioning. As in so many clinical manifestations of OXPHOS diseases, the cognitive abnormalities are heterogenous. In infants and children, the features are often nonspecific, consisting of developmental delays, mental retardation, and, in some cases, developmental regression. Many patients with OXPHOS disease will have normal intelligence when initially evaluated. It is unclear whether normal intellectual function is maintained throughout the lives of most of these patients because long-term follow-up data are limited.

Frequent neurologic manifestations include abnormal tone, seizures, extrapyramidal movements, and autonomic dysfunction. Isolated cognitive defects in patients with OXPHOS disease are exceptional. Features resembling autistic spectrum disorders can also be observed in the rare patients who harbor mtDNA mutations (12). Although most patients with autistic spectrum disorders do not have OXPHOS disease, our experience suggests that atypical patients who have evidence of multiorgan involvement, abnormal brain MRIs, or metabolic abnormalities should be carefully investigated.

Adult patients also display a large array of cognitive symptoms that can include mental retardation, psychosis (13), and dementia. Like pediatric patients, adults usually demonstrate a complex, multisystem disorder. The age of onset of cognitive symptoms in adults with OXPHOS diseases is usually <50 years. Although OXPHOS defects do not appear to play a primary role in most neurodegenerative diseases with onset after 50 years of age, secondary defects may be important contributors to disease pathogenesis. The most likely role of mitochondrial dysfunction in most late onset neurodegenerative diseases such as Alzheimer's disease is to predispose individuals to neurodegeneration or to accelerate cell dysfunction and loss (14–17).

Metabolic Testing

Screening for evidence of an OXPHOS disease is difficult. Metabolic testing can be helpful in obtaining data that support further investigation. Abnormalities in oxidative phosphorylation can produce identifiable defects in related metabolic pathways such as glycolysis, pyruvate metabolism, the tricarboxylic acid cycle, protein catabolism, and fatty acid oxidation. Although the quantitation of organic acids and amino acids in blood, urine, and cerebrospinal fluid can provide useful diagnostic information, normal test results do not exclude the diagnosis. Metabolic acidosis as well as elevations of lactate, pyruvate, lactate/pyruvate ratio (>20), alanine, tricarboxylic acid cycle intermediates, dicarboxylic acids, and/or a generalized amino aciduria can be important diagnostic indicators of an OXPHOS disease. Excretion of carnitine esters or decreased dietary carnitine intake may be associated with reduced blood and tissue carnitine levels.

CSF lactate and pyruvate levels are an important aspect of patient diagnosis (18). Patients can have lactate and pyruvate abnormalities confined to CSF. A 24-h urine collection is useful because it can provide an integrated evaluation of organic and amino acids as well as insight into the function of the highly OXPHOS-dependent proximal renal tubules. Although this is easily accomplished in adults, it is difficult in pediatric patients and spot urine collection is used. In order to enhance the accuracy of quantitative organic and amino acid analysis as well as the ability to compare serial determinations reliably, the blood is collected as a morning sample after an overnight fast. Arterial blood may also be used when questions arise concerning the validity of the venous measurement.

Skeletal Muscle Pathology in OXPHOS Diseases

Most patients suspected of having an OXPHOS disease will require a muscle biopsy. Pathological analysis of the muscle biopsy by histochemistry and immunohistochemistry can be helpful in supporting the diagnosis of such a disease. In most cases, electron microscopy is of little diagnostic utility. The modified Gomori stain, succinate dehydrogenase reaction, and cytochrome c oxidase reaction are essential to the histochemical investigation of patients for oxidative phosphorylation diseases.

Ragged-red fibers are identified by using the modified Gomori stain and are characterized by proliferation of subsarcolemmal and intermyofibrillar mitochondria plus myofibril degeneration. This histological finding almost always signifies the presence of mtDNA. Although this finding can be highly suggestive of an OXPHOS disease, the majority of individuals with such diseases will not have ragged-red fibers, reflecting the large percentage of patients who harbor nDNA mutations in OXPHOS genes. In children who harbor mtDNA mutations that produce ragged-red fibers, this change rarely develops in muscle before about 5 years of age. Ragged-red fibers accumulate with advancing age. The histology of most muscle biopsies will show nonspecific abnormalities that include neurogenic changes, internal nuclei, fiber splitting, myofiber hypertrophy or hypotrophy involving type I or type II fibers, accumulations of lipid, or mild increases in glycogen (19,20). In some individuals the muscle histology may even be normal.

In patients harboring mtDNA mutations, the percentage of ragged-red fibers shows large interindividual variation, ranging from approximately 2 to 70% of the total fibers. Most ragged-red fibers also show an increased succinate dehydrogenase reaction and a decreased cytochrome c oxidase (COX) reaction (COX-negative or COX-deficient fibers). Frequently, the number of COX-negative or COX-deficient fibers is larger than the number of ragged-red fibers. These changes are observed in a wide variety of OXPHOS diseases that include mtDNA depletion diseases, mtDNA deletions, and mt transfer RNA (tRNA) mutations.

The mt tRNA A3243G mutation in the tRNA[leucine(UUR)] gene, which is the most common cause for mitochondrial myopathy, encephalopathy, lactic acidosis, and stroke-like episodes (MELAS), is an important exception. The ragged-red fibers may be COX deficient or show a normal COX reaction. In addition, the blood vessels often show an increased succinate dehydrogenase reaction. It is important to note that COX-deficient fibers, increased succinate dehydrogenase reaction, and ragged-red fibers may be observed in a variety of conditions, including normal aging, zidovudine myopathy, myotonic dystrophy, limb-girdle dystrophy, inclusion body myositis, inflammatory myopathies, and nemaline myopathy.

Because few patients with OXPHOS disease will show characteristic histochemical changes such as cytochrome c oxidase-deficient fibers and ragged-red fibers, more sensitive approaches are required for patient diagnosis. OXPHOS subunit immunohistochemistry is helpful in identifying patients with OXPHOS diseases and assessing which individuals may harbor an mtDNA mutation (21–26). For example, patients who harbor mtDNA mutations that impair mitochondrial protein synthesis will show scattered myofibers, with a decrease in the mtDNA coded subunits I and II of cytochrome c oxidase (complex IV) and a decrease in the closely associated nuclear DNA-coded subunit VIc. Nuclear DNA-coded subunits in other parts of the enzyme will appear normal or mildly reduced.

OXPHOS Biochemistry

OXPHOS enzymology is essential to patient diagnosis. In order to perform accurate assessments of this delicate enzyme system, immediate isolation of mitochondria from fresh muscle biopsies is the most useful approach (27,28). This approach avoids artifacts in OXPHOS enzyme analysis that can be associated with freezing the biopsy prior to mitochondrial isolation, as well as delays in processing the muscle biopsy. As with all the steps of patient evaluation, it is important to exclude other classes of diseases that can mimic OXPHOS diseases by producing increases in lactate, cytochrome c oxidase-deficient fibers, and OXPHOS enzyme defects. Examples of diseases that can produce secondary defects in OXPHOS are Krabbe disease (galactosylceramide β-galactosidase deficiency), molybdemum cofactor deficiency, Alexander disease, and carbohydrate-deficient glycoprotein disease (29,30).

To determine the specific activities of OXPHOS enzymes, the complex I, complex III, and complex IV assays are used to assess electron flow across single OXPHOS complexes; the complex I+III and complex II+III assays assess the movement of electrons between complexes. The specificity of these assays is demonstrated by using specific respiratory inhibitors. The proper functioning of the reagents used in these assays is ensured by performing each enzyme assay in mitochondria isolated from mouse muscle, rat skeletal muscle, or human muscle when available in parallel with the patient assays.

Complex I and complex V (ATP synthase) are the most difficult OXPHOS enzymes to assay. Distinguishing between pathogenic defects and those produced by technical factors can be challenging. In order to assess complex I function, our laboratory uses three complex I assays, which require different electron acceptors:

- n-DecylCoQ as the electron acceptor
- coQ1 as the electron acceptor
- The traditional complex I+III assay

The first two assays are the most specific for mitochondrial complex I activity, but CoQ reduction probably occurs at different sites due to the more hydrophilic nature of CoQ1 and the more lipophilic nature of n-decylCoQ (31). Quality control measures must be rigorously followed for the n-decyl-CoQ assay because this reagent is not widely available and preparations that yield low specific activity measurements can be encountered. The complex I+III assays measure the rate of electron flow between complexes I and III. However, approximately 50% of the observed activity is nonmitochondrial and must be accounted for in the interpretation. Accurate assessment of complex V (ATP synthase) activity requires isolation of intact mitochondria. This assay is not commonly performed due to the frequency of false positives encountered during clinical testing. Maintaining the proton gradient across the mitochondrial inner membrane can be difficult and can produce false positive results.

Due to the complexities associated with oxidative phosphorylation assessment, corroborative data are sought by testing for secondary abnormalities in skin fibroblast β-oxidation. Abnormal OXPHOS function in fibroblasts is associated with impairment of β-oxidation (32). β-Oxidation interfaces with oxidative phosphorylation by donating electrons to coenzyme Q_{10} via the electron transfer flavoprotein (ETF). In addition, functional assemblies between β-oxidation and oxidative phosphorylation may exist (33).

In patients with OXPHOS diseases, β-oxidation of longer chain fatty acids such as palmitate (C16:0) and myristate (C14:0) is often reduced. We observed β-oxidation defects in conjunction with OXPHOS defects (usually complex I defects) in about 24% of our patients. Abnormal metabolism of long chain fatty acids is detrimental to mitochondrial function (34–37) and may contribute to the disease manifestations in patients with OXPHOS disease. In patients with abnormalities in long chain fatty acid oxidation, reductions of these fatty acids in the diet may be helpful.

Genetic Testing for OXPHOS Diseases

At the time of muscle biopsy, a small portion of the biopsy is frozen in liquid nitrogen for DNA isolation. An enormous number of mtDNA mutations (see Appendix 2) are known, along with a growing list of nDNA mutations. Muscle mtDNA is first tested for mtDNA deletions and duplications. Depending on the phenotype, only the most frequently encountered mtDNA mutations are tested. Most mtDNA or nDNA mutations are private or semiprivate mutations (i.e., occurring in relatively few families). To exclude an mtDNA mutation, sequencing of the mtDNA is often necessary, thus permitting assignment of the patient's disease manifestations to the nDNA or the mtDNA. This is a time-consuming process.

Due to complexities associated with recognition of heteroplasmic mtDNA mutations during sequencing, as well as the large array of mtDNA polymorphisms present, the mtDNA sequence must be viewed by an individual with extensive experience in mtDNA sequence interpretation. Pathogenic point mutations in mitochondrial transfer RNA genes appear to be the most commonly encountered mutations in pediatric as well as adult patients (38–40). The yield of genetic testing depends on the age of the patient at presentation, clinical features, family history, and OXPHOS enzyme abnormality. However, assessment of the likelihood of identifying a mutation can be difficult to assign precisely in a clinical setting because maternal transmission may be difficult to delineate conclusively and the phenotype can be nonspecific.

OXIDATIVE PHOSPHORYLATION DISEASE CLASSIFICATION

Due to the extreme clinical heterogeneity inherent in OXPHOS diseases, classification is based on the disease mechanism. Figure 4.2 summarizes the major sites of cellular dysfunction that produce OXPHOS diseases. The major categories are produced by mutations in nDNA genes or mtDNA genes. Mutations in mtDNA produce defects in OXPHOS enzyme polypeptides or in mitochondrial protein synthesis. Nuclear gene mutations produce defects in

- OXPHOS enzyme subunits
- Proteins responsible for OXPHOS enzyme assembly
- Proteins that incorporate metals into OXPHOS enzymes
- Proteins that regulate intergenomic communication
- Cofactors that transfer electrons between OXPHOS enzymes
- Proteins responsible for movement of mitochondria within cells
- Proteins that deliver ATP to the cytoplasm

MTDNA MUTATIONS AND OXPHOS DISEASE

Defects in OXPHOS Enzyme Polypeptides Coded by mtDNA

A large number of mutations with diverse phenotypes are caused by defects in OXPHOS enzyme polypeptides coded by the mtDNA. Examples of this class of mutations include Leber hereditary optic neuropathy (LHON), Leigh disease, and some types of mitochondrial myopathies.

The first mtDNA point mutation was discovered in a maternally inherited disorder called LHON (3). About 95% of LHON cases are caused by mutations involving complex I subunits:

- G11,778A (ND4 subunit) accounts for approximately 69% of cases.
- G3,460A (ND1 subunit) accounts for approximately 13% of cases.
- T14,484C (ND6) accounts for approximately 14% of cases.

Rare mutations involving mtDNA-coded subunits ND5 and ND6 account for the majority of the remaining cases. For unclear reasons, only about 50% of males and about 10% of females develop vision loss, suggesting that nuclear genes or environmental factors alter disease penetrance.

Leigh disease or subacute necrotizing encephalopathy is suspected when cranial nerve abnormalities, breathing abnormalities, and ataxia are observed in conjunction with bilateral hyperintense signals on T2-weighted MRI images in the basal ganglia, thalamus, cerebellum, or brainstem. The Leigh disease phenotype represents a severe defect in ATP generation by OXPHOS and can be caused by a variety of mechanisms. Gene mutations can occur in mitochondrial transfer RNA genes, mtDNA, or nDNA coded polypeptide subunits of OXPHOS enzymes; nDNA genes responsible for intergenomic communication; and nDNA genes responsible for OXPHOS enzyme assembly (39–64). MtDNA mutations that cause Leigh disease occur in mtDNA-coded polypeptides for complex I (ND5 and ND6 subunits) (49,59,65); complex IV (cytochrome c oxidase subunit III) (61); and complex V (ATP6 subunit) (66,67). Mutations in mtDNA appear to account for approximately 7 to 20% of Leigh disease cases (reference 55 and unpublished results).

Mitochondrial myopathies can be caused by mutations in OXPHOS polypeptides coded by the mtDNA. This group of patients is highly heterogeneous and often develops symptoms in other organs over time. However, in order to classify a patient as one with mitochondrial myopathy, myopathic features must dominate the clinical picture. Patients present with muscle weakness that is usually most prominent in proximal muscles, fatigability, exercise intolerance, and abnormal anaerobic threshold measurements. Some individuals will experience myoglobinuria.

Complex III defects due to mutations in the mtDNA-coded cytochrome b gene are a particularly interesting class of polypeptide mutation that usually produces manifestations only in muscle (68,69). Patients harboring this class of mtDNA mutations can be mistakenly diagnosed as having chronic fatigue syndrome or fibromyalgia. Numerous mutations in the cytochrome b gene can cause mitochondrial myopathy and include point mutations (G14846A, G15059A, G15084A, G15168A, G15761A, G15762A, G15723A) and intragenic deletions (24-bp deletion from nucleotide 15498 to 15521) (70–73). In most cases, these cytochrome b mutations are sporadic and appear to be restricted to muscle, thus suggesting that they are somatic mutations in myogenic stem cells after germ-layer differentiation.

DEFECTS IN MITOCHONDRIAL PROTEIN SYNTHESIS

MtDNA Mutations Involving Polypeptide and Protein Synthesis Genes: mtDNA Deletions and Duplications

Large deletions were the first class of mtDNA mutations discovered (1). The mtDNA deletion mutation has the simplest structure, consisting of an mtDNA molecule missing contiguous tRNA and protein-coding genes, thus yielding an mtDNA molecule smaller than the normal 16.6-kb mtDNA. Shortly after the discovery of mtDNA deletion mutants, mtDNAs with duplicated genes were found to coexist with the deleted mtDNAs (74). The structurally more complex mtDNA duplication mutation produces an mtDNA molecule larger than the normal mtDNA and contains two tandemly arranged mtDNA molecules consisting of a full-length 16.6-kb mtDNA coupled to an mtDNA deletion. The duplicated mtDNA appears to be a precursor to the deleted mtDNA and is found in lower concentrations within patient tissues (75). In most cases, the deletions/duplications occur sporadically during oogenesis or during early embryogenesis. In rare cases, the mtDNA deletion/duplication mutants are maternally transmitted (76,77).

Because the mutant mtDNAs are distributed to essentially all tissues in a clonal fashion, a multisystem disorder is produced. For reasons that are not clear, patients who harbor mtDNA deletions and duplications have the most easily recognized and consistent phenotypes among patients with OXPHOS diseases. The most common clinical features are ptosis, ophthalmoplegia, and ragged-red fiber myopathy. This triad of manifestations is highly predictive for the presence of a large mtDNA deletion. Patients with these manifestations can be classified into one of three groups according to their age of onset and the severity of their clinical symptoms.

The most severe variant is Kearns–Sayre syndrome, which is characterized by infantile, childhood, or adolescent onset of disease manifestations and significant multisystem involvement that can include cardiac abnormalities (cardiomyopathies and cardiac conduction defects), diabetes mellitus, cerebellar ataxia, deafness, and evidence of multifocal neurodegeneration. In almost all patients with Kearns–Sayre syndrome, significant increases in CSF protein are present — often over 100 mg/dl. The CSF protein originates from plasma and is due to increased permeability of the blood–brain barrier at the choroids plexus (78). Abnormalities in the choroids plexus are also responsible for the decreased CSF folate observed in these patients.

Like most OXPHOS diseases, cognitive abnormalities occur across a broad phenotypic spectrum. Cognitive features include normal intellectual function, developmental delays in children, mental retardation, learning disorders, autistic features, and dementia. Rarely, patients will present in infancy with an atypical variant called Pearson's syndrome. These individuals manifest anemia, leukopenia, and thrombocytopenia, resulting in frequent transfusions due to high concentrations of the mtDNA deletions in the hematopoietic lineages. Exocrine pancreas dysfunction may also be present. Patients with Pearson's syndrome may have severe systemic manifestations or may be oligosymptomatic. However, if patients survive infancy and early childhood, Kearns–Sayre syndrome develops.

Chronic progressive external ophthalmoplegia (CPEO) plus refers to a disorder of intermediate severity that has an adolescent or adult onset and variable involvement of tissues other than the eyelids and eye muscles. The mildest variant is isolated CPEO in which clinical signs and symptoms develop during adulthood and are limited to the eyelids and eye muscles. In each of these classification groups, patients worsen with age. Individuals who are initially classified as isolated CPEO can progress to CPEO plus.

Mitochondrial Protein Synthesis Defects: Transfer RNA and Ribosomal RNA Mutations

The mtDNA has a unique genetic code. In the nuclear DNA genetic code, 61 of the 64 codons specify the 20 amino acids, and 3 codons are used for termination signals. In the mtDNA, the codon–anticodon pairing is simplified so that only 22 tRNAs are required to read the mitochondrial genetic code. One mitochondrial tRNA can recognize an entire four-member codon family rather than the two tRNAs per codon family required by the nDNA genetic code. These alterations in the mtDNA genetic code render the mtDNA genes incompatible with the nucleus–cytosol system. Consequently, the mtDNA must code tRNAs and ribosomal RNAs (rRNA) that can accommodate these changes. Mutations in the mitochondrial tRNAs and rRNAs produce a highly diverse array of phenotypes. For a detailed discussion of these phenotype-genotype associations, see Shoffner (11).

Although patients with OXPHOS disease can have mutations in any of the 22 tRNAs, the most frequently encountered point mutations are in tRNA genes at positions 3243 of the tRNA [leucine(UUR)] gene and 8344 of the tRNA[lysine] gene. For reasons that are not understood, tRNA mutations have a tendency to produce certain patterns of clinical manifestations, making their presence clinically recognizable. The two best examples of disease produced by defects in mitochondrial protein synthesis are myoclonic epilepsy and ragged-red fiber disease (MERRF) and mitochondrial myopathy, encephalopathy, lactic acidosis, and stroke-like episodes (MELAS).

MERRF is a maternally inherited disease characterized by progressive myoclonic epilepsy, a mitochondrial myopathy with ragged-red fibers, hearing loss, ataxia, and slowly progressive dementia. Approximately 80 to 90% of MERRF cases are caused by a heteroplasmic G to A point mutation at position 8344 in the transfer RNA[lysine] gene. MELAS is a progressive, neurodegenerative disease characterized by stroke-like episodes, seizures, ataxia, hearing loss, diabetes mellitus, and a ragged-red fiber myopathy. Approximately 80% of individuals with the clinical characteristics of MELAS have an heteroplasmic A to G point mutation at position 3243 of the tRNA[leucine(UUR)] gene.

A small percentage of patients with OXPHOS diseases harbor defects in mitochondrial protein synthesis caused by mutation in the 12S ribosomal RNA subunit. Point mutations at positions 1555 (79) and 961 (80) of the 12S ribosomal RNA gene are responsible for increased sensitivity to aminoglycosides. Patients exposed to aminoglycosides experience acute onset of severe high-frequency hearing loss. These mutations appear to enhance the binding of aminoglycosides to the 12S rRNA by increasing the structural similarity to the homologous region of the bacterial rRNA, thus impairing mitochondrial protein synthesis. Family members who are not exposed to aminoglycosides generally show a gradual onset of hearing loss with advancing age. An unusual phenotype caused by the mutation at position 1555 includes a maternally inherited cardiomyopathy (81) and Parkinsonism, deafness, and neuropathy (82).

NUCLEAR DNA MUTATIONS AND OXPHOS DISEASE

Nuclear DNA Mutations in OXPHOS Subunits

Autosomal recessive mutations in polypeptide subunits of OXPHOS enzymes are described only for complex I and complex II (succinate dehydrogenase). Complex I defects are the most commonly encountered OXPHOS defect. Composed of approximately 43 subunits (36 nuclear-coded polypeptides and 7 mtDNA-coded polypeptides), complex I is the largest of the OXPHOS enzyme complexes. Complex I can be fragmented into a hydrophobic group of proteins, a flavoprotein group, and an iron–sulfur group.

In a small number of patients studied to date who harbor complex I defects, mutations in nuclear-coded complex I subunits were identified in about 40% and mutations in subunits coded by the mtDNA in about 5 to 10% (83). These mutations were identified in patients with severe OXPHOS disease with a neurodegenerative course such as Leigh disease. A mitochondrial encephalopathy with profound defects in cognitive and motor development was present in all patients. A defect in complex I was present in skeletal muscle. The enzymology did not distinguish between patients with nDNA and mtDNA mutations. Metabolic testing yielded variable results; some patients showed characteristic increases in lactate in blood, urine, and CSF, but others showed normal levels of these metabolites.

Mutations in the flavoprotein group gene, NDUFV1, disrupt electron flow by impairing the NADH- and FMN-binding sites of complex I. Patients present with manifestations that include Leigh disease, macrocephaly, blindness, spasticity, regression, and myoclonic epilepsy (41,84). Mutations in the iron–sulfur protein genes, NDUFS1 and NDUFS2, produce complex I defects by impairing electron flow through the iron–sulfur groups. Patient manifestations included Leigh disease, leukodystrophy, growth retardation, macrocytic anemia, neonatal lactic acidosis, plus cardiomyopathy (41,52). Mutations in genes that code proteins for the hydrophobic group were found in the NDUFS4 gene (43,53,63,85–87), the NDUFS7 gene (63), and the NDUFS8 gene (88,89). Clinical manifestations included Leigh disease and Leigh disease plus hypertrophic cardiomyopathy.

Complex II (succinate dehydrogenase) occupies a unique position in cellular energetics. This enzyme is an integral part of OXPHOS because it functions to transfer electron to ubiquinone for OXPHOS as well as serving as an integral part of the Krebs cycle (tricarboxylic acid cycle). It catalyzes the oxidation of succinate to fumarate in the cycle. Complex II contains four subunits coded by the nDNA; these are referred to as the flavoprotein subunit (SDHA), the iron–sulfur protein subunit (SDHB), and the two integral membrane proteins, SDHC and SDHD. Mutations in this enzyme are rare. Mutations that produce mitochondrial encephalomyopathies have been found in the flavoprotein (SDHA) and iron–sulfur subunits (SDHB) (90–94). These mutations are transmitted in an autosomal recessive pattern. Some of the clinical presentations are entirely consistent with OXPHOS disease and include Leigh disease, leukodystrophy, and late onset neurodegenerative disorders with ataxia, optic atrophy, and myopathy (44,92,95,96).

Complex II and another Krebs cycle enzyme, fumerase, can harbor mutations that predispose individuals to the development of certain types of tumors. Pheochromocytomas arising in adrenal or extra-adrenal sites and paragangliomas of the head and neck, most commonly involving the carotid bodies, can occur sporadically or be hereditary. Germline mutations in the SDHB, SDHC, and SDHD subunits are the most common cause of familial pheochromocytomas; they account for as many as 70% of cases, whereas mutations in these subunits are identified in about 8% of sporadic tumors (97–99). Most familial paragangliomas are caused by SDHD mutations (100,101). The mechanism by which SDH subunit mutations predispose to pheochromocytomas and paragangliomas has not been defined in detail, but dysregulation of hypoxia-responsive genes and impairment of mitochondria-mediated apoptosis are possible.

OXPHOS Cofactor Defects

Coenzyme Q_{10} transfers electrons from complex I and complex II to complex III. Coenzyme Q_{10} deficiency is an important cause of autosomal recessive mitochondrial encephalomyopathies, which can be treated by oral supplementation of coenzyme Q_{10}. Patients present with a variety of symptoms ranging from predominantly myopathic forms (with recurrent myoglobinuria) to predominantly encephalopathic forms (with ataxia and cerebellar atrophy) (102–105). Rarely, patients with adult onset Leigh disease also can have a primary CoQ_{10} deficiency (106). In patients with ataxia, the cerebellum is usually atrophic. In many patients, the defect is tissue specific. Serum and cultured fibroblasts show normal CoQ_{10} levels. A significant decrease in muscle coenzyme Q_{10} is diagnostic. A muscle biopsy shows a ragged-red fiber myopathy, cytochrome c oxidase deficiency, and increased myofiber lipid.

DEFECTS IN INTERGENOMIC COMMUNICATION

The mechanisms that control intergenomic communication between the nDNA and the mtDNA are beginning to be elucidated by diseases that produce quantitative and qualitative defects in the mtDNA. Quantitative defects are characterized by changes in the copy number of mtDNAs within a cell (mtDNA depletion diseases). Qualitative defects alter the gene structure of the mtDNAs in a random way (multiple mtDNA deletions). Overlap between these two categories occurs with some mutations.

Replication and maintenance of the mtDNA require a large number of nuclear-encoded enzymes and balanced nucleotide pools (Figure 4.3). Proper functioning of mitochondrial nucleotide synthesis is critical because mtDNA replication continues throughout the cell cycle. The deoxynucleoside triphosphate (dNTP) pool is maintained by cytoplasmic dNTPs that enter the mitochondria by specific transporters or by a salvage pathway that uses intramitochondrial precursors.

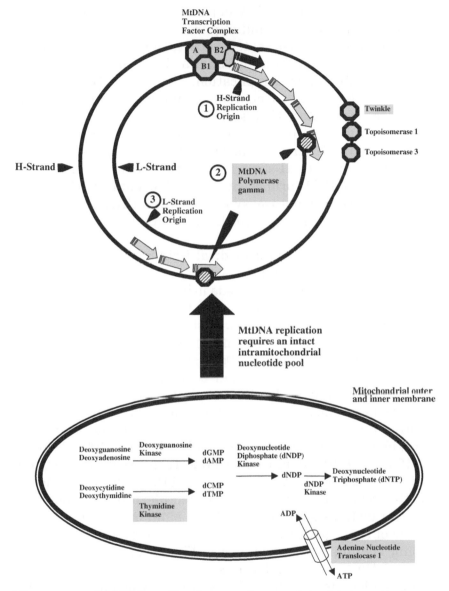

Figure 4.3 MtDNA replication requires an intact intramitochondrial nucleotide pool for proper function. Multiple mtDNA deletions can be produced by altering the nucleotide/nucleoside pool homeostasis or by impairing mtDNA replication. See text for a detailed description of mtDNA replication.

Replication of mtDNA depends on a complex array of mitochondrial proteins and requires a functional dNTP pool. Multiple mtDNAs associated with various proteins forming nucleoids are found in mitochondria. These nucleoids appear important for mtDNA replication. The two mtDNA strands are replicated in an asynchronous manner (Figure 4.3) (107) and are designated H-strand (heavy strand) and L-strand (light strand) due to differences in their guanine and thymidine contents. H-strand replication begins first, using the L-strand as template. H-strand replication begins with the synthesis of an RNA primer by the mitochondrial RNA polymerase. The polymerase synthesizes an RNA primer with subsequent cleavage by mitochondrial RNA processing endonuclease.

Binding of the mitochondrial RNA polymerase requires mitochondrial transcription factors (Tfam, TFB1M, TFB2M); mtDNA synthesis then proceeds using polymerase γ. Once the H-strand synthesis is about two-thirds complete, L-strand synthesis begins. The parent H-strand appears to be maintained in a single-stranded conformation by protein complexes that include the Twinkle (C10orf2) protein and topoisomerase 1 and 3. The parent H-strand is held in a single-stranded conformation and used as the template for L-strand synthesis.

Gene mutations known to produce quantitative or qualitative defects in the mtDNA are found in the thymidine phosphorylase gene (108), thymidine kinase (109), Twinkle (C10orf2) (110), adenine nucleotide translocase 1 (111), and mtDNA-specific DNA polymerase γ (112). Abnormal function of the genes that control the nucleotide pool or genes that control mtDNA synthesis produces similar diseases. A summary of these pathways as they relate to diseases of intergenomic communication is given in Figure 4.3.

MtDNA depletion diseases are an important group of disorders affecting infants and neonates in which a quantitative reduction in mtDNA copy number exists within various tissues. MtDNA depletion is usually an infantile disease characterized by severe muscle weakness, hepatic failure, or renal tubulopathy with fatal outcome. An isolated, progressive myopathy or an encephalomyopathy characterizes the majority of the remaining cases. Most cases have severe mtDNA depletion (>99%) and present shortly after birth and die within the first year of life or may even present prenatally (113). A small percentage survives into adolescence; rarely, patients have been reported who survived into adulthood (114,115). An interesting ethnic predilection, characterized by severe peripheral neuropathy, hepatopathy, and cerebral white matter disease, is found in Navajo populations (116).

Muscle biopsy usually shows ragged-red fibers and cytochrome c oxidase-deficient myofibers. Variable types of OXPHOS enzyme defects are observed. In severe cases, all OXPHOS enzymes are severely affected. The least affected enzyme is complex II (succinate dehydrogenase), which is entirely coded by the nDNA (117). The diagnosis is made using quantitative southern blot analysis that demonstrates that the copy number of the mtDNA is greatly reduced in affected tissues. The mtDNA/nDNA ratio is typically <30%. Interestingly, the unaffected tissues of some patients may show normal levels of mtDNA. In all cases, exclusion of other diseases and drugs that reduce the quantity of mtDNA within tissues like the liver is important and includes such factors as viral hepatitis, iron storage disease, chronic cirrhosis, and antiviral nucleoside analogs.

MtDNA depletion syndromes are caused by autosomal recessive mutations in the nuclear genes that code for mt thymidine kinase (TK2) and deoxyguanosine kinase. Defects in these genes characterize only a small number of patients with mtDNA depletion syndromes. Thymidine kinase and deoxyguanosine kinase deficiencies are recognizable by their tissue-specific manifestations. Decreased thymidine kinase activity causes a fatal skeletal myopathy with little effect on other tissues (118). The factors associated with disease expression are incompletely understood as demonstrated by the rarely encountered patients with milder phenotypes who show reversion of the mtDNA depletion (119). Decreased deoxyguanosine kinase activity causes hepatopathy, encephalopathy, and mtDNA depletion in liver but not in other tissues.

Thymidine phosphorylase mutations cause mitochondrial neurogastrointestinal encephalomyopathy (MNGIE)

(108,120). Thymidine phosphorylase converts thymidine to 2-deoxy D-ribose 1-phosphate and may function to regulate thymidine availability for DNA synthesis. MNGIE is an autosomal recessive disorder characterized by a progressive external ophthalmoplegia, dementia with a progressive leukodystrophy, mitochondrial myopathy, peripheral neuropathy, and prominent involvement of the gastrointestinal tract.

The gastrointestinal manifestations are heralded by significant diarrhea, malabsorption, and weight loss with normal pancreatic function. Radiologic investigations may show marked thickening of the small intestines, which reflects the pathological findings of extensive mural thickening and fibrosis of the submucosa and subserosa. Metabolic testing shows increased blood thymidine levels — a characteristic feature of the disease. MNGIE is easily diagnosed by determining leukocyte thymidine phosphorylase activity. Thymidine phosphorylase deficiency produces quantitative and qualitative defects in the mtDNA that can be detected by mtDNA southern blot analysis.

Qualitative mtDNA defects are produced by mutations in two genes that participate in mtDNA replication: the Twinkle gene and the mitochondrial DNA polymerase γ gene, as well as one gene that helps regulate the mitochondrial nucleotide pool: the adenine nucleotide translocator 1 gene. The Twinkle gene codes a protein that appears to have a helicase function (110). Twinkle is found in nucleosides in association with the transcription factor TFAM and the mitochondrial single-stranded binding protein (121). The mtDNA polymerase γ is responsible for synthesizing new mtDNAs. The adenine nucleotide translocase 1 is important to maintaining the mitochondrial nucleotide pool by transporting ATP out of the mitochondria to the cytoplasm and ADP into the mitochondria.

The most characteristic clinical finding in patients with multiple mtDNA deletions is ophthalmoplegia. Individuals with autosomal dominant variants also have proximal limb weakness, peripheral neuropathy, sensorineural hearing loss, cataracts, endocrine dysfunction, and neuropsychiatric symptoms such as severe depression (122,123). Autosomal recessive progressive variants present differently. Although patients have external ophthalmoplegia, they also manifest cardiomyopathy (124) or peripheral neuropathy, gastrointestinal dysmotility, and leukoencephalopathy (mt neurogastrointestinal encephalomyopathy) (120).

DEFECTS IN OXPHOS ENZYME ASSEMBLY AND PROCESSING

A large number of proteins coded by the nDNA are required for the proper assembly of OXPHOS enzymes. Little is known about these assembly mechanisms for most of the OXPHOS enzymes. Discovery of mutations in genes that control complex IV (cytochrome c oxidase) assembly is beginning to elucidate this process. Complex IV in humans contains three polypeptide subunits coded by the mtDNA and ten subunits coded by the nDNA. In yeast, assembly of this enzyme is associated with over 30 complementation groups, underscoring the complexity of this process (125). In humans, mutations in four genes are known to produce OXPHOS disease by interfering with complex IV assembly. SURF1, COX10, SCO1, and SCO2. All are transmitted in an autosomal recessive fashion.

The first mutation in an assembly factor was identified in patients with Leigh disease. This disease is characterized by early onset hypotonia, ataxia, brainstem abnormalities, regression, and characteristic bilateral basal ganglia lesions that show an increased signal on T2-weighted MRI images. Mutations in a highly evolutionarily conserved gene, the SURF1 gene, were recognized as a cause for systemic complex IV (cytochrome c oxidase) deficiency (62,126). Autosomal recessive mutations in the SURF1 gene, consisting of large deletions, nonsense mutations, and donor–splice site mutants, are heterogeneous. Compound heterozygotes are common.

In yeast, two related COX assembly genes, SCO1 and SCO2, enable the mtDNA-coded subunits I and II and copper to be incorporated into complex IV (cytochrome c oxidase) (127,128). Interestingly, the copper handling for complex IV maturation appears to occur catalytically due to the thioreductase activity of these proteins (129,130). Heme A is an essential component of complex IV (cytochorme c oxidase). The COX10 gene encodes heme A:farnesyltransferase, an enzyme that catalyzes the first step in the conversion of protoheme (heme B) to the heme A prosthetic groups of complex IV. COX15 also decreases heme A synthesis by impairing the formation of the heme A precursor, heme O (131,132).

In humans, these proteins appear to have similar functions. Mutations in COX 10, COX 15, SCO 1, and SCO 2 result in distinct clinical presentations with tissue-specific consequences. SCO 2 and COX 15 mutations produce a fatal cardioencephalopathy and complex IV deficiency (86,133,134). Immunohistochemical studies indicated that the enzymatic deficiency is due to failure to incorporate mtDNA-encoded complex IV subunits. Cardiac failure is manifested as a hypertrophic cardiomyopathy. COX 10 mutations seem to produce more phenotypic variability. Complex IV deficiency due to COX 10 mutations can be associated with renal tubulopathy plus leukoencephalopathy, with anemia, sensorineural deafness, and fatal infantile hypertrophic cardiomyopathy, and with anemia plus Leigh disease (135,136). SCO 1 mutations are rare and patients present with neonatal complex IV deficiency, hepatopathy, ketoacidosis, and encephalopathy (137).

An abnormality in an assembly factor is known for one other enzyme, complex III. The BCS1L protein acts as a chaperone for assembly of complex III polypeptides. Mutations in the BCS1L protein produce two major phenotypes:

- Neonatal proximal tubulopathy, hepatopathy, and encephalopathy
- GRACILE syndrome

GRACILE syndrome was initially described in Finnish patients and is characterized by fetal growth retardation; Fanconi type aminoaciduria; cholestasis; iron overload (liver hemosiderosis, hyperferritinemia, hypotransferrinemia, increased transferrin iron saturation, and free plasma iron); profound lactic acidosis; and early death (138–140). In both presentations, liver and kidney function are significantly impaired. An interesting, unexplained feature of these patients is that complex III activity is not decreased in all patients. British and Turkish patients had complex III deficiency, but Finnish patients had normal complex III activity (140).

Because iron–sulfur proteins are critical for electron transport in OXPHOS enzymes, their proper assembly and incorporation are critical. Two diseases are known to be caused by abnormal iron–sulfur protein processing: Friedreich ataxia (the most common hereditary ataxia) and X-linked sideroblastic anemia with ataxia. Friedreich ataxia is caused by a GAA trinucleotide repeat expansion in the first intron of the frataxin gene that is transmitted in an autosomal recessive fashion. Frataxin is a mitochondrial protein involved in mitochondrial iron homeostasis. Frataxin gene mutations result in impaired activity of the iron–sulfur-containing enzymes within the mitochondria: complex I, complex II, complex III, and aconitase (141). Clinical manifestations are systemic and include hypoactive or absent deep tendon reflexes, ataxia, corticospinal tract dysfunction, impaired vibratory and prioprioceptive function, hypertrophic caridiomyopathy, and diabetes mellitus.

X-linked sideroblastic anemia and ataxia (XLSA/A) is an X-linked recessive disorder caused by mutations in the ABC 7 gene (142). The disorder is characterized by an infantile to early childhood onset of nonprogressive cerebellar ataxia and mild anemia with hypochromia and microcytosis. ABC 7 is an ortholog of the yeast ATM 1 gene whose product localizes to the mitochondrial inner membrane and is involved in iron homeostasis. The ABC 7 gene positively regulates the expression of extramitochondrial thioredoxin and an intramitochondrial iron–sulfur-containing protein, ferrochelatase (143). The thioredoxin has numerous functions, including defense against oxidative stress, control of growth and apoptosis, cocytokine activity, and chemokine activity (144,145).

Heat shock proteins play an important role in the transfer of proteins into the mitochondria. Heat shock protein 60 (hsp60) is a mitochondrial matrix protein involved in the folding and correct assembly of polypeptides into complex mitochondrial enzymes. A deficiency of hsp60 was identified in two patients (146,147). One was characterized by facial dysmorphic features, breathing difficulties at birth, hypotonia, and heart failure; the second had neonatal lactic acidosis, hypotonia, unusual facial features, feeding difficulties, and failure to thrive. Metabolic abnormalities included metabolic acidosis with lactic acidemia, hyperammonaemia, and intermittent ketosis. Consistent with these metabolic changes, defects in various mitochondrial enzymes were observed in both patients and included OXPHOS enzymes, Krebs cycle enzymes, the pyruvate dehydrogenase complex, and the mitochondrial biotin-dependent carboxylases. The molecular defect in these cases has not been identified. However, a mutation in hsp60 was recently identified in a French family with an autosomal-dominant form of hereditary spastic paraplegia (SPG13) (148).

A second type of hereditary spastic paraplegia is caused by a defect in a mitochondrial protein that appears to have chaperonin function. In the autosomal recessive form, this paraplegia is accompanied by a ragged-red fiber myopathy (149). Patients experience progressive weakness, spasticity, and mild decreases in vibratory sensation as their major manifestations. Dysphagia, scoliosis, and optic nerve atrophy may be present. This unique form of hereditary spastic paraplegia is caused by mutations in the paraplegin gene that produces a protein targeted to the mitochondria. Paraplegin has a high degree of homology with a subclass of ATPases called the AAA family. This group of ATPases comprises metalloproteases with proteolytic and chaperonin functions. Patients have ragged-red fibers, cytochrome c oxidase-deficient fibers, and abnormal OXPHOS enzymology in skeletal muscle (150). Precise classification by mechanism is difficult for the hsp60 and paraplegin defects because the pathophysiologic mechanisms are still unclear.

The TIMM8A/DDP1 gene codes for a protein that is localized to the space between the inner and outer mitochondrial membrane and functions to import proteins into the mitochondria. Mutations in this gene produce Mohr–Tranebjaerg syndrome, which is characterized by postlingual progressive sensorineural deafness as the first presenting symptom in early childhood, followed by progressive dystonia, spasticity, dysphagia, mental deterioration, paranoia, and cortical blindness (151–153). Jensen syndrome has also been associated with mutations in the TIMM8A gene and is characterized by profound sensorineural hearing loss with onset in infancy, followed in adolescence by progressive optic nerve atrophy with loss of vision and in adulthood by progressive dementia (154). Extensive CNS calcification can be found in all structures, including meninges, vessels, and neurons.

These disorders are inherited in an X-linked fashion. Mutations impair assembly of the TIMM8A protein (DDP1) with the TIMM 13 protein in the intermembrane space. This TIMM8A/TIMM13 protein complex is responsible for facilitating the import of nuclear-encoded precursor proteins into the mitochondrial inner membrane. This disorder has been proposed to be a mitochondrial disease with a defect in oxidative phosphorylation (155). OXPHOS enzymology results are conflicting, with normal enzyme activities in one family (156) and abnormal OXPHOS enzyme activities in a number of families studied (154).

DEFECTS IN INTRACELLULAR MOVEMENT OF MITOCHONDRIA

Mitochondria are constantly moving within cells. In nerves, mitochondria are moved long distances down axons; this movement is guided by proteins called dynamins and is conducted along cytoskeletal microtubules. Mutations in a gene encoding a mitochondrial dynamin-related guanosine triphosphatase (OPA1) cause an autosomal-dominant form of optic atrophy (Kjer type) (157,158). Patients present with an insidious onset of bilateral visual loss and they characteristically have temporal optic nerve pallor, centrocaecal visual field scotoma, and a color vision deficit that is frequently blue–yellow. OPA1 is anchored to the mitochondrial cristae inner membrane, facing the intermembrane space (159). OPA1 is essential to organization in the mitochondrial inner membrane from which the maintenance of the cristae integrity depends. In addition, OPA1 appears to be important for committing cells to apoptosis (160).

Mitochondrial Membrane Defects

X-linked cardioskeletal myopathy, or Barth syndrome, is characterized by a congenital dilated cardiomyopathy and mitochondrial myopathy with growth retardation. The dilated cardiomyopathy may be associated with endocardial fibroelastosis; ultrastructural studies of the cardiac muscle show mitochondria with abnormally shaped cristae, but no paracrystalline inclusions. Skeletal muscle shows increased lipid on histochemistry and abnormal mitochondrial cristae. Abnormal mitochondria are also found in bone marrow, liver, and renal tubule epithelial cells. Additional features linked to this disease include decreased plasma free carnitine and muscle carnitine, increased urinary 3-methylglutaconic acid and 2-ethylhydracrylic acid, and reduced levels of cytochromes $c+c_1$ in skeletal muscle. Barth syndrome is localized to Xq28 and is caused by mutations in the tafazzin gene (G4.5 gene) (161–164). Tafazzin gene mutations cause a reduction in cardiolipin in the mitochondrial membranes (165–167). Cardiolipin is exclusively found in the inner mitochondrial membrane and is required for optimal function of OXPHOS enzymes.

OXPHOS DISEASE MANAGEMENT

The management of patients with OXPHOS diseases can be difficult. Frequently, multiple specialties must cooperatively manage the patient. Because many of the phenotypes display multiorgan involvement, early recognition and treatment is important. For example, screening of fasting blood glucose and glucose tolerance testing can detect diabetes mellitus. Once diabetes is detected, treatment can be instituted, thus decreasing the risk of complications. Early recognition and treatment of many of the complications of OXPHOS disease is an important measure for reducing the cost of health care for these patients.

Avoidance of certain classes of medications that worsen the patient's OXPHOS defect can also be important. In many patients with OXPHOS diseases, medication side effects are enhanced. Close follow-up of OXPHOS disease patients is necessary when new medications are instituted. For example, valproate is frequently considered as a treatment option for patients with epilepsy. Epileptic seizures are the first recognized symptom in about 53% of patients with mitochondrial encephalomyopathies. Many adults (43%) and most infants (70%) in these groups of patients have atypical mitochondrial encephalomyopathies. Partial seizures, mainly with elementary motor symptoms, and focal or multifocal EEG epileptiform activities characterized the epileptic presentation in 71% of the patients (168). An important adverse effect of valproate treatment is the impairment of fatty acid oxidation, which can further compromise mitochondrial ATP production. Mitochondrial diseases are considered as a risk factor for valproate-induced liver failure and in many cases should be excluded before treatment with valproate (169). Patients with MELAS have experienced worsening of seizures when taking valproate (170).

Metabolic therapies for OXPHOS diseases attempt to increase mitochondrial ATP and decrease free radical production. Metabolic therapies that have been reported to produce a positive therapeutic effect include coenzyme Q_{10}, uridine, idebenone, triacylglycerol, phylloquinone, menadione, succinate, ascorbate, and riboflavin. However, assessment of the efficacy of these treatments has been difficult, due to the clinical and genetic heterogeneity of OXPHOS diseases. The therapeutic efficacy of any reported compounds is limited. In this context, it is important to advise patients to avoid alternative medicine approaches to disease management. Laboratories used by alternative medicine practitioners are often nonmedical, yield unreliable results, and prescribe numerous supplements and herbs without regard to the patient's disease process. These approaches can be dangerous to the health of some patients.

Dicarboxylic aciduria and secondary impairment of long chain fatty acid oxidation occur in patients with OXPHOS diseases. Avoidance of fasting, cornstarch supplementation, and decreased dietary intake of long chain fatty acids may be helpful in selected patients. However, the long-term benefits of dietary manipulations are unknown.

The effects of mild degrees of aerobic activity are a frequent concern voiced by patients and physicians. In ten adults with mitochondrial myopathies, moderate treadmill training over 8 weeks resulted in a 30% improvement of aerobic capacity, a 30% drop in resting and postexercise lactate levels, and a 60% improvement in adenosine triphosphate recovery as measured by ^{31}P-NMR testing (171).

SUMMARY

Physicians in all specialties are increasingly aware of OXPHOS disease. Although the prevalence of OXPHOS disease in the general population is unknown, the number of requests for pediatric and adult evaluations is increasing rapidly. A basic awareness of OXPHOS disease phenotypes as well as of the essential elements of patient evaluation is important for appropriate patient management and referrals. Centers that specialize in OXPHOS disease evaluations can be instrumental in working with referring physicians to develop a cost-effective diagnostic plan individualized to suit the patient's needs.

MtDNA and nDNA sequencing can be important in delineating the inheritance. After a complete evaluation, genetic counseling based on Mendelian principles or mtDNA principles of inheritance can be applied. Although approaches that assess patients for mtDNA mutations are evolving rapidly, significant ambiguity in patient diagnosis often remains, even after detailed testing is complete. Advances in understanding of mutations in nuclear OXPHOS genes will provide a powerful addition to the ability to diagnose, manage, and counsel patients with these disorders.

REFERENCES

1. IJ Holt, AE Harding, JA Morgan–Hughes. Deletions of muscle mitochondrial DNA in patients with mitochondrial myopathies. *Nature* 331: 717–719, 1988.

2. JM Shoffner, MT Lott, AM Lezza, P Seibel, SW Ballinger, DC Wallace. Myoclonic epilepsy and ragged-red fiber disease (MERRF) is associated with a mitochondrial DNA tRNA(Lys) mutation. *Cell* 61: 931–937, 1990.

3. DC Wallace, G Singh, MT Lott, JA Hodge, TG Schurr, AM Lezza. Mitochondrial DNA mutation associated with Leber's hereditary optic neuropathy. *Science* 242: 1427–1430, 1988.

4. S Anderson, AT Bankier, BG Barrell, MH de Bruijn, AR Coulson, J Drouin. Sequence and organization of the human mitochondrial genome. *Nature* 290: 457–465, 1981.

5. R Parvari, I Brodyansky, O Elpeleg, S Moses, D Landau, E Hershkovitz. A recessive contiguous gene deletion of chromosome 2p16 associated with cystinuria and a mitochondrial disease. *Am J Hum Genet* 69: 869–875, 2001.

6. JM Cummins, H Kishikawa, D Mehmet, R Yanagimachi. Fate of genetically marked mitochondrial DNA from spermatocytes microinjected into mouse zygotes. *Zygote* 7: 151–156, 1999.

7. M Schwartz, J Vissing. Paternal inheritance of mitochondrial DNA. *N Eng J Med* 347: 576–580, 2002.

8. M Schwartz, J Vissing. New patterns of inheritance in mitochondrial disease. *Biochem Biophys Res Commun* 310: 247–251, 2003.

9. J St John, D Sakkas, K Dimitriadi, A Barnes, V Maclin, J Ramey. Failure of elimination of paternal mitochondrial DNA in abnormal embryos. *Lancet* 355: 200, 2000.

10. BJ Battersby, JC Loredo–Osti, EA Shoubridge. Nuclear genetic control of mitochondrial DNA segregation. *Nat Genet* 33: 183–186, 2003.

11. JM Shoffner. Oxidative phosphorylation disease. In: Scriver, Beaudet, Valle, Sly, Childs, Kinzler and Vogelstein (Eds.), *The Metabolic and Molecular Bases of Inherited Diseases* (Vol. II, pp. 2367–2423). McGraw-Hill, New York, 2001.

12. WD Graf, J Marin–Garcia, HG Gao, S Pizzo, RK Naviaux, D Markusic. Autism associated with the mitochondrial DNA G8363A transfer RNA(Lys) mutation. *J Child Neurol* 15: 357–361, 2000.

13. S Amemiya, M Hamamoto, Y Goto, H Komaki, I Nishino, I Nonaka. Psychosis and progressing dementia: presenting features of a mitochondriopathy. *Neurology* 55: 600–601, 2000.

14. E Byrne. Does mitochondrial respiratory chain dysfunction have a role in common neurodegenerative disorders? *J Clin Neurosci* 9: 497–501, 2002.

15. PF Chinnery, GA Taylor, N Howell, RM Andrews, CM Morris, RW Taylor. Mitochondrial DNA haplogroups and susceptibility to AD and dementia with Lewy bodies. *Neurology* 55: 302–304, 2000.

16. SM de la Monte, J Chiche, A von dem Bussche, S Sanyal, SA Lahousse, SP Janssens. Nitric oxide synthase-3 overexpression causes apoptosis and impairs neuronal mitochondrial function: relevance to Alzheimer's-type neurodegeneration. *Lab Invest* 83: 287–298, 2003.

17. JM Shoffner. Oxidative phosphorylation defects and Alzheimer's disease. *Neurogenetics* 1: 13–19, 1997.

18. PW Stacpoole, ST Bunch, RE Neiberger, LA Perkins, R Quisling, AD Hutson. The importance of cerebrospinal fluid lactate in the evaluation of congenital lactic acidosis. *J Pediatr* 134: 99–102, 1999.

19. B Koo, LE Becker, S Chuang, F Merante, BH Robinson, D MacGregor. Mitochondrial encephalomyopathy, lactic acidosis, stroke-like episodes (MELAS): clinical, radiological, pathological, and genetic observations. *Ann Neurol* 34: 25–32, 1993.

20. H Reichmann, R Gold, B Meurers, M Naumann, P Seibel, U Walter. Progression of myopathology in Kearns–Sayre syndrome: a morphological follow-up study. *Acta Neuropathol (Berl)*, 85: 679–681, 1993.

21. BJ Hanson, RA Capaldi, MF Marusich, SW Sherwood. An immunocytochemical approach to detection of mitochondrial disorders. *J Histochem Cytochem* 50: 1281–1288, 2002.

22. MF Marusich, BH Robinson, JW Taanman, SJ Kim, R Schillace, JL Smith. Expression of mtDNA and nDNA encoded respiratory chain proteins in chemically and genetically derived ρ0 human fibroblasts: a comparison of subunit proteins in normal fibroblasts treated with ethidium bromide and fibroblasts from a patient with mtDNA depletion syndrome. *Biochim Biophys Acta* 1362: 145–159, 1997.

23. S Rahman, BD Lake, JW Taanman, MG Hanna, JM Cooper, AH Schapira. Cytochrome oxidase immunohistochemistry: clues for genetic mechanisms. *Brain* 123 Pt 3: 591–600, 2000.

24. M Sparaco, T Cavallaro, G Rossi, N Rizzuto. Immunohistochemical demonstration of spinal ventral horn cells involvement in a case of "myoclonus epilepsy with ragged red fibers" (MERRF). *Clin Neuropathol* 19: 200–207, 2000.

25. M Sparaco, EA Schon, S DiMauro, E Bonilla. Myoclonic epilepsy with ragged-red fibers (MERRF): an immunohistochemical study of the brain. *Brain Pathol* 5: 125–133, 1995.

26. K Tanji, TH Vu, EA Schon, S DiMauro, E Bonilla. Kearns–Sayre syndrome: unusual pattern of expression of subunits of the respiratory chain in the cerebellar system. *Ann Neurol* 45: 377–383, 1999.

27. AJ Janssen, JA Smeitink, LP van den Heuvel. Some practical aspects of providing a diagnostic service for respiratory chain defects. *Ann Clin Biochem* 40(Pt 1): 3–8, 2003.

28. XX Zheng, JM Shoffner, AS Voljavec, DC Wallace. Evaluation of procedures for assaying oxidative phosphorylation enzyme activities in mitochondrial myopathy muscle biopsies. *Biochim Biophys Acta* 1019: 1–10, 1990.

29. P Briones, MA Vilaseca, MT Garcia–Silva, M Pineda, J Colomer, I Ferrer. Congenital disorders of glycosylation (CDG) may be underdiagnosed when mimicking mitochondrial disease. *Eur J Pediatr Neurol* 5: 127–131, 2001.

30. PB Kang, JV Hunter, EM Kaye. Lactic acid elevation in extramitochondrial childhood neurodegenerative diseases. *J Child Neurol* 16: 657–660, 2001.

31. M Finel, AS Majander, J Tyynela, AM De Jong, SP Albracht, M Wikstrom. Isolation and characterization of subcomplexes of the mitochondrial NADH:ubiquinone oxidoreductase (complex I). *Eur J Biochem* 226: 237–242, 1994.

32. BS Jakobs, C van den Bogert, G Dacremont, RJ Wanders. β-oxidation of fatty acids in cultured human skin fibroblasts devoid of the capacity for oxidative phosphorylation. *Biochim Biophys Acta* 1211: 37–43, 1994.

33. A Parker, PC Engel. Preliminary evidence for the existence of specific functional assemblies between enzymes of the beta-oxidation pathway and the respiratory chain. *Biochem J* 345 Pt 3: 429–435, 2000.

34. RD Holmes, KH Moore, JP Ofenstein, P Tsatsos, FL Kiechle. Lactic acidosis and mitochondrial dysfunction in two children with peroxisomal disorders. *J Inherit Metab Dis* 16: 368–380, 1993.

35. FV Ventura, JP Ruiter, L Ijlst, IT Almeida, RJ Wanders. Inhibition of oxidative phosphorylation by palmitoyl-CoA in digitonin permeabilized fibroblasts: implications for long-chain fatty acid β-oxidation disorders. *Biochim Biophys Acta* 1272: 14–20, 1995.

36. FV Ventura, JP Ruiter, L Ijlst, IT de Almeida, RJ Wanders. Inhibitory effect of 3-hydroxyacyl-CoAs and other long-chain fatty acid β-oxidation intermediates on mitochondrial oxidative phosphorylation. *J Inherit Metab Dis* 19: 161–164, 1996.

37. L Wojtczak, P Schonfeld. Effect of fatty acids on energy coupling processes in mitochondria. *Biochim Biophys Acta* 1183: 41–57, 1993.

38. M Jaksch, S Kleinle, C Scharfe, T Klopstock, D Pongratz, J Muller–Hocker. Frequency of mitochondrial transfer RNA mutations and deletions in 225 patients presenting with respiratory chain deficiencies. *J Med Genet* 38: 665–673, 2001.

39. MJ Absalon, CO Harding, DR Fain, L Li, KJ Mack. Leigh syndrome in an infant resulting from mitochondrial DNA depletion. *Pediatr Neurol* 24: 60–63, 2001.

40. M Akagi, K Inui, H Tsukamoto, N Sakai, T Muramatsu, M Yamada. A point mutation of mitochondrial ATPase 6 gene in Leigh syndrome. *Neuromuscul Disord* 12: 53–55, 2002.

41. P Benit, D Chretien, N Kadhom, P de Lonlay–Debeney, V Cormier–Daire, A Cabral. Large-scale deletion and point mutations of the nuclear NDUFV1 and NDUFS1 genes in mitochondrial complex I deficiency. *Am J Hum Genet* 68: 1344–1352, 2001.

42. RM Brown, GK Brown. Complementation analysis of systemic cytochrome oxidase deficiency presenting as Leigh syndrome. *J Inherit Metab Dis* 19: 752–760, 1996.

43. SM Budde, LP van den Heuvel, AJ Janssen, RJ Smeets, CA Buskens, L DeMeirleir. Combined enzymatic complex I and III deficiency associated with mutations in the nuclear encoded NDUFS4 gene. *Biochem Biophys Res Commun* 275: 63–68, 2000.

44. M Burgeois, F Goutieres, D Chretien, P Rustin, A Munnich, J Aicardi. Deficiency in complex II of the respiratory chain, presenting as a leukodystrophy in two sisters with Leigh syndrome. *Brain Dev* 14: 404–408, 1992.

45. Y Campos, MA Martin, JC Rubio, LG Solana, C Garcia–Benayas, JL Terradas. Leigh syndrome associated with the T9176C mutation in the ATPase 6 gene of mitochondrial DNA. *Neurology* 49: 595–597, 1997.

46. M Capkova, H Hansikova, C Godinot, H Houst'kova, J Houstek, J Zeman. A new missense mutation of 574C>T in the SURF1 gene — biochemical and molecular genetic study in seven children with Leigh syndrome. *Cas Lek Cesk* 141: 636–641, 2002.

47. A Chakrapani, L Heptinstall, J Walter. A family with Leigh syndrome caused by the rarer T8993C mutation. *J Inherit Metab Dis* 21: 685–686, 1998.

48. RM Chalmers, PJ Lamont, I Nelson, DW Ellison, NH Thomas, AE Harding. A mitochondrial DNA tRNA(Val) point mutation associated with adult-onset Leigh syndrome. *Neurology* 49: 589–592, 1997.

49. M Chol, S Lebon, P Benit, D Chretien, P de Lonlay, A Goldenberg. The mitochondrial DNA G13513A MELAS mutation in the NADH dehydrogenase 5 gene is a frequent cause of Leigh-like syndrome with isolated complex I deficiency. *J Med Genet* 40: 188–191, 2003.

50. DD de Vries, BG van Engelen, FJ Gabreels, W Ruitenbeek, BA van Oost. A second missense mutation in the mitochondrial ATPase 6 gene in Leigh's syndrome. *Ann Neurol* 34: 410–412, 1993.

51. DM Kirby, SG Kahler, ML Freckmann, D Reddihough, DR Thorburn. Leigh disease caused by the mitochondrial DNA G14459A mutation in unrelated families. *Ann Neurol* 48: 102–104, 2000.

52. J Loeffen, O Elpeleg, J Smeitink, R Smeets, S Stockler–Ipsiro-glu, H Mandel. Mutations in the complex I NDUFS2 gene of patients with cardiomyopathy and encephalomyopathy. *Ann Neurol* 49(2): 195–201, 2001.

53. V Petruzzella, R Vergari, I Puzziferri, D Boffoli, E Lamantea, M Zeviani. A nonsense mutation in the NDUFS4 gene encoding the 18 kDa (AQDQ) subunit of complex I abolishes assembly and activity of the complex in a patient with Leigh-like syndrome. *Hum Mol Genet* 10: 529–535, 2001.

54. FM Santorelli, M Barmada, R Pons, LL Zhang, S DiMauro. Leigh-type neuropathology in Pearson syndrome associated with impaired ATP production and a novel mtDNA deletion. *Neurology* 47: 1320–1323, 1996.

55. FM Santorelli, S Shanske, A Macaya, DC DeVivo, S DiMauro. The mutation at nt 8993 of mitochondrial DNA is a common cause of Leigh's syndrome. *Ann Neurol* 34: 827–834, 1993.

56. L Santoro, R Carrozzo, A Malandrini, F Piemonte, C Patrono, M Villanova. A novel SURF1 mutation results in Leigh syndrome with peripheral neuropathy caused by cytochrome c oxidase deficiency. *Neuromuscul Disord* 10: 450–453, 2000.

57. A Shtilbans, S Shanske, S Goodman, CM Sue, C Bruno, TL Johnson. G8363A mutation in the mitochondrial DNA transfer ribonucleic acid Lys gene: another cause of Leigh syndrome. *J Child Neurol* 15(11): 759–761, 2000.

58. CM Sue, C Karadimas, N Checcarelli, K Tanji, LC Papadopou-lou, F Pallotti. Differential features of patients with mutations in two COX assembly genes, SURF-1 and SCO2. *Ann Neurol* 47: 589–595, 2000.

59. RW Taylor, AA Morris, M Hutchinson, DM Turnbull. Leigh disease associated with a novel mitochondrial DNA ND5 mutation. *Eur J Hum Genet* 10: 141–144, 2002.

60. M Teraoka, Y Yokoyama, S Ninomiya, C Inoue, S Yamashita, Y Seino. Two novel mutations of SURF1 in Leigh syndrome with cytochrome c oxidase deficiency. *Hum Genet* 105: 560–563, 1999.

61. V Tiranti, P Corona, M Greco, JW Taanman, F Carrara, E Lamantea. A novel frameshift mutation of the mtDNA COIII gene leads to impaired assembly of cytochrome c oxidase in a patient affected by Leigh-like syndrome. *Hum Mol Genet* 9: 2733–2742, 2000.

62. V Tiranti, K Hoertnagel, R Carrozzo, C Galimberti, M Munaro, M Granatiero. Mutations of SURF-1 in Leigh disease associated with cytochrome c oxidase deficiency. *Am J Hum Genet* 63: 1609–1621, 1998.

63. RH Triepels, LP van den Heuvel, JL Loeffen, CA Buskens, RJ Smeets, ME Rubio Gozalbo. Leigh syndrome associated with a mutation in the NDUFS7 (PSST) nuclear encoded subunit of complex I. *Ann Neurol* 45: 787–790, 1999.

64. M Tulinius, AR Moslemi, N Darin, B Westerberg, LM Wiklund, E Holme. Leigh syndrome with cytochrome-c oxidase deficiency and a single T insertion nt 5537 in the mitochondrial tRNATrp gene. *Neuropediatrics* 34: 87–91, 2003.

65. JM Shoffner, MD Brown, C Stugard, AS Jun, S Pollock, RH Haas. Leber's hereditary optic neuropathy plus dystonia is caused by a mitochondrial DNA point mutation. *Ann Neurol* 38: 163–169, 1995.

66. JM Shoffner, PM Fernhoff, NS Krawiecki, DB Caplan, PJ Holt, DA Koontz. Subacute necrotizing encephalopathy: oxidative phosphorylation defects and the ATPase 6 point mutation. *Neurology* 42: 2168–2174, 1992.

67. Y Tatuch, J Christodoulou, A Feigenbaum, JT Clarke, J Wherret, C Smit. Heteroplasmic mtDNA mutation (T----G) at 8993 can cause Leigh disease when the percentage of abnormal mtDNA is high. *Am J Hum Genet* 50(4): 852–858, 1992.

68. NG Kennaway, NR Buist, VM Darley–Usmar, A Papadimitriou, S Dimauro, RI Kelley. Lactic acidosis and mitochondrial myopathy associated with deficiency of several components of complex III of the respiratory chain. *Pediatr Res* 18: 991–999, 1984.

69. H Reichmann, R Rohkamm, M Zeviani, S Servidei, K Ricker, S DiMauro. Mitochondrial myopathy due to complex III deficiency with normal reducible cytochrome b concentration. *Arch Neurol* 43: 957–961, 1986.

70. AL Andreu, C Bruno, TC Dunne, K Tanji, S Shanske, CM Sue. A nonsense mutation (G15059A) in the cytochrome b gene in a patient with exercise intolerance and myoglobinuria. *Ann Neurol* 45: 127–130, 1999.

71. AL Andreu, C Bruno, S Shanske, A Shtilbans, M Hirano, S Krishna. Missense mutation in the mtDNA cytochrome b gene in a patient with myopathy. *Neurology* 51: 1444–1447, 1998.

72. AL Andreu, MG Hanna, H Reichmann, C Bruno, AS Penn, K Tanji. Exercise intolerance due to mutations in the cytochrome b gene of mitochondrial DNA. *N Engl J Med* 341: 1037–1044, 1999.

73. M Mancuso, M Filosto, JC Stevens, M Patterson, S Shanske, S Krishna. Mitochondrial myopathy and complex III deficiency in a patient with a new stop-codon mutation (G339X) in the cytochrome b gene. *J Neurol Sci* 209: 61–63, 2003.

74. J Poulton, ME Deadman, RM Gardiner. Duplications of mitochondrial DNA in mitochondrial myopathy. *Lancet* 1(8632): 236–240, 1989.

75. B Fromenty, R Carrozzo, S Shanske, EA Schon. High proportions of mtDNA — duplications in patients with Kearns–Sayre syndrome occur in the heart. *Am J Med Genet* 71: 443–452, 1997.

76. SW Ballinger, JM Shoffner, EV Hedaya, I Trounce, MA Polak, DA Koontz. Maternally transmitted diabetes and deafness associated with a 10.4 kb mitochondrial DNA deletion. *Nat Genet* 1: 11–15, 1992.

77. A Rotig, JL Bessis, N Romero, V Cormier, JM Saudubray, P Narcy. Maternally inherited duplication of the mitochondrial genome in a syndrome of proximal tubulopathy, diabetes mellitus, and cerebellar ataxia. *Am J Hum Genet* 50: 364–370, 1992.

78. K Tanji, EA Schon, S DiMauro, E Bonilla. Kearns–Sayre syndrome: oncocytic transformation of choroid plexus epithelium. *J Neurol Sci* 178: 29–36, 2000.

79. TR Prezant, JV Agapian, MC Bohlman, X Bu, S Oztas, WQ Qiu. Mitochondrial ribosomal RNA mutation associated with both antibiotic-induced and non-syndromic deafness. *Nat Genet* 4: 289–294, 1993.

80. RA Casano, DF Johnson, Y Bykhovskaya, F Torricelli, M Bigozzi, N Fischel–Ghodsian. Inherited susceptibility to aminoglycoside ototoxicity: genetic heterogeneity and clinical implications. *Am J Otolaryngol* 20: 151–156, 1999.

81. FM Santorelli, K Tanji, P Manta, C Casali, S Krishna, AP Hays. Maternally inherited cardiomyopathy: an atypical presentation of the mtDNA 12S rRNA gene A1555G mutation. *Am J Hum Genet* 64: 295–300, 1999.

82. D Thyagarajan, S Bressman, C Bruno, S Przedborski, S Shanske, T Lynch. A novel mitochondrial 12SrRNA point mutation in Parkinsonism, deafness, and neuropathy. *Ann Neurol* 48: 730–736, 2000.

83. J Smeitink, R Sengers, F Trijbels, L van den Heuvel. Human NADH:ubiquinone oxidoreductase. *J Bioenerg Biomembr* 33: 259–266, 2001.

84. M Schuelke, J Smeitink, E Mariman, J Loeffen, B Plecko, F Trijbels. Mutant NDUFV1 subunit of mitochondrial complex I causes leukodystrophy and myoclonic epilepsy. *Nat Genet* 21: 260–261, 1999.

85. S Papa, S Scacco, AM Sardanelli, R Vergari, F Papa, S Budde. Mutation in the NDUFS4 gene of complex I abolishes cAMP-dependent activation of the complex in a child with fatal neurological syndrome. *FEBS Lett* 489: 259–262, 2001.

86. S Scacco, V Petruzzella, S Budde, R Vergari, R Tamborra, D Panelli. Pathological mutations of the human NDUFS4 gene of the 18-kDa (AQDQ) subunit of complex I affects the expression of the protein and the assembly and function of the complex. *J Biol Chem* 278: 44161–44167, 2003.

87. L van den Heuvel, W Ruitenbeek, R Smeets, Z Gelman–Kohan, O Elpeleg, J Loeffen. Demonstration of a new pathogenic mutation in human complex I deficiency: a 5-bp duplication in the nuclear gene encoding the 18-kD (AQDQ) subunit. *Am J Hum Genet* 62: 262–268, 1998.

88. M Ito. Complex I deficiency due to NDUFS8 gene mutation. *Nippon Rinsho* 60 (Suppl 4): 441–443, 2002.

89. J Loeffen, J Smeitink, R Triepels, R Smeets, M Schuelke, R Sengers. The first nuclear-encoded complex I mutation in a patient with Leigh syndrome. *Am J Hum Genet* 63: 1598–1608, 1998.

90. T Bourgeron, P Rustin, D Chretien, M Birch–Machin, M Bourgeois, E Viegas–Pequignot. Mutation of a nuclear succinate dehydrogenase gene results in mitochondrial respiratory chain deficiency. *Nat Genet* 11: 144–149, 1995.

91. Y Goto. [Optic atrophy and ataxia (complex II deficiency–mutation in Fp subunit gene of succinate dehydrogenase)]. *Nippon Rinsho* 60 (Suppl 4): 376–377, 2002.

92. M Ito. Complex II deficiency due to Fp gene mutation. *Nippon Rinsho* 60 (Suppl 4): 444–445, 2000.

93. B Parfait, D Chretien, A Rotig, C Marsac, A Munnich, P Rustin. Compound heterozygous mutations in the flavoprotein gene of the respiratory chain complex II in a patient with Leigh syndrome. *Hum Genet* 106: 236–243, 2000.

94. R Van Coster, S Seneca, J Smet, R Van Hecke, E Gerlo, B Devreese. Homozygous Gly555Glu mutation in the nuclear-encoded 70 kDa flavoprotein gene causes instability of the respiratory chain complex II. *Am J Med Genet* 120A: 13–18, 2003.

95. MA Birch–Machin, RW Taylor, B Cochran, BA Ackrell, DM Turnbull. Late-onset optic atrophy, ataxia, and myopathy associated with a mutation of a complex II gene. *Ann Neurol* 48: 330–335, 2000.

96. I Nonaka. Complex II (succinate–ubiquinone reductase) deficiency. *Ryoikibetsu Shokogun Shirizu* 36: 132–134, 2001.

97. RC Aguiar, G Cox, SL Pomeroy, PL Dahia. Analysis of the SDHD gene, the susceptibility gene for familial paraganglioma syndrome (PGL1), in pheochromocytomas. *J Clin Endocrinol Metab* 86: 2890–2894, 2001.

98. BE Baysal, JE Willett–Brozick, EC Lawrence, CM Drovdlic, SA Savul, DR McLeod. Prevalence of SDHB, SDHC, and SDHD germline mutations in clinic patients with head and neck paragangliomas. *J Med Genet* 39: 178–183, 2002.

99. DE Benn, MS Croxson, K Tucker, CP Bambach, AL Richardson, L Delbridge. Novel succinate dehydrogenase subunit B (SDHB) mutations in familial phaeochromocytomas and paragangliomas, but an absence of somatic SDHB mutations in sporadic phaeochromocytomas. *Oncogene* 22: 1358–1364, 2003.

100. AP Gimenez–Roqueplo, J Favier, P Rustin, JJ Mourad, PF Plouin, P Corvol. The R22X mutation of the SDHD gene in hereditary paraganglioma abolishes the enzymatic activity of complex II in the mitochondrial respiratory chain and activates the hypoxia pathway. *Am J Hum Genet* 69: 1186–1197, 2001.

101. PE Taschner, AH Brocker–Vriends, AG van der Mey. [From gene to disease; from SDHD, a defect in the respiratory chain, to paragangliomas and pheochromocytomas]. *Ned Tijdschr Geneeskd* 146: 2188–2190, 2002.

102. C Lamperti, A Naini, M Hirano, DC De Vivo, E Bertini, S Servidei. Cerebellar ataxia and coenzyme Q10 deficiency. *Neurology* 60: 1206–1208, 2003.

103. O Musumeci, A Naini, AE Slonim, N Skavin, GL Hadjigeorgiou, N Krawiecki. Familial cerebellar ataxia with muscle coenzyme Q10 deficiency. *Neurology* 56: 849–855, 2001.

104. S Ogasahara, AG Engel, D Frens, D Mack. Muscle coenzyme Q10 deficiency in familial mitochondrial encephalomyopathy. *Proc Natl Acad Sci USA* 86: 2379–2382, 1989.

105. C Sobreira, M Hirano, S Shanske, RK Keller, RG Haller, E Davidson. Mitochondrial encephalomyopathy with coenzyme Q10 deficiency. *Neurology* 48: 1238–1243, 1997.

106. L Van Maldergem, F Trijbels, S DiMauro, PJ Sindelar, O Musumeci, A Janssen. Coenzyme Q-responsive Leigh's encephalopathy in two sisters. *Ann Neurol* 52: 750–754, 2002.

107. DA Clayton. Structure and function of the mitochondrial genome. *J Inherit Metab Dis* 15: 439–447, 1992.

108. I Nishino, A Spinazzola, M Hirano. Thymidine phosphorylase gene mutations in MNGIE, a human mitochondrial disorder. *Science* 283: 689–692, 1999.

109. A Saada, A Shaag, H Mandel, Y Nevo, S Eriksson, O Elpeleg. Mutant mitochondrial thymidine kinase in mitochondrial DNA depletion myopathy. *Nat Genet* 29: 342–344, 2001.

110. JN Spelbrink, FY Li, V Tiranti, K Nikali, QP Yuan, M Tariq. Human mitochondrial DNA deletions associated with mutations in the gene encoding Twinkle, a phage T7 gene 4-like protein localized in mitochondria. *Nat Genet* 28: 223–231, 2001.

111. J Kaukonen, JK Juselius, V Tiranti, A Kyttala, M Zeviani, GP Comi. Role of adenine nucleotide translocator 1 in mtDNA maintenance. *Science* 289: 782–785, 2000.

112. G Van Goethem, B Dermaut, A Lofgren, JJ Martin, C Van Broeckhoven. Mutation of POLG is associated with progressive external ophthalmoplegia characterized by mtDNA deletions. *Nat Genet* 28: 211–212, 2001.

113. S Arnon, R Aviram, T Dolfin, R Regev, I Litmanovits, R Tepper. Mitochondrial DNA depletion presenting prenatally with skin edema and multisystem disease immediately after birth. *Prenat Diag* 22: 34–37, 2002.

114. TH Vu, M Sciacco, K Tanji, C Nichter, E Bonilla, S Chatkupt. Clinical manifestations of mitochondrial DNA depletion. *Neurology* 50: 1783–1790, 1998.

115. TH Vu, K Tanji, H Valsamis, S DiMauro, E Bonilla. Mitochondrial DNA depletion in a patient with long survival. *Neurology* 51: 1190–1193, 1998.

116. TH Vu, K Tanji, SA Holve, E Bonilla, RJ Sokol, RD Snyder. Navajo neurohepatopathy: a mitochondrial DNA depletion syndrome? *Hepatology* 34: 116–120, 2001.

117. P Hargreaves, S Rahman, P Guthrie, JW Taanman, JV Leonard, JM Land. Diagnostic value of succinate ubiquinone reductase activity in the identification of patients with mitochondrial DNA depletion. *J Inherit Metab Dis* 25: 7–16, 2002.

118. A Saada, A Shaag, O Elpeleg. mtDNA depletion myopathy: elucidation of the tissue specificity in the mitochondrial thymidine kinase (TK2) deficiency. *Mol Genet Metab* 79: 1–5, 2003.

119. MR Vila, T Segovia–Silvestre, J Gamez, A Marina, AB Naini, A Meseguer. Reversion of mtDNA depletion in a patient with TK2 deficiency. *Neurology* 60: 1203–1205, 2003.

120. I Nishino, A Spinazzola, A Papadimitriou, S Hammans, I Steiner, CD Hahn. Mitochondrial neurogastrointestinal encephalomyopathy: an autosomal recessive disorder due to thymidine phosphorylase mutations. *Ann Neurol* 47: 792–800, 2000.

121. N Garrido, L Griparic, E Jokitalo, J Wartiovaara, AM van der Bliek, JN Spelbrink. Composition and dynamics of human mitochondrial nucleoids. *Mol Biol Cell* 14: 1583–1596, 2003.

122. S Servidei, M Zeviani, G Manfredi, E Ricci, G Silvestri, E Bertini. Dominantly inherited mitochondrial myopathy with multiple deletions of mitochondrial DNA: clinical, morphologic, and biochemical studies. *Neurology* 41: 1053–1059, 1991.

123. A Suomalainen, A Majander, M Wallin, K Setala, K Kontula, H Leinonen. Autosomal dominant progressive external oph-thalmoplegia with multiple deletions of mtDNA: clinical, bio-chemical, and molecular genetic features of the 10q-linked disease. *Neurology* 48: 1244–1253, 1997.

124. S Bohlega, K Tanji, FM Santorelli, M Hirano, A al-Jishi, S DiMauro. Multiple mitochondrial DNA deletions associated with autosomal recessive ophthalmoplegia and severe cardi-omyopathy. *Neurology* 46: 1329–1334, 1996.

125. A Tzagoloff, CL Dieckmann. PET genes of *Saccharomyces cer-evisiae*. *Microbiol Rev* 54: 211–225, 1990.

126. Z Zhu, J Yao, T Johns, K Fu, I De Bie, C Macmillan. SURF1, encoding a factor involved in the biogenesis of cytochrome c oxidase, is mutated in Leigh syndrome. *Nat Genet* 20: 337–343, 1998.

127. J Beers, DM Glerum, A Tzagoloff. Purification and character-ization of yeast Sco1p, a mitochondrial copper protein. *J Biol Chem* 277: 22185–22190, 2002.

128. DM Glerum, A Shtanko, A Tzagoloff. SCO1 and SCO2 act as high copy suppressors of a mitochondrial copper recruitment defect in *Saccharomyces cerevisiae*. *J Biol Chem* 271: 20531–20535, 1996.

129. E Balatri, L Banci, I Bertini, F Cantini, S Ciofi–Baffoni. Solu-tion structure of Sco1. A thioredoxin-like protein involved in cytochrome c oxidase assembly. *Structure* (Camb) 11: 1431–1443, 2003.

130. YV Chinenov. Cytochrome c oxidase assembly factors with a thioredoxin fold are conserved among prokaryotes and eukaryotes. *J Mol Med* 78: 239–242, 2000.

131. MH Barros, CG Carlson, DM Glerum, A Tzagoloff. Involvement of mitochondrial ferredoxin and Cox15p in hydroxylation of heme O. *FEBS Lett* 492: 133–138, 2001.

132. MH Barros, FG Nobrega, A Tzagoloff. Mitochondrial ferredoxin is required for heme A synthesis in *Saccharomyces cerevisiae*. *J Biol Chem* 277: 9997–10002, 2002.

133. H Antonicka, A Mattman, CG Carlson, DM Glerum, KC Hoffbuhr, SC Leary. Mutations in COX15 produce a defect in the mitochondrial heme biosynthetic pathway, causing early-onset fatal hypertrophic cardiomyopathy. *Am J Hum Genet* 72: 101–114, 2003.

134. LC Papadopoulou, CM Sue, MM Davidson, K Tanji, I Nishino, JE Sadlock. Fatal infantile cardioencephalomyopathy with COX deficiency and mutations in SCO2, a COX assembly gene. *Nat Genet* 23: 333–337, 1999.

135. H Antonicka, SC Leary, GH Guercin, JN Agar, R Horvath, NG Kennaway. Mutations in COX10 result in a defect in mitochondrial heme A biosynthesis and account for multiple, early-onset clinical phenotypes associated with isolated COX deficiency. *Hum Mol Genet* 12: 2693–2702, 2003.

136. I Valnot, JC von Kleist–Retzow, A Barrientos, M Gorbatyuk, JW Taanman, B Mehaye. A mutation in the human heme A:farnesyltransferase gene (COX10) causes cytochrome c oxidase deficiency. *Hum Mol Genet* 9: 1245–1249, 2000.

137. I Valnot, S Osmond, N Gigarel, B Mehaye, J Amiel, V Cormier–Daire. Mutations of the SCO1 gene in mitochondrial cytochrome c oxidase deficiency with neonatal-onset hepatic failure and encephalopathy. *Am J Hum Genet* 67: 1104–1109, 2000.

138. P de Lonlay, I Valnot, A Barrientos, M Gorbatyuk, A Tzagoloff, JW Taanman. A mutant mitochondrial respiratory chain assembly protein causes complex III deficiency in patients with tubulopathy, encephalopathy and liver failure. *Nat Genet* 29: 57–60, 2001.

139. V Fellman. The GRACILE syndrome, a neonatal lethal metabolic disorder with iron overload. *Blood Cells Mol Dis* 29: 444–450, 2002.

140. I Visapaa, V Fellman, J Vesa, A Dasvarma, JL Hutton, V Kumar. GRACILE syndrome, a lethal metabolic disorder with iron overload, is caused by a point mutation in BCS1L. *Am J Hum Genet* 71: 863–876, 2002.

141. A Rotig, P de Lonlay, D Chretien, F Foury, M Koenig, D Sidi. Aconitase and mitochondrial iron–sulphur protein deficiency in Friedreich ataxia. *Nat Genet* 17: 215–217, 1997.

142. R Allikmets, WH Raskind, A Hutchinson, ND Schueck, M Dean, DM Koeller. Mutation of a putative mitochondrial iron transporter gene (ABC7) in X-linked sideroblastic anemia and ataxia (XLSA/A). *Hum Mol Genet* 8: 743–749, 1999.

143. S Taketani, K Kakimoto, H Ueta, R Masaki, T Furukawa. Involvement of ABC7 in the biosynthesis of heme in erythroid cells: interaction of ABC7 with ferrochelatase. *Blood* 101: 3274–3280, 2003.

144. ES Arner, A Holmgren. Physiological functions of thioredoxin and thioredoxin reductase. *Eur J Biochem* 267: 6102–6109, 2000.

145. A Miranda–Vizuete, AE Damdimopoulos, G Spyrou. The mitochondrial thioredoxin system. *Antioxid Redox Signal* 2: 801–810, 2000.

146. E Agsteribbe, A Huckriede, M Veenhuis, MH Ruiters, KE Niezen–Koning, OH Skjeldal. A fatal, systemic mitochondrial disease with decreased mitochondrial enzyme activities, abnormal ultrastructure of the mitochondria and deficiency of heat shock protein 60. *Biochem Biophys Res Commun* 193: 146–154, 1993.

147. P Briones, MA Vilaseca, A Ribes, A Vernet, M Lluch, V Cusi. A new case of multiple mitochondrial enzyme deficiencies with decreased amount of heat shock protein 60. *J Inherit Metab Dis* 20: 569–577, 1997.

148. JJ Hansen, A Durr, I Cournu–Rebiex, C Georgopoulos, D Ang, MN Neilsen. Hereditary spastic paraplegia SPG13 is associated with a mutation in the gene encoding the mitochondrial chaperonein Hsp60. *Am J Hum Genet* 70: 1328–1332, 2002.

149. G Casari, M De Fusco, S Ciarmatori, M Zeviani, M Mora, P Fernandez. Spastic paraplegia and OXPHOS impairment caused by mutations in paraplegin, a nuclear-encoded mitochondrial metalloprotease. *Cell* 93: 973–983, 1998.

150. CJ McDermott, RW Taylor, C Hayes, M Johnson, KM Bushby, DM Turnbull. Investigation of mitochondrial function in hereditary spastic paraparesis. *Neuroreport* 14: 485–488, 2003.

151. H Jin, E Kendall, TC Freeman, RG Roberts, DL Vetrie. The human family of deafness/dystonia peptide (DDP) related mitochondrial import proteins. *Genomics* 61: 259–267, 1999.

152. H Jin, M May, L Tranebjaerg, E Kendall, G Fontan, J Jackson. A novel X-linked gene, DDP, shows mutations in families with deafness (DFN-1), dystonia, mental deficiency and blindness. *Nat Genet* 14: 177–180, 1996.

153. K Roesch, SP Curran, L Tranebjaerg, CM Koehler. Human deafness dystonia syndrome is caused by a defect in assembly of the DDP1/TIMM8a–TIMM13 complex. *Hum Mol Genet* 11: 477–486, 2002.

154. L Tranebjaerg, PK Jensen, M Van Ghelue, CL Vnencak–Jones, S Sund, K Elgjo. Neuronal cell death in the visual cortex is a prominent feature of the X-linked recessive mitochondrial deafness–dystonia syndrome caused by mutations in the TIMM8a gene. *Ophthalmic Genet* 22: 207–223, 2001.

155. DC Wallace, DG Murdock. Mitochondria and dystonia: the movement disorder connection? *Proc Natl Acad Sci USA* 96: 1817–1819, 1999.

156. J Binder, S Hofmann, S Kreisel, JC Wohrle, H Bazner, JK Krauss. Clinical and molecular findings in a patient with a novel mutation in the deafness–dystonia peptide (DDP1) gene. *Brain* 126 (Pt 8): 1814–1820, 2003.

157. C Alexander, M Votruba, UE Pesch, DL Thiselton, S Mayer, A Moore. OPA1, encoding a dynamin-related GTPase, is mutated in autosomal dominant optic atrophy linked to chromosome 3q28. *Nat Genet* 26: 211–215, 2000.

158. C Delettre, G Lenaers, JM Griffin, N Gigarel, C Lorenzo, P Belenguer. Nuclear gene OPA1, encoding a mitochondrial dynamin-related protein, is mutated in dominant optic atrophy. *Nat Genet* 26: 207–210, 2000.

159. A Olichon, LJ Emorine, E Descoins, L Pelloquin, L Brichese, N Gas. The human dynamin-related protein OPA1 is anchored to the mitochondrial inner membrane facing the inter-membrane space. *FEBS Lett* 523: 171–176, 2002.

160. A Olichon, L Baricault, N Gas, E Guillou, A Valette, P Belenguer. Loss of OPA1 perturbates the mitochondrial inner membrane structure and integrity, leading to cytochrome c release and apoptosis. *J Biol Chem* 278: 7743–7746, 2003.

161. LC Ades, AK Gedeon, MJ Wilson, M Latham, MW Partington, JC Mulley. Barth syndrome: clinical features and confirmation of gene localization to distal Xq28. *Am J Med Genet* 45: 327–334, 1993.

162. S Bione, P D'Adamo, E Maestrini, AK Gedeon, PA Bolhuis, D Toniolo. A novel X-linked gene, G4.5. is responsible for Barth syndrome. *Nat Genet* 12: 385–389, 1996.

163. SB Bleyl, BR Mumford, V Thompson, JC Carey, TJ Pysher, TK Chin. Neonatal, lethal noncompaction of the left ventricular myocardium is allelic with Barth syndrome. *Am J Hum Genet* 61: 868–872, 1997.

164. S Vesel, M Stopar–Obreza, K Trebusak–Podkrajsek, J Jazbec, T Podnar, T Battelino. A novel mutation in the G4.5 (TAZ) gene in a kindred with Barth syndrome. *Eur J Hum Genet* 11: 97–101, 2003.

165. M Schlame, JA Towbin, PM Heerdt, R Jehle, S DiMauro, TJ Blanck. Deficiency of tetralinoleoyl-cardiolipin in Barth syndrome. *Ann Neurol* 51: 634–637, 2002.

166. F Valianpour, RJ Wanders, H Overmars, P Vreken, AH Van Gennip, F Baas. Cardiolipin deficiency in X-linked cardioskeletal myopathy and neutropenia (Barth syndrome, MIM 302060): a study in cultured skin fibroblasts. *J Pediatr* 141: 729–733, 2002.

167. P Vreken, F Valianpour, LG Nijtmans, LA Grivell, B Plecko, RJ Wanders. Defective remodeling of cardiolipin and phosphatidylglycerol in Barth syndrome. *Biochem Biophys Res Commun* 279: 378–382, 2000.

168. L Canafoglia, S Franceschetti, C Antozzi, F Carrara, L Farina, T Granata. Epileptic phenotypes associated with mitochondrial disorders. *Neurology* 56: 1340–1346, 2001.

169. S Krahenbuhl, S Brandner, S Kleinle, S Liechti, D Straumann. Mitochondrial diseases represent a risk factor for valproate-induced fulminant liver failure. *Liver* 20: 346–348, 2000.

170. CW Lam, CH Lau, JC Williams, YW Chan, LJ Wong. Mito-chondrial myopathy, encephalopathy, lactic acidosis and stroke-like episodes (MELAS) triggered by valproate therapy. *Eur J Pediatr* 156: 562–564, 1997.

171. T Taivassalo, EA Shoubridge, J Chen, NG Kennaway, S DiMauro, DL Arnold. Aerobic conditioning in patients with mitochondrial myopathies: physiological, biochemical, and genetic effects. *Ann Neurol* 50: 133–141, 2001.

5

Using Mitochondrial DNA in Population Surveys

CAROLYN D. BERDANIER

CONTENTS

INTRODUCTION

Early in this century, the concept of the gene as the unit of inheritance became widely accepted. At about the same time, it was also recognized that the gene could spontaneously change. It could be altered or mutated so that the appearance or behavior of the organism carrying the gene would also change. Once mutated, the gene would become stable and give rise to subsequent generations of that organism with this "new" characteristic.

Originally, geneticists were concerned with genes located in the nucleus. However, during the last few decades, the mitochondrial genome (mtDNA) has received attention with respect to the metabolic characteristics of an organism. Many species have been studied and their mt genome sequenced and mapped. Table 5.1 is a partial list of the many species that have been sequenced (1–19). Variation within a species is known and most has occurred as spontaneous base substitutions, deletions, or additions in areas of the genome that are not critical to the function of the gene product. For example, mt polymorphism in the human has been found to associate with different responses to exercise (20). Instances in which base substitutions, deletions, or additions have effects on the active sites of the gene products usually (but not always) result in organisms whose functions are compromised. Mitochondrial disease due to a mutation in a critical part of the mtDNA is a particular area of interest in the human (21). Chapter 4 and Chapter 6 through Chapter 8 describe these disorders.

Table 5.1   Mitochondrial DNA in a Variety of Species

Species	# Bases	Remarks	Ref.
Yeast	~50 kb	Some may be linear	1
Red alga	25.8 kb	51 genes	2
Amoeboid protozoan	41.6 kb	Contains open reading frames	3
Pythuim (fungus)	50–150 kb	Linear and circular forms	4
Xenopus laevis	17.553 kb		5
Japanese pond frog	18.4–19.1 kb		6,7
Honey bee		Paternal DNA present	8
Sheep	~16.5	Wild and domestic sheep differ in control region length	9
Cows	16.338 kb	Many polymorphisms reported	10–12
Horses	~16.5	Very diverse	13,14
Donkey	16.67 kb	Length can vary	15
Rat	16.298		16
Gibbon	16.472 kb	Four genes lack complete stop codons	17
Man	16.569 kb	Polymorphisms are common	18,19

Table 5.2 Amino Acid Codons in mtDNA[a]

Amino Acid	Codon
Lysine	AAA, AAG
Asparagine	AAC, AAU
Threonine	ACA, ACC, ACG, ACU
Arginine	AGA, AGG, CGA, CGC, CGG, CGU
Serine	AGC, AGU, UCA,UCC, UCG, UCU
Isoleucine	AUA, AUC, AUU
Methionine	AUG
Glutamine	CAA
Histidine	CAC, CAU
Proline	CCA, CCC, CCG, CCU
Leucine	CUA,CUC, CUG, CUU, UUA, UUG
Glutamate	GAA, GAG
Aspartate	GAC, GAU
Alanine	GCA, GCC, GCG, GCU
Glycine	GGA, GGC, GGG, GGU
Valine	GUA, GUC, GUG, GUU
Tyrosine	UAC, UAU
Cysteine	UGC, UGU
Tryptophan	TGA (different from the nuclear codon for this amino acid)
Phenylalanine	UUC, UUU

[a] Stop codons are UAA, UAG, and UGA; initiator codons are AUG, AUA, AUU, or AUC. Many of the reading frames lack termination codons and only code a T or TA after the last sense codon. The genome of most mammals has a GC content between 36 and 44%.

Base substitutions can occur when the codon for a specific amino acid is not affected. Some amino acids have more than one codon (see Table 5.2). Other substitutions can occur so that the function of the gene product is unaffected because the amino acid substitution is in a peripheral portion of the resultant gene product. In other words, the amino acid substitution has little effect on the functionality of the protein. Still other substitutions can occur in the noncoding bases in the genome. The mt genome has very few of these bases; less than 6% of its total number of bases are noncoding. Of these, 97% are found in the two controller regions: the 30-bp origin of the replication for the light strand and the 898-bp D-loop. The D-loop is sandwiched between the genes for the tRNA[Phe], which is downstream, and the upstream tRNA^{Pro}. Base substitution in these noncoding regions as well as inconsequential polymorphisms within the sequence of the whole genome serve as markers of the genome and can be used to trace related generations through time.

Analysis of human mtDNA has become a very useful tool in tracking population movement throughout the world. Much has been learned using this small genome inherited almost entirely from the mother (22–25). Estimates vary as to the contribution of sperm mitochondria to the resultant progeny. Mitochondria in the sperm are located in the midpiece of the tail. Although this portion of the tail enters the egg during fertilization, it disappears by some unknown mechanism (23).

Many copies of mtDNA are in every cell and heteroplasmy (the presence of mutant and wild-type mtDNA in the same individual) is common in persons with mt disease (see Chapter 4 and Chapter 6 through Chapter 8, this volume). In persons with disease, the percentage of mutant DNA present determines the seriousness of the disease. At 90% or higher mutation, serious tissue malfunction can be expected. In contrast, those with a lower percentage of heteroplasmy are symptom free. If they are female, they can transmit their heteroplasmy to their offspring. The random distribution of the mitochondria to the fertilized egg determines whether the offspring will have a higher or lower percent mutant DNA and mt disease symptoms. Siblings of the same mother can differ in their percent of mutant DNA and can also differ with respect to mt disease severity.

The continuous range of possible gene dosages results in a wide range of phenotypes. This adds an additional complexity to the transmission of inherited mt disease as well as a complexity to tracing inheritance patterns within and between populations originating from the same part of the world. Tracing migration patterns is of interest to anthropologists and evolutionists as well as geneticists.

Although the mutation rate for mt DNA is high (roughly ten times faster than nuclear DNA mutation), most individuals are homoplasmic for their mother's DNA; homoplasmy (absence of mutant mtDNA) is more common than heteroplasmy. The surprising observation is that so much homoplasmy exists given the high rate of mutation. The reason for the homoplasmy has been attributed to a so-called bottleneck whereby a single copy (or several identical copies) of the mtDNA ultimately populates the whole organism (24–32).

Several reports of animal populations have shown that when a single point mutation occurred between the mother and offspring, often the mtDNA switched completely in a single generation. That is, each (the mother and the offspring) was homoplasmic with regard to that base substitution. The mother had one base at that particular location; the offspring had another base at that same position. Some process occurred that ensured that each (mother and offspring) had an identical and different base sequence. The oocyte has ~100,000 copies of the mtDNA, yet the mutation occurred in the same place only once or twice. Therefore, Laipis et al. and others (28–30) suggested that a bottleneck in mitogenesis had occurred, with only one to six copies of the mtDNA serving as the template for all the rest of the DNA. This, in turn, suggested that at some point in oogenesis, mitogenesis ceased or was quiescent and then resumed, using a mutant or normal DNA as its primary template to maintain homoplasmy (31).

However, the bottleneck hypothesis is not without flaw (24,31–33). If homoplasmy was indeed the common event, then homoplasmy with mutant DNA is just as possible as homoplasmy with a normal mtDNA sequence. This does not occur or, if it does, it is a lethal event. So then, how does heteroplasmy occur? How is it maintained?

Bendall and Sykes have reported a heteroplasmic polymorphism that escaped the hypothesized bottleneck (30). This mutation at bp 16189 is frequently accompanied by a heteroplasmic variation of between 9 to 13 bp in the length of the C tract. In these researchers' study, each individual had up to four variants in the number of cytosines nearby. They suggested that this might contribute to an instability at this position due to a loss in thymine. If a C⇒T transition occurs, instability is reduced. These investigators suggested that several, rather than one or two, mtDNAs served as the templates for subsequent copies of the genome. They also suggested that the unit of inheritance is the mitochondrion that contains eight to ten copies of the genome that, in turn, could have more than one sequence available for replication. Other studies likewise showed a skewed segregation of mtDNA in human oocytes with respect to the distribution of mutant vs. normal or wild-type DNA (32,33). This skewed segregation could not occur if the bottleneck theory proposed by Marchington and others (34) is indeed the operating mechanism to explain homoplasmy.

That homoplasmy exists with a variety of polymorphisms suggests that these base substitutions, deletions, and rearrangements are in noncritical regions of the DNA that have little to do with DNA replication or mitogenesis. Once a part of the genome, they remain resident for subsequent generations. As long as the DNA change has little to do with survival and normal function, it continues to be promulgated through the maternal line. How the substitution, deletion, rearrangement, or addition occurred and whether the environment induced these changes will continue to pique the curiosity of population scientists for years to come.

EVOLUTION

The molecular biology of the mitochondrion has provided a new path for the study of evolution of the human species. Mutation in the nucleus accumulates slowly and, indeed, because the nucleus has such an efficient repair system, spontaneous damage or mutation to the nuclear genome is quickly repaired and does not remain resident. In contrast, repair of the mt genome is less efficient. Some repair does take place (35,36), but it is not as efficient as that which occurs in the nucleus.

In addition to the problem of repair to the nucleus with respect to nuclear mutation accumulation, nuclear genes are inherited from both parents and mix in every generation. This mixing obscures the history of subsequent generations because, at each generation, mixing and recombination occurs. Recombination makes it hard to trace particular historical events in the nuclear DNA unless tightly linked segments of this DNA are considered.

As described earlier, mtDNA is maternal with less than 1% coming from the father (22,23,37). Relative to nuclear mutation accumulation, mt mutation accumulates faster and in so doing provides a magnified view of human diversity in the gene pool. MtDNA does not recombine as readily as the nuclear DNA, so it is an excellent tool to define the relationships between individuals. Because homoplasmy is more common than heteroplasmy and because this genome is small compared to the nuclear genome, base sequence differences can be identified more easily than sequence differences in the nuclear genome.

The bottleneck hypothesis, although flawed as discussed earlier, does serve to restrict the genetically effective size of the population within each oocyte, assuring that the inheritance pattern is maternal and haploid. In turn, this means that mtDNA is more sensitive than nuclear DNA to severe reductions in the number of individuals in a population of organisms. Offspring of a pair of individuals carry only one type of mtDNA, but can carry up to four types of nuclear DNA at each gene — all of which are transmissible to their offspring. The pattern of transmission of the mtDNA provides a window into human population movements throughout the world because their particular mtDNA sequence can be detected and relationship maps can be drawn. Among the first to use this genetic information in this way were Avise (38) and Stoneking et al. (39,40).

Cann et al. studied the mtDNA from 147 people (39) drawn from five different areas of the world and deduced that all of these people came from one woman who lived about 200,000 years ago, probably in Africa. They based this deduction on their observation that all except the people from Africa had multiple base substitutions in areas other than that common to all five areas. The mtDNA from the African peoples had the fewest substitutions, deletions, rearrangements, and additions.

Using restriction mapping techniques, they reported that the 147 mtDNAs could be segregated into 133 distinct types specific to a particular region of the world. These clusters were then compared and an average age of the clusters estimated. The mtDNA from the African subjects (identified on the basis of unique nuclear gene characteristics) had only one cluster with an age estimate of 90 to 180 thousand years. Age estimates were based on the assumption that mtDNA divergence occurs at the rate of 2 to 4% per million-year period. They assumed that because there was only one cluster in this DNA, the African peoples were the origin of all the other peoples. They hypothesized that the number of different clusters found in the mtDNA of people from other regions of the world (Europe, Asia, Australia, Papua–New Guinea) represented changes in the DNA that occurred after their predecessors left the African place of origin.

In a subsequent publication Stoneking (40) expanded on this hypothesis, modifying it to address the criticism of the original idea of a single original female in Africa. Using newer methods of mtDNA analysis and developments in the understanding of mtDNA and its inheritance, Stoneking modified his African female origin theory. He indicated that, although it is possible that more than one female served as the common ancestor, the data collected and interpreted to date suggested the migration of people to areas of the world that differed environmentally. This could have resulted in mtDNA change and it was possible that these changes could have spontaneously occurred in the same place.

We do know that the mt base sequence provides several areas that are more vulnerable to mutation than other places. A multiple base repeat, for example, is a vulnerable mutation site; the controller region is one of these sites. Wakely (41) has identified 29 such sites in this region. The variation is not clustered, but occurs wherever there are multiple base repeats. Because the mt genome is so small, it is altogether likely that sequence divergence could happen in the same place many times and yet give the appearance of identical origin.

Because this can happen, the time estimate of Stoneking et al. was questioned by Arnason et al. (42). They suggested that the divergence of different populations from the original progenitor female was probably much earlier than 200,000 years ago. Using fossil evidence, they suggested that hominids evolved much more slowly and that the time line began ~60 million years ago.

Nei (43) and Pesole et al. (44) suggest yet other origin times based on estimations using simulations. An estimate of ~6 million years was proposed by Pesole and coworkers; Nei proposed a shorter period. Models using amino acid substitutions to study the evolution of mt-encoded proteins have not been shown to be informative with respect to evolution and time origin (45). Nonetheless, these models are of interest on a mechanistic level when trying to describe functional differences (or polymorphisms) in the 13 proteins encoded by the mtDNA within a population. They may also be useful in describing differences and similarities between populations, especially when questions about environmental change and mt mutation are posed.

POPULATION VARIANCES IN THE WORLD

The literature on population variance is large. For the purposes of this chapter only a few of these reports will be discussed. The most extensively studied area of the mtDNA has been the noncoding and variable D-loop followed by the noncoding region that follows the gene for cytochrome oxidase V. The D-loop intergenic region is composed of a 9-bp core sequence (CCC-CCTCTA) repeated twice in a tandem array. As mentioned earlier, this base repeat is quite vulnerable to mutation or base substitution. Length substitution has also been reported for this region. The length mutation is of two types: one short and the other long. People of East Asian origin commonly have the short type. An exception to this generality is an instance of an individual from Nepal who had the long type.

An unusual mtDNA sequence variant in the region following the cytochrome oxidase V gene was reported from an Egyptian mummy (46). The mummy had a GGG to GGA transition after the end of the cytochrome oxidase gene. It is hoped that this unusual transition could be used as a marker to trace the movements of people from the Nile to (or from) other parts of Africa and to the Mediterranean region as well. This region is a very small noncoding region in the mtDNA referred to as region V and is sometimes used for identification purposes in population studies. The most well-known variant of this region is the Asian-specific 9-bp deletion found in prehistoric ancient remains of Pacific Islanders. This variant has been used to track the migrations of people from the Melanesian archipelago (47–49). South American aboriginals were found to have this deletion as did the Pima Indians and the Navajos and the Mayas from the Yucatan Peninsula (50–53). The deletion was not found in the Ticuna people of the Amazon region (52,53).

Another deletion that has been found useful in studying Amerindian evolution is a 6-bp deletion between positions 106 and 111 in the D-loop (50,52–54). This deletion, called the Huetar deletion, occurred about 3000 to 5000 years ago, based on its presence in the Chibchan Panamanian Kuna and the estimated divergence time for proto-Talamancan, proto-Guaymi, and proto-Kuna of ~5000 years. The deletion found in the Bribris people suggests that these people descended from Kunas.

Actually, there is some argument about when the ancestors of the Amerinds entered the New World (55). Some argue that their entry occurred more than 30,000 years ago while others maintain that entry occurred less than 13,000 years ago. Segregating populations on the basis of their language and mtDNA sequence, approximately four immigration groups were identified who entered the New World over a 5000-year span of time. Whether some arrived by foot across the Bering Strait or by boat is subject to speculation.

An example of population diversity with respect to origin and time of entry is shown in the study of the people who live on two small islands in the Gulf of Guinea (56). Although the islands are close to each other, their people are very different genetically. One island, Biko, was colonized around 10,000 years ago by migrants from nearby islands; these people show very little DNA diversity. The other island, Sao Tome, was first settled by the Portuguese who brought African slaves with them. The people on this island are quite diverse. Some show clear evidence of their African heritage and others show markers of European people.

Although isolated people tend to be similar genetically, this is not always true. An analysis of 92 unrelated individuals in Galicia, Spain, revealed that there were 50 different sequences defined by 56 variable positions (57). These findings placed this isolated population at the edge of European variation. They were strikingly similar to the Basque people; these observations suggest that they were migrants to this isolated region of Spain during the Upper Paleolithic age.

Genetic diversity has also been reported within the Eskimos. Those from Siberia and those from Alaska were studied and found to differ (58). The individuals could be divided into four haplotypes (A through D). Based on the prevalence of these haplotypes in the people studied, it was suggested that migration to the New World by Eskimos occurred in several stages. The first occurred ~34,000 years ago and subsequent migrations occurred between 13,000 and 16,000 years ago. The 16192C→T transition identifies people from the northern Pacific Rim. It is prevalent in the Alaskan Eskimos and can be found in several Native American groups (58).

Linguistic barriers between populations may not only reflect but also determine evolutionary processes. Rates of gene-frequency change across language boundaries are often significantly higher than those at random locations and linguistically related groups are often genetically related as well. The explanation of this similarity is that the groups must share a common ancestry (59,60).

This hypothesis was tested by Barbujani et al. (61), who investigated the mtDNA sequence across linguistic and geographic boundaries in Italy. They sequenced the 255-bp segment of the D-loop in nine groups of people living in an area of Northern Italy divided by a river. They reported that the most sequence diversity occurred within, rather than between, the different groups. Furthermore, the groups living on the western side of the river seemed to have evolved in relative isolation compared to the groups living on its eastern side. Probably, more intermixing of these people with people outside the region occurred.

A recent survey of European mtDNA has demonstrated the presence of haplotypes A through D as well as another called haplotype X (59). Haplotype X represents ~4% of European mtDNA and has been further characterized by C→T transitions at nucleotide position 16223 and np 16278. Haplotype X appears to be essentially restricted to northern Amerindian groups including the Sioux, Yakima, Ojibwa, and Nuu-Chah-Nulth. This haplotype was also observed in the Navajo. People with this haplotype are considered the original founders of the Native American populations and it has not been found in any Asian populations. This suggests that some of the Native American founders were of Caucasian origin.

As population biologists have looked around the world, they have found numerous similarities and differences among mtDNA sequence. By carefully assessing this diversity, they have been able to speculate about the migratory patterns of humans as well as the degree in intermixing of the respective gene pools. Gene exchange is a common feature of populations when they come into contact with each other and studies showing this have been published (60–63 and others not listed). Without a doubt, using the mtDNA sequence diversity with specific characteristics attributed to specific groups has allowed geographers to map the movement and development of cultural groups. These are powerful tools that can help in understanding the origin of man and his subsequent development.

REFERENCES

1. MA Jacobs, SR Paynes, AJ Bendick. Moving pictures and pulsed-field gel electrophoresis show only linear mitochondrial DNA molecules from yeasts with linear mapping and circular-mapping mitochondrial genomes. *Curr Genet* 30:3–11; 1996.

2. C Leblanc, C Boyen, O Richards, G Bonnard, J-M Grienenberger, B Kloarg. Complete sequence of the mitochondrial DNA of rhodophyte, *Chondrus crispus* (gigartinales). Gene content and genome organization. *J Mol Biol* 250:484–495; 1995.

3. G Burger, I Plante, KM Lonergan, MW Gray. The mitochondrial DNA of the amoeboid protozoan, *Acanthamoeba castellanii*: complete sequence, gene content, and genome organization. *J Mol Biol* 245:522–537; 1995.

4. FN Martin. Linear mitochondrial genome organization *in vivo* in the genus *Pythium*. *Curr Genet* 28:225–234; 1995.

5. BA Roe, D-P Ma, RK Wilson, JF-H Wong. The complete nucleotide sequnece of the *Xenopus laevis* mitochondrial genome. *J Biol Chem* 260:9759–9774; 1985.

6. M Sumida. Mitochondrial DNA differentiation in the Japanese brown frog *Rana japonica* as revealed by restriction endonuclease analysis. *Genes Genet Syst* 72:79–90; 1997.

7. M Sumida. Inheritance of mitochondrial DNAs and allozymes in the female hybrid lineage of two Japanese pond frog species. *Zoo Sci* 14:277–286; 1997.

8. MS Meusel, RFA Moritz. Transfer of paternal mitochondrial DNA during fertilization of honeybee (*Apio mellifera*) eggs. *Curr Genet* 24:539–543; 1993.

9. S Heindleder, K Mainz, Y Plante, H Lewalski. Analysis of mitochondrial DNA indicates that domestic sheep are derived from two different ancestral maternal sources: no evidence for contribution from Urial and Argali sheep. *J Heredity* 89:113–120; 1998.

10. R Steinborn, M Muller, G Brem. Genetic variation in functionally important domains of the bovine mitochondrial DNA control region. *Biochem Biophys Acta* 1397:295–304; 1998.

11. RT Loftus, DE MacHugh, LO Ngere, DS Balaian, AM Bachi, DG Bradley, EP Cunningham. Mitochondrial genetic variation in European, African, and Indian cattle populations. *Animal Genet* 25:265–271; 1994.

12. S Anderson, HL DeBruijn, AR Coulson, IC Eperon, F Sanger, IG Young. Complete sequence of bovine mitochondrial DNA: conserved features of the mammalian mitochondrial genome. *J Mol Biol* 156:683–717; 1982.

13. S Marklund, R Chaudhary, L Marklund, K Sandberg, L Andersson. Extensive mitochondrial DNA diversity in horses revealed by PCR–SSCP analysis. *Animal Genet* 26:193–196; 1995.

14. W Wang, A-H Liu, S-Y Lin, H Lan, B Su, D-W Xie, L-M Shi. Multiple genotypes of mitochondrial DNA within a horse population from a small region in Yunnan province of China. *Biochem Genet* 32:371–379; 1994.

15. X Xu, A Gullberg, U Arnason. The complete mitochondrial DNA of donkey and mitochondrial DNA comparisons among four closely related mammalian species-pairs. *J Mol Evol* 43:438–446; 1996.

16. G Gadaleta, G Pepe, G DeCandia, C Quagliariello, E Sbisa, C Saccone. The complete nucleotide sequence of the *Rattus norvegicus* mitochondrial genome: cryptic signals revealed by comparative analysis between vertebrates. *J Mol Evol* 28:497–516; 1989.

17. U Arnason, A Gullberg, X Xu. A complete mitochondrial DNA molecule of the white handed gibbon, *Hylobates lar*, and comparison among individual mitochondrial genes of all hominoid genera. *Hereditas* 124:185–189; 1996.

18. JE Hixson, WM Brown. A comparison of the small ribosomal RNA genes from mitochondrial DNA of the great apes and humans. Sequences structure, evolution, and phylogenetic implications. *Mol Biol Evol* 3:1–18; 1986.

19. S Anderson, AT Bankier, BG Barrell, MHL Bruijin, AR Coulsen, J Drouin, IC Eperon, DP Nierlich, BA Roe, F Sanger, PH Schreier, AJH Smith, R Staden, IG Young. Sequence and organization of the human mitochondrial genome. *Nature* 290:457–465; 1981.

20. FT Dionne, L Turcotte, C Thibault, MR Boulay, JS Skinner, C Bouchard. Mitochondrial DNA sequence polymorphism, VO_{2max} and response to endurance training. *Med Sci Sports Exercise* 23:177–185; 1991.

21. DC Wallace. Diseases of the mtDNA. *Annu Rev Biochem* 61:175–1212; 1992.

22. RE Giles, H Blanc, HM Cann, DC Wallace. Maternal inheritance of human mitochondrial DNA. *Proc Natl Acad Sci USA* 77:6715–6719; 1980.

23. H Shitara, J-I Hayashi, S Takahama, H Kaneda, H Yonekawa. Maternal inheritance of mouse mtDNA in interspecific hybrids: segregation of the leaked paternal mtDNA followed by the prevention of subsequent paternal leakage. *Genetics* 148:851–854; 1998.

24. J Poulton. Transmission of mtDNA: cracks in the bottleneck. *Am J Hum Genet* 57:224–226; 1995.

25. DR Marchington, GM Hartshorne, D Barlow, J Poulton. Homopolymeric tract heteroplasmy in mtDNA from tissues and single oocytes: support for a genetic bottleneck. *Am J Hum Genet* 60:408–416; 1997.

26. W Hauswirth, P Laipis. Transmission genetics of mammalian mitochondria: a molecular model and experimental evidence. In: E Quagliarello, Ed. *Biogenesis*, vol. 2 in *Achievements and Perspectives of Mitochondrial Research*. Elsevier Biomedical, Amsterdam. 49–59; 1985.

27. W Hauswirth, DA Clayton. Length heterogeneity of a conserved displacement loop sequence in human mitochondrial DNA. *Nucleic Acids Res* 13:8093–8014; 1985.

28. P Laipis, W Hauswirth, T O'Brian, G Michaels. Unequal partitioning of bovine mitochondrial genotypes among siblings. *Proc Natl Acad Sci USA* 85:8107–8110; 1988.

29. NG Larsson, MH Tulinius, E Holmes, A Oldfors, O Andersen, J Wahlstrom, J Aasly. Segregation and manifestations of the mtDNA tRNA(Lys)A&G(8344) mutation of myoclonis epilepsy and ragged red fibers (MERF) syndrome. *Am J Hum Genet* 51:1201–1212; 1992.

30. KE Bendall, BC Sykes. Length heteroplasmy in the first hypervariable segment of the human mtDNA control region. *Am J Hum Genet* 57:248–256; 1995.

31. FV Meirelles, LC Smith. Mitochondrial genotype segregation during preimplantation development in mouse heteroplasmic embryos. *Genetics* 148:877–883; 1998.

32. J Poulton, V Macaulay, DR Marchington. Mitochondrial genetics '98. Is the bottleneck cracked? *Am J Hum Genet* 62:752–757; 1998.

33. RB Blok, DA Gook, DR Thorburn, H-HM Dahl. Skewed segregation of the mtDNA nt 8993(T&G) mutation in human oocytes. *Am J Hum Genet* 60:1495–1501; 1997.

34. DR Marchington, V Macaulay, GM Hartshorne, D Barlow, J Poulton. Evidence from human oocytes for a genetic bottleneck in mtDNA disease. *Am J Hum Genet* 63:769–775; 1998.

35. SP LeDoux, GL Wilson, EJ Beecham, T Stevnsner, K Wassermann, VA Bohr. Repair of mitochondrial DNA after various types of DNA damage in Chinese hamster ovary cells. *Carcinogenesis* 13:1967–1973; 1992.

36. CC Pettepher, SP LeDoux, VA Bohn, GL Wilson. Repair of alkali-labile sites within the mitochondrial DNA of RINr 38 cells after exposure to the nitosourea streptozotocin. *J Biol Chem* 266:3113–3117; 1991.

37. M Schwartz, J Vissing. Paternal inheritance of mitochondrial DNA. *N Engl J Med* 347:576–580; 2002.

38. JC Avise, RM Ball, J Arnold. Current vs. historical population sizes in vertebrate species with high gene flow: a comparison based on mitochondrial DNA lineages and inbreeding theory for neutral mutations. *Mol Biol Evol* 5:331–344; 1988.

39. RL Cann, M Stoneking, AC Wilson. Mitochondrial DNA and human evolution. *Nature* 325:31–36; 1987.

40. M Stoneking. Mitochondrial DNA and human evolution. *J Bioenergetics Biomembranes* 26:251–259; 1994.

41. J Wakeley. Substitution rate variation among sites in hypervariable region 1 of human mitochondrial DNA. *J Mol Evol* 37:613–623; 1993.

42. U Arnason, A Gullberg, A Janke, X Xu. Pattern and timing of evolutionary divergences among hominoids based on analyses of complete mtDNAs. *J Mol Evol* 43:6650–6661; 1996.

43. M Nei. Age of the common ancestor of human mitochondrial DNA. *Mol Biol Evol* 9:1176–1178; 1992.

44. G Pesole, C Gissi, A De Chirico, C Saccone. Nucleotide substitution rate of mammalian mitochondrial genomes. *J Mol Evol* 48:427–434; 1999.

45. Z Yang, R Nielsen, M Hasegawa. Models of amino acid substitution and applications to mitochondrial protein evolution. *Mol Biol Evol* 15:1600–1611; 1998.

46. C Hanni, V Laudet, J Coll, D Stehelin. An unusual mitochondrial DNA sequence variant from an Egyptian mummy. *Genomics* 22:487–489; 1994.

47. E Hagelberg, JB Clegg. Genetic polymorphisms in prehistoric Pacific Islanders determined by analysis of ancient bone. *Proc R Soc London* B 252:163–170; 1993.

48. SW Ballinger, TG Schurr, A Torroni, YY Gan, JA Hodge, K Hassan, KH Chen, DC Wallace. Southeast Asian mitochondrial DNA analysis reveals genetic continuity of ancient mongoloid migrations. *Genetics* 130:139–152; 1992.

49. G Passarino, O Semino, G Modiano, S Santachiara–Benerecetti. COII?tRNALYS intergenic 9-bp deletion and other markers reveal that the Tharus (Southern Nepal) have oriental affinities. *Am J Hum Genet* 53:609–618; 1993.

50. MV Monsalve, H Groot de Restrepo, A Espinel, G Correal, DV Devine. Evidence of mitochondrial DNA diversity in South American aboriginals. *Ann Hum Genet* 58:265–273; 1994.

51. RL Cann, AC Wilson. Length mutations in human mitochondrial DNA. *Genetics* 104:699–711; 1983.

52. TG Schurr, SW Ballinger, Y Gan, JA Hodge, DA Merriwether, DN Laurence, WC Knowler, KM Weiss, DC Wallace. Amerindian mitochondrial DNAs have rare Asian mutations at high frequencies, suggesting they derived from four primary maternal lineages. *Am J Hum Genet* 46:613–623; 1990.

53. A Torroni, TG Schurr, C Yang, EJE Szathmary, RC Williams, MS Schanfield, GA Troup, WC Knowler, DN Lawrence, KM Weiss, DC Wallace. Native American mitochondrial DNA analysis indicates that the Amerind and Nadene populations were founded by two independent migrations. *Genetics* 130:153–162; 1992.

54. M Santos, R Barrantes. Direct screening of a mitochondrial deletion valuable for Amerindian evolutionary research. *Hum Genet* 93:435–436; 1994.

55. A Torrini, JV Neel, R Barrantes, TC Schurr, DC Wallace. Mito-chondrial DNA "clock" for the Amerinds and its implications for timing their entry into North America. *Proc Natl Acad Sci USA* 91:1162–1198; 1994.

56. E Mateu, D Comas, F Calafell, A Perez–Lezaun, A Abade. A tale of two islands: population history and mitochondrial DNA sequence variation of Bioko and Sao Tome, Gulf of Guinea. *Ann Hum Genet* 61:507–518; 1997.

57. A Salas, D Comas, MV Lareu, J Bertranpetit, A Carracedo. mtDNA analysis of the Galician population: a genetic edge of European variation. *Eur J Hum Genet* 6:365–375; 1998.

58. VB Starikovskaya, RI Sukernik, TG Schurr, AM Kogelnik, DC Wallace. mtDNA diversity in Chukchi and Siberian Eskimos: implications for the genetic history of ancient Beringia and the peopling of the new world. *Am J Hum Genet* 63:1473–1491; 1998.

59. MD Brown, SH Hosseini, A Torroni, H-J Bandelt, JC Allen, TG Schurr, R Scozzari, F Cruciani, DC Wallace. mtDNA haplogroup X: an ancient link between Europe/Western Asia and North America? *Am J Hum Genet* 63:1852–1861; 1998.

60. D Comas, F Calafell, E Mateu, A Perez–Lezaun, E Bosch, R Martinez–Arias, J Clarimon, F Facchini, G Fiori, D Luiselli, D Pettener, J Bertranpetit. Trading genes along the silk road: mtDNA sequences and the origin of central Asian populations. *Am J Hum Genet* 63:1824–1838; 1998.

61. G Barbujani, M Stenico, L Excoffier, L Nigro. Mitochondrial DNA sequence variation across linguistic and geographical boundaries in Italy. *Hum Biol* 68:201–215; 1996.

62. M Stenico, L Nigro, G Bertorelle, F Calafell, M Capitanio, C Corrain, G Barbujani. High mitochondrial sequence diversity in linguistic isolates of the Alps. *Am J Hum Genet* 59:1363–1375; 1996.

63 P Laherno, A Sajantila, P Sistonen, M Lukka, P Aula, L Pel-tonen, M-L Savontaus. The genetic relationship between Finns and the Finnish Saami (Lapps): analysis of nuclear DNA and mtDNA. *Am J Hum Genet* 58:1309–1322; 1996.

6

Mitochondrial DNA Mutation, Oxidative Stress, and Alteration of Gene Expression in Human Aging

HSIN-CHEN LEE AND YAU-HUEI WEI

CONTENTS

ABSTRACT

A decline in respiratory function and increase in oxidative stress in mitochondria have been proposed as important contributors to human aging. A wide spectrum of alterations in mitochondria and mitochondrial DNA (mtDNA) has been observed in aged individuals and senescent cells. These include:

- Decline in mitochondrial respiratory function
- Accumulation of mtDNA mutations
- Alteration in the expression of the mitochondrial genes
- Increase in the rate of production of reactive oxygen species (ROS)
- Increase in the extent of oxidative damage to DNA, proteins, and lipids

Responses to oxidative stress and their subsequent consequences in affected tissues play an important role in the deleterious effects of ROS on cellular function and on the apoptotic process, which culminate in aging and degenerative diseases. In this review, we focus on the roles that ROS play in aging-associated oxidative damage to mtDNA and proteins and on the oxidative stress responses of human cells at the molecular and cellular levels. The alterations of gene expression profiles elicited by oxidative stress in aging animals are discussed. Taking recent findings from this and other laboratories together, we suggest that the decline in respiratory function and increase in mitochondrial production of ROS and subsequent accumulation of mtDNA mutations and alteration in the expression of a few clusters of genes play an important role in the aging process.

INTRODUCTION

Aging is a universal but complex biological process. In the past few decades, several molecular mechanisms of aging have been proposed, including cumulative damage caused by reactive oxygen species (ROS), telomere shortening in cultured cells of replicative senescence, genome instability, mutation or altered expression of specific genes, and apoptotic cell death (1).

In 1956, Harman first proposed that free radicals are likely to be the key factor that contributes to human aging (2). He contended that oxygen free radicals produced during aerobic metabolism cause cumulative oxidative damage that eventually results in aging and age-related diseases in humans and in other animals. Subsequently, he extended this notion and suggested that mitochondria are the major target as well as the major source of the free radicals that lead to human aging (3,4). In the past few decades, this so-called "free radical theory of aging" has been tested by many approaches and substantial support has been gained from molecular and cellular biological studies.

Although pro-oxidants may come from a variety of intracellular sources, mitochondria have been established as the major generator and direct target of ROS in animal and human cells (5). Age-related increase in oxidative stress and oxidative damage in mitochondria have thus been proposed to play a critical role in the aging process (5,6). However, whether oxidative damage is a causal factor of aging or merely a correlate with aging is a question that remains to be answered. It has been an increasingly popular notion that the accumulation of oxidative damage during aging is a result of increased production of ROS in mitochondria.

In this review, we discuss aging-associated increase of mitochondrial production of ROS and oxidative damage as well as alteration in cellular responses to oxidative stress in aging. In addition, oxidative damage to mitochondria, impairment of the mechanisms for the removal/disposal of oxidative damage to mtDNA, and alterations in gene expression and apoptotic process of human cells during aging are discussed.

MITOCHONDRIAL OXPHOS FUNCTION DECLINES WITH AGE

The efficiency and capacity of oxidative phosphorylation (OXPHOS) in human and animal tissues and cultured cells has been demonstrated to decline with age. Immunohistochemical staining showed that cytochrome c oxidase (COX)-negative cardiomyocytes and muscle fibers are present in heart, limb, diaphragm, and extraocular muscle of normal subjects and that their number increased with age of the human subject (7–9). Moreover, along with several other investigators, we demonstrated in the late 1980s that bioenergetic functions of isolated mitochondria and electron transport activities of respiratory enzyme complexes gradually decline with age in the liver (10) and skeletal muscle (11,12) of normal human subjects. The age-related decline in mitochondrial respiratory function was also observed in skeletal muscle, heart, and liver tissues of the rat (13–15) and dogs (14). These observations have led us to the same conclusion that bioenergetic function of mitochondria in animal and human tissues declines with age.

Further studies revealed that the respiratory control, OXPHOS efficiency, rates of resting (state 4) and ADP-stimulated (state 3) respiration, and activities of the respiratory enzyme complexes decline with age in various human tissues, although to different extents (10,16). However, it remains unclear whether the amounts of the respiratory enzyme complexes in tissue cells of aged individuals change. Although it was found that the stability of mitochondrial proteins was not changed with age in somatic tissues of the rat (17), a reduction in the rate of protein synthesis was observed in *Drosophila melanogaster*, mouse liver and kidney (18,19), and human skeletal muscle (20). These findings suggest that the decrease in the rates of mitochondrial protein synthesis may contribute to the age-related decline in the capacity of aerobic metabolism and mitochondrial function.

We first observed that the age-dependent decline of glutamate–malate-supported respiration was more dramatic than that of the succinate-supported respiration in human liver (10). This finding quickly led to speculation that mutations in the seven genes of mtDNA coding for NADH dehydrogenase subunits might be involved in this aging-associated respiratory function decline (6). In addition, it was observed that mtDNA rearrangements are abundant in COX-negative fibers in the muscle of elderly subjects and that the proportion of mutated mtDNA is correlated with the progressive decrease of cytochrome c oxidase activity (9). These findings have been confirmed in later studies on various human tissues (16,21) and, together, suggest that mtDNA mutations and alteration of gene expression in mitochondria contribute, at least in part, to the decreased mitochondrial OXPHOS function observed in aged tissue cells.

Direct evidence for the aging-associated decline of mtDNA function has been provided by the observation that steady-state levels of mtRNA were decreased in old *D. melanogaster* (22) and in various aging tissues of the human and animals (23–26). The reduction in the mtRNA levels will result in decline of the mitochondrial protein synthesis and respiratory functions generally observed in old animal tissues. Although the abundance of available mtDNA template is lower and the efficiency of mitochondrial transcription declines, the stability of mRNA is not noticeably decreased in aged tissues of the rat (23,24).

The abundance of the mRNA-encoding mitochondrial RNA polymerase essential for the transcription of mtDNA was found to change significantly in the platelets of old human subjects (27). The expression of mitochondrial transcription factor A (mtTFA), which is a key regulatory protein in the transcription of mtDNA, was found to be up-regulated even in some organs of old rats and in the skeletal muscle of elderly subjects (28). The reduction in the efficiency of mitochondrial transcription may be one of the key factors responsible for the decline of protein synthesis and mitochondrial respiration with age. However, the molecular basis for these changes remains unclear.

It was observed that a reduction in the levels of mtRNA transcripts in older animal tissues is not associated with a decrease in the copy number of mtDNA. In brain and heart of the rat and in *D. melanogaster*, the age-related decline in the levels of mtRNA transcripts was not associated with decrease in the copy number of mtDNA. Indeed, mtDNA copy number was found to increase with age in *D. melanogaster* (29) and in various tissues of the rat (30) and the human

(25,26,31,32). It was suggested that, although the copy number of mtDNA was not decreased in parallel with the mtRNA transcripts, the template quality of mtDNA was significantly decreased in aged animal tissues, probably due to accumulation of oxidative damage (32,33).

Indeed, mutation and oxidative modification of mtDNA have been shown to increase with age in various tissues of the human (34,35) and animals (30). Accumulation of point mutations in the control region of mtDNA, which regulates the replication and transcription of the mitochondrial genome, was observed to occur in skin fibroblasts and in skeletal muscles of aged individuals (36,37). These findings suggest that the decrease in quality of the mtDNA template and the efficiency of mitochondrial transcription contribute to the reduced steady-state levels of mitochondrial RNA transcripts. This, in turn, leads to reduction in the rate of mitochondrial protein synthesis and OXPHOS function in aging human tissues.

MITOCHONDRIAL DNA MUTATIONS ACCUMULATE WITH AGE

Each human cell contains several hundreds to more than tens of thousands of mitochondria, each of which carries two to ten copies of mtDNA (38). Human mtDNA encodes 13 polypeptides, which are essential for the assembly of the mitochondrial energy-generating OXPHOS system, and 12S and 16S rRNAs and 22 tRNAs essential for protein synthesis in mitochondria (39). Thus, any mutation in the coding region of mtDNA could alter the capacity of mitochondria in the production of energy in the form of ATP, the development of a mitochondrial membrane potential, the matrix accumulation of Ca^{2+}, and any other subsidiary function of the respiratory chain.

Mutant mtDNA may coexist with the wild-type mtDNA within a cell — a common condition termed "heteroplasmy" (40). The degree of mtDNA heteroplasmy often varies greatly in different tissues or cells of the same individual (40,41). It has been established that many of these mtDNA mutations start to occur after age reaches the mid-30s and accumulate

Table 6.1   Aging-Associated mtDNA Mutation in Human Tissues

	Nucleotide Position	Tissues	Ref.
	Deletions		
4977 bp	8483 to 13459	Liver, skeletal muscle, brain, heart, lung, spleen, testis, diaphragm, kidney, adrenal gland, skin, oral	42–49
7436 bp	8649 to 16084	Heart, skeletal muscle, liver, skin	50, 51
Multiple deletions		Skeletal muscle, skin, testis, oral	49, 51–53
	Point Mutations		
A3243G	3243	Skeletal muscle	54
A8344G	8344	Extraocular muscle	55
T414G	414	Skin fibroblasts	36
A189G	189	Skeletal muscle	37
T408A	408	Skeletal muscle	37
	Duplications		
260 bp	−567/301	Skeletal muscle, skin, testis	56–58
200 bp	−493/301	Skeletal muscle, skin, testis	56–58

Note: The A3243G transition of mtDNA was found not to be associated with aging in human lung (48) and skin (59).

with age in postmitotic tissues of the human (Table 6.1) (36,37,42–59). Most of these aging-associated mtDNA mutations were originally observed in the affected tissues of patients with mitochondrial diseases. Among them, the most common one is a 4977-bp deletion with a 13-bp direct repeat flanking the 5′- and 3′-end breakpoints at nucleotide positions (np) 8470/8482 and np 13447/13459, respectively (60,61).

This mtDNA deletion was first observed in the skeletal muscle of patients with chronic progressive external ophthalmoplegia (CPEO), Kearns–Sayre syndrome (KSS), or

Pearson's syndrome (60–62). Multiple large-scale deletions of mtDNA can be easily detected, although at rather low levels (<1%), in most tissues of elderly subjects (45,51–53). Two A-to-G point mutations at np 8344 and np 3243 of mtDNA, which are respectively associated with MERRF and MELAS syndromes (63,64), have been shown to accumulate exponentially in the muscle tissues of aged individuals (54,55).

It is important to note that most aging-associated mtDNA rearrangements and point mutations rarely exceed 1% in aging human tissues (35). Thus, it has been questioned whether such low levels of mtDNA mutation can have a significant effect on the bioenergetic function of mitochondria in aging. Although this is a sound argument, it is noteworthy that many mutations, each rare in itself, coexist in aging tissue cells and that the total mtDNA mutation load may become significant. A number of investigators screened large-scale deletions of mtDNA in human tissues using long-range-PCR (65) or a comprehensive PCR-based detection system (66); the results revealed that a broad spectrum of different mtDNA rearrangements are accumulated with age in all the tissues examined.

In this line of studies, we have established that the total amount of mtDNA mutations, including point mutations, length mutations, and rearrangements, may reach such a high level that it could result in significant impairment of mitochondrial respiration and OXPHOS diseases (65). Moreover, the mutated mtDNA molecules may be distributed in an uneven manner among the cells of affected tissues and result in a mosaic pattern of mutant mtDNA. It is equally plausible that mutated mtDNA molecules are distributed in the affected muscle fibers in a segmental manner, with mutations clustered in the affected segment (67–69). It has been shown that the respiratory function of skeletal muscle mitochondria was severely impaired in the fiber segments harboring high proportions of mutated mtDNAs (67). This has been confirmed by recent findings that mtDNA rearrangements are extensive and the levels of full-length mtDNA are reduced in the cytochrome *c* oxidase-negative fibers of the skeletal muscle of aged individuals (68,69).

On the other hand, point mutations in the D-loop region of mtDNA were reported to accumulate with age in human tissues and cultured cells. One somatic mutation in the control region of mtDNA, a T-to-G transversion at np 414 (T414G), was found in skin fibroblasts cultured from individuals over 60 years of age with the level of the mutant mtDNA at 15 to 50% (36). Two other mtDNA mutations in the control region, A189G and T408A, have also been reported to accumulate with age in skeletal muscle (37).

Conversely, we found six aging-associated tandem duplications in the D-loop region of mtDNA from skeletal muscle, skin, and testis tissues of elderly subjects; the incidence and abundance of mtDNA with one of these tandem duplications increased with age (56–58). Because these mutations are located in or near the control region of mtDNA, they may impair the replication and transcription of mtDNA, and thus compromise the bioenergetic function of mitochondria in affected tissue cells of the elderly subjects. This is supported by recent observations that somatic mutations located near the replication origins of mtDNA are associated with reduced mtDNA copy number in human liver tissues (70).

AGING-ASSOCIATED INCREASE IN MITOCHONDRIAL ROS PRODUCTION

It has been well established that ROS and free radicals (e.g., ubisemiquinone and flavosemiquinone) are generated and maintained at a relatively high steady-state level in mitochondria of the tissue cells under normal physiological conditions (71,72). Respiratory enzyme complex I and protonmotive Q cycle operating in complex III are the major sites that generate ROS in the mitochondrial respiratory chain (72). It was reported that approximately 1 to 5% of the oxygen consumed by mitochondria is converted to ROS, including superoxide anions, hydrogen peroxide, and hydroxyl radicals in normal human tissues (72). A number of enzymes in the cytosol, such as NADPH cytochrome b_5 reductase, and oxidases, also generate small amounts of ROS; however, it is established that more than 90% of intracellular ROS are generated at the respiratory chain in the inner mitochondrial membrane during aerobic metabolism (5).

As mentioned earlier, age-dependent decline of mitochondrial OXPHOS function and accumulation of somatic mtDNA mutations can reduce the efficiency of electron transfer, resulting not only in decreased production of ATP but also in increased generation of ROS in mitochondria. It has long been understood that mitochondrial respiration and ATP synthesis are controlled by the ATP demand of energy-utilizing reactions in the tissue cell (73). It is thus plausible that the decline in the efficiency of mitochondria in aging tissue may stimulate electron flow. This could perhaps lead to an increase in ROS production due to increased electron leakage from the defective respiratory chain of mitochondria.

It was demonstrated in the early 1990s that oxidative damage and mutation of mtDNA are increased with age more conspicuously in tissues with higher energy demand (35). It was concurrently shown that the rates of production of superoxide anions and hydrogen peroxide in mitochondria are increased with age in animal tissues (74). Moreover, it was demonstrated in an animal model that the mitochondria isolated from skeletal muscle, heart, and brain of mice lacking the heart/muscle isoform of adenine nucleotide translocase (ANT) produced greater amounts of ROS (75). These results are consistent with previous observations that, when the respiration is blocked by mtDNA mutation or respiratory inhibitors, the rate of ROS production in mitochondria is significantly elevated (76,77).

Several recent studies also demonstrated that human cells harboring mutated mtDNA and/or defective mitochondria were defective in respiratory function and had higher rates of production of superoxide anions and H_2O_2 (75,78). Therefore, aging-related decline in the respiratory chain function can result in an increase of oxidative stress (35).

In addition to the increased production of mitochondrial ROS, oxidative stress can also be elicited by a decline in the capacity of intracellular antioxidant systems. Under normal physiological conditions, cells can cope with and dispose of ROS by an array of antioxidant enzymes, including manganese superoxide dismutase (MnSOD); copper/zinc superoxide dismutase (Cu/ZnSOD); glutathione peroxidases; and catalase. MnSOD and Cu/ZnSOD convert superoxide anion to H_2O_2, which is then transformed to water by glutathione peroxidase or catalase. Together with other small-molecular-weight antioxidants, these enzymes can dispose of ROS and free radicals. When ROS production and sublethal oxidative stress increase, most cells can increase the expression level or activity of these free radical scavenging enzymes to dispose of ROS (75,79,80). However, in aging tissue cells, a fraction of ROS may escape the antioxidant defense system and cause oxidative damage to various cellular constituents including DNA, RNA, proteins, and lipids.

It was demonstrated that complete or partial deficiency in the mitochondrial MnSOD gene increased mitochondrial proton leakage, inhibition of respiration, early and rapid accumulation of mitochondrial oxidative damage, and premature induction of apoptosis (81,82). Moreover, we found that in human skin fibroblasts, the activities of Cu/ZnSOD, catalase, and glutathione peroxidase are decreased with age, but that of MnSOD increases with age up to 65 years and decreases thereafter (83). The decrease of antioxidant capacity and the imbalance in the expression of free radical scavenging enzymes have been suggested to cause increased oxidative stress and damage to tissue cells in the aging process (84,85).

On the other hand, an age-related increase in mitochondrial mass may provide an extra source of oxidants. Overproliferation of abnormal mitochondria generally occurs in the muscle of aged individuals and of patients with mitochondrial myopathy. An increase in the copy number of mtDNA has been observed in aging human tissues (25,26,31,32). In addition, mitochondrial mass and mtDNA copy number are increased during the *in vivo* aging process and *in vitro* replicative senescence of human cells (86,87).

It was demonstrated that mild oxidative stress, elicited by sublethal levels of H_2O_2, caused an increase in mitochondrial mass and mtDNA copy number of human MRC-5 lung cells (88). It was also shown that ROS production was elevated in human cells with higher density of mitochondria (87,89).

These observations led to the conclusion that the age-dependent increase in the production of superoxide anions and H_2O_2 from defective mitochondria is one of the factors contributing to the decline of respiratory function in the aging process.

ENHANCED OXIDATIVE STRESS AND DAMAGE IN AGING TISSUES

Because they are the major intracellular producer of ROS, mitochondria are subjected to direct attack by ROS in animal and human cells. In fact, relatively high levels of oxidized proteins, lipids, and nucleic acids can be detected in mitochondria of tissue cells undergoing normal aerobic metabolism. Because mtDNA is situated on the inner mitochondrial membrane and thus in close proximity to the sites of electron leakage from the respiratory chain, it is highly susceptible to oxidative damage. This was confirmed by the observations that oxidative modification to mtDNA is much more extensive than that to nuclear DNA (90,91). Moreover, it was demonstrated that when the antioxidant defense systems are insufficient for efficient disposal of the excess ROS generated from defective mitochondria, the resultant oxidative stress did lead to increased oxidative damage of mtDNA in affected tissue cells (75).

Damage to mtDNA may appear in the form of base modification, abasic sites, strand breaks, and various other types of DNA lesions. Studies using HPLC and electrochemical detection of the modified nucleosides revealed that the 8-hydroxy 2'-deoxyguanosine (8-OHdG) content in mtDNA of human diaphragm or heart muscle increases with age (34,92). A study using a gene-specific DNA damage assay based on quantitative PCR revealed that damaged nucleotides block the progression of the DNA polymerase, resulting in decreased amplification of the target sequence. This suggests that mtDNA is more susceptible to oxidative damage than nDNA is (91) and that the mtDNA damage accumulates with age in postmitotic tissues (93).

Similarly, the levels of proteins with oxidative modifications and lipid peroxides in mitochondria are increased with age (74,94,95). These observations are consistent with the report that mitochondrial glutathione is markedly oxidized in aging tissues of the rat and mouse (96). The ratio between the oxidized and reduced glutathione and the 8-OHdG content of mtDNA was found to increase concurrently with age in the liver, kidney, and brain of the animals. Interestingly, age-related accrual of oxidative damage to housefly was observed at high levels in a few mitochondrial proteins (97,98).

Mitochondrial aconitase and ANT were found to be the preferred targets of oxidative damage during aging of the animal (97,98). In fact, aconitase has been found to be susceptible to superoxide anions, which cause the release of one iron atom from the iron–sulfur cluster in its active site (99). Inactivation of aconitase may block normal electron flow to oxygen and can cause an increase in ROS production, thus further increasing oxidative damage to biological molecules in mitochondria.

On the other hand, oxidation and functional inactivation of ANT may affect an efficient supply of ADP and impair state-3 respiration of mitochondria as well as the transport of ATP from mitochondria to the cytoplasm. Thus, the increase in ROS production from defective mitochondria can result in increased oxidative damage and increased mutation of mtDNA, further impairing mitochondrial respiration and oxidative phosphorylation and producing more ROS. This leads the cell to enter a vicious cycle in aging (64,84,89,100).

It was demonstrated that treatment of human skin fibroblasts with sublethal levels of oxidative stress results in the formation and accumulation of the common 4977-bp deletion in mtDNA (101). Moreover, environmental insult such as UV irradiation was reported to induce large-scale deletions of mtDNA in human skin (46,53,58) and in human skin fibroblasts (102). Fahn et al. demonstrated that the frequency of occurrence and abundance of mtDNA deletions in the lungs of smokers were higher than in those of nonsmokers (103). Moreover, we demonstrated that the proportion of 4977-bp deleted mtDNA in lung tissues was increased with the smoking index of smokers (48). Furthermore, Mansouri et al. reported that multiple mtDNA deletions are accumulated in the liver of alcoholics, especially in patients who suffered from microvesicular steatosis (104).

Recently, we found that betel quid chewing enhances the occurrence and accumulation of mtDNA mutations in human oral tissue (49). These observations clearly suggest that ROS and free radicals generated by environmental insults (e.g., exposure to UV light and noxious air pollutants); metabolism of drugs and xenobiotics; cigarette smoking; betal quid chewing; and alcohol drinking may induce or increase the formation and accumulation of mtDNA mutations in human tissues during the aging process. Moreover, exogenous oxidative stress may also impair the replication and maintenance of mtDNA and result in a decrease of the copy number (depletion) of mtDNA. Indeed, Attardi and coworkers found that the mtDNA/nuclear DNA ratios of the cybrids established from skin fibroblasts of old donors was significantly lower than those of young donors (105). These changes in the quality and quantity of mtDNA may affect, in a synergistic manner, the bioenergetic function of human mitochondria in the aging process (85,100).

Bioenergetic function of mitochondria was found to decline sharply in human skin fibroblasts following treatment with 200 μM H_2O_2 for 1 h (91). Besides exogenous oxidative insults, endogenous oxidative stress elicited by electron leakage of the mitochondrial electron transport chain was suggested to cause a loss of mitochondrial function (76). Moreover, it was recently shown that an increase in mitochondrial ROS production is associated with decline in the activities of NADH dehydrogenase and succinate dehydrogenase in skeletal muscle and heart of the MnSOD-deficient mice (81). This is consistent with previous findings that the respiratory function of mitochondria is impaired in human tissues and cells afflicted with oxidative damage (6,100).

Taken together, these observations suggest a close relationship between oxidative stress (indicated by glutathione oxidation and oxidative damage to DNA) and mitochondrial dysfunction in human and animal cells during the aging process (6,48). This is the fundamental concept of the mitochondrial role in aging in the scenario of the "vicious cycle"

(6,35,89,100). In essence, oxidative damage or mutation to mtDNA causes defects in the respiratory chain; in turn, this causes increased production of ROS and free radicals, and the cycle is repeated over and over again, with an increased damage to the function of the respiratory chain.

DNA REPAIR SYSTEMS IN MITOCHONDRIA

A steady accumulation of damage can be caused by increased rate of formation or by a decrease in the rate of removal of damaged biological molecules. Many lines of research have suggested that the rate of damage to DNA exceeds the DNA repair capacity in mitochondria, which ultimately leads to the accumulation, with age, of mutation and oxidative damage to mtDNA in somatic tissues.

Due to high susceptibility to free radical attack and inefficient repair of damage, the mitochondrial genome accumulates DNA damage and mutations with age. For example, the 8-OHdG in mtDNA was found to accumulate with age (35,51,59,83). Oxidative damage and mutation to mtDNA were estimated to occur approximately 20 times more frequently than in the nuclear DNA (41,106). Thus, one may expect that mitochondria will possess efficient DNA repair mechanisms to remove oxidative damage from their own DNA. An early report that UV-induced pyrimidine dimers were not removed from mtDNA led to the widely accepted notion that mammalian mitochondria lack efficient DNA repair systems (107). However, a subsequent study showed that mitochondria do have a rather sophisticated DNA repair enzyme system (although with a lower efficiency and limited capacity) to remove various types of oxidative damage on DNA (108).

In vitro repair studies have shown that mitochondria contain all the enzymes required for base excision repair (BER), an important mechanism for the removal of oxidative DNA damage (108). Various nuclear DNA-encoded mitochondrial BER genes have been identified that indicate that mitochondria contain the BER system such as uracil- or 8-OHdG DNA glycosylase (109); apurinic/apyrimidinic (AP) endonucleases (110–112); 8-OHdGTPase (113); DNA polymerase γ; and mitochondrial DNA ligase (114). Moreover, the BER system in mitochondria functions in a way very similar to that in the nucleus (114).

In the process of DNA repair by BER, DNA glycosylase recognizes a damaged nucleobase and then cleaves the N-glycosyl bond between the sugar and the nucleobase to generate an AP site. Some of the glycosylases have an associated AP lyase activity that cleaves the DNA phosphate backbone; others rely on AP endonucleases for DNA strand cleavage. A phosphodiesterase then excises the unsaturated sugar derivatives at the 3′-end of the DNA, and the gap of one nucleotide thus generated is then bridged by a DNA polymerase and the ends are sealed by a DNA ligase.

Because many types of oxidative damage to DNA are continually accumulated with age in mitochondria, the capacity of DNA repair in mitochondria is clearly not enough to cope with the ever increasing damage in mtDNA of somatic tissues with age. It is therefore of great interest and importance to ask whether the DNA repair system undergoes degradation and thereby contributes to the vicious cycle in aging. Hirano et al. examined the age-related change of the DNA repair activity for 8-OHdG and found that the activity of the excision repair for 8-OHdG is decreased with age in human skin fibroblasts (115). It is possible that some of the DNA repair enzymes may be deficient or absent in mitochondria of tissue cells under some pathological conditions.

Driggers and coworkers reported a defect in the DNA repair system of oxidative damage to mtDNA in a human cell line (116). Contrary findings were reported by Souza–Pinto et al., who demonstrated that 8-OHdG glycosylase/AP lyase activity was increased with age in heart and liver mitochondria (112). The results led them to suggest that the repair of oxidative DNA damage could be induced in mitochondria. In contrast, Chen et al. reported that mitochondrial BER activity was markedly declined with age in rat brain (117). This was attributed to the decreased expression of DNA repair enzymes such as 8-OHdG glycosylase and DNA polymerase γ and the reduced activity at the steps of lesion-base incision, DNA repair synthesis, and DNA ligation in the BER pathway (117).

These changes with age in various tissues suggest tissue-specific differences in the capacity or efficiency of repair of DNA damage (118). Indeed, it was found that DNA damage accumulates at different rates in different tissues of the rodents (119). Because successful repair of damaged DNA relies on coordinated action of many repair enzymes, age-dependent decline or imbalance of the activities of the afore-mentioned DNA repair enzymes will result in a compromise of the efficiency and capacity for repair of damaged DNA molecules in aging human and animal tissues. This may account for the aging-associated accumulation of oxidative damage to mtDNA in somatic tissues of the human (100,119).

MITOCHONDRIAL ROLE IN APOPTOSIS DURING AGING

It has been well established that mitochondria play a key role in the initiation and regulation of apoptosis of mammalian cells (84). Accumulation of DNA damage and mutations in the mitochondrial genome leads to mitochondrial function decline with age. Thus, it is comprehensible that apoptosis tends to occur in aging tissues with defective mitochondrial function (119). Moreover, aging-associated increase in the mitochondrial production of ROS may also promote the initiation and execution of apoptosis in somatic tissue cells.

Several lines of evidence suggest that aging is accompanied by alterations in the apoptotic process (119). However, apoptosis is an ATP-dependent process, and it is not clear whether aging will accelerate apoptosis or retard it. Many studies have demonstrated that apoptosis is enhanced in the liver, brain, heart, skeletal muscle, and oocytes of old mice or rats (120–128). This type of cell loss via apoptosis may be triggered by mitochondrial dysfunction elicited by chronic exposure to pro-oxidants; in turn, this increases the likelihood of the opening of mitochondrial permeability transition pores. However, several investigators have reported that senescent cells are less susceptible to apoptosis under the conditions of oxidative stress or bioenergetic breakdown (129,130).

Studies on mtDNA-less ρ^o cells revealed that human cells with mitochondrial respiratory chain dysfunction are more susceptible to staurosporine-induced apoptosis and that the release of cytochrome c from their mitochondria could be prevented by Bcl-2 overexpression like that seen in cells containing mtDNA (131). Moreover, a mouse model of mitochondrial disease caused by knocking out the mitochondrial transcription factor (mtTFA) showed that defects in the respiratory chain are associated with massive apoptosis of affected cells in the heart (132). These observations led to the suggestion that mitochondrial respiratory function may not, after all, be essential for apoptosis. It was reported that, although mtDNA-less ρ^o cells are more susceptible to apoptosis (132,133), they undergo apoptosis with a kinetics slower than that of the parental osteosarcoma cells (134,135).

In addition, it was demonstrated that human skin fibroblasts harboring point mutations mtDNA — such as the G11778A mutation that is the primary genetic defect for Leber's hereditary optic neuropathy (LHON) — were all more sensitive to apoptotic stimuli than were human fibroblasts carrying only wild-type mtDNA (136–138). We have examined the effect of MELAS-associated A3243G mutation and MERRF-associated A8344G mutation on apoptosis response of human cells to apoptotic stimuli. The results revealed that depletion, large-scale deletion, and point mutation of mtDNA render human skin fibroblasts and cybrids more susceptible to apoptosis triggered by UV irradiation and staurosporine (133). These findings suggest that the susceptibility of human cells to apoptotic stimuli can be enhanced by pathogenic mutations in mtDNA.

It has been demonstrated that overproduction of pro-oxidants or ROS can induce oxidative stress and cell death (5,133). High levels of pro-oxidants produced from mitochondria can induce apoptosis by changing cellular redox status, depleting reduced glutathione, reducing the ATP level, and decreasing reducing equivalents such as NADH and NADPH (139). It has been shown that oxidative damage to the ANT in mitochondria increases with age in the flight muscle of houseflies (98). The ANT is a component of mitochondrial permeability transition (MPT) pore, which is a critical regulator of apoptosis (139). In addition, it was shown that oxidative damage to phospholipids in mitochondrial membranes of tissue cells is increased with age (85,93,100).

Located mainly in the inner mitochondrial membrane, cardiolipin has been found to participate in the regulation of the activities of a few mitochondrial enzymes such as ANT and mitochondrial ATP synthase, which are prone to oxidative damage (140). Cardiolipin is an acidic phospholipid that has been demonstrated to bind with cytochrome c (a basic protein) through electrostatic interactions under normal cell physiological conditions. Pro-oxidants and ROS have been shown to facilitate lipid peroxidation and the opening of permeability transition pores and lead to the subsequent release of cytochrome c from mitochondria. It was observed that mitochondrial oxidant production is increased and intracellular levels of Ca^{2+} ions are elevated in myocytes of old animals and human subjects (141,142). Moreover, it has been reported that the threshold for Ca^{2+}-induced MPT decreases with age in mouse lymphocytes, brain, and liver (143). This leads to a lower threshold for the release of apoptogenic proteins into the cytosol.

These events contribute to the age-related increase in the tendency of opening of MPT and release of proapoptotic proteins from the intermembrane space of mitochondria. This was supported by a recent study on skin fibroblasts harboring C8993T mutation in the ATPase 6 gene of mtDNA associated with neurogenic muscle weakness, ataxia, and retinitis pigmentosa (NARP). This mtDNA mutation causes a decrease in the rate of ATP synthesis, displays elevated superoxide anions, and results in massive apoptosis of mutant cells triggered by a metabolic stress upon replacing glucose in the culture medium with galactose (144). Moreover, Kokoszka et al. reported that chronic oxidative stress in mice with a partial deficiency in MnSOD resulted in an increased sensitization of opening the mitochondrial permeability transition pore and premature induction of apoptosis (145). Therefore, aging-associated respiratory function decline, increased mitochondrial ROS production, and oxidative stress may contribute to aging human and animal cells' increased susceptibility to apoptosis.

It has been shown that a reduction in the total number of myocytes was associated with an accelerated decline in cardiac function of the aging heart (146,147). It is thought that apoptosis is a major process contributing to the age-associated loss of cardiomyocytes. Moreover, it was also observed that apoptosis is more prevalent at the late stages of aging (122). Recently, it has been shown that the cytosolic content of cytochrome c was significantly elevated in heart cells of 16- and 24-month-old male Fischer 344 rats when compared with that of 6-month-old rats (125).

It was also reported that the content of the antiapoptotic protein, Bcl-2, decreases with age, but the amount of the proapoptotic protein, Bax, remains unchanged in mitochondrial membranes (125). The decrease in Bcl-2 may result in the opening of the permeability transition pore and release of cytochrome c from mitochondria. The alteration in the relative amounts of apoptotic and antiapoptotic proteins may be one of the molecular mechanisms involved in the increase of apoptosis in aging human tissues.

In a recent study, splenocytes and thymocytes were found to undergo apoptosis with aging in rats and that this was associated with an increase in the expression of p53, Bax, and caspase-3 (127). Moreover, the expressions of Bcl-2 family proteins and caspase family proteins were altered in aging rat brain (148,149). Bcl-2 protein level reached the highest level on embryonic day 19 (E19) and decreased after birth. Bax, Bak, and Bad levels were high from E19 to 2 weeks and decreased significantly at 4 weeks; however, the Bcl-x_L levels remained high from early stage to 96 weeks of age (148).

In addition, the protein expression profiles of caspases-3 and -7 were similar to those of Bax, Bak, and Bad, but those of caspases-8 and -10 were similar to that of Bcl-2. The constitutive expressions of caspases-6 and -9 and Apaf-1 were similar to that of Bcl-x_L (149). However, the biological significance of the differential alterations of these enzymes and proteins involved in the execution and regulation of apoptosis during aging warrants further investigation.

On the other hand, it was demonstrated that aging attenuates apoptosis in the colonic mucosa of Fischer 344 rats (150). In a separate study, the livers of old rats were found to be more resistant to apoptosis in response to a moderate dose of genotoxic stress compared with those of younger rats (151). These observations suggest that an age-related downregulation of apoptosis might be involved in the increase of DNA damage and mutations with age and could explain, at least partially, the increase in the incidence of cancer during aging. However, it remains to be seen whether these results are tissue specific and characteristic of some mitotic but not postmitotic tissues. The causes of the alterations in the sensitivity of human cells to proapoptotic stimuli observed during aging and whether apoptosis per se participates in the normal aging process of the human have remained unclear.

ALTERATIONS IN GENE EXPRESSION DURING AGING PROCESS

To identify the molecular events associated with aging, a number of investigators have examined the age-related, genome-wide changes in the gene expression profile in *D. melanogaster* (152); in skeletal muscle (153), brain (154,155), heart (156–158), and liver (159) of the mouse; in liver of the rat (160); and in skeletal muscle of the rhesus monkey (161). Age-related alterations in mRNA levels may reflect changes in gene expression, in mRNA stability, or in both. It has been expected that analysis of genes that exhibit significant age-dependent changes in their mRNA levels may provide new insights into the long-standing human quest of how we age.

In addition, it has been reported that the majority of the aging-related changes in gene expression profiles in the tissues of animals can be reversed by caloric restriction (153,154,156, 159,162). Thus, it is hoped that elucidation of genes with altered expression in aging and under caloric restriction may provide novel information for a better understanding of the mechanisms by which humans age and the life span as determined and changed with diet.

Weindruch and coworkers (153,154) used oligonucle-otide-based microarrays to analyze the transcriptional alteration in the aging process in gastrocnemius muscle and neocortex and cerebellum of the mouse. They examined the expression patterns of 6347 genes, which represent 5 to 10% of the expressed genes of the mouse genome, in the aging muscle and brain. Age-related changes in the transcript levels were analyzed by comparing the mRNA profiles of 5-month-old mice with those of 30-month-old mice.

They found that aging resulted in a differential gene expression pattern indicative of a marked stress response and lower expression of metabolic and biosynthetic genes in skeletal muscle of the mouse. They proposed that an induction of stress response genes during aging is associated with an increase of damage to proteins and other macromolecules. A decline in the enzyme systems required for the turnover of damaged molecules can result from an energetic deficit in aged cells. The observed alterations in the transcription of genes associated with energy metabolism and mitochondrial function clearly indicate a decrease in mitochondrial biogenesis or turnover secondary to cumulative ROS-induced mitochondrial damage.

On the other hand, the gene expression profile of aging brain tissues indicated an increase of inflammatory response, oxidative stress, and reduced neurotrophic support in neocortex and cerebellum of the mouse (154). Induction of the genes involved in stress response is consistent with a state of higher oxidative stress and accumulation of damaged protein present in the neocortex and cerebellum of aging animals. These include genes encoding chaperone proteins, such as the genes encoding homologues of the heat-shock factors Hsp40 and Hsp59, and cathepsins D, S, and Z, which are the major components of the protein degradation system in the lysosome. Interestingly, a stress response characterized by the induction of heat-shock proteins and other oxidative stress-induced transcripts was also found to occur in aged brain and skeletal muscle of the mouse.

Furthermore, it was shown that most age-related alterations of gene expression in the skeletal muscle of the mouse could be completely or partially prevented by caloric restriction (153). Similarly, caloric restriction selectively attenuated the aging-associated induction of genes encoding proteins involved in inflammatory and stress responses in the mouse brain (154). Indeed, caloric restriction has been shown to (156)

- Slow down the intrinsic rate of aging in mammals
- Retard age-related decline in psychomotor function and ability to fulfill spatial memory tasks
- Decrease the age-associated loss of dendritic spines
- Reduce neuronal degeneration in the animal models of Parkinson's disease

Therefore, the findings that caloric restriction attenuates the increase and decrease of gene expression in aging animals have validated the use of cDNA microarray for analysis of aging-associated transcriptional alterations. These results have provided further support for the hypothesis that oxidative stress is an important cause of the aging of postmitotic tissues.

Recently, Zou et al. used a similar genome-wide approach to examine aging-associated changes in gene expression in *D. melanogaster* (152). They examined the expression patterns of 7829 expressed sequence tags (ESTs), which cover 30 to 40% of the estimated total number of expressed genes in the *Drosophila* genome. This group of researchers further analyzed age-related changes in gene expression by comparing the transcript levels of 3-day-old male flies with those of other males of six different ages. Moreover, they treated 3-day-old flies with paraquat, a commonly used herbicide that generates free radicals, and analyzed oxidative stress response of the flies.

The results showed that the aging process in *D. melanogaster* was accompanied by a reduction in transcript levels of the genes involved in reproduction, metabolism, and protein turnover and that one-third of the aging-regulated genes showed significant changes in response to oxidative stress. The responses to oxidative stress that are altered in aging

Table 6.2   Aging-Associated Alteration in Gene Expression of Mammalian Tissues

Gene	Alteration	Tissue	Ref.
		Stress Response	
Heat shock response	Increase	Skeletal muscle (mouse, monkey), neocortex (mouse), liver (mouse)	153, 154, 159, 161
Heat shock response	Decrease	Heart (mouse), cortex (mouse)	155, 158
DNA damage-inducible genes	Increase	Skeletal muscle (mouse), neocortex (mouse)	153, 154
Oxidative stress-inducible genes	Increase	Skeletal muscle (mouse, monkey), neocortex (mouse)	153, 154, 161
Lysosomal proteases	Increase	Neocortex and cerebellum (mouse), liver (mouse)	154, 159
		Inflammatory Response	
Complement cascade	Increase	Neocortex and cerebellum (mouse), liver (mouse), skeletal muscle (mouse, monkey)	153, 154, 159, 161
MHC molecule	Increase	Neocortex and cerebellum (mouse), skeletal muscle (monkey)	154, 161
Microglia activation factors	Increase	Neocortex (mouse), skeletal muscle (monkey)	154, 161
Inflammatory peptides	Increase	Neocortex and cerebellum (mouse), skeletal muscle (monkey)	154, 161
		Energy Metabolism	
Glycolysis	Decrease	Skeletal muscle (mouse)	153
Glycolysis	Increase	Heart (mouse), neocortex (mouse), liver (rat)	154, 156, 160
Mitochondrial OXPHOS	Decrease	Skeletal muscle (mouse, monkey), heart (mouse), liver (rat)	153, 155, 154, 161
Mitochondrial OXPHOS	Increase	Hypothalamus (mouse)	155
Fatty acid transport	Decrease	Heart (mouse)	156

(continued)

Table 6.2 (continued)   Aging-Associated Alteration in Gene
Expression of Mammalian Tissues

Gene	Alteration	Tissue	Ref.
Mitochondrial β-oxidation	Decrease	Heart (mouse), skeletal muscle (monkey)	156, 161
Mitochondrial β-oxidation	Increase	Liver (rat)	160
Creatine kinase	Increase	Skeletal muscle (mouse), neocortex (mouse), heart (mouse)	153, 154, 156
Protein Turnover			
Protein degradation	Decrease	Skeletal muscle (mouse), neocortex and cerebellum (mouse)	153, 154
Protein degradation	Increase	Heart (mouse), hypothalamus and cortex (mouse)	155, 156
Protein degradation	No effect	Liver (rat)	160
Protein synthesis	Decrease	Heart (mouse), cerebellum (mouse)	154, 156

fruit flies include up-regulation of the genes to counteract
oxidative damage and down-regulation of genes whose expression is increased with endogenous oxidant levels in the flies.

By comparing the gene expression profiles for various
tissues of mice (Table 6.2) (153–156,158–161), we found that
most aging-associated changes in gene expression profiles are
tissue specific. It was noted that aging induces the expression
of a number of stress response genes in skeletal muscle, neocortex, cerebellum, and liver of the laboratory animals
(153,154,159). In contrast, aging reduces the expression of
other stress response genes in the cortex and hypothalamus
(155). Likewise, it was found that aging induced expression
of several inflammatory genes in the neocortex, cerebellum,
and liver, but not in skeletal muscle, cortex, or hypothalamus.

These results also suggest that tissues are subjected to
different stresses during aging, and all the results indicate
that oxidative stress plays a critical role in the aging process

(Table 6.2). Additionally, the analysis of aging-associated alterations in gene expression profile in the mouse and fruit flies has led to the important conclusion that aging reduces expression of the genes involved in energy metabolism and mitochondrial respiratory function (Table 6.2) (152–161). This may be a result of oxidative damage and mutation of mtDNA accumulated in tissue cells during the aging process, which leads to reduced mitochondrial function or biogenesis.

Moreover, the notion that the aging-associated reduction in the efficiency and capacity of oxidative phosphorylation is caused by altered mitochondrial gene expression (Table 6.2) was supported by the induction of free radical scavenging enzymes and inflammatory response proteins in tissue cells or cultured cells of old animals. However, it is not clear whether the observed aging-related changes in the mRNA levels are a causal factor or just a consequence of aging. Further functional studies of the proteins encoded by aging-related genes and their roles in the biology of aging are warranted.

CONCLUDING REMARKS

Aging is a natural, complex, and multifactorial biological process. Many of the studies conducted on cultured human cells and animals have revealed that aging is associated with impairment of bioenergetic functions, decreased ability to respond to stresses, and increased risk of contracting age-associated diseases. Most of these characteristics and phenomena gradually occur in advanced age in organs and tissue cells, which is usually correlated with accumulation of mtDNA mutations, oxidative damage, aberrant proteins, and defective mitochondria in all somatic tissues. These findings suggest that the endogenous and exogenous factors that result in oxidative stresses, mitochondrial defects, impairment of the antioxidant and DNA repair enzyme systems, and decline of protein degradation activities may contribute, in a synergistic manner, to the aging-associated phenotypes.

It has been increasingly appreciated that mitochondria are not only the major metabolic energy supplier but also the main intracellular source and target of ROS in the cell. In mammalian cells, the proper assembly and functioning of mitochondria depend on coordination between a wide array of proteins encoded by genes in the nucleus and the mitochondrial genome (162,163). Communication between the two genomes is essential for the regulation of the synthesis of proteins in the cytoplasm and their subsequent import into mitochondria. Regulation of the activities of specific transcription factors and transcription of gene coding for polypeptides constituting the respiratory chain have been proposed to be crucial for communication between mitochondria and the nucleus in mitochondrial biogenesis (163).

In addition to the effect of the expression of genes in nuclear DNA on the expression of genes in mtDNA, gene expression of mtDNA can also affect the expression of some nuclear DNA-encoded mitochondrial proteins (163,164). Oxygen tension in tissue cells, exercise, and hormone levels have been shown to be able to regulate the mRNA levels of cytochrome *c* oxidase subunits in the mammal (165,166). Thus, mitochondria of the human and animal cells may act as a sensor in regulating energy metabolism and expression of respiratory genes in response to extracellular stimuli.

Experimental data from this and other laboratories have provided ample evidence to support the notion that mutation and oxidative damage to mtDNA and decline of mitochondrial respiratory function in tissue cells are important contributors to human aging (6,35,89). Although it has been generally recognized that modification and mutation of mtDNA play a role in age-dependent mitochondrial function decline, the detailed mechanisms by which these molecular and biochemical events cause human aging remain to be established.

Functional genomics and proteomics approaches to study aging on the genome-wide basis will provide novel information to enable a deeper understanding of the age-related alterations in the structure and function of mitochondria in the aging process. This is critical for elucidation of the molecular mechanism of aging and for a better management of aging and age-related diseases in the human.

ACKNOWLEDGMENTS

Part of the work described in this review was supported by a research grant, NSC93-2811-B010-027, from the National Science Council and a grant, NHRI-EX93-9120BN, from National Health Research Institutes, Taiwan, Republic of China.

REFERENCES

1. FB Johnson, DA Sinclair, L Guarente. Molecular biology of aging. *Cell* 96:291–302, 1999.

2. D Harman. A theory based on free radical and radiation chemistry. *J Gerontol* 11:298–300, 1956.

3. D Harman. The biological clock: the mitochondria. *J Am Geriatr Soc* 20:145–147, 1972.

4. D Harman. Free radical theory of aging: Consequences of mitochondrial aging. *Age* 6:86–94, 1983.

5. KB Beckman, BN Ames. The free radical theory of aging matures. *Physiol Rev* 78:547–581, 1998.

6. YH Wei. Oxidative stress and mitochondrial DNA mutations in human aging. *Proc Soc Exp Biol Med* 217:53–63, 1998.

7. J Müller–Höcker. Cytochrome *c* oxidase deficient cardiomyocytes in the human heart-an age-related phenomenon: a histochemical ultracytochemical study. *Am J Pathol* 134:1167–1173, 1989.

8. J Müller–Höcker. Cytochrome *c* oxidase deficient fibers in the limb muscle and diaphragm of man without muscular disease: an age–related alteration. *J Neurol Sci* 100:14–21, 1990.

9. J Müller–Höcker, P Seibel, K Schneiderbanger, B Kadenbach. Different *in situ* hybridization patterns of mitochondrial DNA in cytochrome *c* oxidase-deficient extraocular muscle fibers in the elderly. *Virchows Arch* 422:7–15, 1993.

10. TC Yen, YS Chen, KL King, SH Yeh, YH Wei. Liver mitochondrial respiratory functions decline with age. *Biochem Biophys Res Commun* 165:994–1003, 1989.

11. JM Cooper, VM Mann, AH Schapira. Analyses of mitochondrial respiratory chain function and mitochondrial DNA deletion in human skeletal muscle: effect of ageing. *J Neurol Sci* 113:91–98, 1992.

12. I Trounce, E Byrne, S Marzuki. Decline in skeletal muscle mitochondrial respiratory chain function: possible factor in ageing. *Lancet* i:637–639, 1989.

13. K Torii, S Sugiyama, K Takagi, T Satake, T Ozawa. Age-related decrease in respiratory muscle mitochondrial function in rats. *Am J Respir Cell Mol Biol* 6:88–92, 1992.

14. S Sugiyama, M Takasawa, M Hayakawa, T Ozawa. Changes in skeletal muscle, heart and liver mitochondrial electron transport activities in rats and dogs of various ages. *Biochem Mol Biol Int* 30:937–944, 1993.

15. M Takasawa, M Hayakawa, S Sugiyama, K Hattori, T Ito, T Ozawa. Age-associated damage in mitochondrial function in rat heart. *Exp Gerontol* 28:269–280, 1993.

16. RH Hsieh, JH Hou, HS Hsu, YH Wei. Age-dependent respiratory function decline and DNA deletions in human muscle mitochondria. *Biochem Mol Biol Int* 32:1009–1022, 1994.

17. RA Menzies, PH Gold. The turnover of mitochondria in a variety of tissues of young adult and aged rats. *J Biol Chem* 246:2425–2429, 1971.

18. PJ Bailey, GC Webster. Lowered rates of protein synthesis by mitochondria isolated from organisms of increasing age. *Mech Ageing Dev* 24:233–241, 1984.

19. DL Marcus, NG Ibrahim, ML Freedman. Age-related decline in the biosynthesis of mitochondrial inner membrane proteins. *Exp Gerontol* 17:333–341, 1982.

20. OE Rooyackers, DB Adey, PA Ades, KS Nair. Effect of age on *in vivo* rates of mitochondrial protein synthesis in human skeletal muscle. *Proc Natl Acad Sci USA* 93:15364–15369, 1996.

21. JM Cooper, VM Mann, AH Schapira. Analyses of mitochondrial respiratory chain function and mitochondrial DNA deletion in human skeletal muscle: effect of ageing. *J Neurol Sci* 113:91–98, 1992.

22. M Calleja, P Pena, C Ugalde, C Ferreiro, R Marco, R Garesse. Mitochondrial DNA remains intact during Drosophila aging, but the levels of mitochondrial transcripts are significantly reduced. *J Biol Chem* 268:18891–18897, 1993.

23. MN Gadaleta, V Petruzzella, M Renis, F Fracasso, P Cantatore. Reduced transcription of mitochondrial DNA in the senescent rat. Tissue dependence and effect of L-carnitine. *Eur J Biochem* 187:501–506, 1990.

24. SP Fernandez–Silva, V Petruzzela, F Fracasso, MN Gadaleta, P Cantatore. Reduced synthesis of mtRNA in isolated mitochondria of senescent rat brain. *Biochem Biophys Res Commun* 176:645–653, 1991.

25. A Barrientos, J Casademont, F Cardellach, E Ardite, X Estivill, A Urbano–Marquez, JC Fernandez–Checa, V Nunes. Qualitative and quantitative changes in skeletal muscle mtDNA and expression of mitochondrial-encoded genes in the human aging process. *Biochem Mol Med* 62:165–171, 1997.

26. A Barrientos, J Casademont, F Cardellach, X Estivill, A Urbano–Marquez, V Nunes. Reduced steady-state levels of mitochondrial RNA and increased mitochondrial DNA amount in human brain with aging. *Mol Brain Res* 52:284–289, 1997.

27. RM Kapsa, AF Quigley, TF Han, MJ Jean–Francois, P Vaughan, E Byrne. mtDNA replicative potential remains constant during ageing: polymerase gamma activity does not correlate with age related cytochrome oxidase activity decline in platelets. *Nucleic Acids Res* 26:4365–4373, 1998.

28. AM Lezza, V Pesce, A Cormio, F Fracasso, J Vecchiet, G Felzani, P Cantatore, MN Gadaleta. Increased expression of mitochondrial transcription factor A and nuclear respiratory factor-1 in skeletal muscle from aged human subjects. *FEBS Lett* 501:74–78, 2001.

29. F Morel, F Mazet, S Touraille, S Alziari. Changes in the respiratory chain complexes activities and in the mitochondrial DNA content during ageing in *D. subobscura*. *Mech Aging Dev* 84:171–181, 1995.

30. MN Gadaleta, G Rainaldi, AMS Lezza, F Milella, F Fracasso, P Cantatore. Mitochondrial DNA copy number and mitochondrial DNA deletion in adult and senescent rats. *Mutat Res* 275:181–193, 1992.

31. HC Lee, CY Lu, HJ Fahn, YH Wei. Aging- and smoking-associated alteration in the relative content of mitochondrial DNA in human lung. *FEBS Lett* 441:292–296, 1998.

32. V Pesce, A Cormio, F Fracasso, J Vecchiet, G Felzani, AM Lezza, P Cantatore, MN Gadaleta. Age-related mitochondrial genotypic and phenotypic alterations in human skeletal muscle. *Free Radic Biol Med* 30:1223–1233, 2001.

33. SA Kovalenko, G Kopsidas, MM Islam, D Heffernan, J Fitzpatrick, A Caragounis, AE Gingold, AW Linnane. The age-associated decrease in the amount of amplifiable full-length mitochondrial DNA in human skeletal muscle. *Biochem Mol Biol Int* 46:1233–1241, 1998.

34. M Hayakawa, K Hattori, S Sugiyama, T Ozawa. Age-associated oxygen damage and mutations in mitochondrial DNA in human heart. *Biochem Biophys Res Commun* 189:979–985, 1992.

35. YH Wei, HC Lee. Oxidative stress, mitochondrial DNA mutation, and impairment of antioxidant enzymes in aging. *Exp Biol Med* 227:671–682, 2002.

36. Y Michikawa, F Mazzucchelli, N Bresolin, G Scarlato, G Attardi. Aging-dependent large accumulation of point mutations in the human mtDNA control region for replication. *Science* 286:774–779, 1999.

37. Y Wang, Y Michikawa, C Mallidis, Y Bai, L Woodhouse, KE Yarasheski, CA Miller, V Askanas, WK Engel, S Bhasin, G Attardi. Muscle-specific mutations accumulate with aging in critical human mtDNA control sites for replication. *Proc Natl Acad Sci USA* 98:4022–4027, 2001.

38. D Bogenhagen, DA Clayton. The number of mitochondrial deoxyribonucleic acid genomes in mouse L and human HeLa cells. *J Biol Chem* 249:7991–7995, 1974.

39. S Anderson, AT Bankier, BG Barrell, MH de Bruijn, AR Coulson, J Drouin, IC Eperon, DP Nierlich, BA Roe, F Sanger, PH Schreier, AJ Smith, R Staden, IG Young. Sequence and organization of the human mitochondrial genome. *Nature* 290:457–465, 1981.

40. J Smeitink, L van den Heuvel, S DiMauro. The genetics and pathology of oxidative phosphorylation. *Nature Rev Genet* 2:342–352, 2001.

41. DC Wallace. Mitochondrial DNA sequence variation in human evolution and disease. *Proc Natl Acad Sci USA* 91:8739–8746, 1994.

42. GA Cortopassi, N Arnheim. Detection of a specific mitochondrial DNA deletion in tissues of older humans. *Nucleic Acids Res* 18:6927–6933, 1990.

43. TC Yen, JH Su, KL King, YH Wei. Ageing-associated 5-kb deletion in human liver mitochondrial DNA. *Biochem Biophys Res Commun* 178:124–131, 1991.

44. K Torii, S Sugiyama, M Tanaka, K Takagi, Y Hanaki, K Iida, M Matsuyama, N Hirabayashi, H Uno, T Ozawa. Ageing-associated deletions of human diaphragmatic mitochondrial DNA. *Am J Respir Cell Mol Biol* 6:543–549, 1992.

45. HC Lee, CY Pang, HS Hsu, YH Wei. Differential accumulations of 4977-bp deletion in mitochondrial DNA of various tissues in human ageing. *Biochim Biophys Acta* 1226:37–43, 1994.

46. JH Yang, HC Lee, KJ Lin, YH Wei. A specific 4977-bp deletion of mitochondrial DNA in human ageing skin. *Arch Dermatol Res* 286:386–390, 1994.

47. HJ Fahn, LS Wang, RH Hsieh, SC Chang, SH Kao, MH Huang, YH Wei. Age-related 4977-bp deletion in human lung mitochondrial DNA. *Am J Respir Crit Care Med* 154:1141–1145, 1996.

48. HC Lee, ML Lim, CY Lu, VW Liu, HJ Fahn, C Zhang, P Nagley, YH Wei. Concurrent increase of oxidative DNA damage and lipid peroxidation together with mitochondrial DNA mutation in human lung tissues during aging — smoking enhances oxidative stress on the aged tissues. *Arch Biochem Biophys* 362: 309–316, 1999.

49. HC Lee, PH Yin, TN Yu, YD Chang, WC Hsu, SY Kao, CW Chi, TY Liu, YH Wei. Accumulation of mitochondrial DNA deletions in human oral tissues — effects of betel quid chewing and oral cancer. *Mutat Res* 493:67–74, 2001.

50. K Hattori, M Tanaka, S Sugiyama, T Obayashi, T Ito, T Satake, Y Hanaki, J Asai, M Nagano, T Ozawa. Age-dependent increase in deleted mitochondrial DNA in the human heart: possible contributory factor to presbycardia. *Am Heart J* 121:1735–1742, 1991.

51. YH Wei, SH Kao, HC Lee. Simultaneous increase of mitochondrial DNA deletions and lipid peroxidation in human aging. *Ann NY Acad Sci* 786:24–43, 1996.

52. C Zhang, A Baumer, RJ Maxwell, AW Linnane, P Nagley. Multiple mitochondrial DNA deletions in an elderly human individual. *FEBS Lett* 297:34–38, 1992.

53. CY Pang, HC Lee, JH Yang, YH Wei. Human skin mitochondrial DNA deletions associated with light exposure. *Arch Biochem Biophys* 312:534–538, 1994.

54. C Zhang, AW Linnane, P Nagley. Occurrence of a particular base substitution (3243 A to G) in mitochondrial DNA of tissues of ageing humans. *Biochem Biophys Res Commun* 195:1104–1110, 1993.

55. C Münscher, T Rieger, J Müller–Höcker, B Kadenbach. The point mutation of mitochondrial DNA characteristic for MERRF disease is found also in healthy people of different ages. *FEBS Lett* 317:27–30, 1993.

56. HC Lee, CY Pang, HS Hsu, YH Wei. Ageing-associated tandem duplications in the D-loop of mitochondrial DNA of human muscle. *FEBS Lett* 354:79–83, 1994.

57. YH Wei, CY Pang, BJ You, HC Lee. Tandem duplications and large-scale deletions of mitochondrial DNA are early molecular events of human aging process. *Ann NY Acad Sci* 786:82–101, 1996.

58. JH Yang, HC Lee, YH Wei. Photoageing-associated mitochondrial DNA length mutations in human skin. *Arch Dermatol Res* 287:641–648, 1995.

59. VWS Liu, C Zhang, CY Pang, HC Lee, CY Lu, YH Wei, P Nagley. Independent occurrence of somatic mutations in mitochondrial DNA of human skin from subjects of various ages. *Hum Mutat* 11:191–196, 1998.

60. IJ Holt, AE Harding, JA Morgan–Hughes. Deletions of mitochondrial DNA in patients with mitochondrial myopathies. *Nature* 331:717–719, 1988.

61. JM Shoffner, MT Lott, AS Voljavec, SA Soueidan, DA Costigan, DC Wallace. Spontaneous Kearns–Sayre/chronic progressive external ophthalmoplegia plus syndrome associated with a mitochondrial DNA deletion: a slip-replication model and metabolic therapy. *Proc Natl Acad Sci USA* 86:7952–7956, 1989.

62. M Zeviani, CT Moraes, S DiMauro, H Nakase, E Bonilla, EA Schon, LP Rowland. Deletions of mitochondrial DNA in Kearns–Sayre syndrome. *Neurology* 38:1339–1346, 1988.

63. DC Wallace. Diseases of the mitochondrial DNA. *Annu Rev Biochem* 61:1175–1212, 1992.

64. YH Wei, HC Lee. Mitochondrial DNA mutations and oxidative stress in mitochondrial diseases. *Adv Clin Chem* 37:83–128, 2003.

65. P Nagley, YH Wei. Ageing and mammalian mitochondrial genetics. *Trends Genet* 14:513–517, 1998.

66. M Hayakawa, K Katsumata, M Yoneda, M Tanaka, S Sugiyama, T Ozawa. Age-related extensive fragmentation of mitochondrial DNA into minicircles. *Biochem Biophys Res Commun* 226:369–377, 1996.

67. AW Linnane, MD Esposti, M Generowicz, AR Luff, P Nagley. The universality of bioenergetic disease and amelioration with redox therapy. *Biochim Biophys Acta* 1271:191–194, 1995.

68. SA Kovalenko, G Kopsidas, JM Kelso, AW Linnane. Deltoid human muscle mtDNA is extensively rearranged in old age subjects. *Biochem Biophys Res Commun* 232:147–152, 1997.

69. G Kopsidas, SA Kovalenko, JM Kelso, AW Linnane. An age-associated correlation between cellular bioenergy decline and mtDNA rearrangements in human skeletal muscle. *Mutat Res* 421:27–36, 1998.

70. HC Lee, SH Li, JC Lin, CC Wu, DC Yeh, YH Wei. Somatic mutations in the D-loop and decrease in the copy number of mitochondrial DNA in human hepatocellular carcinoma. *Mutat Res* 547:71–78, 2004.

71. YH Wei, CP Scholes, TE King. Ubisemiquinone radicals from the cytochrome $b–c_1$ complex of mitochondrial electron transport chain — demonstration of QP-S radical formation. *Biochem Biophys Res Commun* 99:1411–1419, 1981.

72. B Chance, H Sies, A Boveris. Hydroperoxide metabolism in mammalian organs. *Physiol Rev* 59:527–605, 1979.

73. GC Brown. Control of respiration and ATP synthesis in mammalian mitochondria and cells. *Biochem J* 284:1–13, 1992.

74. RS Sohal, HH Ku, S Agarwal, MJ Forster, H Lal. Oxidative damage, mitochondrial oxidant generation, and antioxidant defenses during aging and in response to food restriction in the mouse. *Mech Ageing Dev* 74:121–133, 1994.

75. LA Esposito, S Melov, A Panov, BA Cottrell, DC Wallace. Mitochondrial disease in mouse results in increased oxidative stress. *Proc Natl Acad Sci USA* 96:4820–4825, 1999.

76. C Garcia–Ruiz, A Colell, A Morales, N Kaplowitz, JC Fernandez–Checa. Role of oxidative stress generated from the mitochondrial electron transport chain and mitochondrial glutathione status in loss of mitochondrial function and activation of transcription factor nuclear factor-κB: studies with isolated mitochondria and rat hepatocytes. *Mol Pharmacol* 48:825–834, 1995.

77. C Richter, V Gogvadze, R Laffranchi, R Schlapbach, M Schnizer, M Suter, P Walter, M Yaffee. Oxidants in mitochondria: from physiology to disease. *Biochim Biophys Acta* 1271:67–74, 1995.

78. X Luo, S Pitkanen, S Kassovska–Bratinova, BH Robinson, D Lehotay. Excessive formation of hydroxyl radicals and aldehydic lipid peroxidation products in cultured skin fibroblasts from patients with complex I deficiency. *J Clin Invest* 99:2877–2882, 1997.

79. TC Yen, KL King, HC Lee, SH Yeh, YH Wei. Age-dependent increase of mitochondrial DNA deletions together with lipid peroxides and superoxide dismutase in human liver mitochondria. *Free Radic Biol Med* 16:207–214, 1994.

80. L Brambilla, C Gaetano, P Sestili, V O'Donnel, A Azzi, O Cantoni. Mitochondrial respiratory chain deficiency leads to overexpression of antioxidant enzymes. *FEBS Lett* 418:247–250, 1997.

81. S Melov, P Coskun, M Patel, R Tuinstra, B Cottrell, AS Jun, TH Zastawny, M Dizdaroglu, SI Goodman, TT Huang, H Miziorko, CJ Epstein, DC Wallace. Mitochondrial disease in superoxide dismutase 2 mutant mice. *Proc Natl Acad Sci USA* 96:846–851, 1999.

82. DC Wallace. Animal models for mitochondrial disease methods. *Mol Biol* 197:3–54, 2002.

83. CY Lu, HC Lee, HJ Fahn, YH Wei. Oxidative damage elicited by imbalance of free radical scavenging enzymes is associated with large-scale mtDNA deletions in aging human skin. *Mutat Res* 423:11–21, 1999.

84. HC Lee, YH Wei. Mitochondrial role in life and death of the cell. *J Biomed Sci* 7:2–15, 2000.

85. YH Wei, CY Lu, CY Wei, YS Ma, HC Lee. Oxidative stress in human aging and mitochondrial disease — consequences of defective mitochondrial respiration and impaired antioxidant enzyme system. *Chin J Physiol* 44:1–11, 2001.

86. RJ Shmookler Reis, S Goldstein. Mitochondrial DNA in mortal and immortal human cells. *J Biol Chem* 258:9078–9085, 1983.

87. HC Lee, PH Yin, CW Chi, YH Wei. Increase of mitochondrial mass in human fibroblasts under oxidative stress and during replicative cell senescence. *J Biomed Sci* 9:517–526, 2002.

88. HC Lee, PH Yin, CY Lu, CW Chi, YH Wei. Increase of mitochondria and mitochondrial DNA in response to oxidative stress in human cells. *Biochem J* 348:425–432, 2000.

89. HC Lee, YH Wei. Mitochondrial alterations, cellular response to oxidative stress and defective degradation of proteins in aging. *Biogerontology* 2:231–244, 2001.

90. BN Ames, MK Shigenaga, TM Hagen. Oxidants, antioxidants, and the degenerative diseases of aging. *Proc Natl Acad Sci USA* 90:7915–7922, 1993.

91. FM Yakes, B van Houten. Mitochondrial DNA damage is more extensive and persists longer than nuclear DNA damage in human cells following oxidative stress. *Proc Natl Acad Sci USA* 94:514–519, 1997.

92. M Hayakawa, K Torii, S Sugiyama, M Tanaka, T Ozawa. Age-associated accumulation of 8-hydroxydeoxyguanosine in mitochondrial DNA of human diaphragm. *Biochem Biophys Res Commun* 179:1023–1029, 1991.

93. PH Lin, SH Lee, CP Su, YH Wei. Oxidative damage to mitochondrial DNA in atrial muscle of patients with atrial fibrillation. *Free Radic Biol Med* 35:1310–1318, 2003.

94. AM Hruszkewycz. Lipid peroxidation and mtDNA degeneration. A hypothesis. *Mutat Res* 275:243–248, 1992.

95. S Agarwal, RS Sohal. Differential oxidative damage to mitochondrial proteins during aging. *Mech Ageing Dev* 85:55–63, 1995.

96. J Garcia de la Asuncion, A Millán, R Pla, L Bruseghini, A Esteras, FV Pallardó, J Sastre, J Viña. Mitochondrial glutathione oxidation correlates with age-associated oxidative damage to mitochondrial DNA. *FASEB J* 10:333–338, 1996.

97. LJ Yan, RL Levine, RS Sohal. Oxidative damage during aging targets mitochondrial aconitase. *Proc Natl Acad Sci USA* 94:11168–11172, 1997.

98. LJ Yan, RS Sohal. Mitochondrial adenine nucleotide translocase is modified oxidatively during aging. *Proc Natl Acad Sci USA* 95:12896–12901, 1998.

99. PR Gardner, DD Nguyen, CW White. Aconitase is a sensitive and critical target of oxygen poisoning in cultured mammalian cells and in rat lungs. *Proc Natl Acad Sci USA* 91:12248–12252, 1994.

100. YH Wei, HC Lee, CY Pang, CY Lu, YS Ma. Oxidative damage and mutation to mitochondrial DNA and age-dependent decline of mitochondrial respiratory function. *Ann NY Acad Sci* 854:155–170, 1998.

101. P Dumont, M Burton, QM Chen, ES Gonos, C Frippiat, JB Mazarati, F Eliaers, J Remacle, O Toussaint. Induction of replicative senescence biomarkers by sublethal oxidative stresses in normal human fibroblast. *Free Radic Biol Med* 28:361–373, 2000.

102. M Berneburg, S Grether–Beck, V Kurten, T Ruzicka, K Briviba, H Sies, J Krutmann. Singlet oxygen mediates the UVA-induced generation of the photoaging-associated mitochondrial common deletion. *J Biol Chem* 274:15345–15349, 1999.

103. HJ Fahn, LS Wang, SH Kao, SC Chang, MH Huang, YH Wei. Smoking-associated mitochondrial DNA mutations and lipid peroxidation in human lung tissues. *Am J Respir Cell Mol Biol* 19:901–909, 1998.

104. A Mansouri, B Fromenty, A Berson, MA Robin, S Grimbert, M Beaugrand, S Erlinger, D Pessayre. Multiple hepatic mitochondrial DNA deletions suggest premature oxidative aging in alcoholic patients. *J Hepatol* 27:96–102, 1997.

105. KA Laderman, JR Penny, F Mazzucchelli, N Bresolin, G Scarlato, G Attardi. Aging-dependent functional alterations of mitochondrial DNA (mtDNA) from fibroblasts transformed into mtDNA-less cells. *J Biol Chem* 271:15891–15897, 1996.

106. C Richter, JW Park, BN Ames. Normal oxidative damage to mitochondrial and nuclear DNA is extensive. *Proc Natl Acad Sci USA* 85:6465–6467, 1988.

107. DA Clayton, JN Doda, EC Friedberg. The absence of a pyrimidine dimer repair mechanism in mammalian mitochondria. *Proc Natl Acad Sci USA* 71:2777–2781, 1974.

108. KG Pinz, DF Bogenhagen. Efficient repair of abasic sites in DNA by mitochondrial enzymes. *Mol Cell Biol* 18:1257–1265, 1998.

109. M Takao, H Aburatani, K Kobayashi, A Yasui. Mitochondrial targeting of human DNA glycosylases for repair of oxidative DNA damage. *Nucleic Acids Res* 26:2917–2922, 1998.

110. AE Tomkinson, RT Bonk, S Linn. Mitochondrial endonuclease activities specific for apurinic/apyrimidinic sites in DNA from mouse cells. *J Biol Chem* 263:12532–12537, 1988.

111. DL Croteau, CMJ ap Rhys, EK Hudson, GL Dianov, RG Hansford, VA Bohr. An oxidative damage-specific endonuclease from rat liver mitochondria. *J Biol Chem* 272:27338–27344, 1997.

112. NC Souza–Pinto, DL Croteau, EK Hudson, RG Hansford, AA Bohr. Age-associated increase in 8-oxo-deoxyguanosine glycosylase/AP lyase activity in rat mitochondria. *Nucleic Acids Res* 27:1935–1942, 1999.

113. D Kang, J Nishida, A Iyama, Y Nakabeppu, M Furuichi, T Fujiwara, M Sekiguchi, K Takeshige. Intracellular localization of 8-oxo-dGTPase in human cells, with special reference to the role of the enzyme in mitochondria. *J Biol Chem* 270:14655–14659, 1995.

114. DF Bogenhagen, KG Pinz, RM Perez–Jannotti. Enzymology of mitochondrial base excision repair. *Prog Nucleic Acids Res Mol Biol* 68:257–271, 2001.

115. T Hirano, Y Yamaguchi, H Hirano, H Kasai. Age-associated change of 8-hydroxyguanine repair activity in cultured human fibroblasts. *Biochem Biophys Res Commun* 214:1157–1162, 1995.

116. WJ Driggers, VI Grishko, SP LeDoux, GL Wilson. Defective repair of oxidative damage in the mitochondrial DNA of a *Xeroderma pigmentosum* group A cell line. *Cancer Res* 56:1262–1266, 1996.

117. D Chen, G Cao, T Hastings, Y Feng, W Pei, C O'Horo, J Chen. Age-dependent decline of DNA repair activity for oxidative lesions in rat brain mitochondria. *J Neurochem* 81:1273–1284, 2002.

118. B Karahalil, BA Hogue, NC de Souza–Pinto, VA Bohr. Base excision repair capacity in mitochondria and nuclei: tissue-specific variations. *FASEB J* 16:1895–1902, 2002.

119. JG de la Asuncion, A Millan, R Pla, L Bruseghini, A Esteras, FV Pallardo, J Sastre, J Vina. Mitochondrial glutathione oxidation correlates with age-associated oxidative damage to mitochondrial DNA. *FASEB J* 10:333–338, 1996.

120. L Muskhelishvili, RW Hart, A Turturro, SJ James. Age-related changes in the intrinsic rate of apoptosis in livers of diet-restricted and *ad libitum*-fed B6C3F1 mice. *Am J Pathol* 147:20–24, 1995.

121. JH Morrison, PR Hof. Life and death of neurons in the aging brain. *Science* 278:412–419, 1997.

122. J Kajstura, W Cheng, R Sarangarajan, P Li, B Li, JA Nitahara, S Chapnick, K Reiss, G Olivetti, P Anversa. Necrotic and apoptotic myocyte cell death in the aging heart of Fischer 344 rats. *Am J Physiol Heart Circ Physiol* 271:H1215–H1228, 1996.

123. P Anversa, T Palackal, EH Sonnenblick, G Olivetti, LG Meggs, JM Capasso. Myocyte cell loss and myocyte cellular hyperplasia in the hypertrophied aging rat heart. *Circ Res* 67:871–885, 1990.

124. G Olivetti, M Melissari, JM Capasso, P Anversa. Cardiomyopathy of the aging human heart. Myocyte loss and reactive cellular hypertrophy. *Circ Res* 68:1560–1568, 1991.

125. S Phaneuf, C Leeuwenburgh. Cytochrome *c* release from mitochondria in the aging heart: a possible mechanism for apoptosis with age. *Am J Physiol Regulat Integrat Comp Physiol* 282:R423–R430, 2002.

126. A Dirks, C Leeuwenburgh. Apoptosis in skeletal muscle with aging. *Am J Physiol Regulat Integrat Comp Physiol* 282:R519–R527, 2002.

127. AA Kapasi, PC Singhal. Aging splenocyte and thymocyte apoptosis is associated with enhanced expression of p53, Bax, and caspase-3. *Mol Cell Biol Res Commun* 1:78–81, 1999.

128. Y Fujino, K Ozaki, S Yamamasu, F Ito, I Matsuoka, E Hayashi, H Nakamura, S Ogita, E Sato, M Inoue. DNA fragmentation of oocytes in aged mice. *Hum Reprod* 11:1480–1483, 1996.

129. D Monti, S Salvioli, M Capri, W Malorni, E Straface, A Cossarizza, B Botti, M Piacentini, G Baggio, C Barbi, S Valensin, M Bonafe, C Franceschi. Decreased susceptibility to oxidative stress-induced apoptosis of peripheral blood mononuclear cells from healthy elderly and centenarians. *Mech Ageing Dev* 121:239–250, 2000.

130. M Steinkamp, DO Schachtschabel. Energetic stress in combination with tumor necrosis factor-alpha (TNF-alpha) induces apoptosis of human fibroblasts (WI-38) *in vitro*: reduced responsiveness of senescent cells. *Z Gerontol Geriatr* 34:437–440, 2001.

131. MD Jacobson, JF Burne, MP King, T Miyashita, JC Reed, MC Raff. Bcl-2 blocks apoptosis in cells lacking mitochondrial DNA. *Nature* 361:365–369, 1993.

132. J Wang, JP Silva, CM Gustafsson, P Rustin, NG Larsson. Increased *in vivo* apoptosis in cells lacking mitochondrial DNA gene expression. *Proc Natl Acad Sci USA* 98:4038–4043, 2001.

133. CY Liu, CF Lee, CH Hong, YH Wei. Mitochondrial DNA mutation and depletion increase the susceptibility of human cells to apoptosis. *Ann NY Acad Sci* 1011:133–145, 2004.

134. P Marchetti, SA Susin, D Decaudin, S Gamen, M Castedo, T Hirsch, N Zamzami, J Naval, A Senik, G Kroemer. Apoptosis-associated derangement of mitochondrial function in cells lacking mitochondrial DNA. *Cancer Res* 56:2033–2038, 1996.

135. R Dey, CT Moraes. Lack of oxidative phosphorylation and low mitochondrial membrane potential decrease susceptibility to apoptosis and do not modulate the protective effect of Bcl-x(L) in osteosarcoma cells. *J Biol Chem* 275:7087–7094, 2000.

136. A Wong, G Cortopassi. MtDNA mutations confer cellular sensitivity to oxidant stress that is partially rescued by calcium depletion and cyclosporin A. *Biochem Biophys Res Commun* 239:139–145, 1997.

137. SR Danielson, A Wong, V Carelli, A Martinuzzi, AH Schapira, GA Cortopassi. Cells bearing mutations causing Leber's hereditary optic neuropathy are sensitized to Fas-Induced apoptosis. *J Biol Chem* 277:5810–5815, 2002.

138. A Ghelli, C Zanna, AM Porcelli, AH Schapira, A Martinuzzi, V Carelli, M Rugolo. Leber's hereditary optic neuropathy (LHON) pathogenic mutations induce mitochondrial-dependent apoptotic death in transmitochondrial cells incubated with galactose medium. *J Biol Chem* 278:4145–4150, 2003.

139. G Kroemer, B Dallaporta, M Resche–Rigon. The mitochondrial death/life regulator in apoptosis and necrosis. *Annu Rev Physiol* 60:619–642, 1998.

140. G Petrosillo, FM Ruggiero, M Pistolese, G Paradies. Reactive oxygen species generated from the mitochondrial electron transport chain induce cytochrome *c* dissociation from beef-heart submitochondrial particles via cardiolipin peroxidation. Possible role in the apoptosis. *FEBS Lett* 509:435–438, 2001.

141. JM Nitahara, W Cheng, Y Liu, B Li, A Leri, P Li, D Mogul, SR Gambert, J Kajatura, P Anversa. Intracellular calcium, DNase activity and myocyte apoptosis in aging Fischer 344 rats. *J Mol Cell Cardiol* 30:519–535, 1998.

142. EG Lakatta. Cardiovascular regulatory mechanisms in advanced age. *Physiol Rev* 73:413–467, 1993.

143. M Mather, H Rottenberg. Aging enhances the activation of the permeability transition pore in mitochondria. *Biochem Biophys Res Commun* 273:603–608, 2000.

144. V Geromel, N Kadhom, I Cebalos–Picot, O Ouari, A Polidori, A Munnich, A Rotig, P Rustin. Superoxide-induced massive apoptosis in cultured skin fibroblasts harboring the neurogenic ataxia retinitis pigmentosa (NARP) mutation in the ATPase-6 gene of the mitochondrial DNA. *Hum Mol Genet* 10:1221–1228, 2001.

145. JE Kokoszka, P Coskun, LA Esposito, DC Wallace. Increased mitochondrial oxidative stress in the Sod2 $^{(\pm)}$ mouse results in the age-related decline of mitochondrial function culminating in increased apoptosis. *Proc Natl Acad Sci USA* 98:2278–2283, 2001.

146. P Anversa, T Palackal, EH Sonnenblick, G Olivetti, LG Meggs, JM Capasso. Myocyte cell loss and myocyte cellular hyperplasia in the hypertrophied aging rat heart. *Circ Res* 67:871–885, 1990.

147. G Olivetti, E Cigola, RM Maestri, D Corradi, C Lagrasta, SR Gambert, P Anversa. Aging, cardiac hypertrophy and ischemic cardiomyopathy do not affect the proportion of mononucleated and multinucleated myocytes in the human heart. *J Mol Cell Cardiol* 28:1463–1477, 1996.

148. S Shimohama, S Fujimoto, Y Sumida, H Tanino. Differential expression of rat brain Bcl-2 family proteins in development and aging. *Biochem Biophys Res Commun* 252:92–96, 1998.

149. S Shimohama, H Tanino, S Fujimoto. Differential expression of rat brain caspase family proteins during development and aging. *Biochem Biophys Res Commun* 289:1063–1066, 2001.

150. ZQ Xiao, L Moragoda, R Jaszewski, JA Hatfield, SEG Fligiel, APN Majumdar. Aging is associated with increased proliferation and decreased apoptosis in the colonic mucosa. *Mech Ageing Dev* 122:1849–1864, 2001.

151. Y Suh, KA Lee, WH Kim, BG Han, J Vijg, SC Park. Aging alters the apoptotic response to genotoxic stress. *Nature Med* 8:3–4, 2002.

152. S Zou, S Meadows, L Sharp, LY Jan, YN Jan. Genome-wide study of aging and oxidative stress response in *Drosophila melanogaster. Proc Natl Acad Sci USA* 97:13726–13731, 2000.

153. CK Lee, RG Klopp, R Weindruch, TA Prolla. Gene expression profile of aging and its retardation by caloric restriction. *Science* 285:1390–1393, 1999.

154. CK Lee, R Weindruch, TA Prolla. Gene-expression profile of the ageing brain in mice. *Nat Genet* 25:294–297, 2000.

155. CH Jiang, JZ Tsien, PG Schultz, Y Hu. The effects of aging on gene expression in the hypothalamus and cortex of mice. *Proc Natl Acad Sci USA* 98:1930–1934, 2001.

156. CK Lee, DB Allison, J Brand, R Weindruch, TA Prolla. Transcriptional profiles associated with aging and middle age-onset caloric restriction in mouse hearts. *Proc Natl Acad Sci USA* 99:14988–14993, 2002.

157. N Bodyak, PM Kang, M Hiromura, I Sulijoadikusumo, N Horikoshi, K Khrapko, A Usheva. Gene expression profiling of the aging mouse cardiac myocytes. *Nucleic Acids Res* 30:3788–3794, 2002.

158. MG Edwards, D Sarkar, R Klopp, JD Morrow, R Weindruch, TA Prolla. Age-related impairment of the transcriptional responses to oxidative stress in the mouse heart. *Physiol Genomics* 13:119–127, 2003.

159. SX Cao, JM Dhahbi, PL Mote, SR Spindler. Genomic profiling of short- and long-term caloric restriction effects in the liver of aging mice. *Proc Natl Acad Sci USA* 98:10630–10635, 2001.

160. P Tollet–Egnell, A Flores–Morales, N Stahlberg, RL Malek, N Lee, G Norstedt. Gene expression profile of the aging process in rat liver: normalizing effects of growth hormone replacement. *Mol Endocrinol* 15:308–318, 2001.

161. T Kayo, DB Allison, R Weindruch, TA Prolla. Influences of aging and caloric restriction on the transcriptional profile of skeletal muscle from rhesus monkeys. *Proc Natl Acad Sci USA* 98:5093–5098, 2001.

162. RC Scarpulla. Nuclear control of respiratory chain expression in mammalian cells. *J Bioenerg Biomembr* 29:109–119, 1997.

163. HC Lee, YH Wei. Mitochondrial biogenesis and mitochondrial DNA maintenance in mammalian cells under oxidative stress. *Int J Biochem Cell Biol* 37:822–834, 2005.

164. RO Poyton, JE McEwen. Crosstalk between nuclear and mitochondrial genomes. *Annu Rev Biochem* 65:563–607, 1996.

165. RS Williams, S Salmons, EA Newsholme, RE Kaufman, J Mellor. Regulation of nuclear and mitochondrial gene expression by contractile activity in skeletal muscle. *J Biol Chem* 261:376–380, 1986.

166. Y Xia, LM Buja, RC Scarpulla, JB McMillin. Electrical stimulation of neonatal cardiomyocytes results in the sequential activation of nuclear genes governing mitochondrial proliferation and differentiation. *Proc Natl Acad Sci USA* 94:11399–11404, 1997.

7

Mitochondrial DNA in Cardiomyopathies

CONTENTS

INTRODUCTION

A lot of new evidence has been reported about various cardiovascular diseases due to recent progress in molecular biology. Mitochondrial DNA mutation is one of the themes and has been investigated in patients with cardiomyopathy.

Mitochondria possess an energy-producing system composed of NADH dehydrogenase, cytochrome, and cytochrome oxidase that is encoded and regulated by mitochondrial and nuclear DNA. Human mitochondrial DNA is double stranded and circular, consisting of 16,569 base pairs and containing 11 structural genes for subunits of respiratory enzyme complexes and 2 genes for two subunits of F_0 ATPase, as well as for 22 tRNA molecules and two rRNA molecules (1). Each cardiac myocyte contains 2000 to 3000 mitochondria and each mitochondrion possesses two or three circular DNAs.

Mitochondrial DNA mutations are inherited from the mother (2,3). This DNA is easily damaged because it is continually exposed to free radicals and contains neither histones nor introns, and the mitochondrial DNA repair system is relatively primitive. Oxygen-derived free radicals generated by the mitochondrial inner membrane convert deoxyguanosine (dG) in mitochondrial DNA to 8-hydroxy-dG, which is misread as another base during duplication. Therefore, an increase of 8-hydroxy-dG is synonymous with the accumulation of point mutations.

Ozawa and colleagues have succeeded in determining the entire mitochondrial DNA sequence by using the direct sequencing technique (4,5). Mitochondrial DNA mutation may lead to impairment of mitochondrial function, e.g., deficiency of mitochondrial respiratory chain enzymes or ATPase related to the subunits encoded by mitochondrial DNA. In light of the evidence described later, the term "mitochondrial cardiomyopathy" has been coined to describe cardiomyopathy induced by mitochondrial DNA mutations (6).

IDIOPATHIC CARDIOMYOPATHY

Idiopathic cardiomyopathy consists of hypertrophic, dilated, restrictive cardiomyopathies and arrhythmogenic right ventricular dysplasia. The etiologies of these conditions are still unknown, but recent developments in molecular biology and immunology have provided suggestive evidence, such as the detection of mutations of the genes for myosin, actin, tropomyosin, and troponin. Myocardial mitochondrial DNA mutations have also been detected by Ozawa et al. in cardiomyopathy patients (7–9), and reports concerning mitochondrial DNA mutations in cardiomyopathy patients and their potential role in cardiomyopathy have been produced (10–22).

A point mutation that alters adenine (A) to guanine (G) within the mitochondrial tRNA[Leu(UUR)] gene is common in patients with the syndrome of mitochondrial myopathy, encephalopathy, lactic acidosis, and stroke-like episodes (MELAS); it is also found in patients with cardiomyopathy (9). PCR and southern blot analysis have revealed multiple mitochondrial DNA deletions in a pedigree of inherited dilated cardiomyopathy (23), but the extent to which these mitochondrial DNA mutations are involved in the etiology of idiopathic cardiomyopathy remains to be elucidated.

All humans appear to have the potential to develop cardiomyopathy because the myocardium degenerates with age due to the accumulation of free radical-induced damage to the mitochondrial DNA. Abnormal acceleration of mitochondrial DNA mutations, especially those related to mitochondrial protein synthesis, can induce premature ageing and severe mitochondrial cardiomyopathy (24). An A to G point mutation at position 3260 of the mitochondrial tRNA[Leu(UUR)] gene was found in a maternally inherited disorder that manifested as a combination of adult-onset myopathy and cardiomyopathy (10). An A to G substitution was found at position 4269 in the tRNA[Ile] gene of a patient with fatal cardiomyopathy (12), and an A to G mutation at position 15,923 of the mitochondrial tRNA[Thr] gene is associated with neonatal cardiomyopathy (25).

Other patients with fatal infantile cardiomyopathy have also been reported (26,27), as has the development of severe mitochondrial cardiomyopathy in young people with an A to G transition at position 827 of the mitochondrial 12S rRNA gene (28). It has also been suggested that a point mutation of G to A at 12192 in the tRNAHis gene may be an evolutionary risk factor for cardiomyopathy (29). Therefore, the detection of mitochondrial DNA deletions has been proposed as a new method for investigating sudden cardiac death in which ischemic damage is the primary cause (30).

As yet, no clear correlation has been found between the severity of clinical manifestations and the mutations detected

by conventional analysis of limited regions of the entire mitochondrial DNA. In contrast, comprehensive analysis of mitochondrial DNA by the direct base sequencing technique has
revealed a close correlation between the mitochondrial DNA
genotype and clinical phenotype (5,28). Each cardiac myocyte
contains 2000 to 3000 mitochondria, and each mitochondrion
possesses two or three circular mitochondrial DNAs; thus,
cells can contain normal and mutant mitochondrial DNA in
varying proportions (heteroplasmy). This may result in
marked differences of energy production among myocardial
cells, which may induce arrhythmias.

If the mitochondrial DNA mutation is extensive, oxidative phosphorylation will be depressed, leading to a decrease
of energy production and the development of heart failure.
Mutations in nuclear DNA encoding the mitochondrial respiratory enzyme complex subunits can also affect energy production. Furthermore, it has been suggested that
mitochondrial DNA mutations might activate the mitochondrial apoptotic pathway and thus cause dilated cardiomyopathy (31). The investigation of mitochondrial DNA mutations
may therefore yield various clues about the etiology of
arrhythmias and cardiac dysfunction.

Diagnosis of mitochondrial cardiomyopathy can be confirmed through detecting mitochondrial DNA mutations in
myocardial biopsies. Electron microscopy can show changes in
the size or number of mitochondria. Mitochondrial DNA mutations are observed in hypertrophic and dilated cardiomyopathy.
In fact, it has been reported that about 3% of dilated cardiomyopathies were induced by mitochondrial DNA mutations.

MITOCHONDRIAL MYOPATHY

An etiological relationship between myopathy and cardiomyopathy has long been suspected because cardiomyopathic
patients may show skeletal muscle abnormalities and
patients with myopathy may display cardiomyopathic
changes (32). Mitochondrial myopathies — which comprise
Kerns–Sayre syndrome, including the incomplete type:
chronic progressive external ophthalmoplegia, myoclonic

epilepsy with ragged red fibers (MERFF), and MELAS — are often associated with cardiac disorders. Kerns–Sayre syndrome is characterized by chronic progressive external ophthalmoplegia, heart block, and pigmented retinopathy. The main cause is thought to be mitochondrial DNA deletions (33–38), although mitochondrial dysfunction induced by abnormal nuclear DNA may also be involved (39). The cardiac manifestations of this disease are arrhythmias, such as atrioventricular block, premature ventricular contractions, supraventricular or ventricular tachycardia, sinus dysrhythmia, ST segment and T wave changes, and cardiac dilation and failure (40–46).

In many patients, implantation of a pacemaker is required to prevent sudden death, and some patients may first need a pacemaker at an advanced age. MELAS usually occurs at a young age and is characterized by headache, vomiting, and stroke-like episodes such as hemiplegia (47). This disease is induced by point mutations of the mitochondrial tRNA$^{Leu(UUR)}$ gene, from A to G at position 3243 (48,49) or from thymine (T) to cytosine (C) at position 3271 (50). Cardiac involvement results in the onset of cardiomyopathy (51,52). Mitochondrial DNA mutations have also been reported in Leigh syndrome and Leber's hereditary optic neuropathy (26,53). The mechanisms by which point mutations of the tRNA gene induce pathological changes are described elsewhere (54,55).

DIABETES MELLITUS

Maternally transmitted diabetes mellitus associated with deafness and a mitochondrial DNA deletion of 10.4 kb was reported in 1992 (56); some of these patients also had cardiomyopathy. Two pedigrees of maternally transmitted type 2 (non-insulin-dependent) diabetes mellitus and deafness associated with an A to G transition at position 3243 of mitochondrial DNA were also reported in 1992. One pedigree included individuals with mitral valve prolapse (57) and the other included individuals with cardiomyopathy (58). Therefore, a point mutation at position 3243 of mitochondrial DNA may be a common abnormality leading to diabetes mellitus associated with deafness and/or cardiomyopathy.

ACQUIRED MUTATIONS

Mitochondrial DNA mutations can be induced after birth. An age-dependent increase of deletions to mitochondrial DNA has been observed in the human and rat myocardium (59–64); mitochondrial DNA deletion was found in myocardial autopsy specimens of patients with diabetes, with myocardial infarction (65,66), and patients treated with doxorubicin (66). Autopsy and biopsy specimens of ischemic hearts are reported to show an increase of mitochondrial DNA damage compared with normal hearts (67). In addition, doxorubicin-induced myocardial mitochondrial DNA damage and deletions have been demonstrated in experimental animals (68,69). Therefore, free radicals may play a role in inducing mitochondrial DNA mutations related to the pathological conditions described earlier.

REFERENCES

1. S Anderson, AT Bankier, BG Barrell, MHL de Bruijin, AR Coulson, J Deouin, IC Eperson, DP Nierlich, BA Roe, F Sanger, PH Schreier, AJH Smith, R Staden, IG Young. Sequence and organization of the human mitochondrial genome. *Nature* 290:457–465, 1981.

2. RE Giles, H Blanc, HM Cann, DC Wallace. Maternal inheritance of human mitochondrial DNA. *Proc Natl Acad Sci USA* 77:6715–6719, 1980.

3. T Ozawa, M Yoneda, M Tanaka, K Ohno, W Sato, H Suzuki, M Nishikimi, M Yamamoto, I Nonaka, S Horai. Maternal inheritance of deleted mitochondrial DNA in a family with mitochondrial myopathy. *Biochem Biophys Res Commun* 154:1240–1247, 1988.

4. M Tanaka, W Sato, K Ohno, T Yamamoto, T Ozawa. Direct sequencing of deleted mitochondrial DNA in myopathic patients. *Biochem Biophys Res Commun* 164:156–163, 1989.

5. T Ozawa, K Katsumata, M Hayakawa, M Yoneda, M Tanaka, S Sugiyama. Mitochondrial DNA mutations and survival rate. *Lancet* 345:189, 1995.

6. T Ozawa. Mitochondrial cardiomyopathy. *Herz* 19:105–118, 1994.

7. T Ozawa, M Tanaka, S Sugiyama, K Hattori, T Ito, K Ohno, A Takahashi, W Sato, G Takada, B Mayumi, K Yamamoto, K Adachi, Y Koga, H Toshima. Multiple mitochondrial DNA deletions exist in cardiomyocytes of patients with hypertrophic or dilated cardiomyopathy. *Biochem Biophys Res Commun* 170:830–836, 1990.

8. K Hattori, T Ogawa, T Kondo, M Mochizuki, M Tanaka, S Sugiyama, T Ito, T Satake, T Ozawa. Cardiomyopathy with mitochondrial DNA mutations. *Am Heart J* 122:866–869, 1991.

9. T Obayashi, K Hattori, S Sugiyama, M Tanaka, T Tanaka, S Itoyama, H Deguchi, K Kawamura, Y Koga, H Toshima, N Takeda, M Nagano, T Ito, T Ozawa. Point mutations in mitochondrial DNA in patients with hypertrophic cardiomyopathy. *Am Heart J* 124:1263–1269, 1992.

10. M Zeviani, C Gellera, C Antozzi, M Rimoldi, L Morandi, F Vilani, V Tiranti, S DiDonato. Maternally inherited myopathy and cardiomyopathy: association with mutation in mitochondrial DNA tRNA(Leu)(UUR). *Lancet* 338:143–147, 1991.

11. T Ozawa, S Sugiyama, M Tanaka, K Httori. Mitochondrial DNA mutations and disturbances of energy metabolism in myocardium. *Jpn Circ J* 55:1158–1164, 1991.

12. M Taniike, H Fukushima, I Yanagihara, H Tsukamoto, J Tanaka, H Fujimura, T Nagai, T Sano, K Yamaoka, K Inui, et al. Mitochondrial tRNA(Ile) mutation in fatal cardiomyopathy. *Biochem Biophys Res Commun* 186:47–53, 1992.

13. C Mariotti, V Tiranti, F Carrara, B Dallapiccola, S DiDonato, M Zaviani. Defective respiratory capacity and mitochondrial protein synthesis in transformant cybrids harboring the tRNA(Leu(UUR)) mutation associated with maternally inherited myopathy and cardiomyopathy. *J Clin Invest* 93:1102–1107, 1994.

14. J Hayashi, S Ohta, Y Kagawa, D Takai, S Miyabayashi, K Tada, H Fukushima, K Inui, S Okada, Y Goto, et al. Functional and morphological abnormalities of mitochondria in human cells containing mitochondrial DNA with pathogenic point mutations in tRNA genes. *J Biol Chem* 269:19060–19066, 1994.

15. M Tanaka, T Obayashi, M Yoneda, SA Kovalenkoo, S Sugiyama, T Ozawa. Mitochondrial DNA mutations in cardiomyopathy: combination of replacements yielding cysteine residues and tRNA mutations. *Muscle Nerve* 3:S165–S169, 1995.

16. M Zeviani, C Mariotti, C Antozzi, GM Fratta, P Rustin, A Prelle. OXPHOS defects and mitochondrial DNA mutations in cardiomyopathy. *Muscle Nerve* 3:S170–S174, 1995.

17. Y Takei, S Ikeda, N Yanagisawa, W Takahashi, M Sekiguchi, T Hayashi. Multiple mitochondrial DNA deletions in a patient with mitochondrial myopathy and cardiomyopathy but not ophthalmoplegia. *Muscle Nerve* 18:1321–1325, 1995.

18. T Ozawa. Mitochondrial DNA mutations in myocardial diseases. *Eur Heart J* 16(Suppl):10–14, 1995.

19. LF Turner, S Kaddoura, D Harrington, JM Cooper, PA Pool–Wilson, AH Schapira. Mitochondrial DNA in idiopathic cardiomyopathy. *Eur Heart J* 19:1725–1729, 1998.

20. SS Khogali, BM Mayosi, JM Beattie, WJ McKenna, WJ Watkins, J Poulton. A common mitochondrial DNA variant associated with susceptibility to dilated cardiomyopathy in two different populations. *Lancet* 357:1265–1267, 2001.

21. YY Li, D Chen, SC Watkins, AM Feldman. Mitochondrial abnormalities in tumor necrosis factor-alpha-induced heart failure are associated with impaired DNA repair activity. *Circulation* 104:2492–2497, 2001.

22. RW Taylor, C Giordano, MM Davidson, G d'Amati, H Bain, CM Hayes, H Leonard, MJ Barron, C Casali, FM Santorelli, M Hirano, RN Lightowlers, S DiMauro, DM Turnbull. A homoplasmic mitochondrial transfer ribonucleic acid mutation as a cause of maternally inherited hypertrophic cardiomyopathy. *J Am Coll Cardiol* 41:1786–1796, 2003.

23. A Suomalainen, A Paetau, H Leinonen, A Majander, L Peltonen, H Somer. Inherited idiopathic dilated cardiomyopathy with multiple deletions of mitochondrial DNA. *Lancet* 340:1319–1320, 1992.

24. K. Katsumata, M Hayakawa, M Tanaka, S Sugiyama, T Ozawa. Fragmentation of human heart mitochondrial DNA associated with premature aging. *Biochem Biophys Res Commun* 202:102–110, 1994.

25. KL Yoon, SG Ernst, C Rasmussen, EC Dooling, JR Aprille. Mitochondrial disorder associated with newborn cardiopulmonary arrest. *Pediatr Res* 33:433–440, 1993.

26. LC Papadopoulou, CM Sue, MM Davidson, K Tanji, I Nishino, JE Sadlock, S Krishna, W Walker, J Selby, DM Glerum, RV Coster, G Lyon, E Scalais, R Leibel, P Kaplan, S Shanske, DC De Vivo, E Bonilla, M Hirano, S DiMauro, EA Schon. Fatal infantile cardioencephalomyopathy with COX deficiency and mutations in SCO2, a COX assembly gene. *Nat Genet* 23:333–337, 1999.

27. Y Akita, Y Koga, R Iwanaga, N Wada, J Tsubone, S Fukuda, Y Nakamura, H Kato. Fatal hypertrophic cardiomyopathy associated with an A8296G mutation in the mitochondrial tRNA(Lys) gene. *Hum Mutat* 15:382, 2000.

28. T Ozawa, K Katsumata, M Hayakawa, M Tanaka, S Sugiyama, T Tanaka, S Itoyama, S Nunoda, M Sekiguchi. Genotype and phenotype of severe mitochondrial cardiomyopathy: a recipient of heart transplantation and the genetic control. *Biochem Biophys Res Commun* 207:613–620, 1995.

29. WS Shin, M Tanaka, J Suzuki, C Hemmi, T Toyo–oka. A novel homoplasmic mutation in mtDNA with a single evolutionary origin as a risk factor for cardiomyopathy. *Am J Hum Genetic* 67:1617–1620, 2000.

30. PJ Fouret, G Nicolas, D Lecomte. Detection of the 4977 base pair mitochondrial DNA deletion in paraffin-embedded heart tissue using the polymerase chain reaction — a new method to probe sudden cardiac death molecular mechanisms? *Forensic Sci* 39:693–698, 1994.

31. D Zhang, JL Mott, P Farrar, JS Ryerse, SW Chang, M Stevens, G Denninger, HP Zassenhaus. Mitochondrial DNA mutations activate the mitochondrial apoptotic pathway and cause dilated cardiomyopathy. *Cardiovasc Res* 57:147–157, 2003.

32. A Dunningan, NA Staley, SA Smith, ME Dierpont, D Judd, DG Benditt, DW Benson, Jr. Cardiac and skeletal abnormalities in cardiomyopathy: comparison of patients with ventricular tachycardia or congestive heart failure. *J Am Coll Cardiol* 10:608–618, 1987.

33. IJ Holt, JM Cooper, JA Morgan–Hughes, AE Harding. Deletions of muscle mitochondrial DNA. *Lancet* 1:1462, 1988.

34. M Zeviani, CT Moraes, S DiMauro, H Nakase, E Bonilla, EA Schon, LP Rowland. Deletions of mitochondrial DNA in Kearns–Sayre syndrome. *Neurology* 38:1339–1346, 1988.

35. CT Moraes, S DiMauro, M Zeviani, A Lombes, S Shenske, AF Miranda, H Nakase, E Bonilla, LC Werneck, S Servidei, et al. Mitochondrial DNA deletions in progressive external ophthalmoplegia and Kerns–Sayre syndrome. *N Engl J Med* 320:1293–1299, 1989.

36. B Obermaier–Kusser, J Müller–Höcker, I Nelson, P Lestienne, C Enter, T Riedele, KD Gerbitz. Different copy numbers of apparently identically deleted mitochondrial DNA in tissues from a patient with Kerns–Sayre syndrome detected by PCR. *Biochem Biophys Res Commun* 169:1007–1015, 1990.

37. C Ponzetto, N Bresolin, A Bordoni, M Moggio, G Meola, L Bet, A Prelle, G Scarlato. Kearns–Sayre syndrome: different amounts of deleted mitochondrial DNA are present in several autoptic tissues. *J Neurol Sci* 96:207–210, 1990.

38. S Bosbach, C Kornblum, R Schröder, M Wagner. Executive and visuospatial deficits in patients with chronic progressive external ophthalmoplegia and Kearns–Sayre syndrome. *Brain* 126:1231–1240, 2003.

39. A Soumalainen, A Majander, M Haltia, H Somer, J Lonnqvist, ML Savontaus, L Peltonen. Multiple deletions of mitochondrial DNA in several tissues of a patient with severe retarded depression and familial progressive external ophthalmoplegia. *J Clin Invest* 90:61–66, 1992.

40. KD Gerbitz, B Obermaier–Kusser, S Zierz, D Pongratz, J Müller–Höcker, P Lestienne. Mitochondrial myopathies: divergences of genetic deletions, biochemical defects and the clinical syndromes. *J Neurol* 237:5–10, 1990.

41. P Melacini, C Angelini, G Buja, G Micaglio, ML Valente. Evolution of cardiac involvement in progressive ophthalmoplegia with deleted mitochondrial DNA. *Jpn Heart J* 31:115–120, 1990.

42. D Kenny, J Wetherbee. Kearns–Sayre syndrome in the elderly: mitochondrial myopathy with advanced heart block. *Am Heart J* 120:440–443, 1990.

43. C Bordarier, C Duyckaerts, O Robain, G Ponsot, D Laplane. Kearns–Sayre syndrome. Two clinicopathological cases. *Neuropediatrics* 21:106–109, 1990.

44. C Tranchant, B Mousson, M Mohr, R Dumoulin, M Welsch, C Weess, G Stepien, JM Warter. Cardiac transplantation in an incomplete Kearns–Sayre syndrome with mitochondrial DNA deletion. *Neuromuscul Disord* 3:561–566, 1993.

45. W Sato, M Tanaka, S Sugiyama, K Hattori, T Ito, H Kawaguchi, H Onozuka, H Yasuda, K Ito, G Takada, et al. Deletion of mitochondrial DNA in a patient with conduction block. *Am Heart J* 125:550–552, 1993.

46. KS Ulicny, FC Detterbeck, CD Hall. Sinus dysrhythmia in Kearns–Sayre syndrome. *PACE* 17:991–994, 1994.

47. P Montagna, R Gallassi, R Medori, E Govoni, M Zeviani, S DiMauro, E Lugaresi, F Andermann. MELAS syndrome: characteristic migrainous and epileptic features and maternal transmission. *Neurology* 38:751–754, 1988.

48. Y Goto, I Nonaka, A Horai. A mutation in the tRNA[Leu(UUR)] gene associated with the MELAS subgroup of mitochondrial encephalomyopathies. *Nature* 348:651–653, 1990.

49. M Tanaka, H Ino, K Ohno, T Obayashi, S Ikebe, T Sano, T Ichiki, M Kobayashi, Y Wada, T Ozawa. Mitochondrial DNA mutations in mitochondrial myopathy, lactic acidosis, and stroke-like episodes (MELAS). *Biochem Biophys Res Commun* 174:861–868, 1991.

50. Y Goto, I Nonaka, S Horai. A new mtDNA mutation associated with mitochondrial myopathy, encephalopathy, lactic acidosis and stroke-like episodes (MELAS). *Biochim Biophys Acta* 1097:238–240, 1991.

51. PM Matthews, J Hopkin, RM Brown, JBP Stephenson, D Hilton–Jones, GK Brown. Comparison of the relative levels of the 3243 (A→G) mtDNA mutation in heteroplasmic adult and fetal tissues. *J Med Genet* 31:41–44, 1994.

52. W Sato, M Tanaka, S Sugiyama, T Nemoto, K Harada, Y Miura, Y Kobayashi, A Goto, G Takada, T Ozawa. Cardiomyopathy and angiopathy in patients with mitochondrial myopathy, encephalopathy, lactic acidosis, ans stroke-like episodes. *Am Heart J* 128:733–741, 1994.

53. V Carelli, C Giordano, G d'Amati. Pathogenic expression of homoplasmic mtDNA mutations needs a complex nuclear–mitochondrial interaction. *Trends Genet* 19:257–262, 2003.

54. JF Hess, MA Parisi, JL Bennett, DA Clayton. Impairment of mitochondrial transcription termination by a point mutation associated with the MELAS subgroup of mitochondrial encephalomyopathies. *Nature* 351:236–239, 1991.

55. EA Schon, Y Koga, M Davidson, CT Moraes, MP King. The mitochondrial tRNA[Leu(UUR)] mutation in MELAS: a model for pathogenesis. *Biochim Biophys Acta* 1101:206–209, 1992.

56. SW Ballinger, JM Schoffner, HV Hedaya. Maternally transmitted diabetes and deafness associated with a 10.4 kb mitochondrial DNA deletion. *Nature Genet* 1:11–15, 1992.

57. JM van den Ouweland, HH Lemkes, W Ruitenbeek, LA Sandkuijl, MF de Vijdev, PA Struyvenberg, JJ van de Kamp, JA Massen. Mutation in mitochondrial tRNA[Leu(UUR)] gene in a large pedigree with maternally transmitted type diabetes mellitus and deafness. *Nature Genet* 1:368–371, 1992.

58. W Reardon, RJM Ross, MG Sweeney, LM Luxon, ME Pembrey, AE Harding, RC Trembath. Diabetes mellitus associated with a pathogenic point mutation in mitochondrial DNA. *Lancet* 340:1376–1379, 1992.

59. GA Cortopassi, N Arnheim. Detection of a specific mitochondrial DNA deletion in tissues of older humans. *Nucleic Acids Res* 18:6927–6933, 1990.

60. S Sugiyama, K Hattori, M Hayakawa, T Ozawa. Quantitaive analysis of age-associated accumulation of mitochondrial DNA with deletion in human hearts. *Biochem Biophys Res Commun* 180:894–899, 1991.

61. K Hattori, M Tanaka, S Sugiyama, T Obayashi, T Ito, T Satake, Y Hanaki, J Asai, M Nagano, T Ozawa. Age-dependent increase in deleted mitochondrial DNA in the human ear: possible contributory factor to presbycardia. *Am Heart J* 121:1735–1742, 1991.

62. M Hayakawa, K Hattori, S Sugiyama, T Ozawa. Age-associated oxygen damage and mutations in mitochondrial DNA in human heart. *Biochem Biophys Res Commun* 189:979–985, 1992.

63. M Hayakawa, S Sugiyama, K Hattori, M Takasawa, T Ozawa. Age-associated damage in mitochondrial DNA in human hearts. *Mol Cell Biochem* 119:95–103, 1993.

64. A Baumer, C Zhang, AW Linnane, P Nagley. Age-related human mtDNA deletions: a heterogeneous set of deletions arising at a single pair of directly repeated sequences. *Am J Hum Genet* 54:618–630, 1994.

65. N Takeda, A Tanamura, T Iwai. Mitochondrial DNA deletion in human myocardium. *Mol Cell Biochem* 119:105–108, 1993.

66. N Takeda, A Tanamura, T Iwai, Y Hayashi, S Nomura. Mutations of myocardial mitochondrial DNA in diabetic patients. In: NS Dhalla, GN Pierce, V Panagia, RE Beamish, Eds. *Heart Hypertrophy and Failure*. Boston: Kluwer Academic Publishers, 1995, 59–66.

67. M Corral–Debrinski, G Stepien, JM Shoffner, MT Lott, K Kanter, DC Wallace. Hypoxemia is associated with mitochondrial DNA damage and gene induction. *JAMA* 266:1812–1816, 1991.

68. CN Ellis, MB Ellis, WS Blakemore. Effect of adriamycin on heart mitochondrial DNA. *Biochem J* 245:309–312, 1987.

69. K Adachi, Y Fujiura, F Mayumi, A Nozuhara, Y Sugiu, T Sakanashi, T Hidaka, H Toshima. A deletion of mitochondrial DNA in murine doxorubicin-induced cardiotoxicity. *Biochem Biophys Res Commun* 195:945–951, 1993.

8

Mitochondria in Diabetes Mellitus

HONG KYU LEE

CONTENTS

DIABETES MELLITUS

What Is Diabetes Mellitus?

Diabetes mellitus is a state of chronic hyperglycemia, defined by more than 126 mg/dl of plasma glucose after an overnight fast of more than 8 h [1]. The diabetic state is morbid because the state of chronic hyperglycemia destroys the organs, particularly those with high capillary density. High blood glucose is particularly toxic to endothelial cells, and kidneys, retina of eyes, and nervous tissues, which are highly vasculized [2,3].

The distribution of fasting blood glucose level among the general public is skewed to the right or higher. People with slightly higher fasting blood glucose than normal, 110 to 125 mg/dl, are defined as having impaired fasting glucose (IFG). These subjects are prone to develop clinical diabetes as they age. Another group of subjects, who assimilate glucose poorly when a standard dose (75 g) of glucose is given orally, is designated as people with impaired glucose tolerance (IGT) [1]. These two prediabetic states are not identical; the former implies an insulin secretion defect, whereas the latter suggests insulin resistance [4].

Blood glucose is regulated by many factors; insulin is the most important hormone and secreted by pancreatic β-cells. Insulin lowers blood glucose level by stimulating its transport into the cells and suppressing its production from the liver. Many other humoral factors increase blood glucose level; glucagon, the glucocorticoids, catecholamines, and growth hormone are major players.

Hyperglycemia does not occur without insulin deficiency; thus, the diabetic state is ultimately a state of relative insulin deficiency. However, insulin sometimes does not act to lower blood glucose appropriately. This is frequently seen in obese/diabetic subjects. In this instance, the target cells are insulin resistant. Insulin resistance is considered more important than insulin deficiency in these circumstances. These basic determinants of blood glucose control were taken as a basis for disease classification: type 1, or insulin-dependent diabetes mellitus (IDDM), and type 2, or non-insulin-dependent diabetes mellitus (NIDDM).

Types of Diabetes Mellitus

Primary diabetes mellitus is classified based on pathogenic mechanism, or type of disease processes: type 1, type 2, and other types of diabetes [1]. Type 1 diabetes is characterized by β-cell destruction due to an autoimmune mechanism, with strong genetic influence, particularly from the HLA genes [5]. When any known cause of insulin deficiency is known, such as pancreatic destruction or diabetogenic gene mutation, it is classified as "other form of diabetes" [6]. Mutation of mitochondrial DNA at 3243 A to G, the so-called MELAS mutation, results in diabetes and thus belongs to the group of other types of diabetes [7].

Pregnancy diabetes is yet another class of diabetes. In this case, major morbidity occurs to the baby when maternal hyperglycemia is left untreated. This level comprises over 95 mg/dl of fasting plasma glucose, which means that the permissible level of ambient blood glucose is lower in the fetus than in the adult. In fact, because one measurement of fasting glucose level is usually unsatisfactory to represent the metabolic state of pregnant women, change of blood glucose level after 100 g of oral glucose load is analyzed for the final diagnosis [8].

The clinical features of type 1 diabetes are more severe and rapid than those of type 2. When insulin deficiency is severe, patients lose weight markedly (wasting of muscle and adipose tissue) and develop severe hyperglycemia, glycosuria, and high free fatty acids, which sometimes lead to ketoacidosis if untreated. However, the process of β-cell destruction is rather slow, taking several years after a trigger. This is a serious disease of children and rather frequent among Americans (about 14 cases per 100000, age under 15). The screening of high-risk children with HLA genotype and anti-islet antibodies is predictive of later development of clinical diabetes [5,9]. Several strategies to prevent disease at this stage, such as treatment with insulin and nicotinamide, are under investigation.

The resistance to insulin action (in muscle and liver) rather than insulin deficiency, is regarded as more important in type 2 diabetes. However, as long as pancreatic β-cells secrete enough insulin, diabetic state will not ensue, but obesity can develop. This concept is best illustrated in Figure 8.1, which is changed to adapt the mitochondrial DNA changes associated with evolution of diabetes. It is appropriate here to note that mitochondrial abnormalities in the pancreas could cause an insulin secretion defect and that such abnormalities in muscle and peripheral blood white cells are associated with insulin resistance [10–12]. Clinical features of type 2 diabetes are milder than those of type 1; frequently, patients do not need insulin for survival.

The most important risk factor for type 2 diabetes is aging, and insulin resistance is a major factor in the pathogenesis of type 2 diabetes of the elderly. Frequently, obesity precedes the onset of type 2 diabetes. After diabetes onset, patients may lose weight if the diabetic state is severe. The weight loss indicates insufficient insulin release relative to the body's requirement.

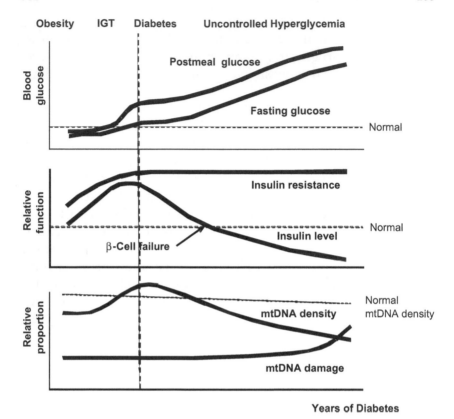

Figure 8.1 General scheme of understanding type 2 diabetes, insulin resistance, and mitochondrial DNA changes. Each line represents a degree of parameter changes. Free radical mechanism is primarily considered as a "cause" in linking these phenotypes and mitochondria.

Metabolic Syndrome and Diabetes Mellitus

Not only are many patients with type 2 diabetes obese, they are also frequently hypertensive and hyperlipidemic. Furthermore, they might have microalbuminuria (slightly increased albumin excretion through kidney), slightly more coagulable blood, high plasma levels of uric acid, and homocysteine. Now this disease of clustered state of metabolic abnormalities is defined as metabolic syndrome.

The World Health Organization (WHO) initially proposed a definition for the metabolic syndrome in 1998 [1]. According to WHO criteria, a participant has the metabolic syndrome if he or she has diabetes, impaired glucose tolerance, impaired fasting glucose, or insulin resistance (based on glucose clamp study) plus two or more of the following abnormalities:

- Blood pressure ≥ 140/90 mmHg
- Triglycerides ≥ 150 mg/dl (1.695 mmol/l) and/or HDL cholesterol < 35 mg/dl (0.9 mmol/l) in men and < 39 mg/dl (1.0 mmol/l) in women
- Central obesity: waist to hip ratio > 0.90 in men and > 0.85 in women and/or BMI > 30 kg/m²
- Microalbuminuria (urinary albumin excretion rate ≥ 20 µg/min or albumin:creatinine ratio ≥ 30 mg/g)

The National Cholesterol Education Program (NCEP) of the United States gives a slightly different definition [13] from that of WHO. NCEP deleted microalbuminuria and insulin resistance, but emphasized hypertension.

In general, the term "metabolic syndrome" is an inclusive clinical term applied to the disease condition characterized by insulin resistance. Many subjects with type 2 diabetes and a prediabetic state of glucose intolerance belong in this category. This author maintains that this condition is better understood as a mitochondrial abnormality.

Complications of Diabetes Mellitus

Diabetes mellitus is defined by the level of plasma glucose — a physiologic state. This definition disregards the state of the vital organs (anatomy). Patients with diabetes usually have normal organs at the onset, as one can easily expect in the early phase of experimental diabetes. However, tissues of diabetic subjects become destroyed as the duration of diabetic state increases. Diabetic complications occur more frequently when hyperglycemia is severe and persists over long periods of time. Complications occur when the vascular supply is rich, such as in the kidneys, retinas, and nerves.

These facts led to the current treatment: tight blood glucose control to prevent complications. Although proved effective in preventing complications, maintaining normoglycemia is a rather difficult task. This difficulty is pressing diabetologists to come up with alternative methods of glycemic control and of preventing complications even when the hyperglycemia persists.

Why should the diabetic state destroy the capillaries? Among the many possibilities are the following [14–16]:

- Nonenzymatic glycosylation of proteins by glucose and accumulation of resultant advanced glycation end-products (AGEs)
- Increased oxidative stress induced by hyperglycemia
- Increased metabolic flux through polyol pathway and secondary changes associated with this altered metabolism
- Alterations in molecules associated with the insulin signaling pathway, particularly DAG-sensitive PKC isoenzymes, Na^+K^+ ATPase activity, and MAP kinase

These changes are combined with the special situation of the endothelial cell, which must produce nitric oxide for vasodilation, rendering these cells particularly vulnerable. However, a recent review presents an alternative view on the pathogenesis of retinopathy and neuropathy [17].

Many agents that target these mechanisms have been developed: polyol pathway (aldose reductase) inhibitors; antioxidants; inhibitors of glycation; and PKC inhibitors [18–21]. Because these metabolic abnormalities are part of the diabetic state, they are interrelated. The diabetic state also affects mitochondrial function. In this book, these topics will be covered as a part of mitochondrial abnormalities and readers should consult specialty textbooks on diabetic complications for more details. In this chapter, we will focus on diabetes mellitus as a special case of mitochondrial disease as a way of understanding diabetic complications, as well as the pathogenesis of diabetes.

MITOCHONDRIAL ABNORMALITIES PRECEDE THE APPEARANCE OF DIABETES MELLITUS

mtDNA Mutation in Diabetes Mellitus: Insulin Secretion

It is now well established that many mitochondrial DNA (mtDNA) base substitutions, duplications, and deletions can cause diabetes mellitus. Approximately 0.5 to 1.5% of all diabetic patients exhibit pathogenic mtDNA mutations such as duplications [22], point mutations [23], and large-scale deletions [24].

Mathews and Berdanier were the first to see diabetes as a disease of mitochondrial origin. They focused on the mutations in the mitochondrial genome that phenotype as diabetes mellitus in human and rats [25]. Recently, Maassen gave a good description on its pathogenesis and clinical feature in a case of a specific mtDNA mutation [26]. Most of the known mtDNA mutations are understood to cause the diabetic state by affecting insulin secretion from β-cells [27]. Wollheim and coworkers showed that a reduction in mitochondrial Ca^{2+} accumulation is responsible for a reduction in insulin secretion, suggesting that modulation of mitochondrial Ca^{2+} uptake may be involved in the desensitization of insulin secretion in response to repeated exposures to metabolic substrates [27]. They also suggested that a mitochondrial factor, distinct from ATP and later identified as glutamate, participated in the triggering of insulin exocytosis.

mtDNA Polymorphism and Type 2 Diabetes: Insulin Resistance

In 1998, Poulton et al. surprised diabetologists by reporting a polymorphism in the first hypervariable region of the mtDNA control region (16189T > C) that was associated with insulin resistance [28]. They screened blood DNA from 251 men born in Hertfordshire from 1920 to 1930 in whom an earlier cohort study had shown that glucose tolerance was inversely related to birth weight. They found that the prevalence of the 16189 variant increased progressively with fasting insulin concentration. The association was independent of age and body mass index and was present after exclusion of the patients with NIDDM or impaired glucose tolerance.

The finding was confirmed in other populations [29]. This group demonstrated a significant association between the 16189C variant and type 2 diabetes in a population-based case-control study involving 932 cases from Cambridgeshire, U.K. The odds ratio was 1.61 in general, but was greatly magnified in individuals with a family history of diabetes from the father's side (odds ratio reached infinity). The prevalence of 16189C variant among individuals who reported having a maternal history of diabetes was not different from those without this history. However, all individuals had diabetes who (1) reported having a paternal family history of diabetes and (2) had the 16189C. This would suggest that nuclear genetic factors inherited from the father increased the background risk of the maternally inherited mitochondrial variant considerably. Such an interaction is biologically plausible because a paternal history of diabetes is associated with low birth weight [29] and the 16189 variant is associated with thinness at birth [30].

This observation was extended to Europeans in New Zealand [31] and Asians. Kim et al. investigated 160 nondiabetic subjects from a 1993 community-based diabetes survey conducted in Yonchon County, Korea, for the T16189C polymorphism [32]. The prevalence of the 16189C variant in Korean adults was 28.8%, and the subjects with the 16189C were found to have higher fasting plasma glucose and higher BMI than those with 16189T. However, fasting insulin, homeostasis model assessment (HOMA) of insulin resistance and β-cell function, cholesterol, and blood pressure were not different between the two groups. A recent analysis found a prevalence of 16189C that was significantly higher among type 2 diabetic patients (unpublished data).

Ji et al. examined 383 unrelated Chinese type 2 diabetics and 151 nondiabetic controls selected by random sampling [33]. The prevalence of the 16189C variant was significantly higher among type 2 diabetics than among controls, particularly in subjects with a maternal family history. In comparison with the type 2 diabetics without the 16189C polymorphism, 16189C variants showed higher fasting serum insulin levels and a lower insulin sensitivity index, suggesting the presence of insulin resistance. Among the control group, however, no significant difference in age, sex, body mass index, waist to hip ratio, serum lipid level, fasting plasma glucose level, fasting serum insulin level, and insulin sensitivity index was found between those with 16189C and those with 16189T.

Although this study confirmed the contribution of the 16189 variant of mtDNA as a genetic predisposing factor for type 2 diabetes in a Chinese population, the absence of association between 16189C and insulin resistance within the control group suggests that it acts with still unknown factors in producing insulin resistance. As noted earlier, the Korean data also showed no difference in insulin sensitivity between two groups.

Furthermore, the length of the C stretch becomes longer when T is changed into C at the 16189 position. Among Koreans, most of those with the 16189C variant have more than ten C repeats, but 85% of those with the wild type had only nine repeats. This could affect transcriptional machinery (data source: NCBI Access No. D84836-D84899). C at the 16189 position is frequently associated with mtDNA 9-bp deletion, a defining character for mitochondrial haplotype B. Because this haplotype is highly prevalent in Indonesia and Southeast Asia, it appears that it might be a genotype for survival in a hot environment. Further studies are needed to clarify the role of this control region because it is also known to be associated with longevity and climate adaptation [34].

mtDNA Density in Diabetes: Insulin Resistance

Lee et al. found that mtDNA density is decreased in subjects that subsequently develop diabetes [10]. They analyzed samples taken from a population-based prospective study and reported that mtDNA density was decreased in the muscles of diabetic subjects. They concluded that the depletion was

due to the diabetic state because the decreased mtDNA density was observed in subjects with type 1 or type 2 diabetes.
These observations will be discussed in more detail later [35].

However, when mtDNA density from peripheral blood
was analyzed with quantitative PCR, the number of mtDNA
copies in 23 subjects who subsequently became diabetic
within 2 years was lower than that of 22 age- and sex-matched
control subjects who remained nondiabetic. The number of
mtDNA copies in subjects who converted was lower than in
controls, even before the development of diabetes (102.8 ± 41.5
vs. 137.8 ± 67.7 copies/pg template DNA). Furthermore,
inverse correlations were noted between mtDNA content and
baseline waist–hip circumference ratio (WHR) ($r = -0.31$) and
fasting glucose level ($r = -0.35$), diastolic blood pressure ($r = -0.36$), and WHR ($r = -0.40$) after development of diabetes,
suggesting that mtDNA density is somehow related to the
insulin resistance syndrome.

A nested case-control design was used to determine
whether mtDNA density change precedes or follows the onset
of diabetes. This was possible because samples were available
from a population-based study from Yonchon County, Korea.
In this study, mtDNA was quantitated with a competitive
PCR. The PCR was designed to produce different PCR products by introducing an internal standard carrying mtDNA
sequence, which competes for the same primers with mtDNA
sequences. By comparing the PCR product density to the
amounts of known internal standard added, mtDNA density
was calculated.

Song et al. extended the observation to diabetic offspring.
Insulin sensitivity in the offspring of type 2 diabetic patients
is known to be lower than that of the offspring of nondiabetic
parents [36]. They selected offspring of type 2 diabetic
patients and 52 age-, sex-, and BMI-matched normal subjects
from an ongoing population-based study at Mokdong, Seoul,
Korea. Of the offspring with a diabetic family history, 52 had
normal glucose tolerance (NGT), 21 had impaired glucose
tolerance (IGT), and 9 had newly diagnosed type 2 diabetes.
The pb-mtDNA density was measured by real-time PCR with
a mitochondria-specific fluorescent probe, normalized by a

nuclear DNA, 28S rRNA gene. The peripheral blood mtDNA/28S rRNA ratio (abbreviated as pb-mtDNA density here) tended to be lower in the 82 offspring of type 2 diabetic patients than in controls (1084 vs. 1304) and was significantly lower in the combined NGT and IGT offspring group (1068) than in the control subjects. In NGT + IGT offspring, the pb-mtDNA density was significantly correlated with logarithmically transformed insulin sensitivity (r = 0.253) and was the main predictor of insulin sensitivity. This study suggested that quantitative pb-mtDNA status might be influenced by a hereditary factor associated with type 2 diabetes.

Similar findings were reported from Hong Kong confirming the major finding of the Mokdong study [37]. In collaboration with Song's group, Ng et al. investigated the mtDNA levels in 152 Chinese type 2 diabetic patients with disease duration ≤ 5 years (51 with maternal family history of diabetes (FH), 51 with paternal FH, and 50 with no parental FH), and 51 age- and sex-matched healthy control subjects. MtDNA levels were measured by quantitative PCR and corrected for 28S rRNA levels. The relative mtDNA levels were similar in diabetic patients and control subjects (547 ± 249 vs. 534 ± 318).

Diabetic patients with FH had similar anthropometric and clinical characteristics compared with the patients with no FH. However, the latter group had significantly higher mtDNA levels than the groups with maternal or paternal FH. In addition, the mtDNA level was inversely correlated with age and blood pressure in control subjects, but mtDNA levels were positively correlated with age, age at diagnosis, and lipid profiles in patients with maternal FH. Multivariate analysis showed that age was the only independent predictor of low mtDNA levels in control subjects. On the other hand, maternal and paternal FH were independent predictors for low level of mtDNA among diabetic patients.

Combined with later studies that showed pb-mtDNA content is inversely correlated with insulin secretion and fuel oxidation rates during glucose clamp studies [38,39], Lee suggested that the mtDNA genome could be considered a thrifty genome, although many known and unknown genetic and environmental factors could be influencing expression of this genome [40]. This hypothesis implies that insulin sensitivity is determined by the qualitative and quantitative state of the mitochondrial genome of a person. Another implication is that a person born with lowered mitochondrial density or set point must secrete more insulin to enhance the glucose utilization. Fat may be the preferred substrate for mitochondrial metabolism when mtDNA density is high. In this state, insulin is less required. This statement needs to be substantiated with further research.

Substantial age-related reductions of mtDNA copy number occur in skeletal muscle and liver, but not in the heart [41]. The decline in mtDNA was not associated with reduced cytochrome c oxidase (COX) transcript levels in tissues with high oxidative capacities such as red soleus muscle or liver; transcript levels were reduced with aging in the less oxidative mixed fiber gastrocnemius muscle. Consistent with transcript levels, COX activity also remained unchanged in aging liver and heart but declined with age in the lateral gastrocnemius. Thus, the effects of aging on mitochondrial gene expression are tissue specific. A compensatory mechanism appears to occur that increases mitochondrial gene expression even though mtDNA copy number has declined.

Similar findings were observed in the human. Barthelemy et al. demonstrated mtDNA depletion in patients with Kearns–Sayre syndrome [42]. Despite the severity of the depletion, *in situ* hybridization using two mtDNA transcripts revealed that the patients have a normal steady-state level of transcription. Such compensation provides clues to the striking contrast between the severity of mtDNA depletion and the clinical manifestations, late onset, and slowly progressive nature of disease.

Whole-body energy expenditure will depend ultimately on the total mitochondrial functional mass of the body. In 1956, Smith reported that the mitochondrial density of the liver correlated with whole-body energy expenditure, and the relative amount of mitochondria in any given tissue was suggested to be the controlling factor in determining the regression of oxygen utilization to the total body size of the species

[43]. Apparently, peripheral blood mtDNA density also seems to correlate with energy utilization, but heart and gastrocnemius muscle do not. This issue will be elaborated further later.

Mitochondrial Function and Insulin Resistance

The linkage between deranged mitochondrial function and insulin resistance was explored only very recently. Peterson et al. studied how insulin resistance arises *in vivo*. They compared elderly people with marked insulin resistance with young controls for insulin-stimulated muscle glucose metabolism [12]. They selected 16 healthy elderly volunteers by screening with a 3-h oral glucose (75 g) tolerance test and with lean body mass (LBM) and fat mass with dual-energy x-ray absorptiometry. They matched these subjects with 13 young volunteers with no family history of diabetes or hypertension, who had similar BMI and habitual physical activity, as assessed by an activity index questionnaire. All participants were in excellent health as determined by a complete medical exam.

Although young and elderly participants had similar fat mass, percent fat mass, and LBM, the elderly participants had slightly higher plasma glucose concentrations and significantly higher plasma insulin concentrations, thus proving that they were indeed relatively insulin resistant. To determine the tissues responsible for the insulin resistance, Peterson and coworkers performed hyperinsulinemic-euglycemic clamp studies, in combination with [6,6-2H2] glucose and [2H5] glycerol tracer infusions. These sophisticated techniques permitted them to measure basal rates of glucose production from liver and the rates of glucose utilization by muscles (derived from the glucose infusion rate required to maintain euglycemia during the clamp). The major difference was that the elderly subjects had ~40% lower insulin-stimulated peripheral glucose uptake than did the young subjects. These investigators suggested an age-related difference in insulin-stimulated muscle glucose metabolism.

Peterson et al. then used muscle and hepatic biopsies and determined intramyocellular lipid content (IMCL) by 1H

NMR spectroscopy. The IMCL content in the soleus muscle was increased by ~45% in the elderly, suggesting that fat had accumulated in muscle. Intrahepatic triglyceride content was also increased by 225% in the elderly, even though no hepatic insulin resistance was detected in the elderly participants during the clamp. These changes were associated with an approximate 40% reduction in mitochondrial oxidative and phosphorylation activity, as assessed by *in vivo* 13C/31P NMR spectroscopy. Furthermore, a strong relationship existed between increased intramuscular fat content and insulin resistance in muscle.

Because increases in the intracellular concentration of fatty acid metabolites have been postulated to activate a serine kinase cascade leading to defects in insulin signaling in muscle and the liver, these results could explain the reduced insulin-stimulated muscle glucose transport activity and reduced glycogen synthesis in muscle. They could account for an impaired suppression of glucose production by insulin in the liver. The authors suggested that an age-associated decline in mitochondrial function, a phenomenon well known to mitochondriologists, contributes to insulin resistance in the elderly. For example, Dicter et al. reported that the glycogen synthesis pathway depended on mitochondrial function because treatment with mitochondrial inhibitors eliminated the majority of glycogen synthesis [44].

Many mitochondrial changes in diabetic tissue are secondary to diabetes [35,45,46]. Mitochondrial change also might precede the diabetes, so interpretation of mitochondrial function and genetic changes warrants great care.

Mitochondria-Related Gene Expression Profile in Insulin Resistance

A recent publication by Patti et al. provides further support for the mitochondrial hypothesis [11]. They analyzed gene expression in skeletal muscle from healthy metabolically characterized nondiabetic (with or without FH negative and positive for diabetes) and diabetic Mexican–American subjects. High-density oligonucleotide arrays and quantitative real-time PCR were used to identify genes differentially expressed in skeletal muscle. Of the 7129 sequences represented on the array, 187 were differentially expressed between control (FH–) and diabetic subjects. By using the MAPP FINDER program [47], they found the top-ranked cellular component terms were mitochondrion, mitochondrial membrane, mitochondrial inner membrane, and ribosome (Z scores of 7.8, 7.2, 7.8, and 6.8, respectively, indicating overrepresentation in FH– vs. diabetic subjects).

Similarly, the top-ranked process term was ATP biosynthesis (Z score of 5.5). Similar results were obtained by using another computer program (ONTOEXPRESS) with multiple correction testing. This program indicated that energy generation, protein biosynthesis/ribosomal proteins, RNA binding, ribosomal structural protein, and ATP synthase complex ontology groups were represented to a significantly greater extent in FH– vs. diabetes than expected, if ontology groups were randomly distributed within the list of top-ranked genes.

The data demonstrated progressive decrease in expression of genes encoding key proteins in oxidative metabolism in insulin-resistant and diabetic subjects. This was associated with reduced expression of multiple nuclear respiratory factor-1 (NRF-1)-dependent genes, encoding key enzymes in oxidative metabolism and mitochondrial function. Although NRF-1 expression is decreased only in diabetic subjects, expression of PPAR γ-coactivator 1-α and -β (PGC1-α/PPARGC1 and PGC1-β/PERC) — coactivators of NRF-1 and PPAR γ-dependent transcription — is decreased in diabetic subjects and FH-positive nondiabetic subjects. Array expression levels for other transcription factors implicated in nuclear-encoded mitochondrial gene transcription, including mitochondrial transcription factor A (Tfam), MEF-2 isoforms, YY1, CREM, CREB1, and CREB3, did not differ but were near background levels. These researchers concluded that the decreased PGC1 expression might be responsible for decreased expression of NRF-dependent genes, leading ultimately to the metabolic disturbances characteristic of insulin resistance and diabetes mellitus.

Interestingly, an inverse relationship was found among levels of mitochondrial complex III subunit, ubiquinol-cytochrome C reductase (UQCRH) gene expression, obesity, and fasting insulin. This raises the possibility that an adipocyte product, nutrient excess, or insulin resistance itself may further contribute to down-regulation of oxidative phosphorylation gene expression. This correlation between the gene expression at the cell level and body mass index is reminiscent of allometry regarding the body scale.

Mitochondrial Abnormalities in Diabetes-Prone Animal Models

Zucker Diabetic Fatty (ZDF) Rat

The ZDF rat is an obese diabetic rat characterized by impaired insulin secretion and insulin sensitivity. The mitochondrial enzymes essential for stimulus-secretion coupling in the pancreatic β-cell are implicated in this rat. Glycerol phosphate dehydrogenase (mGPD) and pyruvate carboxylase (PC) activities were reported to be low in the pancreatic islet [48]. Similar observations were also reported in other rodent models of NIDDM. The key enzyme of the glycerol phosphate shuttle is mGPD, and PC is the key component of the pyruvate malate shuttle. A decrease in these enzyme activities was attributed to a decrease in the net synthesis of thee enzymes rather than by decreased activity. Because enzyme activities were depressed while ZDF rat is euglycemic (6 weeks of age), certain changes associated with diabetic syndrome other than hyperglycemia are suspected as a cause of these alterations.

Bindokas et al. suggested that oxygen free radicals are implicated in the β-cell dysfunction of ZDF rat [49]. They employed the superoxide-mediated oxidation of hydroethidine to ethidium to assess dynamically and directly the relative rates of mitochondrial superoxide anion generation in isolated islets in response to glucose stimulation. Superoxide content of isolated islets increased in response to glucose stimulation. Using digital imaging microfluorometry, they found that the superoxide content of ZDF islets was significantly higher than Zucker lean (ZL) control islets under resting conditions, relatively insensitive to elevated glucose concentrations, and correlated temporally with a decrease in glucose-induced hyperpolarization of the mitochondrial membrane.

Importantly, superoxide levels were elevated in islets from young, prediabetic Zucker diabetic fatty animals. Bindokas and colleagues concluded that overproduction of superoxide was associated with perturbed mitochondrial morphology and may contribute to abnormal glucose signaling found in the Zucker diabetic fatty model of type 2 diabetes mellitus. Combined with the study of MacDonald et al. [48], these observations suggest that certain changes occur in the mitochondria from ZDF rats that precede the onset of diabetes.

Goto–Kakizaki (GK) Rat

GK rat is a genetic model of diabetes characterized by defective insulin secretion as well as by partial maternal inheritance, suggesting mitochondrial involvement in the pathogenesis. The β-cells of adult GK rats have a significantly smaller mitochondrial volume and an increased number of mitochondria per unit tissue volume compared with the β-cells of control islets. No major deletions or restriction fragment polymorphism was detected in mtDNA.

Although mtDNA contents of fetal GK islets and fetal GK liver were not different from those of fetal Wistar rats, adult GK islets contained markedly less mtDNA than the corresponding control islets (the mtDNA contents of adult liver were similar). The lower islet mtDNA contents were paralleled by a decreased content of islet mtRNA (12S ribosomal RNA and cytochrome b messenger RNA). These results suggest that mtDNA damage occurs specifically in islet cells as a consequence of the disturbed metabolic environment of the adult GK rat [50].

Decreased mitochondrial enzyme activities, mGPD and PC, were also observed in these rats. McDonald et al. reported that insulin treatment normalized these activities, suggesting that the changes are results of diabetic syndrome [51]. However, insulin treatment did not restore, but further suppressed, insulin release by the islets [52]. These results strongly indicate that insulin deficiency is not the sole cause behind this impairment and that the glycerol–phosphate shuttle activity is not the rate-limiting step of glucose-induced insulin release in the β-cell of GK rats.

Moreira et al. evaluated the respiratory indexes (respiratory control ratio (RCR) and ADP/O ratio); mitochondrial transmembrane potential; repolarization lag phase; repolarization level; ATP/ADP ratio; and induction of the permeability transition pore mitochondria isolated from GK and normal Wistar rat brain mitochondria of different ages [51]. The effect of amyloid β-peptides on mitochondrial function was also analyzed. Aging induced a decrease in brain mitochondrial RCR, ADP/O, and ATP/ADP ratios and an increase in the repolarization lag phase. Brain mitochondria isolated from older diabetic rats were more prone to induction of the permeability transition pore and accumulated much less Ca^{2+} than those isolated from young rats. Amyloid β-protein potentiated age-related effects on mitochondrial function. These results indicate that mitochondrial dysfunction of GK rat is not limited to the pancreas, but also present in the brain mitochondria.

Otsuka Long–Evans Tokushima Fatty (OLETF) Rat

OLETF rat was derived by selected breeding from Long–Evans strain of rat at Tokushima, Otsuka laboratory (LETO), Japan. This rat is a genetic model of spontaneous development of NIDDM and exhibits hyperglycemia, obesity, and hyperinsulinemia (insulin resistance) similar to that in humans. Among the many genes suspected to cause the phenotype of this animal, defective CCK receptor was thought to cause obesity; the phenotype of CCK knockout mice was normal in weight and glucose tolerance [54].

OLETF develop obesity early on, but diabetes after the 25th week of life. Sato et al. studied the early events of glucose metabolism in the skeletal muscle of the male OLETF rat and found that the activity of the hexokinase II was significantly decreased compared with that of LETO rats [55]; the type I hexokinase activity was not different. The protein content of the hexokinase II was also significantly lower. After insulin stimulation, the intramuscular content of glucose 6-phosphate, which regulates glycogen synthesis in skeletal muscle, was significantly decreased in OLETF rats; however, glycogen synthase activity *in vitro* and intramuscular lactate concentration in these rats did not show significant differences. These results suggested that the molecular defects of skeletal muscle in OLETF rats are similar to those in NIDDM patients. Hexokinase II is closely associated with mitochondrial function and its activity in diabetes will be discussed later.

OLETF rats have a poor capacity for proliferation of pancreatic β-cells after partial pancreatectomy, and this occurs prior to the onset of overt diabetes [56]. Defective morphogenesis is also noted during the culture of the developing OLETF islet, along with a parallel decline in relevant islet hormone contents, when a dramatic increase in β-cell and non-β-cell populations occurs in control rats [57].

In an unpublished study, the author's group found that mtDNA density was lower in the OLETF rat than in LETO rat, in which it was highest in early life (10th week). The mtDNA density declined with age in both groups of rats, but was significantly delayed in LETO rats. This sequence of events is consistent with the idea that alterations in mitochondrial function and mtDNA density change are an early event. It might be related to a poor replication of cells because each cell contains less mtDNA.

The group also found that OLETF rats lose weight when treated with s-adenosylmethionine, probably through a central mechanism. Interestingly, mtDNA density of muscle increased as they lost weight, and the relationship between mtDNA and body mass among OLETF, treated OLETF, and LETO at the 24th week was linear and inverse Although further studies are needed, this inverse relationship might be indicating that the metabolic scale law is operating. The animal is adapting to its energy demand by increasing mitochondrial density and heat generation or vice versa.

Nakaya et al. produced an interesting report on OLETF in regard to the effect of taurine on this rat [58]. The group reported that abdominal fat accumulation, hyperglycemia, and insulin resistance were significantly lower in the taurine-supplemented OLETF rats than in the unsupplemented rats. Serum and liver concentrations of triglyceride and cholesterol were significantly higher in the OLETF rats than in the control rats, which became significantly lower after the taurine supplementation. Interestingly, taurine-supplemented rats also showed higher nitric oxide secretion, as evidenced by increased urinary excretion of nitrite. The importance of nitric oxide (NO) in controlling mitochondrial function will be discussed later.

Taurine is a potent antioxidant and prevents tissue injury. It was reported to improve streptozotocin-induced hyperglycemia in mice [59,60]. Supplementation of the maternal low-protein diet by 2.5% of taurine in the drinking water completely restored the islet cell proliferation as well as the IGF-2 and VEFR content in the islets of fetuses and suckling pups. It also restored the fetal islet vascularization by increasing the number of blood vessels [59]. Moreover, taurine supplementation of the mother's diet normalized the insulin secretion of the fetal islets of dams fed a low-protein diet [60]. The far-reaching consequences of the lower taurine levels in pups born from dams fed a low-protein diet are exemplified by these studies, which have close implication to the thrifty phenotype hypothesis (see below).

A recent report showed that taurine is a constituent of mitochondrial tRNAs and that two novel taurine-containing modified uridines (5-taurinomethyluridine and 5-taurinomethyl-2-thiouridine) were found in mitochondria [61]. Because this modification lacks mutant mitochondrial tRNAs for Leu(UUR) and Lys from pathogenic cells of the mitochondrial encephalomyopathies MELAS and MERRF, respectively, these findings will deepen understanding of the molecular pathogenesis of mitochondrial diseases in the future.

Ob/Ob Mouse

The ob/ob mouse is the most widely studied genetic model of obesity and severe insulin resistance. Vincent et al. compared the patterns of gene expression in skeletal muscle of the obese mouse with that of its thin littermate (ob/+) using the mRNA differential display method [62]. From about 9000 cDNAs displayed, they found eight mRNAs overexpressed in ob/ob muscle:

- Id2 (a negative regulator of the basic helix–loop–helix family of transcription factors)
- Fast skeletal muscle troponin T
- Ribosomal protein L3
- The integral protein of the peroxisomal membrane 22PMP
- The mammalian homolog of geranylgeranyl pyrophosphate synthase
- An mRNA related to phosphatidylinositol-glycan-specific phospholipase D
- Two unknown mRNAs

The level of overexpression of these mRNAs in skeletal muscle varied from a 500% increase to as little as a 25% increase. Two mRNAs were underexpressed: the f-subunit of mitochondrial ATP synthase and a retrovirus-related DNA. Two proteins with multiple transcripts, skeletal muscle α-tropomyosin and one for a repetitive sequence, showed a change in mRNA pattern. Underexpression of the f-subunit of mitochondrial ATP synthase in these mice is consistent with mitochondrial pathogenesis. However, it is not certain at present how this change might be related to leptin resistance, which is considered the primary defect in these mice [63].

Mitochondria in the Malnutrition-Induced Rat: Thrifty Phenotype Hypothesis

Intrauterine growth retardation (IUGR) has been linked to the development of type 2 diabetes in later life. This linkage became famous because of the work of Barker and his group in Southampton, U.K., which has been summarized nicely in a monograph [64]. In summary, low birth weight babies have a predisposition to insulin resistance and metabolic syndrome in general. They may also be more susceptible to breast and colon cancers. This idea was named the "thrifty phenotype hypothesis" and proposed by Hales and Barker as an antithesis to the thrifty genotype hypothesis [65]. The association of low birth weight and the metabolic syndrome has been interpreted as reflecting long-term effects of maternal nutritional factors on fetal growth and the development of tissues regulating glucose metabolism [66,67].

Bertin et al. studied the development of β-cell mass in fetuses of rats deprived of protein and/or energy [68]. Three kinds of diet were used: a low-protein isocaloric diet (5%, instead of 15%; PR); a calorie-restricted diet (50% of the control diet; CER); and a low-protein, calorie-restricted diet (50% of low-protein diet; PER). Only the low-protein diet, independent of total energy intake, led to a lower birth weight. The adult offspring female rats in the three deprived groups exhibited no decrease in body weight and no major impairment in glucose tolerance, glucose utilization, or glucose production (basal state and hyperinsulinemic clamp studies). However, pancreatic insulin content and β-cell mass were significantly decreased in the low-protein isocaloric diet group compared with the two energy-restricted groups.

Further study of the nutrient and energy intake with same protocol revealed that the pancreatic β-cell mass (expressed as absolute value or relatively to fetal body weight) was significantly decreased to a level similar in the CER, PR, and PER fetuses compared with the control fetuses [69]. When the mean insulin content-to-β-cell mass ratio was calculated, this ratio was significantly higher in the low protein-restricted groups than in the control group. Interestingly, an inverse correlation was present between this ratio and fetal insulinemia or insulinemia-to-glycemia ratio (respectively, $r = -0.73$ and $r = -0.65$). The most surprising result of this study was the observation that, among all the parameters analyzed, the lowered fetal plasma level of taurine in the group fed low protein was the main predictor of the fetal plasma insulin level ($r = 0.63$). The importance of taurine was discussed earlier in association with mitochondrial function, and here it was revealed to have an important role for fetal β-cell function.

The data supporting the thrifty phenotype hypothesis were overwhelming, but explanations have been varied. Selak et al. have developed a rat model of uteroplacental insufficiency as a cause of intrauterine growth retardation [70]. By inducing ischemia to a fetus by partially ligating placental blood supply, they made rats become insulin resistant early in life. By 6 months of age, they developed diabetes. Glycogen content and insulin-stimulated 2-deoxyglucose uptake were significantly decreased in muscle.

Mitochondria of IUGR muscle exhibited significantly decreased rates of state-3 oxygen consumption with pyruvate, glutamate, α-ketoglutarate, and succinate. Decreased pyruvate oxidation in IUGR mitochondria was associated with decreased ATP production, decreased pyruvate dehydrogenase activity, and increased expression of pyruvate dehydrogenase kinase 4. Such a defect in IUGR mitochondria leads to a chronic reduction in the supply of ATP available from oxidative phosphorylation. The authors concluded that impaired ATP synthesis in muscle compromises energy-dependent GLUT4 recruitment to the cell surface, glucose transport, and glycogen synthesis; this in turn contributes to insulin resistance and the hyperglycemia of type 2 diabetes.

Park et al. investigated the effect of protein malnutrition during gestation and lactation on mitochondria of the rat liver and skeletal muscle [71]. Female offspring of Sprague–Dawley rats fed a low-protein diet (casein, 80 g/kg) were randomly divided into two groups and weaned onto a low-protein or a control diet (casein, 180 g/kg). The rats in each group were randomly divided into four groups that were killed at 5, 10, 15, and 20 weeks of age. The mitochondrial DNA content of the liver and skeletal muscle was reduced throughout the study period in fetal and early postnatal malnourished rats even if proper nutrition was supplied after weaning. It was accompanied by a decrease in mitochondrial DNA-encoded gene expression, but did not depend on Tfam expression. These findings indicate that poor nutrition in early life causes long-lasting changes in mitochondria that may contribute to the development of insulin resistance in later life.

Park et al. found that these rats also have decreased pancreatic β-cell mass and reduced insulin secretory responses to a glucose load; no difference in glucose tolerance or insulin sensitivity was found until the rats were 20 weeks old [72]. These findings imply that protein malnutrition *in utero* causes the lowering of mtDNA density set point, impairs β-cell development, and leads to an insulin secretion defect. Combined with the report by Selak et al. [70], findings indicate that fetal malnutrition predisposes the fetus to the development of insulin resistance and type 2 diabetes in later life, possibly by a taurine-mitochondria mediated mechanism, i.e., decreasing pancreatic β-cell mass and lowering its set point in muscle and pancreas.

mtDNA Depletion Is Linked to Poor Glucose Utilization

Park et al. investigated the effects of mtDNA depletion on glucose metabolism with mtDNA-depleted (ρ0) human hepatoma SK-Hep1 cells [73]. Established by long-term treatment with ethidium bromide (EtBr), the ρ0 cells have been important tools in investigating the function of mtDNA [74]. EtBr intercalates into circular DNA and inhibits mtDNA replication and transcription at a low concentration (0.1 to 2 μg/ml) without any detectable effect on nuclear DNA division [75]. It has been used to show that inhibition of mitochondrial gene transcription suppresses glucose-stimulated insulin release in a mouse pancreatic β-cell line [76].

The ρ0 SK-Hep1 cells failed to hyperpolarize mitochondrial membrane potential in response to glucose stimulation. However, discrepancies on Δψm changes in ρ0 cells were present, although the impaired oxidative phosphorylation may account for the loss of Δψm. Kennedy et al. reported that Δψm was abolished in ρ0 insulinoma cells [77]; however, Buchet and Godinot reported that Δψm was maintained in ρ0 HeLa cells through the functional F1-ATPase and adenine nucleotide translocator [78]. Whether ρ0 insulinoma or SK-Hep1 cells contain normal activities of F1-ATPase and adenine nucleotide translocator is unknown.

Intracellular ATP content, glucose-stimulated ATP production, glucose uptake, steady-state mRNA and protein levels of glucose transporters, and cellular activities of glucose-metabolizing enzymes were found decreased in $\rho 0$ SK-Hep1 cells, compared with parental cells. GLUT-1 is the most abundant type of glucose transporter in this cell line as in fetal hepatocytes; the major glucose transporter in normal adult hepatocytes, GLUT-2, is not expressed in control SK-Hep1 cells.

Because expression of GLUT-2 is severely decreased throughout hepatocarcinogenesis, the interpretation of data using this cell line needs caution. The specific activities of GAPDH and G6PDH in $\rho 0$ cells were reduced to 75 and 30% of the control cells, respectively. It is interesting that similar changes were noted in human diabetic muscle [35]. The activity of LDH, an indicator for the intracellular NADH accumulation due to dysfunction of the respiratory chain, was enhanced by 53%, implying that minimal energy production for cell survival is achieved via anaerobic metabolism. Interestingly, the activity of hexokinase in $\rho 0$ cells was only 10% of that found in the control cells. These results suggest that the quantitative reduction of mtDNA suppresses the expression of nuclear DNA-encoded glucose transporters and enzymes of glucose metabolism.

A large decrease in hexokinase activity in $\rho 0$SK-Hep1 cells might need special attention. In contrast to other glycolytic enzymes, hexokinase is known to be associated with the mitochondrial outer membrane through its interaction with porin. It uses intramitochondrial ATP supplied by oxidative phosphorylation as a substrate for glucose phosphorylation [79]. Moreover, the activation of hexokinase depends on a contact site-specific structure of the pore, which is voltage dependent and influenced by the electrical potential of the mitochondrial inner membrane [80]. It has been reported that mitochondria lacking a membrane potential induced by a mitochondrial uncoupler such as dinitrophenol decreases the contacts and hexokinase activity in hepatocytes [81].

In western blot analysis, hexokinase protein was significantly decreased in the mitochondrial fraction from $\rho 0$ SK-Hep1 cells (unpublished data). From these data, the authors concluded that the defects in intramitochondrial ATP production and depolarized Δψm in ρ0 cells might have inactivated hexokinase and the decreased hexokinase activity in ρ0 cells may be one of the fundamental causes of the disturbed glucose metabolism. As discussed elsewhere, hexokinase II changes along with the insulin sensitivity; decreased in OLETF rat, an animal model of diabetes; decreased in patients with non-insulin-dependent diabetes mellitus; and increased after exercise.

Thrifty Phenotype, Thrifty Genotype, and Mitochondria

Certain base substitutions in the mitochondrial genome, 3243G and 16189C, for example, predispose the subjects to diabetes and insulin resistance. Those genes that affect the function of the products of the mitochondrial genes, such as UCPs and PGC-1, are also predisposing factors for diabetes and insulin resistance. Furthermore, fetal malnutrition, placental dysfunction, and other factors, which cause low birth weight, are now known to cause mitochondrial dysfunction in animals and human. If one sees the low birth weight baby as a state of low mitochondrial genome density, as determined by nuclear genes, one can understand the insulin resistance associated with mitochondrial dysfunction and why these subjects should develop disease phenotypes later in life (thrifty phenotype hypothesis).

Malnutrition (deficiency of taurine, thus poor mitochondrial development) during repeated famines might have selected those people with genes (mitochondrial genome and genes controlling it) adapted to the environment scarce in nutrients, but susceptible to obesity and diabetes mellitus. Because they had evolved mechanisms to store energy efficiently, if exposed to affluence, they would manifest disease phenotypes. The famous American geneticist/anthropologist Neel gave this explanation when he was asked an opinion for the high frequency of obesity and diabetes found in Pima Indians. Although there was no formal proof for the idea with hard data, his paper became a classic of public health chosen by WHO.

In Wales, U.K., people with diabetes were found to have slightly different mitochondrial genomes than control population, both of them expanded in ancient times. Sherratt et al. have sequenced the 347-bp segment of hypervariable region 1 of mtDNA from leukocytes of 63 subjects with type 2 diabetes and 57 age- and race-matched nondiabetic controls [82]. This region is suitable for investigating maternal ancestry and has been used extensively to study the origins of human racial groups. Consensus sequences for the two study groups were found to be identical. Pairwise sequence analysis showed unimodal distribution of pairwise differences for both groups, suggesting that both populations had undergone expansion in ancient times. The distributions of HV-1 sequence differences were significantly different; mean pairwise differences were 4.7 and 3.8 for the diabetic and control subjects, respectively.

From these data the authors concluded that the diabetic subjects belonged to an ancient maternal lineage that expanded before the major expansion observed in the nondiabetic population. However, phylogenetic trees constructed using several methods failed to show the clustering of all or a subset of the diabetic subjects into one or more distinct lineages. This suggests that random mtDNA changes had occurred and were not haplotype specific. The pressure toward diabetic susceptibility or thriftiness was on the mitochondrial genome in general.

IMPROVEMENT OF GLUCOSE METABOLISM AND MITOCHONDRIAL FUNCTION: EXERCISE

Exercise and Hexokinase II

It is well known that exercise increases insulin sensitivity and prevents the development of type 2 diabetes [83]. A single bout of exercise increases the rate of insulin-stimulated glucose uptake and metabolism in skeletal muscle for 24 to 48 h [84]. This increase has been attributed to an effect of exercise on GLUT-4 glucose transporter, hexokinase activity, or glycogen synthase [85–87]. Muscle glucose uptake (MGU) is controlled by three steps: delivery of glucose to the muscle membrane, transport across the membrane by GLUT4, and intracellular phosphorylation to glucose-6-phosphate by hexokinase (HK).

Koval et al. reported an exercise-induced increase in the functional activity of glycogen synthase by 84% while HK II activity increased about fourfold (in the soluble fraction of the homogenates, not in the particulate fraction) [88]. The HK II mRNA increased twofold. HK I mRNA remained unchanged. As exercise reduces the insulin level and insulin stimulates HK II expression through phosphatidylinositol 3-kinase activation, activation of HK II seems independent of the mitogen-activated protein (MAP) kinase pathway [89]. Becxause HK II overexpression augments exercise-stimulated MGU, a reduction in HK II activity is expected to impair exercise-stimulated MGU. The magnitude of this impairment would be greatest in tissues with the largest glucose requirement [90].

Exercise also exerts its effects on glucose and other metabolism by different pathways, such as MAPK [91]. Ruderman et al. thought that AMPK also plays a major role in regulating lipid metabolism in multiple tissues following exercise and that the net effect of its activation is to increase fatty acid oxidation and diminish glycerolipid synthesis, which in turn might decrease adiposity and increase insulin sensitivity [92]. Then Leary and Shoubridge provided a review on the cellular signals modulating mitochondrial content in response to physiological stimuli such as exercise or cold exposure [93]. Further details will be discussed elsewhere in this book along with the mitochondrial biogenesis. Here, readers are reminded that hexokinase II (HK II) expression and its activity are reduced in patients with non-insulin-dependent diabetes mellitus [94] and that hexokinase binding to mitochondria, a basis for energy metabolism needed in cell proliferation, is stimulated by insulin [95]. The glycogen synthesis pathway depends on mitochondrial function, and treatment with mitochondrial inhibitors eliminates the majority of glycogen synthesis [44].

In a previous section, the marked depression of hexokinase II activity associated with mtDNA depletion in SK Hep-1 cells was described. What happens if mtDNA density is increased?

Exercise and mtDNA Density

Exercise increases maximal exercise capacity as estimated from maximal oxygen uptake (VO_2max). This is strongly correlated with the mitochondrial contents in healthy individuals [96,97]. Tissue mitochondrial content is tightly regulated to ensure adequate, but not excessive, mitochondrial oxidative capacity to meet metabolic demand [98]. Puntschart et al. found that mitochondrially encoded RNAs (COX subunit I, NADH reductase subunit 6, 16S rRNA), as well as nuclear-encoded RNAs (COX subunit IV, succinate dehydrogenase, fumarase), and mtDNA density are increased coordinately in athletes (~1.5-, 1.9-, 1.6-fold higher, respectively) compared to nonathletes [99].

Park et al. showed that the mtDNA density of peripheral blood white cells (pb-mtDNA) is positively correlated with VO_2 max in healthy subjects and inversely with insulin resistance parameters [38,39]. Lim et al. extended these observations to the changes in pb-mtDNA density and the metabolic parameters induced by exercise [100]. A 10-week exercise program, consisting of three sessions a week, each of 1 h, aimed to attain 60 to 80% of VO_2 max. The program decreased the systolic blood pressure and triglyceride, low-density lipoprotein-cholesterol (LDL-C), glucose, and insulin, and increased the high-density lipoprotein-cholesterol (HDL-C) of the young women (mean age of 25). It decreased total cholesterol, LDL-C, and glucose in the older women (mean age of 67).

With these changes, mtDNA density increased significantly after the exercise program in both groups, and the changes in mtDNA density showed a significant positive correlation with the change in VO_2 max. These findings confirm that exercise improves insulin sensitivity and that mitochondrial changes are closely associated with this improvement.

Mechanisms of mtDNA Increase Induced by Exercise

A recent paper by Nisoli et al. provides the first evidence that elevated levels of NO stimulate mitochondrial biogenesis in a number of cell lines via a soluble guanylate–cyclase-dependent signaling pathway that activates PGC_1-α (peroxisome proliferator-activated receptor γ-coactivator-1-α), a master regulator of mitochondrial content [101]. Based on this fact, Leary and Shoubridge provided an excellent review on the intracellular signals modulating mitochondrial content in response to physiological stimuli such as exercise or cold exposure [93]. They link the decrease in the ATP/AMP ratio, rise of intracellular Ca^{2+}, and cAMP stimulation by catecholamines to PGC1 expression. This in turn, could regulate mitochondrial biogenesis.

Whether this signaling cascade represents a widespread mechanism by which mammalian tissues regulate mitochondrial content or how it might integrate with other pathways that control important regulators of mitochondrial biogenesis, such as PGC_1, NRFs, and Tfams, is not clear at this time. Reality seems very complicated because mitochondria play a role in this regulation by taking up calcium through an uptake pathway. Mitochondrial Ca^{2+} accumulation generally increases the ATP production through the activation of mitochondrial enzymes and, in some specific cells, modulates the spatial–temporal dynamics of calcium signals [102]. Figure 8.2 depicts the mechanism of mitochondrial biogenesis following various stimuli.

MECHANISMS OF INCREASED OXIDATIVE STRESS IN DIABETES AND MITOCHONDRIA

Increased Oxidative Stress in Diabetes Mellitus

Considerable evidence has been accumulated to suggest that the production of reactive oxygen species (ROS) is increased in diabetic patients, especially in those with poor glycemic control [104–106]. Excessive ROS can accelerate oxidative

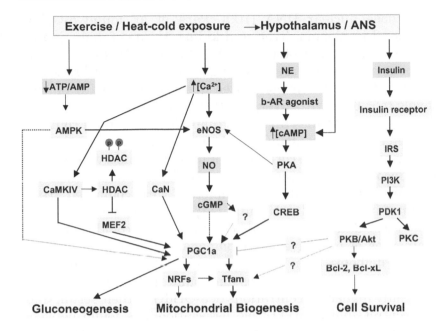

Figure 8.2 Regulation of mitochondrial biogenesis. (Redrawn from Scarpulla, R.C., *Biochim. Biophys. Acta*, 1576:1–14, 2002, and Leary, S.C. and Shoubridge, E.A., *Bioessays*, 25:538–541, 2003.)

damage to macromolecules, including lipids and proteins, as well as to DNA. For example, an increased production of malondialdehyde (MDA), a marker of lipid peroxidation, has been demonstrated in the erythrocyte membranes of diabetic patients, along with a depressed erythrocyte content of reduced glutathione [107]. Moreover, the relationship between markers of lipid peroxidation and metabolic control in type 1 and type 2 diabetic patients is significant [108]. Regarding the oxidation of protein, Suzuki et al. demonstrated local oxidative stress and carbonyl modification of proteins in diabetic glomerulopathy [52].

An ROS-induced modification of a purine residue in DNA produces 8-hydroxydeoxyguanosine (8-OH-dG). It is a sensitive index of oxidative DNA damage; the deoxyguanosine residue in mtDNA is particularly vulnerable to this damage. Mitochondrial DNA sustains almost ten times more damage than does nuclear DNA [109]. Park et al. found 8-OHdG increased in plasma, liver, and kidney tissue of diabetic rats induced by streptozotocin [110]. Plasma 8-OHdG levels in diabetic rats were three times higher than in control rats, and tissue 8-OHdG levels correlated quantitatively with plasma levels (r = 0.64 with liver, r = 0.38 with kidney) — approximately 1.5 to 2 times higher in diabetic rats than in controls. Elevated plasma 8-OHdG levels were normalized by insulin treatment, although correction of hyperglycemia with insulin was incomplete. However, tissue 8-OHdG levels in diabetic rats were lowered almost completely with combined treatment of insulin and antioxidants (probucol or vitamin E).

These data suggest that plasma 8-OHdG could be a useful biomarker of oxidative DNA damage in diabetic subjects. In human diabetes, Shin et al. assessed 8-OHdG as a biomarker of oxidative stress in diabetes and found that patients had significantly higher concentrations of 8-OHdG in their serum than did control subjects [111]. No association was found between the levels of 8-OHdG and HbA1c, suggesting that short periods of glycemic control are not a determinant of oxidative tissue damage. No correlation was found between serum 8-OHdG level and age; however, the duration of diabetes, serum lipids, creatinine or urinary albumin excretion rate, and creatinine clearance rate, all showed a marginal correlation.

Among the diabetic patients, those with severe retinopathy had significantly higher 8-OHdG levels than those with milder retinopathy or without retinal complications. Similarly, 8-OHdG levels in patients with advanced renal disease (azotemia) were higher than in patients without this complication. These findings suggest that oxidative DNA damage may contribute to the development of microvascular complications of diabetes. Based on these findings, treatment of diabetes with antioxidants is suggested [112].

Increased Oxidative Stress in Insulin Resistance

A large database suggests a role for oxidative stress in the pathogenesis of cardiovascular disease. However, information regarding cardiovascular risk factors associated with systemic oxidative stress has largely been derived from highly selected samples with advanced stages of vascular disease; thus, it has been difficult to evaluate the relative contribution of each cardiovascular risk factor to systemic oxidative stress and to determine whether such risk factors act independently and are applicable to the general population.

To determine the clinical conditions associated with systemic oxidative stress, Keany et al. examined 2828 subjects from the Framingham Heart Study and measured urinary creatinine-indexed levels of 8-epi-PGF2-α as a marker of systemic oxidative stress [113]. In age- and sex-adjusted models, increased urinary creatinine-indexed 8-epi-PGF$_{2\alpha}$ levels were positively associated with female sex, hypertension treatment, smoking, diabetes, blood glucose, body mass index, and a history of cardiovascular disease. In contrast, age and total cholesterol were negatively correlated with urinary creatinine-indexed 8-epi-PGF$_{2\alpha}$ levels.

After adjustment for several covariates, decreasing age and total/HDL cholesterol ratio, sex, smoking, body mass index, blood glucose, and cardiovascular disease remained associated with urinary 8-epi-PGF$_{2\alpha}$ levels. Smoking, diabetes, and body mass index were highly associated with systemic oxidative stress as determined by creatinine-indexed urinary 8-epi-PGF$_{2\alpha}$ levels. Translating obesity and diabetes mellitus as a phenotype of the insulin resistant state leads to the fact that this study provides decisive evidence that oxidative stress is operating in this condition.

As presented previously, oxidative stress is present in diabetes mellitus. Why is the diabetic state associated with oxidative stress? Hyperglycemia (the defining character of diabetes mellitus) and high serum free fatty acid levels frequently associated with obesity are known to cause enhanced oxidative stress. These two factors could alter mitochondrial function that, in turn, could cause the insulin resistance state.

Glucose Stimulates Hydrogen Peroxide Production

In the diabetic state, more glucose is shunted through the polyol pathway and results in an elevated cytosolic [NADH].

This alters the NADH/NAD ratio because the oxidation of sorbitol to fructose is mediated by reduction of NAD. This excessive reducing power state of cytosol is similar to the hypoxic state. Williamson and his coworkers thus suggested that reductive stress, rather than oxidative stress, is the appropriate term, which explains the tissue destruction in diabetes mellitus [114]. From the 1990s, evidence emerged suggesting an increased production of superoxide anions occurs in endothelial cells during hyperglycemia [115,116].

Nishikawa et al. studied mechanisms of hyperglycemia-induced superoxide overproduction in detail [117]. Using bovine endothelial cells, they found inhibition of glycolysis-derived pyruvate transport into mitochondria by 4-hydroxy-cyanocinnamic acid completely inhibited hyperglycaemia-induced ROS production. Flux through the TCA cycle, measured by $^{14}CO_2$ production from [U^{14}C] glucose, increased 2.2-fold by 30 mM glucose. These data indicate that glucose exerts a mass action effect on the TCA cycle and that this is the source of the increased ROS-generating substrate induced by hyperglycaemia.

Bovine aortic endothelial cells were then serially incubated with rotenone (an inhibitor of complex I), thenoyltrifluoroacetone (TTFA; an inhibitor of complex II), or carbonyl cyanide m-chlorophenylhydrazone (CCCP; an uncoupler of oxidative phosphorylation that abolishes the mitochondrial membrane proton gradient) and compared with results obtained from incubation with two glucose concentrations. Glucose increased ROS production, but rotenone did not. TTFA and CCCP completely prevented high glucose-induced ROS production, but neither reduced the increased rate of glucose flux through glycolysis; this suggests that complex II and mitochondrial membrane proton gradient are involved. Overexpression of uncoupling protein-1 (UCP1) also prevented the effect of hyperglycaemia, but antisense complementary DNA in the same gene transfer vector did not.

These results show that hyperglycaemia-induced intracellular ROS is produced by the proton electrochemical gradient generated by the mitochondrial electron transport chain. Overexpression of manganese superoxide dismutase (Mn-SOD) also prevented the effect of hyperglycaemia — suggesting that superoxide is the reactive oxygen radical produced.

Is there any reason behind evolution of this physiologic response, induction of superoxide by hyperglycemia, while it is deleterious to the mitochondria? Many years ago, Chevaux et al. suggested that hyperglycemia can be considered a regulatory mechanism favoring glucose uptake and oxidation in patients with diabetes [118]. They studied six cases of chemical diabetes by the steady-state plasma glucose (SSPG) level during a constant infusion of epinephrine, propranolol, glucose, and insulin and found that the rate of carbohydrate oxidation was similar in both groups. The former group had elevated SSPG levels (174 mg/100 ml), compared with six control subjects (96 mg/100 ml). These results suggest that, with the action of insulin in stimulating glucose transport, the elevated plasma glucose concentration in chemical diabetics (an early term for impaired glucose tolerance) allowed for the normalization of carbohydrate oxidation.

A more important function might be the thermic effect of glucose [119], probably through stimulation of uncoupling [120]. More studies are needed to look at whether this effect is generally applicable. It might be. When type 1 diabetes occurs, patients usually lose weight. Weight loss may simply be considered a result of insulin deficiency, but this loss might have an important implication to the metabolic scale law: decreased heat production from decreased muscle mass of diabetes is compensated by the thermic effect of hyperglycemia in maintaining body temperature.

High Glucose Impairs Insulin Secretion and Myocyte Function: Glucose Toxicity

Sakai et al. found that high glucose induces mitochondrial ROS from pancreatic β-cells and that increased ROS suppresses first-phase of GIIS, at least in part, through the suppression of GAPDH activity [121]. Hyperglycemia-induced β-cell and other body dysfunctions are well known as glucose toxicity among diabetologists. This study suggested that hyperglycemia induces a vicious cycle of free radical production that, in turn, results in dysfunctions of the mitochondria in the β-cell.

Cai et al. tested the hypothesis that hyperglycemia per se can induce myocardial dysfunction [122]. Diabetic mice produced by streptozotocin and H9c2 cardiac myoblast cells exposed to high levels of glucose were used. In the hearts of diabetic mice, apoptotic cell death was detected by the terminal TUNEL assay, along with caspase-3 activation and mitochondrial cytochrome c release. Insulin treatment inhibited diabetes-induced myocardial apoptosis as well as suppressed hyperglycemia. When cardiac myoblast H9c2 cells were exposed to high levels of glucose (22 and 33 mmol/l), apoptotic cell death and caspase-3 activation with a concomitant mitochondrial cytochrome c release were observed in the cultures exposed to hyperglycemia. However, they were not seen in the cultures exposed to the same concentrations of mannitol.

Inhibition of caspase-3 with a specific inhibitor, Ac-DEVD-cmk, suppressed apoptosis induced by high levels of glucose. ROS generation was detected in the cells exposed to high levels of glucose. These results suggest that hyperglycemia-induced myocardial apoptosis is mediated, at least in part, by activation of the cytochrome c-activated caspase-3 pathway, which may be triggered by ROS derived from high levels of glucose.

Hyperglycemia Induces a Common mtDNA Deletion *in Vitro*

Egawhary et al. reported that a high-glucose medium combined with an advanced glycation end product of albumin can induce a common mtDNA deletion in cultured human umbilical vein endothelial cells [123,124]. This common 4977-bp mitochondrial deletion has been identified in association with a number of distinct clinical phenotypes, including the Kearns–Sayre syndrome, the Pearson marrow-pancreas syndrome, and chronic progressive external ophthalmoplegia [125].

As discussed earlier, 8-OHdG is increased in body fluid and tissues of diabetic state in human and experimental animals; oxidative DNA damage is an active process of diabetes. DNA damage is an obligatory stimulus for the activation of the nuclear enzyme poly(ADP-ribose) polymerase; thus, activation of this enzyme in turn depletes the intracellular concentration of its substrate NAD(+), slowing the rate of glycolysis, electron transport, and ATP formation, and produces an ADP-ribosylation of the GAPDH. These processes result in acute endothelial dysfunction in diabetic blood vessels that, convincingly, also contributes to the development of diabetic complications. Furthermore, hyperglycemia is not only toxic to endothelium, but also to other tissues, as described in recent works [126–128].

Nonesterified Fatty Acids Stimulate Hydrogen Peroxide Production

It has long been known that increasing the availability of endogenous or exogenous CHO can increase the oxidation of CHO and decrease the oxidation of fat. The opposite is also true (Randle cycle). However, the detailed mechanisms regulating these shifts in fuel use in the face of constant energy demand have not been thoroughly elucidated. These could include:

- Glucose uptake (GLUT4) and phosphorylation (hexokinase)
- Glycogenolysis (glycogen phosphorylase)
- Glycolysis (phosphofructokinase) and conversion to acetyl CoA (pyruvate dehydrogenase) for glucose handling and transport of long chain fatty acids into the cell (fatty acid translocase CD36)
- Release of fatty acids from intramuscular triacylglycerol (hormone sensitive lipase)
- Transport into the mitochondria (carnitine palmitoyl transferase complex) in use of fatty acids.

A recent review by Spriet and Watt contains more detail [129].

Diabetes or insulin deficiency results not only in hyperglycemia, but elevated non-esterified fatty acid (NEFA) levels. Excessive exposure to NEFA is regarded as a potentially important diabetogenic condition by impairing insulin secretion from pancreatic β-cell and target tissues insulin resistant, most likely acting through Randle cycle, by which carbohydrate and fat metabolism interact. The pathophysiologic aspect of diabetes or obesity is called "lipotoxicity."

Recently, formation of ROS in pancreatic β-cells received attention in this process [130,131]. Koshin et al. reported on the effect of oleic acid ROS production in MIN6 cells [132]. Two sites were known to generate ROS in this cell line. Succinate-supported H_2O_2 generation, which reflects membrane potential-dependent ROS production, and H_2O_2 generation supported by glutamate/malate in the presence of antimycin A, reflecting maximal ROS-producing mitochondrial capacity, were increased after 72 h to oleate. With regard to the short-term NEFA effect, an even greater (~fivefold) acceleration of mitochondrial formation of ROS was observed immediately after the addition of 25 to 75 μ*M* oleate to permeabilized β-cells. This acute stimulation of mitochondrial reactive oxygen generation occurred in control cells as well as those exposed for 72 h, suggesting that this response is rather physiologic. Most ROS was generated from complex I.

Metformin is widely used for the treatment of type 2 diabetes and has been proven to reduce diabetic complications and mortality in diabetes, particularly among obese subjects [133]. Recently, Owen et al. provided evidence that its primary site of action is through a direct inhibition of complex 1 of the respiratory chain [134]. Combined with a report that metformin reduces free radical production [135], metformin might exert its beneficial effect by inhibiting ROS production from complex 1 induced by high NEFA in obesity.

NEFAs can act as natural "mild uncouplers" and are thought to prevent mitochondrial ROS production by lowering mitochondrial membrane potential and the degree of reduction of superoxide-producing complexes in the respiratory chain [136]. Indeed, this has been demonstrated in animal mitochondria [137]. However, a recent report demonstrating the opposite effect of arachidonate, palmitate, and oleate on heart mitochondria suggests that FFA can play a dual role in ROS regulation [138]. This was ascribed to the fact that in addition to a protonophoric effect that slows down ROS production, NEFAs also exert a direct inhibiting effect on the mitochondrial respiratory chain. This increases its reduction and thus ROS production.

These findings suggest that NEFAs stimulate mitochondrial formation of ROS and can contribute significantly to NEFA-induced cell damage or lipotoxicity. The combined effects of elevated NEFAs and glucose on mitochondria will be highly complex, and the availability of insulin will further complicate the picture. Few studies have been done on this subject; NEFAs are known to induce uncoupling activity only at sufficiently high membrane potential; the depletion of glucose is expected to lower this potential.

Insulin Acts on Mitochondria and Reduces Free Radical Damage

Mitochondria are the primary site of skeletal muscle fuel metabolism and ATP production. Although insulin is a major regulator of fuel metabolism, its effect on mitochondrial ATP production was not known. Bessman and his group provided evidence that insulin stimulates the TCA cycle and acts by somehow linking hexokinase II and the mitochondrial porin [139]. Wang et al. reported that insulin increases in vastus lateralis muscle mitochondrial ATP production capacity (32 to 42%) in a healthy human, while clamping glucose, amino acids, glucagon, and growth hormone [140]. Increased ATP production occurred in association with increased mRNA levels from mitochondrial (NADH dehydrogenase subunit IV) and nuclear (COX subunit IV) genes (164 to 180%) encoding mitochondrial proteins. In addition, muscle mitochondrial protein synthesis, as well as COX and citrate synthase enzyme activities, was increased by insulin.

Further studies demonstrated no effect of low to high insulin levels on muscle mitochondrial ATP production for people with type 2 diabetes mellitus; however, matched nondiabetic controls increased 16 to 26% when four different substrate combinations were used. The conclusion was that insulin stimulates mitochondrial oxidative phosphorylation in skeletal muscle along with the synthesis of gene transcripts and mitochondrial protein in human subjects. It was also observed that the skeletal muscle of type 2 diabetic patients has a reduced capacity to increase ATP production with high insulin levels.

To evaluate abnormalities in the Tfam function as a cause of mitochondrial dysfunction in diabetes, Kanazawa et al. measured the mitochondrial respiratory chain mRNAs as well as the transcriptional and translational activities in the mitochondria isolated from control and streptozotocin-induced diabetic rat hearts [141]. Tfam is the nuclear-encoded DNA-binding protein containing two high-mobility group domains; it exerts major effects on mtDNA level, mtDNA transcription, and mitochondrial function.

Compared with the control rats, mRNA contents of mito-chondrial-encoded cytochrome b and ATP synthase subunit 6 in diabetic rat hearts were markedly decreased and com-pletely recovered by insulin treatment. The mitochondrial activities of transcription and translation of these genes were decreased significantly in mitochondria isolated from diabetic rats; insulin treatment also completely normalized this also. Interestingly, gel retardation assay showed a reduced binding of Tfam to the D-loop of mitochondrial DNA in diabetic rats, although no difference was found in the Tfam mRNA and protein content between the two groups. On the basis of these findings, Kanazawa et al. [41] suggested that a reduced bind-ing activity of Tfam to the D-loop region in the hearts of diabetic rats might contribute to the decreased mitochondrial protein synthesis.

In this study, this group found an increased basal pro-duction of hydrogen peroxide and an increased lipid peroxide content in the mitochondria isolated from diabetic rat hearts. Furthermore, hydrogen peroxide treatment of the mitochon-dria isolated from the rat hearts decreased transcriptional activity of Tfam. These findings suggest a possibility that decreased Tfam activity is accompanied by abnormal modifi-cations such as protein oxidation. The cause of reduced Tfam binding to the D-loop is not known, although the generation of reactive oxygen species by the diabetic mitochondria might have done so. These findings imply that the reduced binding activity of Tfam in diabetes might be part of the free radical damage that can be reversed by insulin.

"Causal" Treatment of Diabetes

The evidence provided before suggests that the hyperglycemia-derived ROS may act as mediators of diabetic complications. Insulin has the power to reverse it. Overproduction of superoxide by the mitochondrial electron transport chain might be the first and key event in the activation of all other pathways involved in the pathogenesis of diabetic complications. As discussed previously, the traditional explanation for the pathogenesis includes increased polyol pathway flux, increased advanced glycosylation end product formation, activation of protein kinase C, and increased hexosamine pathway flux. However, when mitochondrial superoxide production is blocked (by inhibitors of electron transport chain complex II by an uncoupler of oxidative phosphorylation, or by uncoupling protein-1 and by manganese superoxide dismutase), activation of these pathways of hyperglycemic damage did not occur [117].

These findings suggest new and attractive "causal" antioxidant therapy and may explain why classic antioxidants, such as vitamin E, that work by scavenging already-formed toxic oxidation products have failed to show beneficial effects on diabetic complications [112]. New low-molecular mass compounds that act as SOD or catalase mimetics or L-propionyl-carnitine and lipoic acid that work as intracellular superoxide scavengers, improving mitochondrial function and reducing DNA damage, may be good candidates for such a strategy; preliminary studies support this hypothesis. This causal therapy would also be associated with other promising tools such as LY 333531, PJ34, and FP15, which block the protein kinase β-isoform, poly(ADP-ribose) polymerase, and peroxynitrite, respectively.

Some other options of preventing diabetic complications than glycemic control and intervening glycemic toxicity: thiazolinediones, statins, ACE inhibitors, and angiotensin

1 inhibitors. Thus, it has been suggested that many of their beneficial effects, even in diabetic patients, are due to their property of reducing intracellular ROS generation or more appropriately protecting mitochondria.

SECONDARY MITOCHONDRIAL DNA CHANGES IN DIABETES: DIABETIC COMPLICATIONS

Oxidative mtDNA Damage Is Linked to Diabetic Complications

Mitochondrial DNA is present in an environment with a high risk for exposure to oxidative agents. The free radical theory of aging proposed by Harman states that increased respiration will lead to increased ROS production and damage to cellular lipids, proteins, and nucleic acids [142]. This mechanism is widely believed to cause mitochondrial DNA mutations, which accumulate upon aging [143,144]. The diabetic state is associated with increased oxidative stress and oxidative stress is also increased in the insulin resistance state [106] that precedes type 2 diabetes; thus, it is predictable to find a high rate of mtDNA mutations in diabetic patients.

Suzuki et al. were the first to report mtDNA deletions and 8-OHdG in the muscle DNA of non-insulin-dependent diabetes mellitus (NIDDM) patients [145]. Common mtDNA deletion of 4977 bp (mtDNA4977) and the content of 8-OHdG in the muscle DNA of the NIDDM patients were much higher than those of the control subjects. A significant correlation was found between deleted mtDNA4977 level and the 8-OHdG content; deleted mtDNA and the 8-OHdG content correlated with the duration of diabetes and increased in proportion to the severity of diabetic nephropathy and retinopathy. This kind of damage to mtDNA is also seen in diabetic kidney disease and in vascular endothelial cells, where the most harmful effects of hyperglycemia occur. Keeping in mind the *in vitro* studies of Egawhary et al. [123,124], one can appreciate that hyperglycemia of the diabetic state could damage tissues via oxidative stress.

Kakimoto et al. found that this was indeed the case [147]. They studied streptozotocin-induced diabetic rats for presence of oxidative damages and mtDNA alteration. At 8 weeks after the onset of diabetes, levels of 8-OHdG were significantly increased in mtDNA from kidney of diabetic rats but not in nuclear DNA. Semiquantitative analysis using PCR showed that the frequency of 4834-bp deleted mtDNA was markedly increased in diabetic kidneys at 8 weeks, but not at 4 weeks. Insulin treatment starting at 8 weeks rapidly normalized renal 8-OHdG levels of diabetic rats, but did not reverse the increase in the frequency of deleted mtDNA.

This study demonstrated that oxidative mtDNA damage and subsequent mtDNA deletion accumulates in kidneys of diabetic rats. Consistent with this report, Ha et al. had also observed an increased 8-OHdG level in diabetic rat kidneys [146]. Furthermore, increased levels of 8-OHdG have been reported in urine, mononuclear cells, and skeletal muscles of diabetic patients, suggesting that free radical damage to mtDNA occurs in human diabetes and is rather generalized and not limited to kidneys.

The 4977- and 7436-bp deletions are the most frequent age-associated large-scale deletions seen in humans; they were first reported in the muscle of patients with Kearns–Saye syndrome or CPEO. Because these deletions span a region that encodes five kinds of tRNA, respiratory enzyme subunits in complex I, complex IV, and complex V, the functions of the respiratory enzymes containing deleted mtDNA-encoded proteins subunits may decline gradually in the tissue cells as the deleted mtDNA are accumulated. Ha et al. also reported that the proportion of deleted mtDNA to the total mtDNA reached 0.1% at 8 weeks after the induction of diabetes. This level of deletion observed in kidney tissues of diabetic rats may be too low to affect overall organ function. However, it is likely that this common deletion in mtDNA may be accompanied by other multiple deletions, as shown in aged tissues.

Nomiyama et al. studied the accumulation of somatic mutation in peripheral blood in Japanese diabetic patients [148]. Because only a minor accumulation of the mutant mtDNA was expected in blood cells, they developed a special method for quantifying tiny amounts of it. Under the strict PCR conditions, only mutant mtDNA with np 3243 was amplified; a 100,000-fold excess of wild-type mtDNA was not detected by agarose electrophoresis. The validity of amplified DNA fragment was confirmed by sequencing and not finding a band from template DNA derived from ρ0 HeLa cells. For quantitative analysis, the amplified DNA fragment was monitored with the TaqMan probe in a DNA sequence detection system and estimated using plasmids containing mtDNA fragments with or without the A3243G mutation as the standard.

The results confirmed that nuclear DNA from ρ0 cells did not disturb the measurement of mutant mtDNA in a mixture of the nuclear DNA and the standard plasmid. The authors claimed that this method shows linearity over a wide range of the mutant mtDNA quantity, even at 10,000-fold excess of wild-type mtDNA added. Mutant mtDNA of 0.001 to 0.1% of total mtDNA could be quantified. Further caution was provided by allocating all the samples randomly arranged and by repeating the measurement five times.

The researchers studied 383 diabetic subjects, aged 18 to 80 years, and 290 age-matched, nondiabetic, healthy subjects, aged 0 to 60 years. Although the level of the A3243G mutation was negligible in the newborn group, it increased in healthy subjects aged 20 to 29 and 41 to 60 years. In diabetic patients, the mutation rate increased along with age and the duration of diabetes. In the middle-aged group (41 to 60 years old), the A3243G mutation accumulates fourfold higher in the diabetic patients than in the healthy subjects.

Moreover, multiple regression analysis revealed that the most critical factor associated with this mutation in diabetic patients was the duration of diabetes. Also, the genotypes of DD and DI-CC of ACE-p22phox showed the highest mutational rate and the thickest IMT. These findings suggest that a secondary mtDNA mutation is associated with diabetes and linked to the diabetic complications.

mtDNA Depletion in Diabetic Complications

Direct proof for the presence of mtDNA depletion in diabetes mellitus was first provided by Antonetti et al. [35]. This group screened subtraction libraries from normal and type 2 diabetic human skeletal muscle for genes that were increased in expression in diabetes; they identified four different mitochondrially encoded genes, all of which are located on the heavy strand of the mitochondrial genome: cytochrome oxidase I, cytochrome oxidase III, NADH dehydrogenase IV, and 12s rRNA. As estimated by southern blot analysis, mitochondrial DNA copy number decreased approximately 50%. Northern blot analysis assessed a 1.5- to 2.2-fold increase in the expression of these mRNA molecules relative to total RNA in type 1 and type 2 diabetes.

Mitochondrial gene expression increased approximately 2.5-fold when expressed relative to mitochondrial DNA copy number. For cytochrome oxidase I, similar changes in mitochondrial gene expression were observed in muscle of non-obese diabetic and ob/ob mice — models of type 1 and type 2 diabetes, respectively. By contrast, expression of cytochrome oxidase 7a, a nuclear-encoded subunit of cytochrome oxidase, did not change or decreased slightly; the expression of mitochondrial transcription factor 1 in human skeletal muscle did not change with type 1 or type 2 diabetes. The subjects studied had long history of diabetes; therefore, it appears that mtDNA depletion occurs in long-term uncontrolled diabetes. Increase in mitochondrial gene expression associated with mtDNA depletion may have been compensated by the increase in mitochondrial respiration and oxidative stress.

Abnormal Mitochondrial Function of Diabetes Mellitus Exacerbates Insulin Deficiency

Streptozotocin induces several morphologic changes in the mitochondria that may result in impaired energy production in the cells [149,150]. Streptozotocin-induced diabetic rat hepatocytes showed a markedly decreased oxidation of succinate, suggesting that the mitochondrial TCA cycle was impaired. Three days of insulin treatment did not restore this impairment, although gluconeogenesis was suppressed, suggesting that a longer period of treatment is needed to reverse the damage to the mitochondria caused by the diabetes. A depressed state-3 respiration, oxidative phosphorylation rate, and Mg^{2+}-dependent ATPase activities were also reported in mitochondria from diabetic hearts.

These effects were partially reversible with 2 weeks of insulin treatment and fully reversible after 4 weeks of insulin therapy. The results indicated the presence of a generalized depression in mitochondrial function in chronic diabetes; such a defect could contribute to the development of diabetic complications such as cardiomyopathy.

Recently, Yechoor et al. studied the coordinated patterns of gene expression for substrate and energy metabolism in skeletal muscle of diabetic mice induced by streptozotocin [46]. They used Affymetrix oligonucleotide microarrays to define the full set of alterations in gene expression in skeletal muscle caused by diabetes and the loss of insulin action and studied control and streptozotocin-diabetic mice. Of the genes studied, 235 were identified as changed in diabetes, with 129 genes up-regulated and 106 down-regulated.

The analysis revealed a coordinated regulation at key steps in glucose and lipid metabolism, mitochondrial electron transport, transcriptional regulation, and protein trafficking. The mRNAs for all of the enzymes of the fatty acid β-oxidation pathway were increased, whereas those for GLUT4, hexokinase II, the E1 component of the pyruvate dehydrogenase complex, and subunits of all four complexes of the mitochondrial electron transport chain were coordinately down-regulated. Metabolic abnormalities underlying diabetes are primarily the result of lack of adequate insulin action and associated changes in protein phosphorylation and gene expression, so insulin treatment is expected to restore these abnormalities. However, only about half of the alterations in gene expression in diabetic mice could be corrected after 3 days of insulin treatment and euglycemia. As described earlier, it would take weeks of insulin treatment to reverse the changes.

NO and Vascular Complications of Diabetes

Bian and Murad outlined the functions of NO, which is normally produced from endothelial cells and functions as a signal molecule to vascular smooth muscle cell as vasodilatation factor (endothelium derived vasodilation factor; VDRF) [151]. In a physiologic state, NO reversibly inhibits mitochondrial respiration by competing with oxygen at cytochrome oxidase and exerts some of its main physiological and pathological effects on cell functions by competitively inhibiting this enzyme. NO may also be a physiological regulator of the affinity of mitochondrial respiration for oxygen, enabling mitochondria to act as sensors of oxygen over the physiological range [152].

However, increased superoxide production by hyperglycemia could damage mitochondria and mtDNA. Superoxide overproduction is accompanied by increased nitric oxide generation in endothelial cells due to an NOS and inducible NOS uncoupled state. This phenomenon favors the formation of the strong oxidant peroxynitrite, which in turn would further damage DNA.

Molecular Mechanisms of mtDNA Depletion

The molecular mechanisms leading to mtDNA damages were described previously: excessive superoxide generation from the mitochondrial electron transfer chain could make an innocent molecule like NO into a toxic one, could damage mitochondria and its DNA, and would stimulate more free radical production. Mechanisms of mtDNA depletion in diabetes are poorly understood.

Tfam plays a major role in mitochondrial transcription and its functions [140]. As discussed previously, Kanazawa et al. reported that Tfam binding to the D-loop of mtDNA is decreased in streptozotocin-induced diabetes and that this can be reversed by insulin treatment [141]. It is expected that decreased Tfam would decrease mtDNA level. Choi et al. found that Tfam expression disappears in mtDNA depleted ρ0 SK Hep-1 cells, a line derived from human hepatocellular

carcinoma [153]. Furthermore, in comparison to the parent
ρ+ cells, a decrease was seen in

- Intracellular ATP
- Glucose-stimulated ATP production
- Glucose uptake
- Steady-state mRNA
- Protein levels of glucose transporters
- Cellular activities of glucose-metabolizing enzymes

These results suggest that the quantitative reduction of
mtDNA may down-regulate the expression of glucose transporters and suppress glucose metabolism, resulting in a
decrease of ATP production and glucose utilization, at least
in certain cell lines. These researchers found that exogenous
hydrogen peroxide decreased Tfam-driven gene transactiva-
tion and glucose increased it. Tfam protein exhibits a strong
binding affinity to circular DNA, which might protect mtDNA
from ROS. A cycle between Tfam protein level and the oxidative stress induced by mtDNA damage is suggested.

The acute effects of high glucose on cells appear different
according to the cell types. Choi et al. reported that peripheral
blood mtDNA density slowly increases after the induction of
diabetes with streptozotocin in rats [153]. Because Tfam is a
key regulator involved in mitochondrial DNA transcription
and replication, they studied the effect of overexpression of
NRF-1 and high concentration (50 mM) of glucose on the promoter activity of the rat Tfam in L6 rat skeletal muscle cells.

The expression of Tfam is coordinated by a limited set of
transcription factors. Nuclear respiratory factors (NRF-1 and
-2) are two major *trans*-acting factors known to play a key
role in the transcription of the human Tfam gene. Peroxisome
proliferator-activated receptor-γ coactivaor-1 (PGC-1) stimulates the expression of NRF-1 and coactivates the transcription function of NRF-1 on the promoter of Tfam (12).
Furthermore, genomic footprinting studies in tumor cells
showed a high level of Sp1 binding to the promoter up-regulated Tfam.

The addition of 50 mM glucose for 24 h increased Tfam
promoter activity by up to twofold. The glucose-induced Tfam

expression was dose dependent and cell-type specific. Glucose increased the Tfam promoter-driven transactivity in L6 (skeletal muscle), HIT (pancreatic β-cell), and CHO (ovary) cells, but not in HepG2 (hepatoma), HeLa, and CV1 (kidney) cells. Among various monosaccharides, only glucose and fructose increased the Tfam promoter activity.

Oxidative stress might not be involved in glucose-induced Tfam expression because the treatment with antioxidants such as vitamin C, vitamin E, probucol, or α-lipoic acid did not suppress the induction. None of inhibitors against protein kinase C, MAP kinase, and PI_3 kinase altered the glucose-induced Tfam promoter activity, suggesting that general phosphorylation might not be involved in its signaling. However, a dominant negative of NRF-1, of which 200 amino acids of C-terminus were truncated, completely suppressed the glucose-induced Tfam induction. It was concluded that high glucose-induced Tfam transcription in L6 cells might be mediated by NRF-1.

Although some important factors involved in Tfam expression have been identified, the upstream signaling pathway of these factors and other related factors in various cellular conditions remain largely undetermined. Haraguchi et al. mapped three positive (PR1-3) and one negative (NR1) transcriptional control domains in the promoter of the human F_0F_1-ATP synthase β subunit gene (ATPsyn β) in the context of expression in myogenic cells [154]. OXBOX- and REBOX-specific binding of an OXBOX/REBOX-like region within the conserved sequence block C of the human mitochondrial DNA D-loop sequence are consistent with the idea that OXBOX- and REBOX DNA-binding factors coordinate the expression of mitochondrial energy genes in highly oxidative tissues by working with general transcription factors such as SP1 and CCAAT DNA-binding proteins in the nucleus and Tfam in the mitochondrion.

Then what tells the nuclei to make more mitochondria? One possibility is that the nuclei monitor the redox state of the cell and thus the mitochondria. The mitochondria oxidize NADH derived from carbohydrates and fats with O_2 to generate H_2O and trap the energy released in ATP. Defects in

OXPHOS will reduce ATP production and increase the NADH/NAD+ ratio, thus making the cell more reduced. The more reduced environment of the cell found in diabetes could be monitored by a ubiquitous transcription factor, such as the REBOX-binding factor. This factor would bind to its 5′ *cis*-elements in nuclear DNA (nDNA) bioenergetic genes when in a reducing environment. In turn, this would induce the transcription of nDNA proteins involved in the bioenergetics and biogenesis of mitochondria in chronic hyperglycemia.

WHOLE-BODY ENERGY METABOLISM IN DIABETES AND METABOLIC SYNDROME

Type 1 Diabetes

In patients with insulin dependent diabetes, the capacity to oxidize and store glucose is markedly reduced. After patients are treated with an artificial β-cell unit for 72 h, blood glucose level is normalized. Glucose storage is returned to normal. However, restoration of the capacity to oxidize carbohydrates in response to the test meal was gradual, and the ability to store carbohydrate was normalized within 24 h. These results suggest that an "inducible" process by insulin is involved, but it is not known how this induction occurs [155].

No clear explanation is possible for this normalization of the glucose oxidation process. However, as described previously, Kanazawa et al. reported the normalization of mitochondrial function, i.e., reduced Tfam binding in streptozotocin-induced diabetes, by insulin treatment [141]. Furthermore, Memon et al. have shown that the markedly decreased oxidation of succinate in streptozotocin-induced diabetic rat hepatocyte improved after 3 days of insulin treatment, if not completely [156].

Type 2 Diabetes

In subjects with obesity and type 2 diabetes, lipid oxidation is generally increased as serum NEFA is elevated. Felber and his group have studied the role of fuel oxidation in the pathogenesis of insulin resistance of obesity and type 2 diabetes

[157,158]. In 1987, they reported that, in a postabsorptive state as well as in response to insulin stimulation (during oral glucose tolerance test (OGTT) and during hyperinsulinemic clamp), lipid oxidation is significantly increased in all obese, obese–glucose-intolerant, and obese–diabetic groups in comparison with young or older controls. Basal glucose oxidation is significantly decreased in obese–nondiabetic and obese–glucose-intolerant subjects. During the OGTT and during the insulin clamp, insulin-stimulated glucose oxidation was decreased in all three obese groups.

In contrast, nonoxidative glucose disposal was markedly inhibited in nondiabetic and diabetic obese patients during the euglycemic insulin clamp. Statistical analysis revealed that lipid and glucose oxidation was strongly and inversely related in the basal state during euglycemic hyperinsulinemic clamp and during OGTT. As explained previously, competition between fuels for final oxidation process through TCA cycle, or Randle cycle, was suggested to cause this relation. Because mitochondria are the site of the TCA cycle, one can easily associate this fuel competition with this organelle.

Nonoxidative glucose uptake after glucose ingestion was normal in nondiabetic obese and glucose-intolerant obese subjects and decreased in diabetic obese subjects. Fasting and post-OGTT hyperglycemia were the strongest (negative) correlates of nonoxidative glucose disposal in single and multiple regression models. The negative correlation between lipid oxidation and nonoxidative glucose uptake, although significant, was much weaker. Among the conclusions made by this group were that (1) glucose oxidation is reduced in nondiabetic obese and diabetic obese individuals; and (2) hyperglycemia provides a compensatory mechanism for the defect in nonoxidative glucose disposal in nondiabetic obese subjects.

If fasting hyperglycemia is seen as a manifestation of insulin deficiency and the fact that nonoxidative glucose oxidation is normalized by insulin treatment in patients with insulin-dependent diabetes [155] is taken into consideration, one can see that the defective glucose oxidation is the primary defect associated with insulin resistance, and glycogen storage is associated with hyperglycemia (or insulin deficiency). This conclusion is consistent with the idea that mitochondrial defect is primary in the metabolic syndrome and hyperglycemia plays a compensatory role, as discussed previously.

Obesity

Obesity is defined by an increased amount of adipose tissue in the body. This increased body mass is related to total body metabolic rate. Information on the alterations in body metabolism associated with obesity is huge and conflicting. In obesity, more calories are consumed than are used up — suggesting that obese people should have excess energy. On the other hand, obese people experience the fatigue and decreased physical endurance that indicate diminished energy supply in the body.

Wlodek and Gonzales developed and applied a "knowledge discovery in databases" procedure to metabolic data and concluded that, in the chain of metabolic events leading to obesity, the crucial event is the inhibition of the TCA cycle at the step of aconitase [159]. This enzyme is of special interest because a nonheme iron sulfur cluster participates in the active site of this enzyme. Iron–sulfur center is a well-known target of free radical damage. Inhibition of the TCA cycle disturbs energy metabolism and results in ATP deficiency with simultaneous fat accumulation. These researchers suggested further steps in obesity development are the consequences of diminished energy supply: inhibition of β-oxidation, leptin resistance, increases in appetite and food intake, and a decrease in physical activity.

Thus, this theory shows that obesity does not need to be caused by overeating and sedentary lifestyle, but rather may be the result of the "obese" change in metabolism that forces people to overeat and save energy to sustain metabolic functions of cells. The logic leads this obese change to the environmental factors. This conclusion is in total agreement with mitochondrial hypothesis of insulin resistance, in which a deranged individual cellular mitochondrial unit is considered primary, and lowered mitochondrial function at cell level would lead to the compensatory increase in body mass.

The relationship between resting energy expenditure (REE) and metabolically active fat-free mass (FFM) is a cornerstone in the study of physiological aspects of body weight regulation and human energy requirements. Recent studies and reviews offer further details [160–163]. Decreased metabolic rate has been shown to be a strong predictor of obesity among Pima Indians [164].

In contrast, subjects who gained weight increased energy expenditure after a weight gain, which is appropriate for their fat-free mass [165]. Again, the details explaining these changes will be not discussed here. Instead, an explanation based on the allometric scale law between body mass and metabolic rate will be provided.

COMBINATION OF MITOCHONDRIAL HYPOTHESIS AND BIOLOGICAL ALLOMETRIC SCALING LAW

Biological Allometric Scaling Law

Pioneering work published by Max Rubner in the 1880s reported that mammalian basal metabolic rate (BMR) was proportional to $M^{2/3}$. In accordance with simple geometric and physical principles, it was thought that an animal's rate of metabolic heat production was matched to the rate at which heat was dissipated through its body surface. In 1932, Kleiber concluded that basal metabolic rate scaled not in proportion with surface area, but with an exponent of 3/4 or body mass raised to the power of 3/4, which was supported by Brody's famous mouse-to-elephant curve.

Recent reformulation of allometric scaling laws by West and his group provides a reason why mitochondrial density should correlate with insulin sensitivity [166,167]. This group generated a general model for the origin of the allometric scaling law from three general principles required in living organisms:

- A space-filling, fractal-like branching pattern is necessary to supply the organism with what it needs.

- The final unit of this branching pattern is a size-invariant unit.
- The energy required to distribute resources is minimized.

West et al. extended the relation between body mass and metabolic power, which covers 27 orders of magnitude of body size, from elephant, mice, and mitochondrion to electron transfer chain enzymes [168]. This subject is very complicated and some of its details still hotly debated [169–172].

Whether the allometry of mammalian basal metabolic rate accounting for variation associated with body temperature, digestive state, and phylogeny has a metabolic scaling exponent of 3/4 or 2/3 [170] or whether the proposed general model of West and his group is right, one should accommodate the fact that allometry can extend from the individual mitochondrion to intact mammals to account for differences in the energetic efficiency between individuals and between species.

Mitochondria of a Cell and Whole Body

In 1956, Smith reported that the mitochondrial density of the liver correlated with whole-body energy expenditure; the relative amount of mitochondria in any given tissue was suggested to be the controlling factor in determining the regression of oxygen utilization on total body size of the species. Allometric scaling law between body mass and metabolic rate dictates that a lowered metabolic rate should raise body mass.

It is illuminating to see from work of West et al. [168] that each cultured cell has an identical metabolic power and that its power decreases as the weight of the animal increases [168]. As these researchers discuss in their findings, this argument could be reversed: if one knows the scale of power generation at the molecular level, it will be sufficient to predict the metabolic rate of individual mitochondria and cells (whether *in vitro* or *in vivo*) as well as intact mammals. Work by Lee's group [38,39] suggests that peripheral blood mtDNA density could serve as a marker. Muscle mtDNA is known to correlate with whole-body energy metabolism, but not all muscle seems to reflect it.

Heat Production, Body Temperature Control, and
Biological Scale

Body temperature is another key constraint of metabolic scale law. West and his coworkers believe that it is imperative for organisms to keep the core temperature optimal to make enzyme activities function optimally. Warm-blooded animals were evolved in adaptation to the body temperature of 36.7°C. When an animal decreases its energy-generating capacity, as reflected by lowered mtDNA density in peripheral blood, it will be imperative for that animal to prevent heat loss (by increasing fat tissue) and increase metabolic rate to keep optimum core temperature. Uncoupling of mitochondria and hydrogen peroxide production induced by NEFAs and glucose and the stimulation of their transport by insulin might be considered as a compensatory process.

In other words, metabolic syndrome could be understood as a state resulting from a lowered heat production of individual heat-producing cells of the body. The sensor of body temperature resides in the hypothalamus, which will control sympathetic activity and appetite and set the insulin sensitivity [173]. Hyperinsulinemia would stimulate heat production and adipogenesis.

Metabolic adjustment of homeothermic animals to a cold environment is very complicated. For example, cold-exposed animals present a significantly lower insulin-induced reduction in food consumption compared with animals maintained at room temperature. Animals exposed to cold are resistant at the molecular and functional levels to the actions of insulin in the hypothalamus [174].

More active backward pathway of Krebs cycle (BPKC) found in obese subjects is consistent with this adaptation. Insulin and glucose stimulate BPKC, perhaps by increasing the supply of pyruvate through stimulation of glycolysis [175]. Continuous decline of mitochondrial function in obesity is suggested by work of Golay et al. [176]. They found that the thermogenic responses (by indirect calorimetry) to 100 g oral glucose load in three groups of obese subjects (normal glucose tolerance, impaired glucose tolerance, and frank diabetes) averaged 6.8, 7.0, and 3.1%, respectively.

With the evolution of obesity (i.e., 6 years later), the glucose-induced thermogenesis (GIT) was significantly reduced in the nondiabetic groups to 4.1 and 3.0%, respectively, and was still blunted in the diabetic group (2.1%). The decrease in GIT was accompanied by a reduction in glucose tolerance and insulin response with no change in fasting plasma insulin.

Golay et al. concluded that the decrease in GIT, which accompanies the worsening of glucose tolerance and the occurrence of diabetes, is a mechanism that may have contributed to maintain the obesity state by a reduction of energy expenditure. From the mitochondrial perspective, these studies can be understood as demonstrations of how diabetes mellitus evolved from insulin resistance and obesity; hyperglycemia might have played an additional compensatory role when the heat generation was further reduced.

SUMMARY: NEW PARADIGM FOR UNDERSTANDING DIABETES MELLITUS

In this review, diabetes mellitus is explained as a disease of the mitochondrial genome. Because eukaryotes are of symbiotic origin, the mitochondrion is an essential part of this life form. When the genes coding for mitochondrial components have problems, diseases involving energy utilization will be produced. Diabetes mellitus is a prototype of these diseases. The process of development in early life, particularly malnutrition and placental ischemia, is an important determinant in forming the basic body plan, and this plan persists throughout life. The effect is most apparent in the mitochondria.

Once born, the mitochondria must fulfill the needs of the body plan set by the nuclear genes. Mitochondria must generate ATP and heat simultaneously. If the set point of a mitochondrion is low, it must work harder, which will result in more ROS and insulin resistance and obesity. If more ROS than ATP is produced, as in the case of subjects with MELAS mutations, a subject will not become obese.

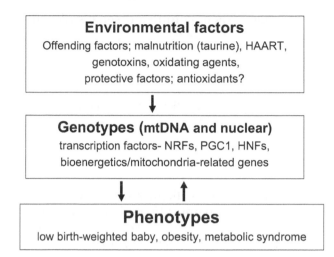

Figure 8.3 General outline of mitochondrial hypothesis. Networks of genotypes and phenotypes are expressed in accordance with the rules of allometric body scale, in response to the environmental challenge.

This mitochondrial hypothesis is just an alternative way of looking at diabetes and interpreting data. It is based on bioenergetics and, more broadly, biophysics — not just on the biochemistry of mitochondria.

Taking partly from Strohman's words, human disease phenotypes are controlled not only by genes, but also by lawful self-organizing networks that display system-wide dynamics [177]. Based on the mitochondrial unit, these networks range from metabolic pathways to signaling pathways that regulate hormone action, the summation of which should equal whole-body energetics. When perturbed, networks alter their output of matter and energy; depending on the environmental context, this can produce a pathological or a normal phenotype.

Strohman proposed studies of the dynamics of these networks by approaches based on metabolic control analysis. West et al. suggested a general model for the origin of allometric scaling law from three general principles required in the living organisms. This chapter discussed a vicious cycle of poor mitochondrial biogenesis and response to correct it (Figure 8.3). These ideas may provide new insights into understanding the pathogenesis of complex diseases like diabetes mellitus.

ACKNOWLEDGMENT

This work was supported by a grant from Ministry of Health & Welfare, Republic of Korea (02-PJ1-PG1-CH04-0001).

REFERENCES

1. Alberti KG, Zimmet PZ: Definition, diagnosis and classification of diabetes mellitus and its complications. Part 1: diagnosis and classification of diabetes mellitus provisional report of a WHO consultation. *Diabet. Med.* 15:539–553, 1998.

2. Ceriello A, dello Russo P, Amstad P, Cerutti P: High glucose induces antioxidant enzymes in human endothelial cells in culture. Evidence linking hyperglycemia and oxidative stress. *Diabetes* 45:471–477, 1996.

3. Giugliano D, Ceriello A, Paolisso G: Oxidative stress and diabetic vascular complications. *Diabetes Care* 19:257–267, 1996.

4. Carnevale Schianca GP, Rossi A, Sainaghi PP, Maduli E, Bartoli E: The significance of impaired fasting glucose vs. impaired glucose tolerance: importance of insulin secretion and resistance. *Diabetes Care* 26:1333–1337, 2003.

5. Thorsby E, Ronningen KS: Role of HLA genes in predisposition to develop insulin-dependent diabetes mellitus. *Ann. Med.* 24:523–531, 1992.

6. Kishimoto M, Hashiramoto M, Araki S, Ishida Y, Kazumi T, Kanda E, Kasuga M: Diabetes mellitus carrying a mutation in the mitochondrial tRNA(Leu(UUR)) gene. *Diabetologia* 38:193–200, 1995.

7. Suzuki S, Hinokio Y, Hirai S, Onoda M, Matsumoto M, Ohtomo M, Kawasaki H, Satoh Y, Akai H, Abe K: Pancreatic β-cell secretory defect associated with mitochondrial point mutation of the tRNA(LEU(UUR)) gene: a study in seven families with mitochondrial encephalomyopathy, lactic acidosis and stroke-like episodes (MELAS). *Diabetologia* 37:818–825, 1994.

8. Classification and diagnosis of diabetes mellitus and other categories of glucose intolerance. National Diabetes Data Group. *Diabetes* 28:1039–1057, 1979.

9. Kimpimaki T, Kulmala P, Savola K, Kupila A, Korhonen S, Simell T, Ilonen J, Simell O, Knip M: Natural history of β-cell autoimmunity in young children with increased genetic susceptibility to type 1 diabetes recruited from the general population. *J. Clin. Endocrinol. Metab.* 87:4572–4579, 2002.

10. Lee HK, Song JH, Shin CS, Park DJ, Park KS, Lee KU, Koh CS: Decreased mitochondrial DNA content in peripheral blood precedes the development of non-insulin-dependent diabetes mellitus. *Diabetes Res. Clin. Pract.* 42:161–167, 1998.

11. Patti ME, Butte AJ, Crunkhorn S, Cusi K, Berria R, Kashyap S, Miyazaki Y, Kohane I, Costello M, Saccone R, Landaker EJ, Goldfine AB, Mun E, DeFronzo R, Finlayson J, Kahn CR, Mandarino LJ: Coordinated reduction of genes of oxidative metabolism in humans with insulin resistance and diabetes: potential role of PGC1 and NRF1. *Proc. Natl. Acad. Sci. U.S.A.* 100:8466–8471, 2003.

12. Petersen KF, Befroy D, Dufour S, Dziura J, Ariyan C, Rothman DL, DiPietro L, Cline GW, Shulman GI: Mitochondrial dysfunction in the elderly: possible role in insulin resistance. *Science* 300:1140–1142, 2003.

13. Executive summary of the third report of The National Cholesterol Education Program (NCEP) Expert Panel on Detection, Evaluation, and Treatment of High Blood Cholesterol in Adults (Adult Treatment Panel III). *JAMA* 285:2486–2497, 2001.

14. Purves T, Middlemas A, Agthong S, Jude EB, Boulton AJ, Fernyhough P, Tomlinson DR: A role for mitogen-activated protein kinases in the etiology of diabetic neuropathy. *FASEB J.* 15:2508–2514, 2001.

15. Mandarino LJ: Current hypotheses for the biochemical basis of diabetic retinopathy. *Diabetes Care* 15:1892–1901, 1992.

16. Stevens MJ, Feldman EL, Greene DA: The aetiology of diabetic neuropathy: the combined roles of metabolic and vascular defects. *Diabet. Med.* 12:566–579, 1995.

17. Barber AJ: A new view of diabetic retinopathy: a neurodegenerative disease of the eye. *Prog. Neuropsychopharmacol. Biol. Psychiatry* 27:283–290, 2003.

18. Di Leo MA, Ghirlanda G, Gentiloni SN, Giardina B, Franconi F, Santini SA: Potential therapeutic effect of antioxidants in experimental diabetic retina: a comparison between chronic taurine and vitamin E plus selenium supplementations. *Free Radic. Res.* 37:323–330, 2003.

19. Agardh CD, Stenram U, Torffvit O, Agardh E: Effects of inhibition of glycation and oxidative stress on the development of diabetic nephropathy in rats. *J. Diabetes Complications* 16:395–400, 2002.

20. Kelly DJ, Zhang Y, Hepper C, Gow RM, Jaworski K, Kemp BE, Wilkinson–Berka JL, Gilbert RE: Protein kinase C β-inhibition attenuates the progression of experimental diabetic nephropathy in the presence of continued hypertension. *Diabetes* 52:512–518, 2003.

21. Obrosova IG, Minchenko AG, Vasupuram R, White L, Abatan OI, Kumagai AK, Frank RN, Stevens MJ: Aldose reductase inhibitor fidarestat prevents retinal oxidative stress and vascular endothelial growth factor overexpression in streptozotocin-diabetic rats. *Diabetes* 52:864–871, 2003.

22. Ballinger SW, Shoffner JM, Gebhart S, Koontz DA, Wallace DC: Mitochondrial diabetes revisited. *Nat. Genet.* 7:458–459, 1994.

23. Kadowaki T, Kadowaki H, Mori Y, Tobe K, Sakuta R, Suzuki Y, Tanabe Y, Sakura H, Awata T, Goto Y: A subtype of diabetes mellitus associated with a mutation of mitochondrial DNA. *N. Engl. J. Med.* 330:962–968, 1994.

24. Ballinger SW, Shoffner JM, Hedaya EV, Trounce I, Polak MA, Koontz DA, Wallace DC: Maternally transmitted diabetes and deafness associated with a 10.4 kb mitochondrial DNA deletion. *Nat. Genet.* 1:11–15, 1992.

25. Mathews CE, Berdanier CD: Noninsulin-dependent diabetes mellitus as a mitochondrial genomic disease. *Proc. Soc. Exp. Biol. Med.* 219:97–108, 1998.

26. Maassen JA: Mitochondrial diabetes: pathophysiology, clinical presentation, and genetic analysis. *Am. J. Med. Genet.* 115:66–70, 2002.

27. Maechler P, Wollheim CB: Mitochondrial function in normal and diabetic β-cells. *Nature* 414:807–812, 2001.

28. Poulton J, Brown MS, Cooper A, Marchington DR, Phillips DI: A common mitochondrial DNA variant is associated with insulin resistance in adult life. *Diabetologia* 41:54–58, 1998.

29. Poulton J, Luan J, Macaulay V, Hennings S, Mitchell J, Wareham NJ: Type 2 diabetes is associated with a common mitochondrial variant: evidence from a population-based case-control study. *Hum. Mol. Genet.* 11:1581–1583, 2002.

30. Casteels K, Ong K, Phillips D, Bendall H, Pembrey M: Mitochondrial 16189 variant, thinness at birth, and type-2 diabetes. ALSPAC study team. Avon Longitudinal Study of Pregnancy and Childhood. *Lancet* 353:1499–1500, 1999.

31. Poulton J, Bednarz AL, Scott–Brown M, Thompson C, Macaulay VA, Simmons D: The presence of a common mitochondrial DNA variant is associated with fasting insulin levels in Europeans in Auckland. *Diabet. Med.* 19:969–971, 2002.

32. Kim JH, Park KS, Cho YM, Kang BS, Kim SK, Jeon HJ, Kim SY, Lee HK: The prevalence of the mitochondrial DNA 16189 variant in nondiabetic Korean adults and its association with higher fasting glucose and body mass index. *Diabet. Med.* 19:681–684, 2002.

33. Ji L, Gao L, Han X: Association of 16189 variant (T→C transition) of mitochondrial DNA with genetic predisposition to type 2 diabetes in Chinese populations. *Zhonghua Yi. Xue. Za Zhi.* 81:711–714, 2001.

34. Coskun PE, Ruiz–Pesini E, Wallace DC: Control region mtDNA variants: longevity, climatic adaptation, and a forensic conundrum. *Proc. Natl. Acad. Sci. U.S.A* 100:2174–2176, 2003.

35. Antonetti DA, Reynet C, Kahn CR: Increased expression of mitochondrial-encoded genes in skeletal muscle of humans with diabetes mellitus. *J. Clin. Invest.* 95:1383–1388, 1995.

36. Song J, Oh JY, Sung YA, Pak YK, Park KS, Lee HK: Peripheral blood mitochondrial DNA content is related to insulin sensitivity in offspring of type 2 diabetic patients. *Diabetes Care* 24:865–869, 2001.

37. Ng MCY, Christley JJH, Cockram CS, Chan JCN: Qualitative and quantitative analysis of mitochondrial DNA in Chinese patients with type 2 diabetes. Abstract # 2073. *Diabetes* 50:A493, 2001.

38. Park KS, Lee KU, Song JH, Choi CS, Shin CS, Park DJ, Kim SK, Koh JJ, Lee HK: Peripheral blood mitochondrial DNA content is inversely correlated with insulin secretion during hyperglycemic clamp studies in healthy young men. *Diabetes Res. Clin. Pract.* 52:97–102, 2001.

39. Park KS, Song JH, Lee KU, Choi CS, Koh JJ, Shin CS, Lee HK: Peripheral blood mitochondrial DNA content correlates with lipid oxidation rate during euglycemic clamps in healthy young men. *Diabetes Res. Clin. Pract.* 46:149–154, 1999.

40. Lee HK: Method of proof and evidences for the concept that mitochondrial genome is a thrifty genome. *Diabetes Res. Clin. Pract.* 54 Suppl 2:S57–S63, 2001.

41. Barazzoni R, Short KR, Nair KS: Effects of aging on mitochondrial DNA copy number and cytochrome c oxidase gene expression in rat skeletal muscle, liver, and heart. *J. Biol. Chem.* 275:3343–3347, 2000.

42. Barthelemy C, Ogier DB, Diaz J, Cheval MA, Frachon P, Romero N, Goutieres F, Fardeau M, Lombes A: Late-onset mitochondrial DNA depletion: DNA copy number, multiple deletions, and compensation. *Ann. Neurol.* 49:607–617, 2001.

43. Smith RE: Quantitative relations between liver mitochondria metabolism and total body weight in mammals. *Ann. NY Acad. Sci.* 62:403–422, 1956.

44. Dicter N, Madar Z, Tirosh O: Alpha-lipoic acid inhibits glycogen synthesis in rat soleus muscle via its oxidative activity and the uncoupling of mitochondria. *J. Nutr.* 132:3001–3006, 2002.

45. Harper ME, Dent R, Monemdjou S, Bezaire V, Van Wyck L, Wells G, Kavaslar GN, Gauthier A, Tesson F, McPherson R: Decreased mitochondrial proton leak and reduced expression of uncoupling protein 3 in skeletal muscle of obese diet-resistant women. *Diabetes* 51:2459–2466, 2002.

46. Yechoor VK, Patti ME, Saccone R, Kahn CR: Coordinated patterns of gene expression for substrate and energy metabolism in skeletal muscle of diabetic mice. *Proc. Natl. Acad. Sci. U.S.A.* 99:10587–10592, 2002.

47. Doniger SW, Salomonis N, Dahlquist KD, Vranizan K, Lawlor SC, Conklin BR: MAPPFinder: using gene ontology and GenMAPP to create a global gene-expression profile from microarray data. *Genome Biol.* 4:R7, 2003.

48. MacDonald MJ, Tang J, Polonsky KS: Low mitochondrial glycerol phosphate dehydrogenase and pyruvate carboxylase in pancreatic islets of Zucker diabetic fatty rats. *Diabetes* 45:1626–1630, 1996.

49. Bindokas VP, Kuznetsov A, Sreenan S, Polonsky KS, Roe MW, Philipson LH: Visualizing superoxide production in normal and diabetic rat islets of Langerhans. *J. Biol. Chem.* 278:9796–9801, 2003.

50. Serradas P, Giroix MH, Saulnier C, Gangnerau MN, Borg LA, Welsh M, Portha B, Welsh N: Mitochondrial deoxyribonucleic acid content is specifically decreased in adult, but not fetal, pancreatic islets of the Goto–Kakizaki rat, a genetic model of noninsulin-dependent diabetes. *Endocrinology* 136:5623–5631, 1995.

51. MacDonald MJ, Efendic S, Ostenson CG: Normalization by insulin treatment of low mitochondrial glycerol phosphate dehydrogenase and pyruvate carboxylase in pancreatic islets of the GK rat. *Diabetes* 45:886–890, 1996.

52. Suzuki N, Aizawa T, Asanuma N, Sato Y, Komatsu M, Hidaka H, Itoh N, Yamauchi K, Hashizume K: An early insulin intervention accelerates pancreatic β-cell dysfunction in young Goto–Kakizaki rats, a model of naturally occurring noninsulin-dependent diabetes. *Endocrinology* 138:1106–1110, 1997.

53. Moreira PI, Santos MS, Moreno AM, Seica R, Oliveira CR: Increased vulnerability of brain mitochondria in diabetic (Goto–Kakizaki) rats with aging and amyloid-β-exposure. *Diabetes* 52:1449–1456, 2003.

54. Noble F, Roques BP: Phenotypes of mice with invalidation of cholecystokinin (CCK(1) or CCK(2)) receptors. *Neuropeptides* 36:157–170, 2002.

55. Sato T, Magata K, Koga N, Mitsumoto Y: Defect of an early event of glucose metabolism in skeletal muscle of the male Otsuka Long–Evans Tokushima fatty (OLETF) rat, a non-insulin-dependent diabetes mellitus (NIDDM) model. *Biochem. Biophys. Res. Commun.* 245:378–381, 1998.

56. Shima K, Zhu M, Mizuno A: Pathoetiology and prevention of NIDDM lessons from the OLETF rat. *J. Med. Invest.* 46:121–129, 1999.

57. Zhu M, Mizuno A, Noma Y, Murakami T, Kuwajima M, Shima K, Lan MS: Defective morphogenesis and functional maturation in fetal islet-like cell clusters from OLETF rat, a model of NIDDM. *Int. J. Exp. Diabetes Res.* 1:289–298, 2001.

58. Nakaya Y, Minami A, Harada N, Sakamoto S, Niwa Y, Ohnaka M: Taurine improves insulin sensitivity in the Otsuka Long–Evans Tokushima fatty rat, a model of spontaneous type 2 diabetes. *Am. J. Clin. Nutr.* 71:54–58, 2000.

59. Aruoma OI, Halliwell B, Hoey BM, Butler J: The antioxidant action of taurine, hypotaurine and their metabolic precursors. *Biochem. J.* 256:251–255, 1988.

60. Trachtman H, Futterweit S, Maesaka J, Ma C, Valderrama E, Fuchs A, Tarectecan AA, Rao PS, Sturman JA, Boles TH: Taurine ameliorates chronic streptozocin-induced diabetic nephropathy in rats. *Am. J. Physiol.* 269:F429–F438, 1995.

61. Boujendar S, Arany E, Hill D, Remacle C, Reusens B: Taurine supplementation of a low protein diet fed to rat dams normalizes the vascularization of the fetal endocrine pancreas. *J. Nutr.* 133:2820–2825, 2003.

62. Cherif H, Reusens B, Ahn MT, Hoet JJ, Remacle C: Effects of taurine on the insulin secretion of rat fetal islets from dams fed a low-protein diet. *J. Endocrinol.* 159:341–348, 1998.

63. Suzuki T, Suzuki T, Wada T, Saigo K, Watanabe K: Taurine as a constituent of mitochondrial tRNAs: new insights into the functions of taurine and human mitochondrial diseases. *EMBO J.* 21:6581–6589, 2002.

64. Vicent D, Piper M, Gammeltoft S, Maratos–Flier E, Kahn CR: Alterations in skeletal muscle gene expression of ob/ob mice by mRNA differential display. *Diabetes* 47:1451–1458, 1998.

65. Zhang Y, Proenca R, Maffei M, Barone M, Leopold L, Friedman JM: Positional cloning of the mouse obese gene and its human homologue. *Nature* 372:425–432, 1994.

66. Hales CN, Barker DJ: The thrifty phenotype hypothesis. *Br. Med. Bull.* 60:5–20, 2001.

67. Metzger BE: Biphasic effects of maternal metabolism on fetal growth. Quintessential expression of fuel-mediated teratogenesis. *Diabetes* 40 Suppl 2:99–105, 1991.

68. Bertinm E, Gangnerau MN, Bailbe D, Portha B: Glucose metabolism and β-cell mass in adult offspring of rats protein and/or energy restricted during the last week of pregnancy. *Am. J. Physiol.* 277:E11–E17, 1999.

69. Poulsen P, Vaag AA, Kyvik KO, Moller JD, Beck–Nielsen H: Low birth weight is associated with NIDDM in discordant monozygotic and dizygotic twin pairs. *Diabetologia* 40:439–446, 1997.

70. Selak MA, Storey BT, Peterside I, Simmons RA: Impaired oxidative phosphorylation in skeletal muscle of intrauterine growth-retarded rats. *Am. J. Physiol. Endocrinol. Metab.* 285:E130–E137, 2003.

71. Park KS, Kim SK, Kim MS, Cho EY, Lee JH, Lee KU, Pak YK, Lee HK: Fetal and early postnatal protein malnutrition cause long-term changes in rat liver and muscle mitochondria. *J. Nut.* 133:3085–3090, 2003.

72. Park HK, Jin CJ, Cho YM, Park do J, Shin CS, Park KS, Kim SY, Cho BY, Lee HK: Changes of mitochondrial DNA content in the male offspring of protein-malnourished rats. *Ann. NY Acad. Sci.* 1011:205–216, 2004.

73. Park KS, Nam KJ, Kim JW, Lee YB, Han CY, Jeong JK, Lee HK, Pak YK: Depletion of mitochondrial DNA alters glucose metabolism in SK-Hep1 cells. *Am. J. Physiol. Endocrinol. Metab.* 280:E1007–E1014, 2001.

74. King MP, Attardi G: Human cells lacking mtDNA: repopulation with exogenous mitochondria by complementation. *Science* 246:500–503, 1989.

75. Zylber E, Vesco C, Penman S: Selective inhibition of the synthesis of mitochondria-associated RNA by ethidium bromide. *J. Mol. Biol.* 44:195–204, 1969.

76. Hayakawa T, Noda M, Yasuda K, Yorifuji H, Taniguchi S, Miwa I, Sakura H, Terauchi Y, Hayashi J, Sharp GW, Kanazawa Y, Akanuma Y, Yazaki Y, Kadowaki T: Ethidium bromide-induced inhibition of mitochondrial gene transcription suppresses glucose-stimulated insulin release in the mouse pancreatic β-cell line β-HC9. *J. Biol. Chem.* 273:20300–20307, 1998.

77. Kennedy ED, Maechler P, Wollheim CB: Effects of depletion of mitochondrial DNA in metabolism secretion coupling in INS-1 cells. *Diabetes* 47:374–380, 1998.

78. Buchet K, Godinot C: Functional F1-ATPase essential in maintaining growth and membrane potential of human mitochondrial DNA-depleted ρ degrees cells. *J. Biol. Chem.* 273:22983–22989, 1998.

79. Cesar MC, Wilson JE: Further studies on the coupling of mitochondrially bound hexokinase to intramitochondrially compartmented ATP, generated by oxidative phosphorylation. *Arch. Biochem. Biophys.* 350:109–117, 1998.

80. Gerbitz KD, Gempel K, Brdiczka D: Mitochondria and diabetes. Genetic, biochemical, and clinical implications of the cellular energy circuit. *Diabetes* 45:113–126, 1996.

81. Brdiczka D: Function of the outer mitochondrial compartment in regulation of energy metabolism. *Biochim. Biophys. Acta* 1187:264–269, 1994.

82. Sherratt EJ, Thomas AW, Gill–Randall R, Alcolado JC: Phylogenetic analysis of mitochondrial DNA in type 2 diabetes: maternal history and ancient population expansion. *Diabetes* 48:628–634, 1999.

83. Knowler WC, Barrett–Connor E, Fowler SE, Hamman RF, Lachin JM, Walker EA, Nathan DM: Reduction in the incidence of type 2 diabetes with lifestyle intervention or metformin. *N. Engl. J. Med.* 346:393–403, 2002.

84. Devlin JT, Hirshman M, Horton ED, Horton ES: Enhanced peripheral and splanchnic insulin sensitivity in NIDDM men after single bout of exercise. *Diabetes* 36:434–439, 1987.

85. Devlin JT, Horton ES: Effects of prior high-intensity exercise on glucose metabolism in normal and insulin-resistant men. *Diabetes* 34:973–979, 1985.

86. Ren JM, Semenkovich CF, Gulve EA, Gao J, Holloszy JO: Exercise induces rapid increases in GLUT4 expression, glucose transport capacity, and insulin-stimulated glycogen storage in muscle. *J. Biol. Chem.* 269:14396–14401, 1994.

87. O'Doherty RM, Bracy DP, Granner DK, Wasserman DH: Transcription of the rat skeletal muscle hexokinase II gene is increased by acute exercise. *J. Appl. Physiol.* 81:789–793, 1996.

88. Koval JA, DeFronzo RA, O'Doherty RM, Printz R, Ardehali H, Granner DK, Mandarino LJ: Regulation of hexokinase II activity and expression in human muscle by moderate exercise. *Am. J. Physiol.* 274:E304–E308, 1998.

89. Osawa H, Sutherland C, Robey RB, Printz RL, Granner DK: Analysis of the signaling pathway involved in the regulation of hexokinase II gene transcription by insulin. *J. Biol. Chem.* 271:16690–16694, 1996.

90. Fueger PT, Hcikkinen S, Bracy DP, Malabanan CM, Pencek RR, Laakso M, Wasserman DH: Hexokinase II partial knockout impairs exercise-stimulated glucose uptake in oxidative muscles of mice. *Am. J. Physiol. Endocrinol. Metab.* 285:E958–E963, 2003.

91. Goodyear LJ, Chang PY, Sherwood DJ, Dufresne SD, Moller DE: Effects of exercise and insulin on mitogen-activated protein kinase signaling pathways in rat skeletal muscle. *Am. J. Physiol.* 271:E403–E408, 1996.

92. Ruderman NB, Park H, Kaushik VK, Dean D, Constant S, Prentki M, Saha AK: AMPK as a metabolic switch in rat muscle, liver and adipose tissue after exercise. *Acta Physiol. Scand.* 178:435–442, 2003.

93. Leary SC, Shoubridge EA: Mitochondrial biogenesis: which part of "NO" do we understand? *Bioessays* 25:538–541, 2003.

94. Vestergaard H, Bjorbaek C, Hansen T, Larsen FS, Granner DK, Pedersen O: Impaired activity and gene expression of hexokinase II in muscle from non-insulin-dependent diabetes mellitus patients. *J. Clin. Invest.* 96:2639–2645, 1995.

95. Golshani–Hebroni SG, Bessman SP: Hexokinase binding to mitochondria: a basis for proliferative energy metabolism. *J. Bioenerg. Biomembr.* 29:331–338, 1997.

96. Hoppeler H, Luthi P, Claassen H, Weibel ER, Howald H: The ultrastructure of the normal human skeletal muscle. A morphometric analysis on untrained men, women and well-trained orienteers. *Pflugers Arch.* 344:217–232, 1973.

97. Lampert E, Mettauer B, Hoppeler H, Charloux A, Charpentier A, Lonsdorfer J: Structure of skeletal muscle in heart transplant recipients. *J. Am. Coll. Cardiol.* 28:980–984, 1996.

98. Holloszy JO, Coyle EF: Adaptations of skeletal muscle to endurance exercise and their metabolic consequences. *J. Appl. Physiol.* 56:831–838, 1984.

99. Puntschart A, Claassen H, Jostarndt K, Hoppeler H, Billeter R: mRNAs of enzymes involved in energy metabolism and mtDNA are increased in endurance-trained athletes. *Am. J. Physiol.* 269:C619–C625, 1995.

100. Lim S, Kim SK, Park KS, Kim SY, Cho BY, Yim MJ, Lee HK: Effect of exercise on the mitochondrial DNA content of peripheral blood in healthy women. *Eur. J. Appl. Physiol.* 82:407–412, 2000.

101. Nisoli E, Clementi E, Paolucci C, Cozzi V, Tonello C, Sciorati C, Bracale R, Valerio A, Francolini M, Moncada S, Carruba MO: Mitochondrial biogenesis in mammals: the role of endogenous nitric oxide. *Science* 299:896–899, 2003.

102. Brini M, Pinton P, King MP, Davidson M, Schon EA, Rizzuto R: A calcium signaling defect in the pathogenesis of a mitochondrial DNA inherited oxidative phosphorylation deficiency. *Nat. Med.* 5:951–954, 1999.

103. Scarpulla RC: Nuclear activators and coactivators in mammalian mitochondrial biogenesis. *Biochim. Biophys. Acta* 1576:1–14, 2002.

104. Armstrong D, al Awadi F: Lipid peroxidation and retinopathy in streptozotocin-induced diabetes. *Free Radic. Biol. Med.* 11:433–436, 1991.

105. Kitahara M, Eyre HJ, Lynch RE, Rallison ML, Hill HR: Metabolic activity of diabetic monocytes. *Diabetes* 29:251–256, 1980.

106. Leinonen J, Lehtimaki T, Toyokuni S, Okada K, Tanaka T, Hiai H, Ochi H, Laippala P, Rantalaiho V, Wirta O, Pasternack A, Alho H: New biomarker evidence of oxidative DNA damage in patients with non-insulin-dependent diabetes mellitus. *FEBS Lett.* 417:150–152, 1997.

107. Nagasaka Y, Fujii S, Kaneko T: Effects of high glucose and sorbitol pathway on lipid peroxidation of erythrocytes. *Horm. Metab. Res.* 21:275–276, 1989.

108. Jain SK, McVie R: Effect of glycemic control, race (white vs. black), and duration of diabetes on reduced glutathione content in erythrocytes of diabetic patients. *Metabolism* 43:306–309, 1994.

109. Lee YS, Lee HS, Park MK, Hwang ES, Park EM, Kasai H, Chung MH: Identification of 8-hydroxyguanine glycosylase activity in mammalian tissues using 8-hydroxyguanine specific monoclonal antibody. *Biochem. Biophys. Res. Commun.* 196:1545–1551, 1993.

110. Park KS, Kim JH, Kim MS, Kim JM, Kim SK, Choi JY, Chung MH, Han B, Kim SY, Lee HK: Effects of insulin and antioxidant on plasma 8-hydroxyguanine and tissue 8-hydroxydeoxyguanosine in streptozotocin-induced diabetic rats. *Diabetes* 50:2837–2841, 2001.

111. Shin CS, Moon BS, Park KS, Kim SY, Park SJ, Chung MH, Lee HK: Serum 8-hydroxy-guanine levels are increased in diabetic patients. *Diabetes Care* 24:733–737, 2001.

112. Ceriello A: New insights on oxidative stress and diabetic complications may lead to a "causal" antioxidant therapy. *Diabetes Care* 26:1589–1596, 2003.

113. Keaney JF, Jr., Larson MG, Vasan RS, Wilson PW, Lipinska I, Corey D, Massaro JM, Sutherland P, Vita JA, Benjamin EJ: Obesity and systemic oxidative stress: clinical correlates of oxidative stress in the Framingham Study. *Arterioscler. Thromb. Vasc. Biol.* 23:434–439, 2003.

114. Williamson JR, Kilo C, Ido Y: The role of cytosolic reductive stress in oxidant formation and diabetic complications. *Diabetes Res. Clin. Pract.* 45:81–82, 1999.

115. Maziere C, Auclair M, Rose–Robert F, Leflon P, Maziere JC: Glucose-enriched medium enhances cell-mediated low density lipoprotein peroxidation. *FEBS Lett.* 363:277–279, 1995.

116. Graier WF, Simecek S, Kukovetz WR, Kostner GM: High D-glucose-induced changes in endothelial Ca2+/EDRF signaling are due to generation of superoxide anions. *Diabetes* 45:1386–1395, 1996.

117. Nishikawa T, Edelstein D, Du XL, Yamagishi S, Matsumura T, Kaneda Y, Yorek MA, Beebe D, Oates PJ, Hammes HP, Giardino I, Brownlee M: Normalizing mitochondrial superoxide production blocks three pathways of hyperglycaemic damage. *Nature* 404:787–790, 2000.

118. Chevaux F, Curchod B, Felber JP, Jequier E: Insulin resistance and carbohydrate oxidation in patients with chemical diabetes. *Diabetes Metab.* 8:105–108, 1982.

119. Pittet P, Gygax PH, Jequier E: Thermic effect of glucose and amino acids in man studied by direct and indirect calorimetry. *Br. J. Nutr.* 31:343–349, 1974.

120. Baines A, Ho P: Glucose stimulates O2 consumption, NOS, and Na/H exchange in diabetic rat proximal tubules. *Am. J. Physiol. Renal Physiol.* 283:F286–F293, 2002.

121. Sakai K, Matsumoto K, Nishikawa T, Suefuji M, Nakamaru K, Hirashima Y, Kawashima J, Shirotani T, Ichinose K, Brownlee M, Araki E: Mitochondrial reactive oxygen species reduce insulin secretion by pancreatic β-cells. *Biochem. Biophys. Res. Commun.* 300:216–222, 2003.

122. Cai L, Li W, Wang G, Guo L, Jiang Y, Kang YJ: Hyperglycemia-induced apoptosis in mouse myocardium: mitochondrial cytochrome C-mediated caspase-3 activation pathway. *Diabetes* 51:1938–1948, 2002.

123. Egawhary DN, Swoboda BE, Chen J, Easton AJ, Vince FP: Diabetic complications and the mechanism of the hyperglycaemia-induced damage to the mtDNA of cultured vascular endothelial cells: (I) Characterization of the 4977 base pair deletion and 13 bp flanking repeats. *Biochem. Soc. Trans.* 23:518S, 1995.

124. Swoboda BE, Egawhary DN, Chen J, Vince FP: Diabetic complications and the mechanism of the hyperglycaemia-induced damage to the mt DNA of cultured vascular endothelial cells: (II) the involvement of protein kinase C. *Biochem. Soc. Trans.* 23:519S, 1995.

125. McDonald DG, McMenamin JB, Farrell MA, Droogan O, Green AJ: Familial childhood onset neuropathy and cirrhosis with the 4977bp mitochondrial DNA deletion. *Am. J. Med. Genet.* 111:191–194, 2002.

126. Santos DL, Palmeira CM, Seica R, Dias J, Mesquita J, Moreno AJ, Santos MS: Diabetes and mitochondrial oxidative stress: a study using heart mitochondria from the diabetic Goto–Kakizaki rat. *Mol. Cell Biochem.* 246:163–170, 2003.

127. Yorek MA: The role of oxidative stress in diabetic vascular and neural disease. *Free Radic. Res.* 37:471–480, 2003.

128. Barber AJ: A new view of diabetic retinopathy: a neurodegenerative disease of the eye. *Prog. Neuropsychopharmacol. Biol. Psychiatry* 27:283–290, 2003.

129. Spriet LL, Watt MJ: Regulatory mechanisms in the interaction between carbohydrate and lipid oxidation during exercise. *Acta Physiol. Scand.* 178:443–452, 2003.

130. Carlsson C, Borg LA, Welsh N: Sodium palmitate induces partial mitochondrial uncoupling and reactive oxygen species in rat pancreatic islets *in vitro*. *Endocrinology* 140:3422–3428, 1999.

131. Barbu A, Welsh N, Saldeen J: Cytokine-induced apoptosis and necrosis are preceded by disruption of the mitochondrial membrane potential (Deltapsi(m)) in pancreatic RINm5F cells: prevention by Bcl-2. *Mol. Cell Endocrinol.* 190:75–82, 2002.

132. Koshkin V, Wang X, Scherer PE, Chan CB, Wheeler MB: Mitochondrial functional state in clonal pancreatic β-cells exposed to free fatty acids. *J. Biol. Chem.* 278:19709–19715, 2003.

133. Effect of intensive blood-glucose control with metformin on complications in overweight patients with type 2 diabetes (UKPDS 34). UK Prospective Diabetes Study (UKPDS) Group. *Lancet* 352:854–865, 1998.

134. Owen MR, Doran E, Halestrap AP: Evidence that metformin exerts its antidiabetic effects through inhibition of complex 1 of the mitochondrial respiratory chain. *Biochem. J.* 348 Pt 3:607–614, 2000.

135. Bonnefont–Rousselot D, Raji B, Walrand S, Gardes–Albert M, Jore D, Legrand A, Peynet J, Vasson MP: An intracellular modulation of free radical production could contribute to the beneficial effects of metformin towards oxidative stress. *Metabolism* 52:586–589, 2003.

136. Skulachev VP: Uncoupling: new approaches to an old problem of bioenergetics. *Biochim. Biophys. Acta* 1363:100–124, 1998.

137. Korshunov SS, Korkina OV, Ruuge EK, Skulachev VP, Starkov AA: Fatty acids as natural uncouplers preventing generation of O2.- and H2O2 by mitochondria in the resting state. *FEBS Lett.* 435:215–218, 1998.

138. Cocco T, Di Paola M, Papa S, Lorusso M: Arachidonic acid interaction with the mitochondrial electron transport chain promotes reactive oxygen species generation. *Free Radic. Biol. Med.* 27:51–59, 1999.

139. Bessman SP, Mohan C, Zaidise I: Intracellular site of insulin action: mitochondrial Krebs cycle. *Proc. Natl. Acad. Sci. U.S.A.* 83:5067–5070, 1986.

140. Wang J, Wilhelmsson H, Graff C, Li H, Oldfors A, Rustin P, Bruning JC, Kahn CR, Clayton DA, Barsh GS, Thoren P, Larsson NG: Dilated cardiomyopathy and atrioventricular conduction blocks induced by heart-specific inactivation of mitochondrial DNA gene expression. *Nat. Genet.* 21:133–137, 1999.

141. Kanazawa A, Nishio Y, Kashiwagi A, Inagaki H, Kikkawa R, Horiike K: Reduced activity of mtTFA decreases the transcription in mitochondria isolated from diabetic rat heart. *Am. J. Physiol. Endocrinol. Metab.* 282:E778–E785, 2002.

142. Harman DA: theory based on free radical and radiation chemistry. *J. Gerontol.* 11:298–300, 1956.

143. Linnane AW, Zhang C, Baumer A, Nagley P: Mitochondrial DNA mutation and the ageing process: bioenergy and pharmacological intervention. *Mutat. Res.* 275:195–208, 1992.

144. Rustin P, Kleist–Retzow JC, Vajo Z, Rotig A, Munnich A: For debate: defective mitochondria, free radicals, cell death, aging-reality or myth-ochondria? *Mech. Ageing Dev.* 114:201–206, 2000.

145. Suzuki S, Hinokio Y, Komatu K, Ohtomo M, Onoda M, Hirai S, Hirai M, Hirai A, Chiba M, Kasuga S, Akai H, Toyota T: Oxidative damage to mitochondrial DNA and its relationship to diabetic complications. *Diabetes Res. Clin. Pract.* 45:161–168, 1999.

146. Ha H, Kim C, Son Y, Chung MH, Kim KH: DNA damage in the kidneys of diabetic rats exhibiting microalbuminuria. *Free Radic. Biol. Med*. 16:271–274, 1994.

147. Kakimoto M, Inoguchi T, Sonta T, Yu HY, Imamura M, Etoh T, Hashimoto T, Nawata H: Accumulation of 8-hydroxy-2′-deoxyguanosine and mitochondrial DNA deletion in kidney of diabetic rats. *Diabetes* 51:1588–1595, 2002.

148. Nomiyama T, Tanaka Y, Hattori N, Nishimaki K, Nagasaka K, Kawamori R, Ohta S: Accumulation of somatic mutation in mitochondrial DNA extracted from peripheral blood cells in diabetic patients. *Diabetologia* 45:1577–1583, 2002.

149. Harano Y, DePalma RG, Lavine L, Miller M: Fatty acid oxidation, oxidative phosphorylation and ultrastructure of mitochondria in the diabetic rat liver. Hepatic factors in diabetic ketosis. *Diabetes* 21:257–270, 1972.

150. Mythili MD, Vyas R, Akila G, Gunasekaran S: *Microsc. Res. Tech.* 63:274–281, 2004.

151. Bian K, Murad F: Nitric oxide (NO) — biogeneration, regulation, and relevance to human diseases. *Front Biosci.* 8:d264–d278, 2003.

152. Brown GC: Nitric oxide regulates mitochondrial respiration and cell functions by inhibiting cytochrome oxidase. *FEBS Lett.* 369:136–139, 1995.

153. Choi YS, Kim S, Pak YK: Mitochondrial transcription factor A (mtTFA) and diabetes. *Diabetes Res. Clin. Pract.* 54 Suppl 2:S3–S9, 2001.

154. Haraguchi Y, Chung AB, Neill S, Wallace DC: OXBOX and REBOX, overlapping promoter elements of the mitochondrial F0F1-ATP synthase β-subunit gene. OXBOX/REBOX in the ATPsyn β-promoter. *J. Biol. Chem.* 269:9330–9334, 1994.

155. Foss MC, Vlachokosta FV, Cunningham LN, Aoki TT: Restoration of glucose homeostasis in insulin-dependent diabetic subjects. An inducible process. *Diabetes* 31:46–52, 1982.

156. Memon RA, Mohan C, Bessman SP: Impaired mitochondrial protein synthesis in streptozotocin diabetic rat hepatocytes. *Biochem. Mol. Biol. Int.* 37:627–634, 1995.

157. Felber JP, Ferrannini E, Golay A, Meyer HU, Theibaud D, Curchod B, Maeder E, Jequier E, DeFronzo RA: Role of lipid oxidation in pathogenesis of insulin resistance of obesity and type II diabetes. *Diabetes* 36:1341–1350, 1987.

158. Golay A, Munger R, Assimacopoulos–Jeannet F, Bobbioni–Harsch E, Habicht F, Felber JP: Progressive defect of insulin action on glycogen synthase in obesity and diabetes. *Metabolism* 51:549–553, 2002.

159. Wlodek D, Gonzales M: Decreased energy levels can cause and sustain obesity. *J. Theor. Biol.* 225:33–44, 2003.

160. Leibel RL, Hirsch J: Diminished energy requirements in reduced-obese patients. *Metabolism* 33:164–170, 1984.

161. Leibel RL, Rosenbaum M, Hirsch J: Changes in energy expenditure resulting from altered body weight. *N. Engl. J. Med.* 332:621–628, 1995.

162. Nelson KM, Weinsier RL, Long CL, Schutz Y: Prediction of resting energy expenditure from fat-free mass and fat mass. *Am. J. Clin. Nutr.* 56:848–856, 1992.

163. Wang Z, Heshka S, Gallagher D, Boozer CN, Kotler DP, Heymsfield SB: Resting energy expenditure-fat-free mass relationship: new insights provided by body composition modeling. *Am. J. Physiol. Endocrinol. Metab.* 279:E539–E545, 2000.

164. Ravussin E, Lillioja S, Knowler WC, Christin L, Freymond D, Abbott WG, Boyce V, Howard BV, Bogardus C: Reduced rate of energy expenditure as a risk factor for body-weight gain. *N. Engl. J. Med.* 318:467–472, 1988.

165. Weyer C, Pratley RE, Salbe AD, Bogardus C, Ravussin E, Tataranni PA: Energy expenditure, fat oxidation, and body weight regulation: a study of metabolic adaptation to long-term weight change. *J. Clin. Endocrinol. Metab.* 85:1087–1094, 2000.

166. West GB, Brown JH, Enquist BJ: A general model for the origin of allometric scaling laws in biology. *Science* 276:122–126, 1997.

167. West GB, Brown JH, Enquist BJ: The fourth dimension of life: fractal geometry and allometric scaling of organisms. *Science* 284:1677–1679, 1999.

168. West GB, Woodruff WH, Brown JH: Allometric scaling of metabolic rate from molecules and mitochondria to cells and mammals. *Proc. Natl. Acad. Sci. U.S.A.* 99 Suppl 1:2473–2478, 2002.

169. Darveau CA, Suarez RK, Andrews RD, Hochachka PW: Allometric cascade as a unifying principle of body mass effects on metabolism. *Nature* 417:166–170, 2002.

170. White CR, Seymour RS: Mammalian basal metabolic rate is proportional to body mass 2/3. *Proc. Natl. Acad. Sci. U.S.A.* 100:4046–4049, 2003.

171. West GB, Savage VM, Gillooly J, Enquist BJ, Woodruff WH, Brown JH: Physiology: why does metabolic rate scale with body size? *Nature* 421:713, 2003.

172. Santillan M: Allometric scaling law in a simple oxygen exchanging network: possible implications on the biological allometric scaling laws. *J. Theor. Biol.* 223:249–257, 2003.

173. Schwartz MW, Woods SC, Porte D, Jr., Seeley RJ, Baskin DG: Central nervous system control of food intake. *Nature* 404:661–671, 2000.

174. Torsoni MA, Carvalheira JB, Pereira–Da-Silva M, Carvalho-Filho MA, Saad MJ, Velloso LA: Molecular and functional resistance to insulin in hypothalamus of rats exposed to cold. *Am. J. Physiol. Endocrinol. Metab.* 285:E216–E223, 2003.

175. Belfiore F, Iannello S: Fatty acid synthesis from glutamate in the adipose tissue of normal subjects and obese patients: an enzyme study. *Biochem. Mol. Med.* 54:19–25, 1995.

176. Golay A, Jallut D, Schutz Y, Felber JP, Jequier E: Evolution of glucose induced thermogenesis in obese subjects with and without diabetes: a 6-year follow-up study. *Int. J. Obes.* 15:601–607, 1991.

177. Strohman R: Maneuvering in the complex path from genotype to phenotype. *Science* 296:701–703, 2002.

9

Drugs, Nutrients, and Hormones in Mitochondrial Function

CAROLYN D. BERDANIER

CONTENTS

INTRODUCTION

Mitochondria (mt) have long been known to be sensitive to a wide variety of compounds that include drugs, nutrients, and hormones. Among the first to be reported are those that influence the respiratory chain and the coupling of respiration to ATP synthesis (oxidative phosphorylation, OXPHOS). These compounds (see Table 1.7, Chapter 1) allowed scientists to learn how the chain worked and how its function was linked to that of ATP synthesis. Of course, some of these compounds were not drugs for the treatment of disease but merely tools of the biochemist. Yet, their use paved the way for subsequent investigations with compounds that had a direct effect on mt OXPHOS or an indirect effect via changes in mt gene expression, changes in mt membrane function, or induction of free radical damage to the genome or the membranes. In addition, many compounds affect mt function secondarily; this effect developed unintentionally and was not the primary purpose of the drug.

A number of contemporary drugs fit this description. Some compounds were originally developed for other purposes but later, when the mechanism of their action was elucidated, were found to act via an effect on mt DNA and/or mt membrane integrity. Nutrient intake and hormonal status likewise can affect mt function. Some hormones have direct effects on mt gene transcription and others affect nuclear transcription. Some hormones affect both genomes. Only a few hormones have been studied in detail with respect to OXPHOS. It is the purpose of this chapter to highlight some of the compounds that affect mt function.

COMPOUNDS GENERATING FREE RADICALS

A number of compounds generate free radicals in such amounts as to have immediate and devastating effects on cells that have a high ATP need. Neuronal, renal, and pancreatic β-cells are typical of cells requiring high ATP. They are particularly vulnerable to destruction if their ability to synthesize ATP is interrupted. Humans or animals exposed to compounds that produce free radicals usually (but not always) show signs of pancreatic dysfunction, renal dysfunction, and neuronal deficits. Some of these compounds have been used in experimental animals to produce type 1 diabetes mellitus. The whole body response to these compounds is dose and age dependent. In many instances, the net result of exposure is seen in the mt compartment.

Mitochondria are a significant source of reactive oxygen species during reductive stress (1). Under normal conditions, mitochondria produce up to 90% of the reactive oxygen in the cell. These organelles generate superoxide anions and convert them to hydrogen peroxide via the enzyme superoxide dismutase. The hydrogen peroxide, in turn, is converted to water (2). It should be noted that unsaturated fatty acids present in the mt membranes are also vulnerable to peroxidation and can serve as a source of free radicals in the mt compartment.

Table 9.1 Compounds That
Affect Mitochondria via the
Production of Free Radicals

Vacor
Streptozotocin
Alloxan
Zidovudine (AZT)

Respiratory inhibitors such as rotenone and rhrein (3) or mercury (4) seem to enhance the production of free radicals. The list of compounds producing free radicals is extensive and is summarized in Table 9.1. Details of their effects follow.

Vacor

Vacor is a rodenticide. Its active ingredient is N-3-pyridylm-ethyl-N'-ρ-nitrophenyl urea. The dinitrophenyl urea portion of the molecule is responsible for its free-radical-generating function, which is potentiated by the pyridylmethyl portion. It came to the attention of scientists when several (~30) humans unsuccessfully attempted suicide by ingesting Vacor. This resulted in severe insulinopenic diabetes mellitus with ketoacidosis as well as a toxic neuropathy affecting peripheral and autonomic nerve tracts.

Beta cells isolated from humans who had consumed Vacor showed evidence of cytotoxicity (5). These cells were unresponsive to the glucose signal for insulin release. This is an ATP-dependent process. Because Vacor is a strong producer of free radicals, it likely induced free radical damage to the mt DNA and the mt membrane; this damage in turn resulted in β-cell destruction. Point mutations in the mt DNA have been reported, as has a reduction in mitochondrial number (6).

Insulin release depends on an intact and fully functional OXPHOS system (7). Vacor specifically inhibits complex I of the respiratory chain (5). This complex has an affinity for ubiquinone (7). The reduction in complex I activity means that the respiratory chain is inhibited with respect to use of substrates that enter at site 1. Fatty acid-supported OXPHOS is not impaired because reducing equivalents produced by fatty acid oxidation enter respiration via complex II. However, most of the metabolic intermediates of glucose and amino acids produce reducing equivalents that enter via complex I, so intermediary metabolism is suppressed. This means that the oxidation of fatty acids is one of the few processes working (somewhat) in the individual poisoned by Vacor.

Complete oxidation is impaired because the final step produces acetate that in turn contributes its reducing equivalents to complex I. Because fatty acids are serving as the main metabolic fuel, the quantity of ketones produced from such oxidation is in excess of that which can be used by muscle and brain. This means that the ketones accumulate and disturb the acid–base balance, resulting in ketoacidosis. This is the typical metabolic pattern of a person with uncontrolled type 1 diabetes mellitus.

Streptozotocin

Streptozotocin is a 1-methyl-1-nitrosourea linked to the C2 position of 2-deoxyglucose. The glucose facilitates its entry into the pancreatic β-cell. It is an unstable compound when in solution and if used in experimental animals must be used within 15 minutes of solution for full activity. It is a broad-spectrum antibiotic produced by *Streptomices achromogenes* that has antitumor, antibacterial, and oncogenic activity (9–11). When it is administered to mice or rats in large doses (200 mg/kg body weight), type 1 diabetes development is immediate (12). Necrotized pancreatic β-cells can be observed within 6 hours of treatment. Cells that produce glucagon and somatostatin are unaffected. Renal tumor cells and tubular lesions also develop with this treatment (10,13).

When streptozotocin is administered in five smaller doses (40 mg/kg body weight), insulitis (inflammation of the β-cells) can be observed (14). The change in glucose metabolism is observed within the first day of treatment, but the onset of diabetes is slower. Ultrastructural studies of the islets have revealed occasional necrotic β-cells, numerous infiltrating lymphocytes and macrophages, and large numbers of type C virus particles within the many intact partially degranulated β-cells (11).

The immune system seems to be involved in the whole-body response to the multiple low-dose streptozotocin treatment (15). Thymus-deficient mice are unaffected by streptozotocin, but thymus-sufficient mice develop diabetes after exposure to the compound. The immune reaction also involves the free radical suppression system (16), suggesting that the mode of action of streptozotocin is via free radical attack on the intracellular mt membranes and the mtDNA. Decreased nuclear and mt gene expression has been reported in β-cells isolated from streptozotocin-treated animals (17). Insulin has been shown to stimulate mt gene transcription (18) and it is possible that drug-induced damage to the islet cells resulted in an ablation of this hormone effect.

In addition to effects on DNA, streptozotocin also affects the fatty acid composition of the membranes within the cell (19). Diminished unsaturated fatty acid synthesis is reflected in a lower P/S ratio in the membranes after treatment (20). This change in unsaturation index would mean that the membranes of the mitochondria are less fluid than those in untreated rats and that this would have a negative effect on ATP synthesis efficiency (21). As well, evidence suggests increased free radical formation in cells from streptozotocin-treated rats (22). Streptozotocin also serves as a noncompetitive inhibitor of succinyl-CoA synthetase *in vitro* (23); however, because its half-life *in vivo* is so short, its action as an inhibitor of this enzyme is very short lived indeed. Altogether, then, streptozotocin has multiple effects on mitochondria: direct destruction, damage to mt DNA, and perturbation of the membrane lipid.

Alloxan

Alloxan is another unstable compound related to uric acid in structure. It is a pyrimidine derivative that was reported by Dunn et al. to produce type 1 diabetes in rabbits (24). Since this first report in 1943, it has been realized that hyperglycemia had been observed by earlier scientists but had not received very much of their attention. The instability of the compound under physiological conditions and its thiol reactivity made for inconsistent results. In addition, age, strain, species, and gender differences in response made it difficult to study the mechanism of action of this compound (25).

Alloxan was reported to be a potent inhibitor of β-cell glucokinase (25). Glucokinase is considered to be the glucose sensor of the β cell and depends on OXPHOS. As a ring structure, alloxan is a potent free radical generator (26,27). When exposed to reducing substances such as ascorbic acid and molecular oxygen, alloxan and its reduced analog, dialuric acid, form a reduction–oxidation cycle in which superoxide anion radicals are formed. These radicals can dismute (via mt superoxide dismutase) to form hydrogen peroxide as part of the free radical suppression system.

Because these radicals are very reactive, cell damage can occur. The membranes and DNA are damaged; the most vulnerable DNA is that found in the mitochondrial compartment because it is in this compartment that ~90% of the molecular oxygen can be found. Proximity of the radicals to membranes and DNA facilitates damage. Just as renal damage was reported in streptozotocin-treated animals, so too has this been reported in animals treated with alloxan. Alloxan seems to be more difficult than streptozotocin to use as a diabetes-producing compound because of its greater instability. Both compounds are unstable, however.

PHARMACEUTICALS

Zidovudine

Zidovudine or AZT is a potent inhibitor of the replication of the HIV virus. It is 3'-azidothymidine,3'azido-2,3-dideoxythymidine and is incorporated into viral DNA, resulting in premature termination of DNA synthesis. It has some major limitations in use, however, because of its serious side effects. This compound is a free radical generator. AZT causes an increase in the urinary excretion of a marker of free radical attack on DNA, 8-oxo-7,8-dihydro-2'-deoxyguanosine (28,29). In addition, AZT attacks mt glutathione and lipid (measured as malondialdehyde) and markers of these reactions have also been found to increase in HIV-infected humans treated with AZT. If supplemental doses of vitamins E and C are given with the AZT, some of the damage to the mitochondria can be prevented (28). Other nutrients, i.e., selenium, copper, iron, and provitamin A (carotene), that function in the free radical scavenging system may also have a beneficial effect, but these have not been studied extensively.

AZT also has effects on mt OXPHOS. A dose-dependent inhibition of state-3 succinate-supported respiration has been reported, as has an inhibition of the hydrolytic activity of the H^+-ATPase (30). The evidence that mt DNA was attacked by AZT-generated free radical damage is lacking; however, the facts that malondialdehyde and the end product of guanosine oxidation appeared in the urine and that respiration was impaired suggest that AZT attacked the mitochondria as well as had effects on the virus.

OXPHOS Stimulants/Inhibitors

The respiratory chain is described in Chapter 1 of this volume. In brief, it consists of four respiratory complexes (I, II, III, and IV) that work to pass reducing equivalents down the chain to be joined with oxygen to make water. In the process, energy is generated that is captured by the F_1F_0ATPase and used to synthesize ATP. Substrates that donate their reducing equivalents to complex I result in the synthesis of three molecules of ATP. Most of the metabolites of glucose and amino acids fit this description. Metabolites that donate their reducing equivalents to the FAD-linked complex II result in the synthesis of two molecules of ATP. A few metabolites donate their reducing equivalents to complex III, resulting in formation of one ATP. The details of this coupling of respiration and ATP synthesis are described in Chapter 1. Suffice it to say that drugs and hormones, as well as nutrients, affect the activity of the respiratory chain as well as its coupling to ATP synthesis. These are listed in Table 9.2.

Table 9.2   OXPHOS Stimulators/Inhibitors

Adriamycin (doxorubicin)	Inhibitor
Sulfonamides, i.e., tolbutamide	Enhancer
Metformin	Enhancer
Thiozolidines	Enhancer
Aspirin	Enhancer
Ethanol	Decreased coupling
Statins	Inhibits production of ubiquinone
Thyroid hormones	Enhancer
Low-protein diet	Reduces OXPHOS function
Essential fatty acid-deficient diets	Uncouples OXPHOS
Sucrose-rich diets	Mildly uncouples OXPHOS
Iodine-deficient diets	Reduces OXPHOS function
Copper-deficient diets	Reduces OXPHOS activity
Menhaden-oil diets	Increases coupling
Arachidonic acid	Inhibitor
Exercise	Stimulates OXPHOS
Glucagon	Stimulator
DHEA	Stimulator *in vivo*, inhibitor *in vitro*
Glucocorticoids	Stimulator
General anesthetics	UNCOUPLER OF OXPHOS
Impramine	UNCOUPLER OF OXPHOS
Theophylline	Inhibitor of respiratory chain
DDT	Inhibitor
Methamphetamine, methylenedioxymethamphetamine	Inhibitor

DRUGS

Adriamycin

The anthracycline antibiotic adriamycin is widely used to treat a broad range of malignancies (31). Unfortunately, it cannot be used over the long term because it results in specific cardiotoxicity, which is cumulative and total dose dependent. Among the early changes in heart tissue are changes in the morphology and function of the mitochondria. In studies of rats treated with adriamycin, mt OXPHOS was impaired at complexes III and IV (32). Complexes I and II were minimally affected. Subsequent studies revealed that adriamycin could be found bound to cardiolipin with a fixed stoichiometry of two molecules of the drug to one molecule of cardiolipin

(33,34). Cardiolipin is a phospholipid found in the inner mt membrane.

Adriamycin also binds mt RNA and this binding can account for a reduction in mt protein synthesis. The binding of adriamycin to cardiolipin results in a peroxidation of this lipid with subsequent alteration in the inner membrane function and mtDNA-directed protein synthesis. The peroxidation is enhanced in copper-deficient rats, suggesting that the Cu–Zn superoxide dismutase may play a role in determining the cumulative dose at which adriamycin induces cell injury via peroxidation (35).

Adriamycin inhibits the cardiac mt calcium pump, probably due to its effect on OXPHOS and ATP production (36). ATP is needed for this calcium pump. If calcium accumulates in the mitochondrion, mt death occurs. The mechanism of action of adriamycin also includes a modification of mt sulfhydryl groups (37) and an induction of riboflavin deficiency (38). Riboflavin is the major component of FAD, a coenzyme importantly involved in the respiratory chain as well as in the appropriate function of the α-glycerophosphate shuttle. This shuttle functions in the transport of reducing equivalents from the cytosol to the mt compartment and its function is impaired in riboflavin deficiency (39). Reducing equivalents brought into the mt compartment via this shuttle enter the respiratory chain via the FAD-linked complex II.

Hypoglycemic Agents

A number of oral glucose-lowering drugs have been developed to help the person with type 2 diabetes mellitus control blood glucose levels. Although tolbutamide (a sulfonourea), metformin (a biguanide), and the thiazolidines (glitizones) act by different pathways, each is effective in lowering blood glucose. Tolbutamide is the oldest of these drugs. It works by increasing the oxidation of glucose to CO_2 and water. In the presence of insulin, it reduces the production of lactate, suggesting that its effect is on mt OXPHOS. When OXPHOS is aberrant, lactate levels rise. Tolbutamide, in the absence of insulin, has the reverse effect in brown fat cells. In these cells, tolbutamide alone increases uncoupling of ATP synthesis from respiration (uncouples respiration from ATP synthesis).

Different cell types respond differently. Liver slices, for example, did not respond to tolbutamide in the presence of insulin in the same way as white or brown fat cells. In the white fat cells, respiration as well as glucose oxidation, was enhanced by tolbutamide in the presence of insulin. This was not observed in the liver slices. Another hypoglycemic drug, phenformin, had no effect on mt metabolism at all. It served to lower blood glucose by enhancing insulin's effect on glucose transport into cells (40). Metformin, on the other hand, increased mt fatty acid oxidation, lactate production, mt mass, and basal lipolysis (41). It stimulated complex II respiration and had an overall stimulatory effect on catalysis.

Troglitazone, one of the thiazolidinediones, also improves insulin action and suppresses hyperlipidemia and glucose utilization (41). The glitazones, in general, are active stimulators of the peroxisome proliferator-activated receptor (PPAR)-γ; they have a gene regulatory function. Abbreviated PPAR-γ seems to stimulate the transcription of a number of nuclear and mt genes involved in lipid and carbohydrate metabolism. Troglitazone treatment of cultured cells resulted in an increase in mitochondrial mass yet inhibited aerobic metabolism and basal lipolysis (41).

Troglitazone has also been shown to suppress thermogenesis in isolated fat cells, suggesting that this drug could improve mt coupling of respiration to ATP synthesis (41,42). This hypothesis was denied in a subsequent study using HepG2 cells (43). When exposed to troglitazone, these cells died; total mortality occurred within 5 hours. The effect was time and dose dependent. One must note here that HepG2 cells are not normal cells. They are derived from a hepatoma and, as such, have poor mt function even in the absence of drugs. Because hepatotoxicity in humans has been reported, troglitazone was withdrawn from the market. Other glitazones remain in use.

NSAID

Aspirin is one of the oldest drugs on the market. It is a common over-the-counter, nonsteroid, anti-inflammatory drug (NSAID). In addition to relieving inflammation, the NSAIDs also relieve the pain of headache and have effects on intermediary metabolism. Aspirin lowers blood glucose and significantly increases oxygen consumption by isolated mitochondria (44). ATP synthesis is suppressed under the same treatment conditions that increase oxygen consumption, so the assumption is that aspirin increases uncoupling. It also increases gluconeogenesis and decreases glycogen levels in the liver.

The role of aspirin is not direct with respect to its effect on OXPHOS but is via its conversion to salicylic acid (45). In some individuals, aspirin sensitivity appears and, if it is ignored, Reye's syndrome develops. This syndrome is characterized by acute encephalopathy, fatty degeneration of the liver, and mitochondrial dysfunction. The dysfunction is as described earlier: an increase in respiration without a corresponding increase in ATP synthesis.

Aspirin, as well as a number of drugs of the phenylproprionic acid family of drugs (e.g., ibuprofen [Advil, Motrin, Nuprin]) and of the pyrrole acetic acid family (e.g., indomethecin [Indocin]), inhibits the cyclooxygenase reaction that is the first step in the production of eicosinoids. In so doing, these drugs reduce inflammation, pain, and blood clotting. None of these functions involves the mitochondria directly.

Ethanol

In the context of this chapter, ethanol is regarded as a drug rather than as a dietary ingredient or beverage. Chronic excessive ethanol consumption has deleterious effects on the liver, heart, skeletal muscle, pancreas, brain, and other tissues (46) and results in large, poorly organized mitochondria with decreased function. Coupling of respiration to ATP synthesis is insufficient (47–49), leading to a disordered lipid metabolism. Lipids accumulate in the liver and it is not uncommon to observe elevated serum lipids as well.

Gluconeogenesis is inhibited when ethanol is added to the incubation medium of isolated hepatocytes (50). Mt superoxide dismutase is elevated in monkeys fed ethanol; this suggests that ethanol stimulates free radical production that, in turn, probably damages membranes and DNA (51). Derangement in membrane function has been reported in animals fed ethanol. The effects of ethanol are likely mediated by free radicals. Ethanol also has a direct effect on the proteins of cell membranes and is known to precipitate proteins in solution or suspension.

Statins

The discovery and development of HMG-CoA reductase inhibitors blazed a new path in the control of blood cholesterol (52,53). According to current thinking, reducing the level of circulating cholesterol through inhibiting *de novo* synthesis of cholesterol reduces the risk of coronary vessel disease. A number of these statin drugs are currently on the market: mcvastatin, lovostatin, simvastatin, and pravastatin (trade names: Lipator, Zocor, etc.). All of these drugs have a doubling-ring structure analogous to rings A and B of cholesterol. On the second ring, another five-carbon open ring or a carbon chain can link this ring to another ring. On the first ring, a four-carbon chain can be linked to the ring by an oxygen atom.

In any case, these structures serve as competitive inhibitors of HMG-CoA reductase, the rate-limiting enzyme in cholesterol synthesis pathway. This means that the product of this reaction (mevolonate) is not produced in as great a quantity for use in the subsequent synthesis of ubiquinone. The product, ubiquinone, is an essential component of the respiratory chain and in the chain is called coenzyme Q_{10}. Over time, chronic treatment with these drugs results in a short supply of CoQ_{10} with the consequence of reduced mt respiration and perhaps mt and cell death (54). This probably explains the side effects of the statins in some individuals. Muscle degeneration is a characteristic sign of mt dysfunction and loss. Loss of mitochondria in the brain is also a possibility, with subsequent losses in neural function.

HORMONES

Thyroid Hormone

Thyroid hormone has a number of effects on mitochondria. It stimulates the initial steps of gluconeogenesis (55,56) by increasing the activity of the malate–aspartate shuttle (57,58). Some of the thyroid hormone effect on cellular metabolism is through its influence on OXPHOS (59–74). Respiration is increased (as is ATP synthesis) in hyperthyroid animals and decreased in hypothyroid animals. In turn, these effects can be explained by its effect on the lipid composition of the mt membranes as well as by an effect on mt and nuclear gene expression (75–84). Hypothyroid rats have significant increases in membrane cardiolipin and more linoleic but less arachidonic acid in their phospholipids (75,76). In another study, the reverse was reported (77). In this report, total cholesterol increased (77) and phospholipids decreased. It should be noted that the mt membranes contain far less cholesterol than does the plasma membrane. Nonetheless, these changes should have effects on mt membrane fluidity that, in turn, should affect the activity of a number of membrane-associated carriers and enzymes.

Hypothyroidism also results in a decrease in the activity of the microsomal desaturases (78); this can explain the altered ratio of polyunsaturated fatty acids to saturated fatty acids (the P/S ratio). The fatty acid composition of the diet fed the hypothyroid rats can determine the effect of hypothyroidism on the membrane fatty acid composition (79). A diet supplemented with unsaturated fatty acids reversed the hypothyroid effect on the membranes. With this diet, there was little need for the desaturases, so the thyroid deficiency effect on the membrane P/S ratio was bypassed.

Hyperthyroid rats, in contrast, show an increase in the oxidation of fatty acids and a decrease in triacylglycerol formation from palmitate and oleate (80). In these rats, hepatic membrane cholesterol content decreases and phospholipid content increases (81). The phospholipids of the hyperthyroid rats had an increase in the conversion of linoleic acid to arachidonic acid (81). In contrast, brain mitochondrial membranes had no significant change in cholesterol content, but phosphatidylethanolamine increased at the expense of phosphatidylserine, phosphatidylinositol, and phosphatidylcholine. Mitochondrial membrane fluidity (82) also increased. As mentioned, these changes in fluidity could affect the activities of membrane-associated systems. Indeed, it has been reported that the microenvironment of the α-glycerophosphate dehydrogenase was dramatically influenced by thyroid status as was the tricarboxylate carrier (83,84).

The components of the respiratory chain and the F_1F_0ATPase are embedded in membrane and, as could be anticipated, OXPHOS is increased in hyperthyroid animals and decreased in hypothyroid animals (59,84–89). Some of the thyroid hormone effect is, as described previously, an effect on membrane fluidity and some an effect on the activity of the shuttle systems responsible for the entry of reducing equivalents into the mitochondrial compartment for use by the respiratory chain. One of these shuttles is the α-glycerophosphate shuttle, which is increased with thyroid treatment because its rate-limiting component, the mt FAD linked α glycerophosphate dehydrogenase, is induced by the hormone. There is more of this enzyme and it is more active (57,83–90).

In addition, thyroid hormone increases the activity of the adenine nucleotide translocase (60,62,91) that transports ATP out of the mitochondrial compartment in exchange for ADP, which is then used to synthesize more ATP. Proton permeability of the mt membranes is also increased in hyperthyroid tissues partly due to the previously described change in fatty acid composition and subsequent fluidity and partly due to differences in the proton flux/mass protein ratio (92).

Membrane permeability differences due to thyroid hormone were also reported with respect to the permeability of ions (K^+, Ca^{++}, phosphate). Mitochondria from hyperthyroid rats showed higher rates of K^+ and phosphate accumulation and required far more Ca^{++} to uncouple OXPHOS and cause mt death (93–96). Hyperthyroidism induces an mt permeability transition that is mediated by reactive oxygen species and membrane protein thiol oxidation (96). Overall, thyroid status affects mitochondrial function by affecting the availability of substrates and the supply of reducing equivalents, by altering composition of the mt membrane, and by altering the use of nutrients that supply the substrates.

Nutrients

Nutritional status can have profound effects on mt function (97–114). Low-protein diets (97), coconut-oil (essential fatty acid-free) diets (98), sucrose-rich diets (99), and iodine-deficient diets (100) have been shown to reduce mt function vis-a-vis OXPHOS. Essential fatty acid-deficient diets reduce membrane fluidity and OXPHOS is somewhat uncoupled. Iodine deficiency results in a deficiency of thyroid hormone synthesis with the result of hypothyroidism as described earlier.

Low-protein diets affect the synthesis of proteins essential to the appropriate functioning of mitochondria. Not all proteins, however, are affected. There seems to be a hierarchy of proteins whose synthesis is impaired; shuttle systems are more readily affected than are proteins of the respiratory chain. High-protein diets result in a proliferation of mitochondria with a parallel increase in gluconeogenesis, urea synthesis, and cytochrome oxidase activity (101). Protein-free feeding of neonatal rats resulted in a significant decrease in OXPHOS and cytochrome and ATPase activities (102).

Copper deficiency results in a reduction of the copper–zinc superoxide dismutase and this, in turn, results in peroxide-induced damage to the mt membrane and perhaps proteins of OXPHOS (103,104). Copper is an essential element in cytochrome c oxidase and one would expect a reduction in OXPHOS in copper-deficient animals. Such has been reported (105,106).

The type of fat fed to an animal also affects OXPHOS. As described previously, hydrogenated coconut oil feeding results in less efficient ATP synthesis (98) and feeding menhaden oil has the reverse effect (107,108). Menhaden oil is rich in long chain polyunsaturated fatty acids and mt membranes of rats fed this oil are more fluid than when rats are fed more saturated fat diets. In the long term, however, this increase in saturation results in formation of more free radicals (109) and subsequent damage to vital organs, in turn reducing life span (110). Studies of specific fatty acids and OXPHOS showed that arachidonic acid irreversibly inhibited complex I activity while having no effect on the other respiratory complexes (111,112). *In vitro*, free fatty acids decouple OXPHOS by dissipating intramembranal protons without inhibiting ATP synthesis driven by the proton electrogradient (112,113).

Exercise

Rats forced to exercise on a treadmill or forced to swim had an increase in mitochondrial number as well as in mitochondrial enzyme activity and OXPHOS (115,116). Muscle mitochondria from endurance-trained rats generated normal transmembrane potentials, ADP/O ratios, and respiratory control ratios (state 3:state 4 respiration) (115). In addition, exercise induces a coordinate increase in muscle oxidative capacity, in insulin sensitivity, and in the expression of some of the mt and nuclear genes involved in mt gene expression (117). Exercise also increases the formation of free radicals that can have deleterious effects on mitochondrial function in the long term (118,119). Studies of the longevity of exercised and sedentary rats showed that exercised rats had a shorter life span than did sedentary rats.

Other Hormones

Glucagon stimulates α-ketoglutarate metabolism in isolated mitochondria (120) as well as stimulates OXPHOS (121). It stimulates the electron transport chain (122) in much the same way as epinephrine. DHEA, a steroid hormone intermediate, inhibits respiration *in vitro* while enhancing it *in vivo* (123). DHEA is a precursor of the steroid sex hormones. The glucocorticoids also have effects on OXPHOS *in vivo* (124). Hepatic mitochondria prepared from glucocorticoid-treated and control rats showed that this hormone treatment resulted in an increase in OXPHOS activity. More ATP was produced, more ATP was exchanged for ADP, and intramitochondrial concentrations of magnesium and potassium (124) increased.

Miscellaneous Compounds

General anesthetics uncouple OXPHOS (125) as does imipramine (a tricyclic antidepressant) (126) and doxorubicin (adriamycin) (127). Doxorubicin also damages the NADH oxidation pathway, resulting in an inactivation of the respiratory chain. Theophylline also reduces the activity of the respiratory chain (128). Lastly, toxins such as the insecticide DDT (129) and methamphetamine or 3,4-methylenedioxymethamphetamine (MDMA or Ecstasy) (130) depress OXPHOS, suppressing the activity of the respiratory chain and of the F_1F_0ATPase.

HORMONES AND NUTRIENTS IN
MITOCHONDRIAL GENE EXPRESSION

The mechanisms for mitochondrial gene expression and its regulation are reviewed in Chapter 3. Transcription depends on a number of nuclear-encoded proteins. Among these are the receptor proteins that bind to specific elements in the displacement loop (D-loop), mt transcription factor A (mtTFA), mt termination factor (mtTERF), DNA polymerase, RNA polymerase, and RNA processing enzymes. Transcription is initiated within the D-loop that serves as the common promoter region for the entire genome. In addition to mtTFA and mtTERF, nutrients and hormones have been shown to influence mt gene transcription and mt gene expression (listed in Table 9.3). In instances when low levels of mtTFA are present, low numbers of mitochondria are also present (131).

The expressions of the OXPHOS genes and of mitogenesis are not isolated events. Because the genes that encode the components of OXPHOS occur in both genomes, the coordinated expression is a necessity. The nuclear respiratory factors (NRF) 1 and 2 and the general transcription factor Sp1 play roles in this coordination by simultaneously regulating mtTFA (132–137). Mitochondrial TFA and number increase during rapid growth and parallel the increased expression of other genes whose products are involved in bioenergetics. Heddi et al., for example, have shown that all

Table 9.3 Hormones and
Nutrients That Affect
Mitochondrial Gene
Expression

Growth factors
Exercise
Thyroid hormone
Glucocorticoids
Insulin
Vitamin A (retinoic acid)
Vitamin D
Vitamin B$_{12}$
Iron
Peroxisomal proliferators

of the OXPHOS genes are expressed at once (138). In this study, these workers showed that patients with mitochondrial DNA mutations overexpressed all of these genes (perhaps as a compensatory mechanism) to overcome the effects of the mutation on OXPHOS. Increased expression of the following was overexpressed in these patients:

*   Nuclear-encoded creatine kinase
*   Several subunits of the F$_1$ATPase
*   The adenine nucleotide translocase
*   Glycogen phosphorylase
*   Hexokinase I
*   Phosphofructokinase
*   The E$_1$ α subunit of pyruvate dehydrogenase
*   The ubiquinone oxidoreductase

The mechanism of this overexpression may be due to redox-sensitive transcription factors, although other explanations are possible as well.

Growth (Factors That Stimulate Growth)

Mt gene expression is regulated by growth and development in a tissue-specific manner. In the heart, regulation occurs at the level of transcription whereas, in the liver, mRNA stability and translational efficiency regulate gene expression during growth (139,140). In rat heart, mitochondrial transcripts increased between 1 day and 3 months of age and then decreased between 3 months and 18 months of age. In neonatal liver, the half-life of mt-mRNA was much longer than that in adult liver and translational efficiency peaked 1 hour after birth. This response to growth was also seen in nuclear-encoded components of OXPHOS (141,142).

The mechanism for this post-transcriptional regulation involves the nuclear-encoded F_1 ATPase β-subunit (143,144). The 3′ untranslated region (UTR) of this gene contains a translational enhancer that functionally resembles an internal ribosome entry site. During fetal development, a protein (3′ βFBP) binds this enhancer and inhibits translation. Within 1 hour of birth, this protein no longer binds the 3′ UTR, unmasking the enhancer. This produces a spike in $βF_1$ ATPase protein levels. In the adult, this protein is present again and its translation is once again inhibited.

Exercise

Another treatment that results in an increase in the number of mitochondria is exercise. The oxidative capacity of mammalian striated muscle can vary markedly over nearly a tenfold range (145), reflecting major differences in the expression of mitochondrial and nuclear genes that encode the components of OXPHOS. Red muscle contains more mitochondria than does white muscle or mixed muscle. As a result, these different muscle types have different numbers of mitochondria. However, when stimulated by exercise, all muscle types responded similarly. All increased their mitochondrial number and their oxidative capacity (146). Similarly, exercise tolerance is variable and seems to be related to mt sequence, especially that of the D-loop (147,148).

Exercise intolerance is a feature of individuals having mutations in the cytochrome b gene in the mtDNA (148). Exercise appears to reverse the age-related loss in mt function and is also associated with fewer mtDNA deletions in rats studied from weaning to 22 months of age (149). As has been pointed out in Chapter 6, age-related changes in mtDNA and mt function can be related to oxidative stress. Detailed studies of cells subjected to oxidative stress showed that these cells increase mt mass and mtDNA content as a way of compensating for the age-related loss in function (150).

Exercise tolerance can be affected by the health status of the subjects. Those with HIV treated with protease inhibitors are less tolerant of exercise than control subjects without HIV. Protease inhibitors have the side effects of lipodystrophy and lactic acidosis and it is thought that they may target mitochondria as well as the HIV virus (151).

Endurance training induces muscle-specific adaptations in mt function that include a tighter coupling of OXPHOS as well as increases in mt mass, number, and mt protein synthesis (152,153). Exercise results in an increase in mtTFA and that in turn elicits an increase in mt number and mt transcripts (154). There is a direct correlation between mtTFA and mt transcripts as well as mt gene expression (155–158).

Exercise, especially strenuous exercise, elicits a whole cascade of hormonal and metabolic adaptations (115,117,119,147). The glucocorticoids, the catecholamines, glucagon, insulin, thyroid hormone, and growth hormone are involved. Each of these hormones acts not only on metabolic fuel flux but also on the expression of genes in the nuclear and mitochondrial compartment that contribute to oxidative capacity. In turn, this is linked to an increase in mt number. Hormones as mt gene transcription agents have not been well studied, but some evidence indicates that mt gene transcription is under the influence of several hormones.

HORMONES

Thyroid Hormone

As described in the preceding section on factors that influence OXPHOS, the thyroid hormones have several mechanisms whereby they influence mitochondrial function. Of interest in this section is their role in mt gene expression. In 1990, Leung and McKee reported that thyroid hormone is important to mt protein synthesis (159). They used the model of cardiac hypertrophy to show that thyroid hormones (mainly T_3) stimulated the synthesis of mitochondrial proteins. Those encoded by the nucleus and those encoded by the mt genome were found to be increased with thyroid treatment. Earlier, Wooten and Cascarano reported that thyroid hormone stimulated mitogenesis that, in turn, led to cellular hyperplasia (160).

Thyroid hormone stimulates OXPHOS, mitogenesis, and mt protein synthesis; it also stimulates the synthesis of uncoupling proteins (UCPS) located in the mitochondrial matrix (160–164). Again, this action of thyroid hormone is related to the regulation of energy balance not only by the cell but also by the whole body (165). The mechanism of action of thyroid hormone on UCP synthesis relates to its action as a stimulator of nuclear UCP gene transcription. Thyroid hormone binds to its cognate thyroid hormone receptor protein (v-erb A); this in turn binds to its element in the nuclear DNA. In so binding, it stimulates transcription along with other transcription agents (166). Nuclear thyroid hormone binding can be inhibited by fatty acids, especially oleic acid (167). Hypothyroidism results in a reduction in mt mass and mt number as well as in the amount of $F_1F_0ATPase$ without having an effect on the expression of the mt-encoded subunits of this complex (168).

Thyroid hormone also binds to a receptor protein found in the mt matrix that binds to the mtDNA (169–171). This receptor protein, c-erb Aα1, is smaller than the nuclear receptor (c-Erb Aβ1, β2, and β3) protein and has some parts that are identical to it. Two mt proteins bind thyroid hormone: one in the inner membrane and one in the matrix. The matrix protein binds to mtDNA in the D-loop and, at the time it was reported, was thought to be the first receptor protein to exist in the mt matrix (170). DNA-binding experiments showed that the 43-kDa protein binds specifically to four mtDNA sequences. These base sequences are similar to the nuclear T_3 response elements.

The interaction of T_3 and v-erb A stimulates myogenic differentiation. T3 stimulates the production of its binding protein and v-erb A stimulates T_3 activity (172). These data from myoblast cultures show that changes in mt activity occurred prior to myoblast terminal differentiation and could be involved in the process of myogenesis. Enriquez et al. have shown that *in vivo* and *in vitro* T_3 regulates transcription of the mt genome. It modulates mt RNA levels and the mRNA/rRNA ratio by influencing transcription rate (173).

In so doing, T_3 has a profound effect on cellular respiration (174). If the thyroid hormone receptor is aberrant due to a mutation in the nuclear gene that dictates its amino acid structure, there will be resistance to the stimulatory effect of the hormone on gene expression (175). The thyroid hormone receptor requires zinc and, if the animal is zinc deficient, the receptor protein will not be functionally active (176). This is true for many of the DNA-binding proteins that contain zinc fingers. Whether all of the DNA-binding receptor proteins in the mt matrix require zinc has yet to be shown. Actually, receptor protein identification is at a primitive state at this time. Future research should be directed towards the identification of additional transcription agents in this compartment.

Glucocorticoids

Although T_3 was the first hormone to be shown to affect mt gene expression, the adrenal steroid, glucocorticoid, was soon discovered to have this effect also. These two hormones were known to interact in the expression of the nuclear genome by activating protein receptors; thus, it was only natural to explore their possible roles in mt gene expression. Demonacos et al. proposed that a glucocorticoid-binding protein existed in the mt matrix (177) and that a response element for this receptor was on the D-loop.

These receptors are nuclear encoded and their synthesis is stimulated by glucocorticoid. Upon synthesis, they rapidly migrate into the mt compartment and bind to putative GR elements (GREs). Glucocorticoids have been shown to stimulate mitogenesis and mt-encoded respiratory chain proteins (178). ACTH, the hormone that stimulates glucocorticoid synthesis and release, seems to play a role as well in mt gene expression (179). It is not known whether this effect is direct or indirect because no one has isolated and identified an mt receptor protein or a response element for this hormone.

Insulin

People with mt disease frequently have disordered glucose metabolism (see Chapter 8). In part, this is due to the important role mitochondria play in the synthesis and release of insulin by the pancreas (180–183). This process is very dependent on adequate supplies of ATP produced by OXPHOS. The synthesis of insulin requires a ready supply of ATP and the release of insulin is via an ATP-dependent process. In addition, insulin plays a role in mt gene expression (184). An insulin response element and an insulin response protein have been imputed from studies of human skeletal muscle. Putative receptors for estrogen have been suggested, but studies that show a role for estrogen in mt gene expression are lacking.

NUTRIENTS

Vitamin A (Retinoic Acid)

Countless genes are regulated by retinoic acid (RA) bound to one of its DNA-binding receptor proteins. A small sampling of these genes is shown in Table 9.4. Several key enzymes in intermediary metabolism are found in this list. The nuclear retinoic acid receptors (RAR and RXR) are similar to the nuclear receptors for the steroid hormones (185,186). Both are members of the so-called super family of steroid receptors. There are three RAR and three RXR genes, designated α, β, and (γ. Each gene has several isoforms that arise from different promoter usage or alternative splicing (187,189–191).

The differences between the receptors occur in their N-terminal region (192). In addition, they have different expression patterns in the embryo and the adult and are regulated differently by vitamin A (193,194). In vitamin A-deficient rats, mRNA for RAR-β is decreased with no effect on mRNA expression of RARα or RARγ in lung, liver, and intestine. In the

Table 9.4 Gene Products
Affected by Retinoic Acid

PEPCK
Glucokinase
$F_1F_0ATPase$
Ornithene decarboxylase
Fructose 2,6-bisphosphatase
Growth hormone
Alcohol dehydrogenase
Insulin
Osteocalcin
Calbindin
Retinoic acid-binding proteins

testes, $RAR\alpha_2$ expression was higher in retinol-deficient rats than in retinol-sufficient rats, and increased with RA feeding. RA also increased $RAR\alpha_2$ expression in the embryo. In addition, RA up-regulates $RAR\alpha_1$ mRNA in the adult testes, but not lung, liver, intestine, or embryo; $RAR\beta_1$ and $RAR\beta_2$ in adult lung, liver, intestine, and embryo, but not testes; and $RAR\gamma_2$ in adult lung, liver, and intestine, but not embryo.

The complex distribution and regulation of these receptor isoforms has led to the hypothesis that each isoform has a specific function. Studies with $RXR\alpha$ knockout mice have revealed that genes involved in energy production, including those encoded by mtDNA, are regulated during embryonic heart development and that $RXR\alpha$ is essential for energy production in the embryonic heart (195). Studies with isoform specific null mutant mice have shown that the signs of vitamin A deficiency are also seen in RAR/RXR single or double null mutant mice (187,196–199).

Nonetheless, redundancy occurs. In some cells types, single receptor isoform deletion mutations had little or no apparent effect (200). Although the exact function of each isoform is still not known, it is clear that RAR/RXRs play a critical role in the function of vitamin A in gene expression. There are several reports of the presence of RAR γ in the mt compartment that strongly suggest a role for RA in mt gene expression (201–203). This isoform was not found in the nuclear compartment.

Retinoic acid also plays a role in mt gene expression as alluded to earlier. It appears to affect the expression of mt genes for NADH dehydrogenase subunit 5 (204), cytochrome c oxidase subunit I mRNA and 16srRNA (203), and the ATPase 6 gene (201,202). It appears that an interaction occurs between retinoic acid and thyroid hormone vis-a-vis mt gene expression.

In primary hepatocytes cultured in a serum-free medium, it was found that protein expression of the mt-encoded ATPase subunit a could be regulated by DHEA (a PPAR-α ligand) or RA alone, but not T_3 alone (205). Interestingly, T_3 synergistically enhanced the DHEA and RA effects. Combining DHEA and RA reduced ATPase subunit a (ATPase 6) gene expression. These data are consistent with the hypothesis that mt-RAR and mt-PPAR compete for binding to the TR/p43 and the DR2 response element in the mtDNA promoter. Binding studies are needed to confirm this explanation.

Retinoic acid works to stimulate mt transcription directly and indirectly through an increase in nuclear-encoded mt transcription factors, i.e., mtTFA (201). In turn, this drives mitogenesis so that an increase in the number of mitochondria can be observed (201).

Vitamin D

Vitamin D serves as an enhancer of the transcription of calcium-binding proteins (206). Unlike retinoic acid, however, its genomic action is tissue specific. A putative response element for the vitamin D receptor has been identified in the D-loop of the mtDNA; however, not all cell types have their mt transcription enhanced by this vitamin. The difference has to do with whether there is a cytosolic vitamin D carrying protein in the cell. Those cells that have one, i.e., bone cells or kidney cells, also appear to bind the vitamin in the mitochondria (206,207). Its function in that organelle has not been completely elucidated. In intestinal cells, the addition of vitamin D to the incubation media up-regulates the expression of six of the mt genes. Studies of the relationship of this up-regulation and mt function have not been conducted.

Vitamin B$_{12}$

Vitamin B$_{12}$ is an important coenzyme in pyrimidine and purine synthesis that also plays a role in mt mRNA synthesis. Reports of decreased transcription activity of the mt genome in vitamin-deficient cells have been made, although OXPHOS seems to be unimpaired (208). Probably, the effect of a deficiency of this vitamin relates to its role in the synthesis of the rapidly turning over RNA. This places a high demand on purine and prymidine synthesis.

Iron

Living mitochondria incubated with the ferrous ion show major changes in the conformations of mt DNA as assessed by southern blotting (209). High-iron loads result in strand breaks in the mt genome, just as in nuclear DNA. Iron plays a role in lipid peroxidation and thus iron overdose can result in damage to the genome (210,211). Iron has a role in heme synthesis and, in one part of heme synthesis, iron liganded to ATP is transferred to an mt receptor (212). Similarly, the synthesis of the cytochromes in the respiratory chain also depends on the presence of iron. Because iron is incorporated into the cytochrome structure, iron deficiency is characterized by an impaired control of mt respiration (213). At this time, no evidence suggests an mt iron receptor or a role of iron in mt transcription or translation.

Peroxisomal Proliferators

Wu et al. found that PPAR γ coactivator-1 (PGC-1) can increase nuclear mRNA levels for NRF-1 and 2 as well as act as a coactivator for NRF-1 on the nuclear mtTFA promoter (214). PGC-1 can be induced by cold and refeeding and is thought to play a key role in thermogenesis. It was originally found to coactivate PPAR-γ on the uncoupling protein 1 promoter in brown fat cells (215). PGC-1 binds the DNA binding/hinge domain of PPAR-γ in a ligand-dependent manner. It also binds the thyroid receptor (TR), estrogen receptor (ER), and retinoic acid receptors (RARs) in varying degrees of ligand dependence. Thus, PGC-1 can coordinate thermogenesis by the regulation of nuclear and mitochondrial OXPHOS genes via NRFs as well as the uncoupling proteins via PPAR-γ and TR.

Transcription is not always coordinated between the nucleus and mitochondria. Several researchers failed to find coordinated transcription in various thyroid hormone states (216–222). Peroxisomal proliferators such as Clofibrate, per-fluorooctanoic acid, and acetylsalicyclic acid alter mRNA levels for mt genes cytochrome c oxidase (subunit 1) and subunit 1 of the NADH dehydrogenase. The mRNAs for adenine nucle-aotide translocase were also selectively increased as was that for malic enzyme. In this instance, these drugs mimicked the effects of thyroid hormone on mt gene expression (223).

In addition to regulation during mitochondrial proliferation and differentiation, mitochondrial transcription may also be directly regulated by hormone receptors (173,224). Figure 9.1 shows where putative glucocorticoid (GRE), vitamin D (VDRE), thyroid hormone (TRE), insulin (IRE), and retinoic acid (RARE) response elements have been found in the D-loop (173,177,180,184,225–230). The glucocorticoid receptor (GR), a variant of the thyroid receptor (TR/p43), and a truncated PPAR-γ (mt-PPAR) have been shown to bind this region of mtDNA *in vitro* (177,231). Casas et al. reported the existence of a 45-kDa protein related to a peroxisome prolif-erator γ_2 in the mitochondrial matrix. They found that this protein bound to the D-loop complexed to the thyroid-binding protein (231). As mentioned earlier, thyroid hormone increases mt transcription through binding to the thyroid binding protein bound to the D-loop; it also affects mitochon-drial transcription by increasing mtTFA (232).

Casas et al. reported that mt-PPAR bound this DR2 in a complex with TR/p43 (231). This element has also been shown to act as an RARE in other genes. Thus, it is possible that retinoic acid bound to its receptor could also directly regulate mitochondrial transcription. It should be noted that the mt-PPAR lacks the C-terminus ligand-binding domain. Mitochondrial PPAR gene expression is regulated by fenofi-brate and fasting in a PPAR-α-dependent manner (231).

mtDNA D-loop

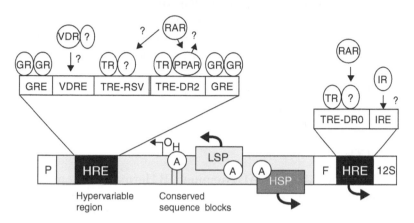

Figure 9.1 The region of the D-loop that contains putative hormone response elements is shown as an expansion of the area upstream of the conserved sequence blocks. There are unknown elements as well.

Therefore, mt gene expression may be regulated with a PPAR-γ-like mechanism, but by PPAR-α-ligands in a nucleus-dependent manner. This mt-PPAR still retains the binding site for PGC-1.

In addition, oxidative stress may serve as a signal for mt biogenesis (232). With oxidative damage to the mitochondria, there may be some way of signaling the nucleus to turn on genes that are important to the manufacture of more mitochondria. The circuitry for this signaling system has not been worked out; however, it does seem logical to suppose that some sort of crosstalk exists between the compartments to sustain mt function.

In primary hepatocytes cultured in a serum-free medium, it was found that protein expression of the mt-encoded ATPase subunit a could be regulated by DHEA (a PPAR-α-ligand) or RA alone, but not T_3 alone (205). Interestingly, T_3 synergistically enhanced the DHEA and RA effects. Combining DHEA and RA reduced ATPase subunit a expression. These data are consistent with the hypothesis that mt-RAR and mt-PPAR compete for binding to the TR/p43 and the DR2 response element in the mtDNA promoter. Binding studies are needed to confirm this explanation. Thus, evidence suggests direct regulation of mt gene transcription by hormones (thyroid hormone, insulin, glucocorticoids, estrogen) and vitamins (vitamins A and D). Some of these actions may involve peroxisome proliferator-activated receptors as well as cognate receptors for the individual compound.

REFERENCES

1. TL Dawson, GJ Gores, AL Nieminen, B Herman, JJ Lemasters. Mitochondria as a source of reactive oxygen species during reductive stress in rat hepatocytes. *Am J Physiol* 264: C961–967; 1993.

2. RS Sohal. Mitochondria generate superoxide anion radicals and hydrogen peroxide. *FASEB J* 11:1269–1270; 1997.

3. M Glinn, L Ernster, CP Lee. Initiation of lipid peroxidation in submitochondrial particles: effects of respiratory inhibitors. *Arch Biochem Biophys* 290:57–65; 1991.

4. B-O Lund, DM Miller, JS Woods. Studies on Hg(II) induced H2O2 formation and oxidative stress *in vivo* and *in vitro* in rat kidney mitochondria. *Biochem Pharmacol* 45:2017–2024; 1993.

5. JH Karam, PA Lewitt, CW Young, RE Nowlain, BJ Frankel, H Fujiya, ZR Freedman, GM Gradsky. Insulinopenic diabetes after rodenticide (Vacor) ingestion. A unique model of acquired diabetes in man. *Diabetes* 29:971–978; 1980.

6. MD Espositi, A Ngo, MA Myers. Inhibition of mitochondrial complex I may account for IDDM induced by intoxication with rodenticide Vacor. *Diabetes* 45:1531–1534; 1996.

7. CB Wolheim. Beta-cell mitochondria in the regulation of insulin secretion: a new culprit in type 2 diabetes. *Diabetologia* 43:265–277; 2000.

8. MD Esposti, A Ngo, GL McMullen, A Ghelle, A Sparla, B Benelli, M Ratta, AW Linnane. The specificity of mitochondrial complex I for ubiquinone. *Biochem J* 313:327–334; 1996.

9. AB Adolfe, ED Glasofer, WM Troetel, J Ziegenfuss, JE Stambaugh, AJ Weiss, RW Manthei. Fate of streptozotocin (NSC-85998) in patients with advanced cancer. *Cancer Chemother Rep* 59: 547–551; 1975.

10. RN Arison, FL Feudale. Induction of renal tumors by streptozotocin in rat. *Nature* 214:1254–1255; 1967.

11. RN Arison, EL Ciaccio, MS Glitzer, JA Cassaro, MP Pruss. Light and electron microscopy of lesions in rats rendered diabetic with streptozotocin. *Diabetes* 16:51–56; 1967.

12. G Brosky, J Logothetopolous. Streptozotocin diabetes in the mouse and guinea pig. *Diabetes* 18:606–611; 1969.

13. R Rasch. Tubular lesions in streptozotoci-diabetic rats. *Diabetologia* 27:32–37, 1984.

14. AA Like, AA Rossini. Streptozotocin-induced pancreatic insulitis: new model of diabetes mellitus. *Science* 193:415–417; 1976.

15. S-G Paik, N Fleischer, S-I Shin. Insulin-dependent diabetes mellitus induced by subdiabetogenic doses of streptozotocin: obligatory role of cell-mediated autoimmune processes. *Proc Nat Acad Sci* 77:6129–6133; 1980.

16. G Gold, M Manning, A Heldt, R Nowlain, JR Pettit, GA Grodsky. Diabetes induced with multiple subdiabetogenic doses of streptozotocin. Lack of protection by exogenous superoxide dismutase. *Diabetes* 30:634–638; 1981.

17. N. Welsh, S Paabo, M Welsh. Decreased mitochondrial gene expression in isolated islets of rats injected neonatally with streptozotocin. *Diabetologia* 34:626–631; 1991.

18. X Huang, K-F Eriksson, A Vaag, M Lehtovirta, M Hansson, E Laurila, T Kanninen, BT Olesen, I Kurucz, L Koranyi, L Groop. Insulin regulated mitochondrial gene expression is associated with glucose flux in human skeletal muscle. *Diabetes* 48:1508–1514; 1999.

19. DL Topping, ME Targ. Time course of changes in blood glucose and ketone bodies, plasma lipids and liver fatty acid composition in streptozotocin-diabetic male rats. *Hormone Res* 6:129–137; 1975.

20. FH Fass, WL Carter. Altered fatty acid desaturation and microsomal fatty acid composition in the streptozotocin-treated diabetic rat. *Lipids* 15:953–961; 1980.

21. TA Brasitus, PK Dudeja. Corrrection of abnormal lipid fluidity and composition of rat ileal microvillus membranes in chronic streptozotocin-induced diabetes by insulin therapy. *J Biol Chem* 260:12405–12409; 1985.

22. DD Gallaher, AS Csallany, DW Shoeman, JM Olson. Diabetes increases excretion of urinary malonaldehyde conjugates in rats. *Lipids* 28:663–666; 1993.

23. L Boquist, I Ericsson. Inhibition by streptozotocin of the activity of succinyl-CoA synthetase *in vitro* and *in vivo*. *FEBS* 196:341–343; 1986.

24. JS Dunn, J Kirkpatrick, NGB McLetchie, SV Telfer. Necrosis of the islets of Langerhans produced experimentally. *J Path Bact* 55:245–257; 1943.

25. S Lenzen, U Panten. Alloxan: history and mechanism of action. *Diabetologia* 31:337–342; 1988.

26. RE Heikkila. Free radical production by alloxan. *Eur J Pharmacol* 44:191–196; 1977.

27. WJ Malaisse. Alloxan toxicity to the pancreatic B-cell. A new hypothesis. *Biochem Pharmacol* 31:3527–3534; 1982.

28. JG Asuncion, ML Olmo, J Sastre, A Millan, A Pellin, FV Pallardo. AZT treatment induces molecular and ultrastructural oxidative damage to muscle mitochondria. *J Clin Invest* 102:4–9; 1998.

29. JG Asuncion, ML Olmo, J Sastre, FV Pallardo, J Vina. Zidovudine (AZT) causes an oxidation of mitochondrial DNA in mouse liver. *Hepatology* 29:985–987; 1999.

30. LF Pereira, MBM Oliveira, EGS Carnieri. Mitochondrial sensitivity to AZT. *Cell Biochem Function* 16:173–181; 1998.

31. RC Young, RF Ozols, CE Meyers. Adriamycin: a chemotherapeutic drug for malignancy treatment. *N Eng J Med* 305:139–153; 1981.

32. K Nicolay, B De Kruijff. Effects of adriamycin on respiratory chain acitivites in mitochondria from rat liver, rat heart and bovine heart. Evidence for a preferential inhibition of complex III and IV. *Biochemica Biophysica Acta* 892:320–330; 1987.

33. K Nicolay, RJM Timmers, E Spoelstra, R Van Der Neut, JJ Fok, YM Huigen, AJ Verkleij, B De Kruijff. The interaction of adriamycin with cardiolipin in model and rat liver mitochondrial membranes. *Biochemica Biophysica Acta* 778:359–371; 1984.

34. E Goormaghtigh, P Huart, M Praet, R Brasseur, J-M Ruysschaert. Structure of the adriamycin-cardiolipin complex. Role in mitochondrial toxicity. *Biophys Chem* 35:247–257; 1990.

35. J Fischer, MA Johnson, RL Tackett. The effect of copper deficiency on adriamycin toxicity. *Toxicol Lett* 57:147–158; 1991.

36. L Moore, EJ Landon, DA Cooney. Inhibition of the cardiac mitochondrial calcium pump by adriamycin *in vitro*. *Biochem Med* 18:131–138; 1977.

37. PM Sokolove. Mitochondrial sulfhydryl group modification by adriamycin aglycones. *FEB* 234:199–202; 1988.

38. R Ogura, H Ueta, Y Hino, T Hidaka, M Sugiyama. Riboflavin deficiency caused by adriamycin. *J Nutr Sci Vitaminol* 37:473–477; 1991.

39. G Wolf, RS Rivlin. Inhibition of thyroid hormone induction of mitochondrial α-glycerophosphate dehydrogenase in riboflavin deficiency. *Endocrinol* 86:1347–1353; 1970.

40. SS Chan, JN Fain. Uncoupling action of sulfonylureas on brown fat cells. *Mol Pharmacol* 6:513–523; 1970.

41. JM Lenhard, SA Kliewer, MA Paulik, KD Plunket, JM Lehmann, JE Weiel. Effects of troglitazone and metformin on glucose and lipid metabolism. *Biochem Pharmacol* 54:801–808; 1997.

42. MA Paulik, RG Buckholz, ME Lancaster, WS Dallas, EA Hull–Ryde, JE Weiel, JM Lenhard. Development of infrared imaging to measure thermogenesis in cell culture: thermogenic effects of uncoupling protein-2, troglitazone and b-adrenergic agents. *Pharmaceutical Res* 15:944–949; 1998.

43. MA Tirmenstein, CX Hu, TL Gales, BE Maleeff, PK Narayanan, E Kurali, TK Hart, HC Thomas, LW Schwartz. Effects of troglitazone on HepG$_2$ viability and mitochondrial function. *Toxicol Sci* 69:131–138; 2002.

44. MA Mehlman, RB Tobin, EM Sporn. Oxidative phosphorylation and respiration by rat liver mitochondria from aspirin treated rats. *Biochem Pharmacol* 21:3279–3285; 1972.

45. L Thomkins, KH Lee. The aspirin effect on respiration is due to its conversion to salicylic acid. *J Pharm Sci* 58:102–106; 1969.

46. JB Hoek. Mitochondrial energy metabolism in chronic alcoholism. *Curr Topics Bioenergetics* 17:197–241; 1994.

47. A Chedid, CL Mendenhall, T Tosch, T Chen, L Rabin, P Garcia–Pont, SJ Goldberg, T Kiernan, LB Seef, M Sorrell, C Tamburro, RE Weesner, R Zetterman. Significance of megamitochondria in alcoholic liver disease. *Gastroenterology* 90:1858–1864; 1986.

48. M Arai, MA Leo, M Nakano, ER Gordon, CS Lieber. Biochemical and morphological alterations of baboon hepatic mitochondria after chronic ethanol consumption. *Hepatology* 4:165–174; 1984.

49. S Wahid, JM Khanna, PJ Carmichael, Y Israel. Mitochondrial function following chronic ethanol treatment: effect of diet. *Res Comm Chem Path Pharmacol* 30:477–491; 1980.

50. HA Krebs, RA Freedland, R Hems, M Stubbs. Inhibition of hepatic gluconeogenesis by ethanol. *Biochem J* 112:117–124; 1969.

51. CL Keen, T Tamura, B Lonnerdal, LS Hurley, CH Halsted. Changes in hepatic superoxide dismutase activity in alcoholic monkeys. *Am J Clin Nutr* 41:929–932; 1985.

52. A Endo. The discovery and development of HMG-CoA reductase inhibitors. *J Lipid Res* 33:1569–1562; 1992.

53. DB Hunninghake, EA Stein, CA Dujovne, WS Harris, EB Feldman, VT Miller, JA Tobert, PM Laskarzewski, E Quiter, J Held, AM Taylor, S Hopper, SB Leonard, BK Brewer. The efficacy of intensive dietary therapy alone or combined with lovostatin in outpatients with hypercholesterolemia. *N Eng J Med* 328:1213–1219; 1993.

54. G De Pinieux, P Chariot, M Ammi–Said, F Louarn, JL Lejone, A Astier, B Jacotot, R Gherardi. Lipid lowering drugs and mitochondrial function: effects of HMG-CoA reductase inhibitors on serum ubiquinone and blood lactate/pyruvate ratio. *Br J Clin Pharmacol* 42:333–337; 1996.

55. MJ Muller, HJ Seitz. Dose dependent stimulation of hepatic oxygen consumption and alanine conversion to CO2 and glucose by 3,5,3'-triiodo-L-thyronine (T$_3$) in isolated perfused liver of hypothyroid rats. *Life Sci* 28:2243–2249; 1981.

56. CD Berdanier. Effects of thyroid hormone on the gluconeogenic capacity of lipemic BHE rats. *Proc Sco Exp Biol Med* 172:187–193; 1983.

57. RB Tobin, CD Berdanier, RE Ecklund, V DeVore, C Caton. Mitochondrial shuttle activities in hyperthyroid and normal rats and guinea pigs. *J Environ Pathol Toxicol* 3:289–305, 1980.

58. CD Berdanier, D Shubeck. Effects of thyroid hormone on mitochondrial activity in lipemic BHE rats. *Proc Soc Exp Biol Med* 166:348–354; 1981.

59. HG Klemperer. The uncoupling of oxidative phosphorylation in rat liver mitochondria. *Biochem J* 60:122–128; 1955.

60. RP Hafner, GC Brown, MD Brand. Thyroid–hormone control of state-3 respiration in isolated rat liver mitochondria. *Biochem J* 265:731–734; 1990.

61. P Venditti, S Di Meo, T De Leo. Effect of thyroid state on characteristics determining the susceptibility to oxidative stress of mitochondrial fractions from rat liver. *Cell Physiol Biochem* 6:283–295; 1996.

62. HJ Seitz, MJ Muller, S Soboll. Rapid thyroid-hormone effect on mitochondrial and cytosolic ATP/ADP ratios in the intact liver cell. *Biochem J* 227:149–153; 1985.

63. K Nishiki, M Erecinska, DF Wilson, S Copper. Evaluation of oxidative phosphorylation in hearts from euthyroid, hypothyroid and hyperthyroid rats. *Am J Physiol* 235:C212–C219; 1978.

64. RB Gregory, MN Berry. On the thyroid hormone-induced increase in respiratory capacity of isolated rat hepatocytes. *Biochem Biophys Acta* 1098:61–67; 1991.

65. S Luvisetto, I Schmehl, E Intravaia, E Conti, GF Azzone. Mechanism of loss of thermodynamic control in mitochondria due to hyperthyroidism and temperature. *J Biol Chem* 267:15348–15355; 1992.

66. ME Harper, JS Ballantyne, M Leach, MD Brand. Effects of thyroid hormones on oxidative phosphorylation. *Biochem Soc Trans* 21:785–792; 1993.

67. WE Thomas, A Crespo–Armas, J Mowbray. The influence of nanomolar calcium ions and physiological levels of thyroid hormone on oxidative phosphorylation in rat liver mitochondria. A possible signal amplification mechanism. *Biochem J* 247:315–320; 1987.

68. S Luvisetto, I Schmehl, E Conti, E Intravaia, GF Azzone. Activation of respiration and loss of thermodynamic control in hyperthyroidism. Is it due to increased slipping of mitochondrial pumps? *FEBS* 291:17–20; 1991.

69. ME Harper, MD Brand. Use of top-down elasticity analysis to identify sites of thyroid hormone-induced thermogenesis. *Proc Soc Exp Biol Med* 208:228–237; 1995.

70. J Bouhnik, J-P Clot, M Baudry, R Michel. Early effects of thyroidectomy and triiodothyronine administration on rat liver mitochondria. *Mol Cell Endocrinol* 15:1–12; 1979.

71. K Sterling. Thyroid hormone action at the cell level. *N Eng J Med* 300:117–123; 173–177; 1979.

72. S Sugiyama, T Kato, T Ozawa, K Yagi. Deterioration of mitochondrial function in heart muscles of rats with hypothyroidism. *J Clin Biochem Nutr* 11:199–204; 1991.

73. I Ezawa, M Yamamoto, S Kimura, E Ogata. Alterations of oxidative phosphorylation reactions in mitochondria isolated from hypothyroid rat liver. *Eur J Biochem* 141:9–13; 1984.

74. S Iossa, A Barletta, G Liverini. The effect of thyroid state and cold exposure on rat liver oxidative phosphorylation. *Mol Cell Endocrinol* 75:15–18; 1991.

75. FL Hoch, C Subramanian, GA Dhopeshwarkar, JF Mead. Thyroid control over biomembranes: VI lipids in liver mitochondria and microsomes of hypothyroid rats. *Lipids* 16:328–335; 1981.

76. D Raederstorff, CA Meier, U Moser, P Walter. Hypothyroidism and thyroxin substitution affect the n-3 fatty acid composition of rat liver mitochondria. *Lipids* 26:781–787; 1991.

77. G Paradies, FM Ruggiero, P Dinoi. The influence of hypothyroidism on the transport of phosphate and on the lipid composition in rat liver mitochondria. *Biochem Biophys Acta* 1070:180–189; 1991.

78. FL Hoch, JW Depierre, L Ernster. Thyroid control over biomembranes. Liver microsomal cytochrome b5 in hypothyroidism. *Eur J Biochem* 109:301–306; 1980.

79. KW Withers, AJ Hubert. The influence of dietary fatty acids and hypothyroidism on mitochondrial fatty acid composition. *Nutr Res* 7:1139–1159; 1987.

80. JA Stakkestad, J Bremer. The metabolism of fatty acids in hepatocytes isolated from triiodothyronine-treated rats. *Biochem Biophys Acta* 711:90–100; 1982.

81. FM Ruggiero, C Landriscina, GV Gnoni, E Quagiariello. Lipid composition of liver mitochondria and microsomes in hyperthyroid rats. *Lipids* 19:171–178; 1984.

82. CS Bangur, JL Howland, SS Kattare. Thyroid hormone treatment alters phospholipid composition and membrane fluidity of rat brain mitochondria. *Biochem J* 305:9–32; 1995.

83. Z Beleznai, E Amler, H Rauchova, Z Drahota, V Janecsik. Influence of thyroid status on the membranes of rat liver mitochondria. Unique localization of L-glycerol-3-phosphate dehydrogenase. *FEB* 243:247–250; 1989.

84. G Paradies, FM Ruggiero. Enhanced activity of the tricarboxylate carrier and modification of lipids in hepatic mitochondria from hyperthyroid rats. *Arch Biochem Biophys* 278:425–430; 1990.

85. JM Tibaldi, N Sahnoun, MI Surks. Response of hepatic mitochondrial α-glycerophosphate dehydrogenase and malic enzyme to constant infusions of L-triiodothyronine in rats bearing the Walker 256 carcinoma. Evidence for the divergent postreceptor regulation of the thyroid hormone response. *J Clin Invest* 74:705–714; 1984.

86. Y-P Lee, AE Takemori, H Lardy. Enhanced oxidation of α-glycerophosphate by mitochondria of thyroid-fed rats. *J Biol Chem* 234:3051–3055; 1959.

87. JH Oppenheimer, E Silva, HL Schwartz, MI Surks. Stimulation of hepatic mitochondrial α-glycerophosphate dehydrogenase and malic enzyme by L-triiodothyronine. Characteristics of the response with specific nuclear thyroid hormone binding sites fully saturated. *J Clin Invest* 59:517–527; 1977.

88. C van Hardeveld, R Rusche, AAH Kassenaar. Sensitivity of mitochondrial α-glycerophosphate dehydrogenase to thyroid hormone in skeletal muscle of the rat. *Horm Metab Res* 8:153–154; 1976.

89. B Bulos, B Saktor, IW Grossman, N Altman. Thyroid control over mitochondrial α-glycerophosphate dehydrogenase in rat liver as a function of age. *J Gerontol* 26:13–19; 1971.

90. S Muller, HJ Seitz. Cloning of a cDNA for the FAD-linked glycerol-3-phosphate dehydrogenase from rat liver and its regulation by thyroid hormones. *Proc Nat Acad Sci USA* 91:10581–10585; 1994.

91. W Hoppner, UB Rasmussen, G Abuerreish, H Wohlrab, HJ Seitz. Thyroid hormone effect on gene expression of the adenine nucleotide translocase in different rat tissues. *Mol Endocrinol* 2:1127–1131; 1988.

92. MD Brand, D Steverding, B Kadenbach, PM Stevenson, RP Hafner. The mechanism of the increase in mitochondrial proton permeability induced by thyroid hormones. *Eur J Biochem* 206:775–781; 1992.

93. SB Shears, JR Bronk. Ion transport in liver mitochondria from normal and thyroxine-treated rats. *J Bioenergetics Biomembranes* 12:379–393; 1980.

94. RF Castilho, AJ Kowaltowski, AE Vercesi. 3,5,3′-triiodothyronine induces mitochondrial permeability transition mediated by reactive oxygen species and membrane protein thiol oxidation. *Arch Biochem Biophys* 354:151–157; 1998.

95. G Paradies, FM Ruggiero. Stimulation of phosphate transport in rat-liver mitochondria by thyroid hormones. *Biochim Biophys Acta* 1019:133–136; 1990.

96. S Soboll, C Horst, H Hummerich, JP Schumacher, HJ Seitz. Mitochondrial metabolism in different thyroid states. *Biochem J* 281:171–173; 1992.

97. RS Tyzbir, AS Kunin, NM Sims, E Danforth. Influence of diet composition on serum triiodothyronine (T$_3$) concentration, hepatic mitochondrial metabolism and shuttle system activity in rats. *J Nutr* 111:252–259; 1981.

98. OE Deaver, RC Wander, RH McCusker, CD Berdanier. Diet effects on mitochondrial phospholipid fatty acids and mitochondrial function in BHE rats. *J Nutr* 116:1148–1155; 1986.

99. RC Wander, CD Berdanier. Effects of dietary carbohydrate on mitochondrial composition and function in two strains of rats. *J Nutr* 115:190–199; 1985.

100. E Danforth, AG Burger. The impact of nutrition on thyroid hormone physiology and action. *Annu Rev Nutr* 9:210–227; 1989.

101. R Didier, C Remesy, C Demigne, P Fafournoux. Hepatic proliferation of mitochopndria in response to a high protein diet. *Nutr Res* 5:1093–1102; 1985.

102. J Ferreira, L Gil. Nutritional effects on mitochondrial bioenergetics. Alterations in oxidative phosphorylation by rat liver mitochondria. *Biochem J* 218:61–67; 1984.

103. JM Matz, JT Saari, AM Bode. Functional aspects of oxidative phosphorylation and electron transport in cardiac mitochondria of copper deficient rats. *J Nutr Biochem* 6:644–652; 1995.

104. REC Wildman, R Hopkins, ML Failla, DM Medeiros. Marginal copper-restricted diets produce altered cardiac ultrastructure in the rat. *Proc Soc Exp Biol Med* 210:43–49; 1995.

105. WT Johnson, SN Dufault, SM Neuman. Altered nucleotide content and changes in mitochondrial energy states associated with copper deficiency in rat platelets. *J Nutr Biochem* 6:551–556; 1995.

106. Y Hoshi, O Hazeki, M Tamura. Oxygen dependence of redox state of copper in cytochrome oxidase *in vitro*. *J Appl Physiol* 74:1622–1627; 1993.

107. M-JC Kim, CD Berdanier. Influence of menhaden oil on mitochondrial respiration in BHE rats. *Proc Soc Exp Biol Med* 192:172–1176; 1989.

108. M-JC Kim, CD Berdanier. Nutrient-gene interactions determine mitochondrial function: effect of dietary fat. *FASEB J* 12:243–248; 1998.

109. K Wickwire, K Kras, C Gunnett, D Hartle, CD Berdanier. Menhaden oil feeding increases potential for renal free radical production in BHE/cdb rats. *Proc Soc Exp Biol Med* 209:397–402; 1995.

110. CD Berdanier, B Johnson, DK Hartle, WA Crowell. Lifespan is shortened in BHE/cdb rats fed a diet containing 9% menhaden oil and 1% corn oil. *J Nutr* 122:1109–1131; 1992.

111. H Morii, Y Takeuchi, Y Watanabe. Selective inhibition of NADH-CoQ oxidoreductase (complex I) of rat brain mitochondria by arachidonic acid. *Biochem Biophys Res Commun* 178:1120–1126; 1991.

112. H Rottenberg, S Steiner–Mordoch. Free fatty acids decouple oxidative phosphorylation by dissipating intramembranal protons without inhibiting ATP synthesis driven by the proton electrochemical gradient. *FEBS* 202:314–318; 1986.

113. H Rottenberg, K Hashimoto. Fatty acid uncoupling of oxidative phosphorylation in rat liver mitochondria. *Biochemistry* 25:1747–1755; 1986.

114. TY Aw, DP Jones. Nutrient supply and mitochondrial function. *Annu Rev Nutr* 9:229–251; 1989.

115. KJA Davies, L Packer, GA Brooks. Biochemical adaptation of mitochondria, muscle and whole animal respiration to endurance training. *Arch Biochem Biophys* 209:539–554; 1981.

116. CA Tate, PE Wolkowicz, J McMillin–Wood. Exercise induced alterations of hepatic mitochondrial function. *Biochem J* 208:695–701; 1982.

117. KR Short, JL Vittone, ML Bigelow, DN Proctor, RA Rizza, JM Coenen–Schimke, K Sreekumaran Nair. Impact of aerobic exercise training on age-related changes in insulin sensitivity and muscle oxidative capacity. *Diabetes* 52:1888–1896; 2003.

118. A Herrero, G Barja. ADP-regulation of mitochondrial free radical production is different with complex I or complex II-linked substrates: Implications for the exercise paradox and brain hypermetabolism. *J Bioenergetics Biomembranes* 29:241–249; 1997.

119. HE Poulson, S Loft, K Vistisen. Extreme exercise and oxidative DNA modification. *J Sports Sci* 14:343–346; 1996.

120. EA Siess, OH Wieland. Glucagon-induced stimulation of 2-oxoglutarate metabolism in mitochondria from rat liver. *FEBS Lett* 93:301–306; 1978.

121. CB Jensen, FD Sistare, HC Hamman, RC Haynes. Stimulation of mitochondrial functions by glucagon treatment. Evidence that effects are not artifacts of mitochondrial isolation. *Biochem J* 210:819–827; 1983.

122. MD Brand, L D'Alessandri, HMGPV Reis, RP Hafner. Stimulation of the electron transport chain in mitochondria isolated from rats treated with mannoheptulose or glucagon. *Arch Biochem Biophys* 283:278–284; 1990.

123. CD Berdanier, WF Flatt. DHEA and mitochondrial respiration. In: *Dehydroepiandrosterone* (M Kalimi and W Regelson, Eds.) Walter de Gruyter & Co, Berlin. 377–391; 2000.

124. EII Allen, AB Chisholm, MA Titheradge. Stimulation of hepatic oxidative phosphorylation following dexamethasone treatment of rats. *Biochem Biophys Acta* 725:71–76; 1983.

125. H Rottenberg. Uncoupling of oxidative phosphorylation in rat liver mitochondria by general anesthetics. *Proc Nat Acad Sci USA* 80:3313–3317; 1983.

126. RR Rajan, SS Katyare. Cooperativity in inhibition of coupled and uncoupled respiration in rat liver and brain mitochondria by imipramine. *Indian J Biochem Biophys* 22:71–77; 1985.

127. O Marcillat, Y Zhang, KJA Davies. Oxidative and nonoxidative mechanisms in the inactivation of cardiac mitochondrial electron transport chain components by doxirubicin. *Biochem J* 259:181–189; 1989.

128. RB Tobin, B Friend, CD Berdanier, MA Mehlman, V DeVore. Metabolic responses of rats to chronic theophylline ingestion. *J Toxicol Environ Health* 2:361–369; 1976.

129. AJM Moreno, VMC Madeira. Mitochondrial bioenergetics as affected by DDT. *Biochem Biophys Acta* 1060:166–174; 1991.

130. KB Burrows, G Gudelsky, BK Yamamoto. Rapid and transient inhibition of mitochondrial function following methamphetamine or 3,4-methylenedioxymethamphetamine administration. *Eur J Pharmacol* 9:11–18; 2000.

131. N-G Larsson, A Oldfors, E Holme, DA Clayton. Low levels of mitochondrial transcription factor A in mitochondrial depletion. *Biochem Biophys Res Comm* 200:1374–1381; 1994.

132. JA Enriquez, P Fernandez–Silva, J Montoya. Autonomous regulation in mammalian mitochondrial DNA transcription. *Biol Chem* 380:737–747; 1999.

133. H Tomura, H Endo, Y Kagawa, S Ohta. Novel regulatory enhancer in the nuclear gene of human mitochondrial ATP synthase β-subunit. *J Biol Chem* 265:6525–6529; 1990.

134. JV Verbasius, RC Scarpulla. Activation of the human mitochondrial transcription factor A gene by nuclear respiratory factors. A potential regulatory link between nuclear and mitochondrial gene expression in organelle biogenesis. *Proc Natl Acad Sci USA* 91:1309–1312; 1994.

135. MJ Evans, RC Scarpulla. NRF-1: a trans-activator of nuclear-encoded respiratory genes in animal cells. *Genes Dev* 4:1023–1034; 1990.

136. CA Chau, MJ Evans, RC Scarpulla. Nuclear respiratory factor 1 activation sites in genes encoding the (subunit of ATP synthase, eukaryotic initiation factor 2α and tyrosine aminotransferase: specific interaction of purified NRF-1 with multiple target genes. *J Biol Chem* 267:6999–7002; 1992.

137. CA Virbasius, JV Virbasius, RC Scarpulla. NRF-1, an activator involved in nuclear-mitochondrial interactions, utilizes a new DNA binding domain conserved in a family of developmental regulators. *Genes Dev* 7:2431–2436; 1993.

138. A Heddi, G Stepien, PJ Benke, DC Wallace. Coordinate induction of energy gene expression in tissues of mitochondrial disease patients. *J Biol Chem* 274:22968–22976; 1999.

139. J Marin–Garcia, R Ananthakrishnan, MJ Goldenthal. Mitochondrial gene expression in rat heart and liver during growth and development. *Biochem Cell Biol* 75:137–142; 1997.

140. LK Ostronoff, JM Izquierdo, JM Cuezva. Mt mRNA stability regulates the expression of the mitochondrial genome during liver development. *Biochem Biophys Res Com* 217:1094–1098; 1995.

141. LK Ostronoff, JM Izquierdo, JA Enriquez, J Montoya, JM Cuezva. Transient activation of mitochondrial translation regulates the expression of the mitochondrial genome during mammalian mitochondrial differentiation. *Biochem J* 316:183–191; 1996.

142. JM Izquierdo, J Ricart, LK Ostronoff, G Egea, JM Cuezva. Changing patterns of transcriptional and post-transcriptional control of β-F_1 ATP synthase gene expression during mitochondrial biogenesis in liver. *J Biol Chem* 270:10342–10350; 1995.

143. AM Luis, JM Izquierdo, LK Ostronoff, M Salinas, JF Santaren, JM Cuezva. Translational regulation of mitochondrial differentiation in neonatal rat liver: specific increase in the translational efficiency of the nuclear-encoded mitochondrial β-F_1-ATPase mRNA. *J Biol Chem* 268:1868–1875; 1993.

144. JM Izquierdo, JM Cuezva. Control of the translational efficiency of β-F_1-ATPase mRNA depends on the regulation of a protein that binds the 3' untranslated region of the mRNA. *Mol Cell Biol* 17:5255–5268; 1997.

145. FT Dionne, L Turcotte, M-C Thibault, MR Boulay, JS Skinner, C Bouchard. Mitochondrial DNA sequence polymorphism, VO_{2max} and response to endurance training. *Med Sci Sport Exercise* 23:177–185; 1991.

146. M Tonkonogi, B Walsh, T Tiivel, V Saks, K Sahlin. Mitochondrial function in human skeletal muscle is not impaired by high intensity exercise. *Pflugers Arch* 437:562–568; 1999.

147. MB Brearley, S Zhou. Mitochondrial DNA and maximum oxygen consumption. *Sport Sci* 5:1–3; 2001.

148. AL Andreu, MG Hanna, H Reichmann, C Bruno, AS Penn, K Tanji, F Pallotti, S Iwata, E Bonilla, B Lach, J Morgan–Hughes, S DiMauro, S Shanske, CM Sue, T Pulkes, A Siddiqui, JB Clark, J Land, M Iwata, J Schaefer. Exercise intolerance due to mutations in the cytochrome b gene of mitochondrial DNA. *N Eng J Med* 341:1037–1044; 1999.

149. S Qingde, Y Xirang, Z Yong, J Jinlei, L Shusen. The effect of aging and aerobic exercise on mitochondrial DNA deletions in rat liver. *Hong Kong KJ Sports Med Sports* 14:1; 2002.

150. H-C Lee, P-H Yin, C-Y Lu, Y-H Wei. Increase of mitochondria and mitochondrial DNA in response to oxidative stress in human cells. *Biochem J* 348:425–432; 2000.

151. BT Rogea, JAL Calbetb, K Mollera, H Ulluma, HW Hendelc, J Gerstofta, BK Pedersena. Skeletal muscle mitochondrial function and exercise capacity in HIV-infected patients with lipodystrophy and elevated p-lactate levels. *AIDS* 16:973–982; 2002.

152. BY Hochachka. Endurance training induces muscle-specific changes in mitochondrial function in skinned muscle fibers. *J Appl Physiol* 92:2429–2438; 2002.

153. EE McKee, BL Grier, GS Thompson, ACF Leung, JD McCourt. Coupling of mitochondrial metabolism and protein synthesis in heart mitochondria. *Am J Physiol* 258:E503–E510; 1990.

154. T Gustafsson, J Bengtsson, U Widegren, N-G Larsson, E Jansson, CJ Sundberg. Increased protein levels of mitochondrial transcription factor A (Tfam) in response to exercise training. *Scan Phys Abstr* 262; 2000.

155. HL Garstka, M Facke, JR Escribano, RJ Wiesner. Stoichiometry of mitochondrial transcripts and regulation of gene expression by mitochondrial transcription factor A. *BBRC* 200:619–626; 1994.

156. J Poulton, K Morten, C Freeman–Emmerson, C Potter, C Sewry, V Dubowitz, H Kidd, J Stephenson, W Whitehouse, FJ Hansen, M Paris, G Brown. Deficiency of the human mitochondrial transcription factor h-mtTFA in infantile mitochondrial myopathy is associated with mtDNA depletion. *Hum Mol Genet* 3:1763–1773; 1994.

157. N-G Larsson, A Oldfors, E Holme, DA Clayton. Low levels of mitochondrial transcription factor A in mitochondrial depletion. *Biochem Biophys Res Commun* 200:1374–1378; 1994.

158. J Montoya, A Perez–Martos, HL Garstka, RJ Wiesner. Regulation of mitochondrial transcription by mitochondrial transcription factor A. *Mol Cell Biochem* 174:227–230; 1997.

159. ACF Leung, EE McKee. Mitochondrial protein synthesis during thyroxine-induced cardiac hypertrophy. *Am J Physiol* 258: E511–E518; 1990.

160. WL Wooten, J Cascarano. The effect of thyroid hormone on mitochondrial biogenesis and cellular hyperplasia. *J Bioenergetics Biomembranes* 12:1–12; 1980.

161. SL Sigurdson, J Hims Hagen. Role of hyperthyroidism in increased thermogenesis in the cold acclimated Syrian hamster. *Can J Physiol Pharm* 66:826–829; 1987.

162. KR Short, J Nygren, R Barazzoni, J Levine, KS Nair. T$_3$ increases mitochondrial ATP production in oxidative muscle despite increased expression of UCP2 and 3. *Am J Physiol Endocrinol Metab* 280:E761–E769; 2001.

163. C Guerra, C Roncero, A Porras, M Fernandez, M Benito. Triiodothyronine induces the transcription of the uncoupling prottein genes and stabilizes its mRNA in fetal brown adipocyte primary cultures. *J Biol Chem* 271:2076–2081; 1996.

164. J Triandafillou, C Gwilliam, J Hims–Hagen. Role of thyroid hormone in cold induced changes in rat brown adipose tissue mitochondria. *Can J Biochem* 60:530–537; 1982.

165. S Soboll. Thyroid hormone action on mitochondrial energy transfer. *Biochem Biophys Acta* 1144:1–16; 1993.

166. P De Nayer. Thyroid hormone action at the cellular level. *Hormone Res* 26:48–57; 1987.

167. FRM van der Klis, WM Wiersinger, JJM de Vijlder. Studies on the mechanism of inhibition of nuclear triiodothyronine binding by fatty acids. *FEB* 246:6–12; 1989.

168. JM Izquierdo, E Jimenez, JM Cuezva. Hypothyroidism affects the expression of the B-F$_1$-ATP synthase gene and limits mitochondrial proliferation in rat liver at all stages of development. *Eur J Biochem* 232:344–350; 1995.

169. K Sterling, PO Milch. Thyroid hormone binding by a component of mitochondrial membrane. *Proc Nat Acad Sci USA* 72:3225–3229; 1975.

170. C Wrutniak, I Cassar–Malek, S Marchal, A Rascle, S Heusser, JM Keller, J Flechon, M Dauca, J Samarut, J Ghysdael. A 43-kDa protein related to c-Erb A α1 is located in the mitochondrial matrix of rat liver. *J Biol Chem* 270:16347–16354; 1995.

171. F Casas, P Rochard, A Rodier, I Cassar–Malek, S Marchal–Victorian, RJ Wiesner, G Cabello, C Wrutniak. A variant form of the nuclear triiodothyronine receptor c-ErbA alpha 1 plays a direct role in regulation of mitochondrial RNA synthesis. *Mol Cell Biol* 19:7913–7924; 1999.

172. P Rochard, I Cassar–Malek, S Marchal, C Wrutniakk, G Cabello. Changes in mitochondrial activity during avian myoblast differentiation: influence of triiodothyronine or v-erb A expression. *J Cell Physiol* 168:239–247; 1996.

173. JA Enriquez, P Fernadez–Silva, N Garrido–Perez, MJ Lopez–Perez, A Perez–Martos, J Montoya. Direct regulation of mitochondrial RNA synthesis by thyroid hormone. *Mol Cell Biol* 19:657–670; 1999.

174. TM Pillar, HJ Seitz. Thyroid hormone and gene expression in the regulation of mitochondrial respiratory function. *Eur J Endocrinol* 136:231–239; 1997.

175. SC Meier–Heusler, X Zhu, C Juge–Aubry, A Pernin, AG Burger, S-Y Cheng, CA Meier. Modulation of thyroid hormone action by mutant thyroid hormone receptors, c-erb Aα2 and peroxisome proliferator-activated receptor: evidence for different mechanisms of inhibition. *Mol Cell Endocrinol* 107:55–66; 1995.

176. HC Freake, KE Govoni, K Guda, C Huang, SA Zinn. Actions and interactions of thyroid hormone and zinc status in growing rats. *J Nutr* 131:1135–1141; 2001.

177. CV Demonacos, N Karayanni, E Hatzoglou, C Tsiriyiotis, DA Spandidos, CE Sekeris. Mitochondrial genes as sites of primary action of steroid hormones. *Steroids* 61:226–232; 1996.

178. F Djoudi, J Bastin, T Gilbert, A Rotig, P Rustin, C Merlet–Benichou. Mitochondrial biogenesis and development of respiratory chain enzymes in kidney cells: role of glucocorticoids. *Am J Physiol* 267:C245–C254; 1994.

179. M Raikhinstein, I Hanukoglu. Mitochondrial-genome-encoded RNAs: differential regulation by corticotropin in bovine adrenocortical cells. *Proc Nat Acad Sci USA* 90:10509–10513; 1993.

180. JP Silva, M Kohler, C Graff, A Oldfors, MA Magnuson, P-O Berggren, N-G Larsson. Impaired insulin secretion and β-cell loss in tissue specific knockout mice with mitochondrial diabetes. *Nat Genet* 26:336340; 2000.

181. CB Wolheim. Beta cell mitochondria in the regulation of insulin secretion: a new culprit in type II diabetes. *Diabetologia* 43:265–277; 2000.

182. T Hayakawa, M Noda, K Yasuda, H Taniguchi, I Miwa, H Sakura, Y Terauchi, J Hayashi, GW Sharp, Y Kanazawa, Y Akanuma, Y Yazaki, T Kadowaki. Ethidium bromide-induced inhibition of mitochondrial gene transcription suppresses glucose-stimulated insulin release in the mouse pancreatic beta-cell line betaHC9. *J Biol Chem* 273:20300–20307; 1998.

183. P Maechler, CB Wolheim. Mitochondrial function in normal and diabetic beta cells. *Nature* 414:807–812; 2001.

184. X Huang, K-F Eriksson, A Vaag, M Lehtovirta, M Hansson, E Laurila, T Kanninen, BT Oleson, I Kurucz, L Koranyi, L Groop. Insulin regulated mitochondrial gene expression is associated with glucose flux in human skeletal muscle. *Diabetes* 48:1508–1514; 1999.

185. M Petkovich, NJ Brand, A Krust, P Chambon. A human retinoic acid receptor which belongs to the family of nuclear receptors. *Nature* 220:444–450; 1987.

186. DJ Mangelsdorf, ES Ong, JA Dyke, RM Evans. Nuclear receptor that identifies a novel retinoic acid response pathway. *Nature* 345:224–229; 1990.

187. P Kastner, A Krust, C Mendelsohn, JM Garnier, A Zelent, P LeRoy, A Staub, P Chambon. Murine isoforms of retinoic acid receptor γ with specific patterns of expression, *Proc Natl Acad Sci USA* 87:2700–2704; 1990.

188. DJ Mandelsdorf, U Borgmeyer, RA Heyman, JY Zhou, ES Ong, AE Oro, A Kakizuka, RM Evans. Characterization of three RXR genes that mediate the action of 9-*cis* retinoic acid. *Genes Dev* 6:329–344; 1992.

189. P LeRoy, H Nakshatri, P Chambon. Mouse retinoic acid receptor γ2 isoform is transcribed from a promoter that contains a retinoic acid response element. *Proc Natl Acad Sci USA*, 88:10138–10142; 1991.

190. V Giguere, M Shago, R Zirngibl, P Tate, J Rossant, S Varmuza. Identification of a novel isoform of the retinoic acid receptor γ expressed in the mouse embryo. *Mol Cell Biol* 10:2335–2340; 1990.

191. JM Lehmann, B Hoffman, M Pfahl. Genomic organization of the retinoic acid receptor gamma gene. *Nucl Acids Res* 19: 573–578; 1991.

192. P Chambon. The retinoid signaling pathway: molecular and genetic analyses. *Cell Biol* 5:115–125; 1994.

193. K Takeyama, R Kojima, T Ohashi, T Sato, H Mano, S Masushige, S Kato. Retinoic acid differentially up-regulates the gene expression of retinoic acid receptor α and β isoforms in embryo and adult rats. *Biochem Biophys Res Commun* 222:395–400; 1996.

194. R-U Haq, M Pfahl, F Chytil. Retinoic acid affects the expression of nuclear retinoic acid receptors in tissues of retinol-deficient rats. *Proc Natl Acad Sci USA* 88:8272–8276; 1991.

195. D Lohnes, P Kastner, A Dierich, M Makr, M LeMeur, P Chambon. Function of retinoic acid receptor α in the mouse. *Cell* 73:643–658; 1993.

196. P Kastner, JM Grondona, M Mark, A Gansmuller, M LeMeur, D Decimo, J-L Vonesch, P Dolle, P Chambon. Genetic analysis of RXR developmental function: convergence of RXR and RAR signaling pathways in heart and eye morphogenesis. *Cell* 78:987–1003; 1994.

197. C Mendelsohn, M Mark, P Dolle, A Dierich, MP Gaub, A Krust, C Lampron, P Chambon. Retinoic acid receptor β2 (RARβ2) null mutant mice appear normal. *Exp Biol* 166:246–258; 1994.

198. JM Grondona, P Kastner, A Gansmuller, D Decimo, P Chambon, M Mark. Retinal dysplasia and degeneration in RAR2/RAR2 compound mutant mice. *Development* 122:2173–2188; 1996.

199. W Krezel, N Ghyselinck, TA Samad, V Dupe, P Kastner, E Borrelli, P Chambon. Impaired locomotion and dopamine signaling in retinoid receptor mutant mice. *Science* 279:863–867; 1998.

200. P Kastner, M Mark, P Chambon. Nonsteroid nuclear receptors: what are genetic studies telling us about their role in real life? *Cell* 83:859–869; 1995.

201. HB Everts, DO Claassen, CL Hermoyian, CD Berdanier. Nutrient–gene interactions: dietary vitamin A and mitochondrial gene expression. *IUBMB Life* 53:295–301; 2002.

202. SJ Ruff, DE Ong. Cellular retinoic acid binding protein is associated with mitochondria. *FEBS Letters* 487:282–286; 2000.

203. G Li, Y Liu, SS Tsang. Expression of a retinoic acid-inducible mitochondrial ND5 gene is regulated by cell density in bovine papillomavirus DNA-transformed mouse C127 cells but not in revertant cells. *Int J Oncol* 5:301–307; 1994.

204. IC Gaemers, AMM Van Pelt, APN, Themmen, DG De Rooij. Isolation and characterization of all-trans-retinoic acid-responsive genes in the rat testis. *Mol Reprod Dev* 50:1–6; 1998.

205. CD Berdanier, HB Everts, E Hermoyian, CE Mathews. Role of vitamin A in mitochondrial gene expression. *Diab Res Clin Prac* 54:511–527; 2001.

206. GK Whitfield, J-C Hsieh, PW Jurutka, SH Selznick, CA Haussler, PN MacDonald, MR Haussler. Genomic actions of 1,25-dihydroxyvitamin D_3. *J Nutr* 125:1690S–1694S; 1995.

207. SY Chou, SS Hannah, KE Lowe, AW Norman, HL Henry. Tissue specific regulation by vitamin D status of nuclear and mitochondrial gene expression in kidney and intestine. *Endocrinology* 136:5520–5526; 1995.

208. P Cantatore, V Petruzzella, C Nicoletti, F Papadia, F Fracasso, P Rustin, MN Gadaleta. Alteration of mitochondrial DNA and RNA levels in human fibroblasts with impaired vitamin B_{12} coenzyme synthesis. *FEBS Lett* 432:173–178; 1998.

209. J Asin, A Perez–Martos, P Silva–Fernandez–Silva, J Montoya, AL Andreu. Iron II induces changes in the conformation of mammalian mitochondrial DNA resulting in a reduction of its transcriptional rate. *FEBS Lett* 480:161–164; 2000.

210. F Lucesoli, CG Fraga. Oxidative damage to lipids and DNA concurrent with decrease of antioxidants in rat testes after acute iron intoxication. *Arch Biochem Biophys* 316:567–571; 1995.

211. G Minotti. Sources and role of iron in lipid peroxidation. *Chem Res Toxicol* 6:134–146; 1992.

212. J Weaver, S Pollack. Two types of receptors for iron in mitochondria. *Biochem J* 271:463–466; 1990.

213. WT Willis, PR Dallman. Impaired control of respiration in iron deficient muscle mitochondria. *Am J Physiol* 257:C1080–C1085; 1989.

214. Z Wu, P Puigserver, U Andersson, C Zhang, G Adelmant, V Mootha, A Troy, S Cinti, B Lowell, RC Scarpulla, BM Spiegelman. Mechanisms controlling mitochondrial biogenesis and respiration through the thermogenic coactivator PGC-1. *Cell* 98:115–124; 1999.

215. P Puigserver, Z Wu, CW Park, R Graves, M Wright, BM Spiegelman. A cold-inducible coactivator of nuclear receptors linked to adaptive thermogenesis. *Cell* 92:829–839; 1998.

216. JM Izquierdo, JM Cueza. Evidence of post-transcriptional regulation in mammalian mitochondrial biogenesis. *Biochem Biophys Res Commun* 196:55–60; 1993.

217. K Luciakova, R Li, BD Nelson. Differential regulation of the transcript levels of some nuclear-encoded and mitochondrial-encoded respiratory-chain components in response to growth activation. *Eur J Biochem* 207:253–257; 1992.

218. RJ Stevens, ML Nishio, DA Hood. Effect of hypothyroidism on the expression of cytochrome c and cytochrome c oxidase in heart and muscle during development. *Mol Cell Biochem* 143:119–127; 1995.

219. BD Nelson, K Luciakova, R Li, S Betina. The role of thyroid hormone and promoter diversity in the regulation of nuclear encoded mitochondrial proteins. *Biochim Biophys Acta* 1271:85–91; 1995.

220. K Luciakova, BD Nelson. Transcript levels for nuclear-encoded mammalian mitochondrial respiratory chain components are regulated by thyroid hormone in an uncoordinated fashion. *Eur J Biochem* 207:247–251; 1992.

221. JM Izquierdo, E Jimenez, JM Cuezva. Hypothyroidism affects the expression of the β-F$_1$-ATP synthase gene and limits mitochondrial proliferation in rat liver at all stages of development. *Eur J Biochem* 232:344–350; 1995.

222. JA Enriquez, P Fernadez–Silva, J Montoya. Autonomous regulation in mammalian mitochondrial DNA transcriptions. *J Biol Chem* 380:737–747; 1999.

223. Y Cai, BD Nelson, R Li, K Luciakova, JW DePierre. Thyromimetic action of the peroxisome proliferators Clofibrate, perfluorooctanoic acid and acetylsalicyclic acid includes changes in mRNA levels for certain genes involved in mitochondrial biogenesis. *Arch Biochem Biophys* 325:107–112; 1996.

224. DD Chang, DA Clayton. Precise assignment of the light-strand promoter of mouse mitochondrial DNA: a functional promoter consists of multiple upstream domains. *Mol Cell Biol* 6:3253–3261; 1986.

225. DD Chang, DA Clayton. Precise assignment of the heavy-strand promoter of mouse mitochondrial DNA: cognate start sites are not required for transcriptional initiation. *Mol Cell Biol* 6:3262–3267; 1986.

226. DF Bogenhagen, EF Applegate, BK Yoza. Identification of a promoter for transcription of the heavy strand mtDNA: *in vitro* transcription and deletion mutagenesis. *Cell* 36:1105–1113; 1984.

227. JFH Wong, DP Ma, RK Wilson, BA Roe. DNA sequence of the Xenopus Laevis mitochondrial heavy and light strand replication origins and flanking tRNA genes. *Nucleic Acids Res* 11:4977–4995; 1983.

228. K Umesono, KK Murakami, CC Thompson, RM Evans. Direct repeats as selective response elements for the thyroid hormone, retinoic acid, and vitamin D$_3$ receptors. *Cell* 65:1255–1266; 1991.

229. AM Naar, JM Boutin, SM Lipkin, VC Yu, JM Holloway, CK Glass, MG Rosenfeld. The orientation and spacing of core DNA-binding motifs dictate selective transcriptional responses to three nuclear receptors. *Cell* 65:1267–1279; 1991.

230. C Wrutniak, I Cassar–Malek, S Marchal, A Rascle, S Heusser, JM Keller, J Flechon, M Dauca, J Samarut, J Ghysdael. A 43-kDa protein related to c-Erb A α1 is located in the mitochondrial matrix of rat liver. *J Biol Chem* 270:16347–16354; 1995.

231. F Casas, L Domenjoud, P Rochard, R Hatier, L Daury, A Bianchi, P Kremarik–Bouillaud, P Becuwe, JM Keller, H Schohn, C Wrutniak–Cabello, G Cabello, M Dauca. A 45-kDa protein related to PPARγ2 induced by peroxisome proliferators located in the mitochondrial matrix. *FEBS Lett* 478:4–8; 2000.

232. S Miranda, R Foncea, J Guerrero, F Leighton. Oxidative stress and upregulation of mitochondria;l biogenesis genes in mitochondrial DNA depleted HeLa cells. *Biochem Biophys Res Commun* 258:44–49; 1999.

233. HL Garstka, M Facke, JR Escribano, RJ Wiesner. Stoichiometry of mitochondrial transcripts and regulation of gene expression by mitochonrial transcription factor A. *Biochem Biophys Res Commun* 200:619–626; 1994.

234. F Casas, T Pineau, P Rochard, A Rodier, L Daury, M Dauca, G Cabello, C Wrutniak–Cabello. New molecular aspects of regulation of mitochondrial activity by fenofibrate and fasting. *FEBS Lett* 482:71–74; 2000.

10

Bacterial F_1F_0 ATP Synthase as a Model for Complex V

BRIAN D. CAIN,
TAMMY BOHANNON GRABAR, AND
DEEPA BHATT

CONTENTS

INTRODUCTION

F_1F_0 adenosine triphosphate (ATP) synthase (complex V) catalyzes the terminal step in oxidative phosphorylation. The carriers of the electron transport chain (complexes I to IV) establish an electrochemical gradient of protons across a membrane. The energy of ion translocation down the gradient is used by F_1F_0 ATP synthases in the phosphorylation of ADP. In view of the central metabolic importance of ATP, it is not surprising that members of the F_1F_0 ATP synthase family are almost ubiquitously distributed throughout biology. With the exceptions of a few organisms, the F_1F_0 ATP synthases are located in the inner mitochondrial membranes of eukaryotes and the cytoplasmic membranes of prokaryotes. In the case of plants, a second F_1F_0 ATP synthase is located in the thylakoid membrane to make use of the proton gradient established via the photosynthetic reaction centers.

The F_1F_0 ATP synthase family members share common catalytic properties and molecular architectures. As a result, mechanistic observations on F_1F_0 ATP synthase from one species are most often applicable to the enzymes found in other organisms. Over the past decade, enormous progress has been made in our understanding of the structure and function of the F_1F_0 ATP synthases (1–5). At the same time, a growing number of human genetic disorders have been directly associated with defects in oxidative phosphorylation (6–8). A subset of the disorders has been attributed to defects in F_1F_0 ATP synthase. The purpose of this chapter is to consider what the basic science says about the nature of defects in F_1F_0 ATP synthase. The human disease mutations are seemingly restricted to the mitochondrially encoded *ATPase6* and *ATPase8* genes, so the focus will be on bacterial F_0 sector stator subunits *a* and *b* of F_1F_0 ATP synthase and the homologous mammalian proteins. To appreciate the defects fully, one must first have an understanding of the structure of the F_1F_0 ATP synthase and the mechanism of ATP synthesis.

STRUCTURE OF F_1F_0 ATP SYNTHASE

The F_1F_0 ATP synthases are multimeric enzyme complexes. Electron micrographs of the enzyme revealed two masses, reflecting the F_1 and F_0 sectors, linked by two narrow stalk structures (9–11). The simplest bacterial enzymes have as few as 8 different types of subunits (Figure 10.1) in comparison to the human mitochondrial enzyme, which has at least 16 subunits (3). All eight of the *Escherichia coli* subunits have homologues in the enzymes of higher organisms. Unfortunately, the subunit nomenclature between experimental systems is not consistent (Table 10.1). The F_0 sector is embedded within the membrane bilayer and functions to conduct protons across the membrane. This drives ATP synthesis at catalytic sites located within F_1. This can be reversed *in vitro* by supplying ATP as a substrate to drive proton pumping across a membrane.

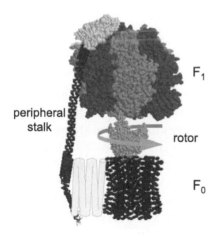

Figure 10.1 Space-filling model of the *Escherichia coli* F_1F_0 ATP synthase. F_1F_0 subunits included in the model were: $\alpha_3\beta_3\delta\gamma\varepsilon ab_2c_{10}$. The arrow shows the direction of rotation of $\gamma\varepsilon c_{10}$ during ATP synthesis. The cylinder representations for the *a* subunit and portions of the *b* subunits reflect the fact that no high-resolution structure exists for these areas of the enzyme for any species.

Table 10.1 F_1F_0 ATP Synthase Subunit

| Bacterium | Chloroplast | Mitochondria | | Function |
		Yeast	Bovine	
a	a	a	a	Catalytic site
b	b	b	b	Catalytic site
g	g	g	g	Rotor
d	d	OSCP	OSCP	Stator
e	e	d	d	Rotor
—	—	e	e	?
a	*a* (or IV)	*a* or 6	a	Proton channel, stator
b	*b* (or I)	*b* or 4	—	Stator?
—	*b'* (or II)	—		Stator?
c	*c* (or III)	*c* or 9	c	Proton channel, rotor
		D	d	Stator?
		8	A6L	Stator?
		E	e	?
		F	f	?
		G	g	?
		H	F6	Stator?
		inh1p	IF1	Regulator
		stf1p		
		I		
		K		

The F_1 and F_0 sectors can be readily separated for study *in vitro* by washing inverted membrane vesicles with low ionic strength buffers. In the bacterial enzyme, the water-soluble F_1 sector consists of the $\alpha_3\beta_3\gamma\delta\varepsilon$ subunits (Figure 10.1). When F_1 is washed from the membrane, it acts as a water-soluble ATPase. Much of the mass of F_1 consists of the $\alpha_3\beta_3$ hexamer. High-resolution structures of the bovine (12), rat (13), and yeast (14) F_1 have been determined by x-ray crystallography. The α and β subunits alternate within the hexamer like the segments of an orange. The three catalytic sites are located at the interfaces between these subunits, but most of the residues in contact with the nucleotides are contributed by the β subunits.

The γ and ε subunits form the central or rotor stalk. Long α-helical segments of the γ subunit extend up through the donut hole in the center of the $\alpha_3\beta_3$ hexamer, reaching near the very top of the F_1 sector. The bacterial δ subunit appeared to cap the top of the peripheral or second stalk (11,15). Recent evidence suggests that subunit δ associates with F_1 by interacting with a single α subunit on the outside of the $α_3β_3$ hexamer (16). The more complex mitochondrial enzymes have additional subunits, and many of these are associated with the stalk structures (2,3). This includes the IF_1 inhibitor subunit thought to regulate the reverse ATP hydrolysis reaction.

Release of F_1 leaves the intrinsic membrane subunits of F_0 sector, ab_2c_{10}, in the membrane as a functional protonophore (Figure 10.1). Only the structure of the c subunit has been determined at a high level of resolution (17,18). The c subunits form hairpin loops spanning the membrane twice with the loop located on the F_1 side of the membrane. Multiple c subunits form a ring oriented front to back (18–21). The actual number of c subunits in the ring ranges from 10 to 14 in the F_1F_0 ATP synthase family. The number is certainly species specific and may indeed vary with physiological conditions within an organism (22). The c subunit houses a highly conserved and mechanistically essential acidic amino acid, *asp61* in the *E. coli* enzyme, that undergoes protonation and deprotonation during proton translocation.

Located to one side of the c subunit ring is a single a subunit. The a subunits are extremely hydrophobic proteins and are thought to span the membrane five times (23,24). The subunit contains conserved amino acids thought to form the F_0 ion translocation channel (25). In bacterial enzymes, a dimer of two b subunits forms the peripheral stalk (26,27). Each has a single membrane span and an extended hydrophilic domain that stretches to near the top of F_1 in the intact enzyme complex. There the b_2 dimer participates in a direct interaction with the δ subunit (28,29). Additional contacts have been detected with α and β subunits.

MECHANISM OF ATP SYNTHESIS

The principles governing catalytic activity by F_1F_0 ATP synthase were described in the "binding change mechanism" by Boyer (1). The substrates are inorganic phosphate (P_i) and

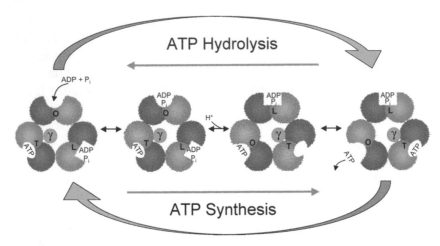

Figure 10.2 A variation on the binding change mechanism. This simplified scheme was derived from the more detailed enzymatic mechanism described by Weber and Senior (5) in which all three catalytic sites are filled with nucleotides transiently during ATP synthesis. F_1 subunits: α_3, β_3, γ.

adenosine diphosphate (ADP), and the reaction requires magnesium (Mg^{+2}). A schematic diagram of a simplified variation of the mechanism (5) is shown in Figure 10.2. Three distinct catalytic sites were defined:

- The tight site containing ATP
- The loose site depicted with ADP and P_i bound
- An empty open site

The reaction starts with binding of substrates at the open site. Then an energy input from proton translocation occurs so that the open site becomes a loose site, the loose site assumes the tight conformation, and the tight site is converted to an open conformation. The translocated protons do not participate in catalysis per se, but provide the energy for the conformational changes. ATP is formed in the tight site, and the original molecule of ATP is released from the open site. Upon ATP release, the starting structure has been regenerated, albeit rotated 120°.

The binding change mechanism made two important predictions. First, the three catalytic sites in the β subunits existed at any moment in time in differing conformations. Indeed, this prediction was satisfied when three distinct conformations were observed in the high-resolution structures of bovine F_1 (12). The second prediction was physical rotation as an integral feature of the enzymatic mechanism. Rotation of the γ subunit relative to the $\alpha_3\beta_3$ hexamer was confirmed by experiments performed in several laboratories (30–32).

The most elegant of these studies was performed by Noji et al., who engineered a fluorescently labeled actin filament onto the γ subunit in F_1 affixed onto a slide via the β subunits (33). Fluorescence microscopy yielded single molecule observations revealing ATP hydrolysis-dependent rotation of the γ subunit-actin filament. Subsequent experiments using variations of the approach have yielded observations of rotation of the ε subunit and the ring of c subunits, indicating that the "rotor" consists of $\gamma\varepsilon c_{10-14}$ functioning as a unit (32). The remaining F_1F_0 subunits are commonly referred to as the "stator" subunits.

Models for the overall mechanism have been developed to describe ATP synthesis in F_1F_0 ATP synthase by proton-driven rotation (Figure 10.3). The current prevailing model suggests that protons, in the form of H_3O^+, enter a half channel housed in the a subunit (34,35). This leads to protonation of *asp61* of a c subunit in the ring. In some way, the protonation event — or perhaps the release of ion from a previously protonated c subunit in the ring into an exit channel also located within subunit a — promotes the generation of torque.

The molecular mechanism of torque generation is presently one of the most actively debated subjects in the field. Whatever the molecular mechanism is, the c subunit ring rotates relative to the stator a subunit. Rotation of the c ring in turn forces rotation of the entire rotor. The rotation of the γ subunit within the $\alpha_3\beta_3$ hexamer results in the changes in conformation at the catalytic sites. Translocation of three to four protons is required to generate enough torque to drive the 120° rotation of the rotor leading to synthesis of one molecule of ATP. The $\alpha_3\beta_3$ hexamer is held in place against

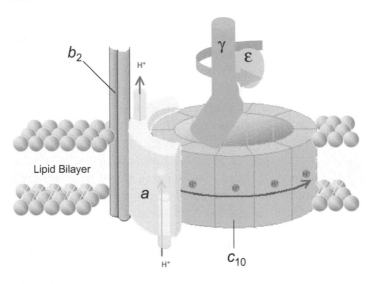

Figure 10.3 Function of F_0. The current model for proton translocation and torque generation in the F_0 sector. F_1F_0 subunits: γ, ε, a, b_2, c_{10}. (Model drawn from schemes proposed by W. Junge, *Proc Natl Acad Sci* 96: 4735–4737, 1999, and S.B. Vik and B.J. Antonio, *J Biol Chem* 269: 30364–30369, 1994.)

the rotation by stator $b_2\delta$ subunits of the peripheral stalk (Figure 10.1).

GENETIC DISEASE AND F_1F_0 ATP SYNTHASE

Defects in energy metabolism have been implicated in complex, multifactorial conditions in humans, including aging, Alzheimer's disease, cancer, and diabetes (36–40). Inborn errors in metabolism affecting the central nervous system, sensory organs, and musculature have been specifically associated with a number of specific defects in the mitochondrial genome (mtDNA) (6,7,41,42). These tissues are most profoundly affected by mitochondrial defects because of their high requirements for energy metabolism. Some of the mitochondrial diseases have resulted from major deletions in the mtDNA (43,44) or mitochondrial DNA rearrangements (45).

However, many of the inherited disorders are the product of single base substitution mutations (46,47). Patients present with a complex pathophysiology because of differences in the defects resulting from differing mutations within a mitochondrial gene, differences in the percentage of defective mitochondrial genomes inherited by individuals, and differences in somatic segregation of defective mitochondria during development within each individual. Although paternal inheritance of mitochondrial disease has been recently reported (48), the vast majority of mitochondria appear to be inherited from the mother. As a result, mitochondrial deficiencies typically display a characteristic pattern of maternal inheritance (6,7).

Many clinically distinct disease syndromes have been reported to be associated with defects in F_1F_0 ATP synthase in humans (6). Examples of inborn errors in metabolism associated with F_1F_0 ATP synthase deficiency include neurogenic muscle weakness ataxia retinitis pigmentosa (NARP); Leigh syndrome; Leber hereditary optic neuropathy (LHON); and familial bilateral striatial necrosis (FBSN). Surprisingly, although the human enzyme has the typical mammalian complement of 16 different subunits — each the product of a different gene — all reported F_1F_0 ATP synthase defects resulted from single base substitution changes in the *ATPase6* (subunit *a*) and *ATPase8* (subunit 8, or alternatively subunit A6L) genes. These genes were the only two F_1F_0 ATP synthase subunit genes found encoded in the human mitochondrial genome. To date, no specific mutation in the 14 nuclear-encoded F_1F_0 ATP synthase subunit genes has been shown to produce a disease phenotype.

The majority of disease mutations studied affected the *ATPase6* gene. At least six specific single base substitution mutations affecting *ATPase6* are known to result in mitochondrial disease. Most of these disease mutations occurred at or in the immediate vicinity of the codons for the conserved amino acids located in the transmembrane spans of the *a* subunit (Figure 10.4). The severity of symptoms resulting from *ATPase6* gene defects appeared to be directly proportional to the degree of heteroplasmy for the mutations. An individual with approximately 80% defective mitochondrial

```
                      TM4
Mitochondria          IQPMALAVRLTANITAGHLLMHLIGGAT
Bacteria              SKPLSLGLRLFGNMYAGELIFILIAGLL
Chloroplasts          TKPLSLSFRLFGNILADELVVGVLISLV


                      TM5
Mitochondria          LLTILEFAVALIQAYVFTLLVSLY
Bacteria              PWAIFHILIITLQAFIFMVLTIVY
Chloroplasts          VIMLLGLFTSAIQALIFATLAGAY
```

Figure 10.4 Primary structure conservation in subunit a. TM4 and TM5 are shown with strongly conserved amino acids shaded. Many of these conserved sites are thought to line the F_0 proton channel.

genomes seems to be clinically discernable, with profound disease apparent above about 90% (6,49,50). For one of the conserved codons (*leu-156* of the a subunit), two different substitutions of mtDNA nt8993 have been reported. In patients with comparable levels of defective mitochondrial DNA, those with the nt8993 T→C mutation *(leu-156→pro)* presented with less severe symptoms than patients with the nt8993 T→G mutation *(leu-156→arg)* (51,52). Therefore, the specific base substitution in *ATPase6* has an impact on the penetrance of the disease phenotype.

Several nuclear-encoded genes have been implicated in mitochondrial diseases (6,42); however, these did not occur in genes for subunits of F_1F_0 ATP synthase or the other respiratory complexes. Although other mutations undoubtedly exist in the 14 nuclear F_1F_0 ATP synthase genes, no diseases associated with these genes have been reported. This may be because of a mitochondrial threshold effect (53). A load of 80% defective *ATPase6* genes was needed to result in clinically important effects that could be diagnosed. In all probability, an individual with a heterozygous defect in a nuclear F_1F_0 ATP synthase gene would appear asymptomatic. The homozygous condition would likely be fatal in early embryogenesis. Therefore, mutations in the nuclear genes are likely to remain undetected by conventional clinical diagnostic approaches.

MUTATIONS AFFECTING THE *a* SUBUNIT

The *a* subunit is the largest of the F$_0$ subunits and is thought to span the membrane five times (TM1-TM5) (23,24). Across species, *a* subunit genes share striking sequence homology in the areas encoding transmembrane spanning segments TM2, TM4, and TM5 (Figure 10.4). A specific constellation of amino acids has been particularly well conserved across all F$_1$F$_0$ ATP synthases in TM4 and TM5. Many of these conserved residues are thought to line the F$_0$ proton channel. The *a* subunits are extremely hydrophobic and, like many membrane proteins, have proven very difficult to study by traditional biochemical and structural approaches. To date, no high-resolution structural data exist on an *a* subunit from any species. Most of the available information on structure and function of the *a* subunit was deduced from extensive mutagenesis studies of the *E. coli uncB* (*a* subunit) gene (21,25).

Single amino acid substitution mutations affecting the conserved residues in TM4 and TM5 often resulted in *E. coli* defective for F$_1$F$_0$ ATP synthase. The most convenient approach for detecting loss of function was to grow a mutant strain on minimal medium supplemented with succinate as the sole carbon source. Because succinate is a tricarboxylic acid cycle intermediate, ATP production depends on oxidative phosphorylation and an active F$_1$F$_0$ ATP synthase. All of the conserved amino acids have been individually replaced by site-directed mutagenesis experiments carried out in many laboratories. Typically, radical substitutions involving replacement of a charged amino acid with a nonpolar residue, or replacing a small side-chain amino acid with a bulky one, resulted in a slow- or no-growth phenotype.

Similarly, human *ATPase6* gene mutations resulting in mitochondrial diseases were also radical amino acid substitutions affecting the conserved regions of the *a* subunit. For the *E. coli* enzyme, the most important amino acid was *arg-210* located in TM4 (25). The arginine has been conserved throughout evolution in the *a* subunits of all F$_1$F$_0$ ATP synthase family members, and in the human enzyme it is *arg-159*. All substitutions for *E. coli arg-210* resulted in a total loss of F_1F_0 ATP synthase function (54–57). Apparently, *arg-210* is essential for F_0-mediated proton translocation and may function in modulating the pKa of *c* subunit *asp-61* for the protonation–deprotonation events. Replacement of the numerous other conserved residues with at least some conservative amino acid substitutions allowed detectable levels of enzyme activity. Although important in proton translocation, none of these sites should be viewed as mechanistically essential for F_0 function.

The biochemical bases for the *a* subunit defects were usually studied using a battery of assays measuring F_1F_0 and F_0 function. Essentially four distinct biochemical phenotypes were observed for the *a* subunit mutants:

- Failure to assemble the F_1F_0 complex
- Impaired F_0-mediated proton translocation with normal F_1 ATP hydrolysis activity
- Impaired proton translocation and inhibition of F_1 activity
- Proton leakage due to uncoupling of a functional F_0 proton channel from F_1 activity

The phenotypes were amino acid-substitution specific. For example, the loss in F_1F_0 ATP synthase in the *arg-210* mutants usually resulted from blocked F_0-mediated proton translocation (54–57). However, the *arg-210→ala* substitution allowed modest passive proton translocation through F_0, suggesting a partially uncoupled phenotype (56).

The clearest examples of mutations affecting assembly and stability were those involving truncation of the *a* subunit (58). Such mutations resulted in decreased steady state levels of intact F_1F_0 ATPase activity in the membranes prepared from the mutant strains. These decreases reflected low levels of *a* and *b* subunits and consequently reduced F_1 binding. Many of the single amino acid substitutions in the *a* subunit displayed slight reductions in steady state levels of F_1F_0 ATP synthase, indicating possible assembly defects. However in general, the single amino acid substitutions yielded levels of intact enzyme complexes in mutant cell membranes at least approaching wild-type levels.

Most of the a subunit mutations yielding a strong growth phenotype displayed impaired proton translocation. Many, like the *arg-210→ile* substitution, allowed assembly of substantial numbers of stable F_1F_0 complexes in the membranes (57). The F_1F_0 ATPases had normal dicyclohexylcarbodiimide (DCCD)-sensitive F_1F_0 ATP hydrolysis activity. DCCD binds with high affinity to the c subunit inhibiting ATP hydrolysis in an intact F_1F_0, so this parameter has been used as an assay of coupling between the F_1 and F_0 sectors. In the case of the *arg-210→ile* substitution, F_1 washed from the membranes had wild-type ATPase activity, but no apparent F_0-mediated proton translocation was left in the membranes. As a result, ATP hydrolysis failed to produce proton-driven ATP synthesis or ATP-driven proton pumping activity by the intact F_1F_0 complexes.

Some amino acid substitutions resulted in impaired proton translocation and marked inhibition of ATP hydrolysis activity of F_1F_0 ATP synthase. The best example of such a mutation affected a strongly conserved site. The *ala-217→arg* substitution yielded intact F_1F_0 complexes that had sharply reduced, DCCD-insensitive ATPase activity (57,59). As expected, the F_1 sector functioned as a normal F_1-ATPase when washed from the membrane. Moreover, determination of the microrate constants for ATP hydrolysis under uni-site catalysis conditions indicated that the catalytic sites retained normal properties in the defective F_1F_0 complexes (59). Rapid ATP hydrolysis associated with normal catalytic cooperativity was lost, and the position of the rotor ε subunit appeared to be locked (57). One plausible explanation for these observations was a suppression of rotor movement as a result of the *ala-217→arg* substitution, thus inhibiting catalytic cooperativity needed for rapid ATP hydrolysis. In contrast, the rotor may be allowed to "free wheel" in the more common class of mutants represented by the *arg-210→ile* substitution.

A few mutations affecting the a, b, and c subunits have a most unusual biochemical phenotype. These amino acid substitutions resulted in the uncoupling of F_1 from a functional F_0 in an intact F_1F_0 ATP synthase complex. Perhaps the most impressive uncoupling phenotype was observed for the mutation a subunit *gly-218→asp* (60). F_1 and F_0 were abundantly active when separated, but the intact F_1F_0 complex was profoundly defective for ATP synthesis. This was a property of the specific substitution for *gly-218* by aspartate. Other amino acid replacements at the same position, such as the *gly-218→lys* substitution, yielded inhibition of F_0 proton translocation (60). The effects of *gly-218→asp* and *glu-218→lys* substitutions were partially suppressed by a second mutation, *his-245→gly,* suggesting a direct interaction important to forming the F_0 proton channel.

In summary, some mutations affecting the a subunit resulted in reduced assembly of F_1F_0 complexes. However, for the majority of single amino acid substitutions, defective a subunits were stably integrated into the F_1F_0 ATP synthase complexes. The capacity to conduct proton translocation and its coupling to catalytic activity varied greatly depending on the specific amino acid substitution. Importantly, the inhibition of F_1F_0 ATPase activity was not necessarily a direct consequence of a defect in proton translocation.

IMPLICATIONS FOR HUMAN COMPLEX V

In view of the conservation of TM4 and TM5 and the central role of the a subunit in F_0 function, observations made on the bacterial F_1F_0 ATP synthase are believed to be applicable to mammalian enzyme. That thought raised the question: if so many missense mutations in the bacterial *uncB(a)* gene result in functionally important defects, then why are so few alleles known in the human *ATPase6* gene? The nature of the biochemical phenotypes of the *E. coli uncB(a)* gene mutations may have important implications for detection and pathophysiology of human complex V disorders.

First, because some amino acid substitutions in the bacterial a subunit severely impaired ATP synthesis without significantly altering F_1F_0 ATP hydrolysis activity, apparently normal ATP hydrolysis activity cannot be considered definitive evidence of a normal F_1F_0 ATP synthase for diagnostic purposes. Unfortunately, ATP hydrolysis is the easiest assay of enzymatic function. The diagnostic problem may be exacerbated because most molecular biology-based approaches for diagnoses of *ATPase6* deficiencies have employed restriction fragment length polymorphism (RFLP) analysis for the nt8993 (*leu*-156) mutations. Most of the possible detrimental *ATPase6* gene mutations would not generate an RFLP and could not be detected by this method. As a result, it is likely that *ATPase6* deficiencies would be missed in the clinical setting. For example, Marin–Garcia et al. reported four cases of reduced skeletal muscle F_1F_0 ATP synthase activity in pediatric cardiomyopathy patients that were not associated with any known mtDNA pathogenic mutations (61). These might well be *ATPase6* gene mutations that did not alter the restriction pattern producing an RFLP.

Second, the observations of efficient assembly of defective *a* subunits into bacterial F_1F_0 ATP synthases suggested a potential problem in human mitochondria. In a mitochondrion with normal as well as defective *ATPase6* genes, high levels of defective *a* subunits might be expected to compete with normal *a* subunits for integration into human F_1F_0 complexes. As a result, in cases with a high degree of heteroplasmy, single amino acid substitution mutations in *ATPase6* genes would exert a dominant negative phenotype. Precedent for this concept can be found in the bacterial literature (62). The *gly-59→asp* substitution was reported to act as an intragenic second site suppressor mutation for *arg210→lys* when overexpressed from a plasmid, but only in the presence of a normal copy of the *uncB(a)* gene in the *E. coli* chromosome. An alternative interpretation of this apparent suppression is that the secondary mutation resulted in improper folding of the overexpressed defective *a* subunit, preventing it from competing with the wild-type gene product.

Finally, if the bacterial F_1F_0 ATP synthase accurately portrays human enzyme, then construction of mutations comparable to the human *ATPase6* disease mutations in the *E. coli uncB(a)* gene should yield an inactive F_1F_0 ATP synthase. The bacterial approach provided access to in-depth biochemical analyses not possible with limited amounts of patient tissue and tissue culture cell lines. We have modeled both of the human *ATPase6* nt8993 substitutions by site-directed mutagenesis of the *E. coli uncB(a)* gene (63,64). The mutant with the *leu-207→arg* failed to grow on succinate-based medium, and *leu-207→pro* mutant grew very slowly, indicating defects in F_1F_0 ATP synthase. Moreover, the *leu-207→arg* and *leu-207→pro* substitutions severely affected the F_1F_0 ATP synthase activity *in vitro*.

The total numbers of intact F_1F_0 ATP synthase complexes were reduced in membranes prepared from these strains, suggesting partial assembly defects, and F_0-mediated proton translocation properties of the intact complexes were sharply curtailed. The defects appeared stronger for the *leu-207→arg* substitution mirroring the relative severity of symptoms in the patients with similar levels of heteroplasmy. Capaldi and coworkers adopted a similar *E. coli* modeling strategy for investigating the effects of several human *ATPase6* gene defects (65–67). They independently observed marked decreases in activity from the *leu-207→pro* substitution. Two disease-associated *ATPase6* mutations altering amino acids in TM5 were also studied. The *E. coli leu-262-pro* enzyme modeling the FSBN syndrome mutation (nt9185 T→C) resulted in a partial F_1F_0 ATP synthase deficiency comparable to *leu-207-pro*. The Leigh syndrome-associated mutation (nt9176 T→G) was modeled by constructing the *leu-259→arg* substitution and the resulting F_1F_0 complex was efficiently assembled but fully inactivated.

Recently, cybrid technology has been employed to study the effects of mtDNA point mutations in *ATPase6* (nt8993) (68,69). In these experiments, enucleated patient cells with a high degree of heteroplasmy for the mtDNA defect were fused to an mtDNA deletion ρ^0 tissue culture cell line. The cybrids were selected for homoplasmic mtDNA carrying the *ATPase6* defect. F_1F_0 ATP synthase deficiency was readily apparent as an energy metabolism defect. Limited restoration of F_1F_0 ATP synthase function was achieved in cybrid cells using excessive ectopic overexpression of a recombinant *a* subunit. For these experiments, the *ATPase6* gene was engineered with the universal genetic code and a mitochondrial targeting signal, and then the construct was transfected into the cybrid cells (70).

MUTATIONS AFFECTING THE *b* SUBUNIT

The principal components of the peripheral stalk of the *E. coli* F_1F_0 ATP synthase are two identical *b* subunits and the δ subunit. A considerable body of evidence from physical and biochemical studies indicates that the *b* subunits adopt an extended conformation forming a parallel homodimer reaching from the periplasmic side of the cytoplasmic membrane to near the top of the F_1 sector (Figure 10.1 and Figure 10.5) (25–27). Formation of the b_2 dimer has been shown to be an early step in the assembly of the F_1F_0 ATP synthase complex (71). A hydrophobic segment at the amino terminal end of each *b* subunit spans the membrane once, and there is a large hydrophilic domain. The *b* subunits have little primary sequence conservation. In some photosynthetic bacterial F_1F_0 ATP synthases and in the chloroplast enzymes, two *b*-like subunits are present, and these proteins are encoded by different genes (Table 10.1).

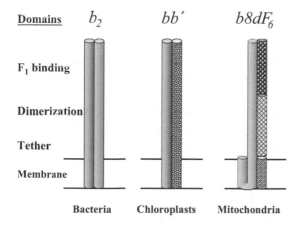

Figure 10.5 Speculative model for the *b*-like subunits. Ample evidence supports the parallel arrangement of the bacterial b_2 homodimer and the chloroplast *bb'* heterodimer. The model shown for the mitochondrial $b8dF_6$ structure is more speculative.

In the mitochondrial F_1F_0 ATP synthase, at least four distinct subunits might play the role of the bacterial b_2 dimer. Figure 10.5 shows a speculative diagram of a mammalian peripheral stalk consisting of the mammalian b, subunit 8 (A6L), d, and F6 subunits. One of the b subunit homologues is the mammalian b subunit. The mitochondrial b subunit is probably anchored to the inner mitochondrial membrane by not one but two membrane-spanning segments and possesses a large hydrophilic domain reminiscent of its bacterial counterpart. Although there is little apparent primary sequence homology, the overall shape and characteristics of the protein favor this interpretation.

Three mammalian proteins may account for the other bacterial b subunit. In this model, the subunit 8 contributes the membrane spanning segment, and a combination of subunit d and subunit F6 forms a hydrophilic domain (Figure 10.5). The assignment of the subunit 8 as a possible homologue of one of the transmembrane segments of the *E. coli* b_2 subunits is based on its membrane orientation and the location of the *ATPase8* gene in the mitochondrial genome of most eukaryotes. The human subunit 8 protein is only 68 amino acids in length and has not been studied in detail, so virtually everything that is known about *ATPase8* has come from studies of the *Saccharomyces cerevisiae* gene and its product subunit, Y8 (72).

The carboxyl terminus of subunit Y8 was determined to be on the F_1 side of the mitochondrial membrane. Unlike the first membrane-spanning segment of the mitochondrial b subunit, the orientation of subunit Y8 is parallel to the orientations of the transmembrane domains of the bacterial b_2 homodimer. In this view, the first of two membrane-spanning domains in the mammalian b subunit has no homologous segment in the bacterial F_1F_0 ATP synthase. The position of the *ATPase8* (subunit 8) gene in the mitochondrial genome of eukaryotes suggests strong conservation of some essential function for a very small protein.

Characterization of *ATPase8* defective yeast strains revealed that subunit Y8 was necessary for assembly of F_1F_0 ATP synthase (72). Indirect evidence from mutant and antibody studies also indicated that the conformation of subunit Y8 had an influence on F_0 proton translocation. Ectopic expression of Y8 subunit engineered with a mitochondrial targeting signal complemented an *ATPase8* deficient yeast strain (73). Perucca–Lostanlen et al. reported that the pathogenesis of maternally inherited diabetes and deafness may be associated with a mutation at nt8381 in *ATPase8* (40). To our knowledge, that study remains the only reported human F_1F_0 ATP synthase subunit disease-related defect mapped outside of the *ATPase6* gene.

The lack of strong sequence homology between the bacterial *b* subunits and the mammalian F_1F_0 ATP synthase subunits implies that the types of mutation modeling experiments conducted on the *a* subunit may not be directly interpretable for *b* subunit studies. Nevertheless, the mammalian enzyme has a peripheral stalk that has the same stator function, so the structure–function relationships should be similar to those of the *E. coli* enzyme.

Circular dichroism spectra suggested that at least 80% of the bacterial *b* subunit assumed an α-helical secondary structure (27). Analytical ultracentrifugation of a polypeptide modeling the *b* subunit hydrophilic domain, referred to as b_{sol}, suggested a highly elongated molecule. However, no high-resolution structures have been obtained for the peripheral stalk within F_1F_0, an isolated b_2 dimer, or for an individual *b* subunit. Four functional domains, called the membrane, tether, dimerization, and F_1-binding domains, were defined by Dunn et al. (26); this concept provided a useful nomenclature for describing the protein.

The membrane domain, *met-1* to *trp-26*, anchors the *b* subunit to the membrane. Dmitriev et al. used nuclear magnetic resonance (NMR) spectroscopy to study the structure of a 34-amino acid polypeptide modeling the membrane domain dissolved in an inorganic solvent (74). The $b_{met1\text{-}glu34}$ peptide formed an α-helix with a 20° bend. Disulfide bridge formation between cysteines inserted into the membrane domain was used to probe for areas in which the two *b* subunits were in close contact in an F_1F_0 complex. A model was developed in which the membrane domains of the two *b* subunits were involved in protein–protein interactions, but then flared apart as the subunit crossed the membrane.

Recently, a systematic mutagenesis strategy was employed as a test of this model and found that it is indeed plausible (75). Multiple mutations affecting sites proximal to the periplasmic side of the membrane had markedly stronger F_1F_0 ATP synthase defective phenotypes in comparison to similar changes in the area approaching the cytoplasmic side of the membrane. The primary defect for these mutations was a failure to assemble the F_1F_0 ATP synthase complex efficiently.

Cysteine cross-linking studies have also detected interactions between the extreme amino terminal end of the *b* subunit and the *a* and *c* subunits (76,77). The efficiency of *b*–*c* cross-linking was never greater than 50%. This suggested that the two individual *b* subunits occupied distinct positions in the bacterial F_1F_0 ATP synthase: one in contact with the *a* subunit and the other located near the ring of *c* subunits. This would be compatible with the membrane domains of the subunit *b* and subunit 8 of mammalian complex V might interact separately with subunits *a* and *c*.

The *b* subunit tether domain, *met-30* to *ser-60*, roughly corresponds to the portion of the peripheral stalk extending from the membrane surface to the bottom of F_1 (Figure 10.1 and Figure 10.5). The structure of the tether domain has not been determined at high resolution. The number of amino acids between two highly conserved and functionally important amino acids, *arg-36* and *ala-79*, has been strictly conserved across the broad spectrum of prokaryotic organisms. Coupled with the functional role of the peripheral stalk, this led to a widely held view that the peripheral stalk was a rigid, rod-like structural feature of F_1F_0 ATP synthase. However, research manipulating the length of the tether domain altered that thinking.

Intact functional F_1F_0 ATP synthases were detected containing *b* subunits with deletions of up to 11 amino acids in the $b_{\Delta 50 \rightarrow 60}$ and $b_{\Delta 54 \rightarrow 64}$ mutants (78). Similarly, active enzyme was present in mutants with insertions of 14 amino acids, $b_{+leu54 \rightarrow lys67}$ (79). The predicted secondary structure of the tether domain was an extended α-helix, so the longest functional deletions and insertions would subtract about 16 Å or add 21 Å of material to the peripheral stalk, respectively. These changes in length were startlingly large in view of the fact that the measured distance from the membrane to the base of F_1 was only 45 Å (9). Therefore, a model in which the peripheral stalk has properties of a flexible, rope-like structure has gained favor over the rigid stalk model.

Recently, it has been possible to construct F_1F_0 complexes with a heterodimer of b subunits differing in the length of the tether domain (80). The flexibility of this domain may be a necessary feature of the F_1F_0 ATP synthase because the enzyme is reversible and the direction of rotation must reverse based on whether the enzyme is catalyzing ATP synthesis or hydrolysis. The peripheral stalk would be expected to reorient its position depending upon rotation, so the flexibility of the tether domain may provide the hinge allowing this repositioning to occur. The tether domain contains a highly conserved amino acid, b_{arg-36}, that has been implicated in a structural role influencing proton translocation through F_0 (81). Amino acid substitutions, such as $b_{arg-36 \to ile}$ or $b_{arg-36 \to glu}$, resulted in assembled F_1F_0 complexes that displayed defects in F_0-mediated proton conduction or disruption of coupling between catalysis and H^+ translocation, respectively. Cross-linking studies place b_{arg-36} in the immediate vicinity of the a subunit (76,82). Therefore, b_{arg-36} probably plays a structural role aligning the proton exit channel.

What does the tether domain of the mammalian enzyme look like? Clearly, the section of the mammalian subunit b as it emerges from the membrane would occupy a comparable position in complex V. Based on cross-linking experiments on eukaryotic enzymes (2), one can speculate that subunit d may contribute the second tether domain in complex V (Figure 10.5). Indeed, subunits d and b have been chemically cross-linked in the bovine enzyme (83).

The bacterial b subunit dimerization domain, *ser-60* to *lys-122,* has been shown to be essential for forming the peripheral stalk (26). Protein–protein interactions between the two b subunits were shown to be important for assembly of F_1F_0

ATP synthase. Mutations affecting *ala-79*, conserved in *b* subunits among bacterial species, resulted in major complex assembly defects by altering dimerization (84,85). However, subunit contacts were not necessarily restricted to b_2 dimer formation because *b*–α cross-linking has been observed at sites with this segment (26).

Recently, Del Rizzo et al. determined the structure of the $b_{thr62\text{-}lys122}$ polypeptide at 1.55-Å resolution by x-ray crystallography (86). The $b_{thr62\text{-}lys122}$ peptide crystallized as a monomer, and the structure revealed an extended 90-Å long protein in an α-helical structure. This seems too long for this one segment of bacterial *b* subunits, so the straight linear structure may not fully reflect the dimerization domain in the context of the F_1F_0 ATP synthase complex. Nevertheless, the comparable segment of mammalian *b* subunit can be expected to have an extended α-helical structure. The *d* subunit seems to be long enough to constitute the dimerization domain as well as the tether domain.

The structure of the F_1-binding domain, *lys-122* to *leu-156*, has not been studied at high resolution. Interactions between the *b* subunits and the α and β subunits have been demonstrated (26). However, the majority of the contacts between b_2 and F_1 are thought to occur through interaction with the δ subunit perched atop the F_1 sector. The b_2δ interaction appears to occur at the extreme carboxyl termini of the proteins (87). Even very small modifications of the *b* subunit within the F_1-binding domain have been shown to disrupt the enzyme. Changing the length of this domain by internal deletion of as few as two amino acids (Bhatt and Cain, unpublished observation) or a two-amino acid truncation of the carboxyl terminus (88) essentially halts F_1F_0 complex assembly. Similarly, truncation of the b_{sol} peptide by four amino acids prevented its reconstitution with subunit δ.

In the bovine enzyme, subunit *b* has a protein–protein interaction with subunit OSCP, the bacterial δ subunit homologue. Subunit F_6 appears to be the best candidate for a homologue of the second bacterial *b* subunit F_1-binding domain. Subunit F_6 has been cross-linked to subunits *b* and α. However, a cautionary note is that to our knowledge, no report has indicated a direct interaction between subunit F_6 and the OSCP subunit.

In summary, the peripheral stalks of bacterial F_1F_0 ATP synthase and human complex V probably consist of $b_2\delta$ and $b8dF_6$OSCP, respectively. The mammalian b subunit is homologous to one of the bacterial b subunits, and mammalian subunits 8, d, and F_6 appear to constitute the other. Only the small subunit 8 is encoded in the human mitochondrial genome; reports of mutations associated with human diseases for the peripheral stalk subunits are very limited.

THE FUTURE

We have learned a great deal about F_0 and the human disease mutations affecting complex V. Most of this information has been gleaned from mutational analysis and traditional biochemistry approaches. Certainly, the most pressing need for improving understanding of the F_0 sector is high-resolution structural information. Obviously, the most useful structure would be intact F_1F_0 ATP synthase, but partial complexes, such as F_0 alone or even isolated a subunit, would be enormously helpful. The major scientific issues remaining are the subunit interactions within F_0; the mechanism of proton translocation; the coupling of proton translocation to the generation of torque; and the biochemical bases of complex V deficiency in inherited diseases and multifactor syndromes. Study of F_1F_0 ATP synthase is a maturing field and all of these questions should be within our grasp in the foreseeable future.

REFERENCES

1. PD Boyer. The ATP synthase — a splendid molecular machine. *Annu Rev Biochem* 66: 717–749, 1997.

2. D Stock, C Gibbons, I Arechaga, AG Leslie, JE Walker. The rotary mechanism of ATP synthase. *Curr Opin Struct Biol* 10: 672–679, 2000.

3. PL Pedersen, YH Ko, S Hong. ATP synthases in the year 2000: evolving views about the structures of these remarkable enzyme complexes. *J Bioenerg Biomemb* 32: 325–332, 2000.

4. RA Capaldi, R Aggeler. Mechanism of the F_1F_0 ATP synthase, a biological rotary motor. *Trends Biochem Sci* 27: 154–160, 2002.

5. J Weber, AE Senior. ATP synthesis driven by proton transport in F_1F_0 ATP synthase. *FEBS Lett* 545: 61–70, 2003.

6. EA Schon. Mitochondrial genetics and disease. *Trends Biochem Sci* 25: 555–560, 2000.

7. S Dimauro, EA Schon. Mitochondrial DNA mutations in human disease. *Am J Med Genet* 106: 18–26, 2001.

8. TH Vu, M Hirano, S DiMauro. Mitochondrial diseases. *Neurol Clin* 20: 809–839, 2002.

9. S Wilkens, RA Capaldi. ATP synthase's second stalk comes into focus. *Nature* 393: 29, 1998.

10. S Wilkens, RA Capaldi. Electron microscopic evidence of two stalks linking the F_1 and F_0 parts of the *Escherichia coli* ATP synthase. *Biochim Biophys Acta* 1365: 93–97, 1998.

11. S Wilkens. F_1F_0-ATP synthase-stalking mind and imagination. *J Bioenerg Biomembr* 32: 333–339, 2000.

12. JP Abrahams, AGW Leslie, R. Lutter, JE Walker. Structure at 2.8-Å resolution of F_1-ATPase from bovine heart mitochondria. *Nature* 370: 621–628, 1994.

13. MA Bianchet, J Hullihen, PL Pedersen, LM Amzel. The 2.8-Å structure of rat liver F_1-ATPase: configuration of a critical intermediate in ATP synthesis/hydrolysis. *Proc Natl Acad Sci* 95: 11065–11070, 1998.

14. D Stock, AGW Leslie, JE Walker. Molecular architecture of the rotary motor in ATP synthase. *Science* 286: 1700–1705, 1999.

15. AJ Rodgers, RA Capaldi. The second stalk composed of the *b* and δ subunits connects F_0 to F_1 via an α subunit in the *Escherichia coli* ATP synthase. *J Biol Chem* 273: 29406–29410, 1998.

16. J Weber, A Muharemagic, S Wilke–Mounts, AE Senior. F_1F_0 ATP synthase. Binding of the δ subunit to a 22-residue peptide mimicking the N-terminal region of α subunit. *J Biol Chem* 278: 13623–13626, 2003.

17. ME Girvin, RH Fillingame. Determination of local protein structure by spin label difference 2D NMR: the region neighboring Asp61 of subunit c of the F_1F_0 ATP synthase. *Biochemistry* 34: 1635–1645, 1995.

18. ME Girvin, VK Rastogi, F Abildgaard, JL Markley, RH Fillingame. Solution structure of the transmembrane H⁺-transporting subunit c of the F_1F_0 ATP synthase. *Biochemistry* 37: 8817–8824, 1998.

19. VK Rastogi, ME Girvin. Structural changes linked to proton translocation by subunit c of the ATP synthase. *Nature* 402: 263–268, 1999.

20. RH Fillingame, W Jiang, O Dmitriev. Coupling H⁺ transport to rotary catalysis in F-type ATP synthases: structure and organization of the transmembrane rotary motor. *J Exper Biol* 203: 9–17, 2000.

21. RH Fillingame, CM Angevine, OY Dmitriev. Coupling proton movements to c-ring rotation in F_1F_0 ATP synthase: aqueous access channels and helix rotations at the a–c interface. *Biochim Biophys Acta* 1555: 29–36, 2002.

22. RA Schemidt, J Qu, JR Williams, WS Brusilow. Effects of carbon source on expression of F_0 genes and on the stoichiometry of the c subunit in the F_1F_0 ATP synthase of *Escherichia coli*. *J Bacteriol* 180: 3205–3208, 1998.

23. FI Valiyaveetil, RH Fillingame. Transmembrane topography of subunit a in the *Escherichia coli* F_1F_0 ATP synthase. *J Biol Chem* 273: 16241–16247, 1998.

24. T Wada, JC Long, D Zhang, SB Vik. A novel labeling approach supports the five-membrane model of subunit a of the *Escherichia coli* ATP synthase. *J Biol Chem* 274: 17353–17357, 1999.

25. BD Cain. Mutagenic analysis of the F_0 stator subunits. *J Bioenerg Biomemb* 32: 365–371, 2000.

26. SD Dunn, M Revington, DJ Cipriano, BH Shilton. The b subunit of *Escherichia coli* ATP synthase. *J Bioenerg Biomembr* 32: 347–355, 2000.

27. J Greie, G Deckers–Hebestreit, K Altendorf. Subunit organization of the stator part of the F_0 complex from *Escherichia coli* ATP synthase. *J Bioenerg Biomembr* 32: 357–364, 2000.

28. AJW Rodgers, S Wilkins, R Aggeler, MB Morris, SM Howitt, RA Capaldi. The subunit δ-subunit *b* domain of the *Escherichia coli* F_1F_0 ATPase. *J Biol Chem* 272: 31058–31064, 1997.

29. SD Dunn, J Chandler. Characterization of a $b_2\delta$ complex from *Escherichia coli* ATP synthase. *J Biol Chem* 273: 8646–8651, 1998.

30. TM Duncan, VV Bulygin, Y Zhou, ML Hutcheon, RL Cross. Rotation of subunits during catalysis by *Escherichia coli* F_1-ATPase. *Proc Natl Acad Sci* 92: 10964–10968, 1995.

31. D Sabbert, W Junge. Stepped vs. continuous rotary motors at the molecular scale. *Proc Natl Acad Sci* 94: 2312–2317, 1997.

32. H Noji, M Yoshida. The rotary machine in the cell, ATP synthase. *J Biol Chem* 276: 1665–1668, 2001.

33. H Noji, R Yasuda, M Yoshida, K Kinosita. Direct observation of the rotation of F_1-ATPase. *Nature* 368: 299–302, 1997.

34. W Junge. ATP synthase and other motor proteins. *Proc Natl Acad Sci* 96: 4735–4737, 1999.

35. SB Vik, BJ Antonio. A mechanism of proton translocation by F_1F_0 ATP synthases suggested by double mutants of the *a* subunit. *J Biol Chem* 269: 30364–30369, 1994.

36. S Dimauro, K Tanji, E Bonilla, F Pallotti, EA Schon. Mitochondrial abnormalities in muscle and other aging cells: classification, causes and effects. *Muscle Nerve* 26: 597–607, 2002.

37. R Castellani, K Harai, G Aliev, KL Drew, A Nunomura, A Takeda, AD Cash, ME Obrenovich, G Perry, MA Smith. Role of mitochondrial dysfunction in Alzheimer's disease. *J Neurosci Res* 70: 357–360, 2002.

38. PF Chinnery, DC Samuels, J Elson, DM Turnbull. Accumulation of mitochondrial DNA mutations in ageing, cancer, and mitochondrial disease: is there a common mechanism? *Lancet* 360: 1323–1325, 2002.

39. JA Maassen. Mitochondrial diabetes, diabetes and the thiamine responsive megalblastic anaemia syndrome and MODY-2. Diseases with common pathology? *Panmin Med* 44: 295–300, 2002.

40. D Perucca–Lostanlen, H Narbonne, JB Hernandez, P Staccini, A Saurieres, V Paquis–Flucklinger, B Vialettes, C Desnuelle. Mitochondrial variations in patients with maternally inherited diabetes and deafness syndrome. *Biochem Biophys Res Commun* 277: 771–775, 2000.

41. JV Leonard, AH Schapira. Mitochondrial respiratory chain disorders I: mitochondrial DNA defects. *Lancet* 355: 299–304, 2000.

42. M Orth, AH Schapira. Mitochondria and degenerative disorders. *Am J Med Genet* 106: 27–36, 2001.

43. S Schanske, Y Tang, M Hirano, Y Nishigaki, K Tanji, E Bonilla, C Sue, S Krishna, JR Carlo, J Willner, EA Schon, S Dimauro. Identical mitochondrial DNA deletion in a woman with ocular myopathy and in her son with Pearson syndrome. *Am J Hum Genet* 71: 679–683, 2002.

44. V Maximo, P Soares, J Lima, J Cameselle–Teijeiro, M Sobrinho–Simoes. Mitochondrial DNA somatic mutations (point mutations and large deletions) and mitochondrial DNA variants in human thyroid pathology. *Am J Pathol* 160: 1857–1865, 2002.

45. O Musumeci, AL Andreu, S Schanske, N Bresolin, GP Comi, R Rothstein, EA Schon, S Dimauro. Intragenic inversion of mtDNA: a new type of pathogenic mutation in a patient with mitochondrial myopathy. *Am J Hum Genet* 66: 1900–1904, 2000.

46. CJ Wilson, NW Wood, JV Leonard, R Surtees, S Rahman. Mitochondrial DNA point mutation T9176C in Leigh syndrome. *J Child Neurol* 15: 830–833, 2000.

47. C Giordano, F Pallotti, WF Walker, N Checcarelli, O Musumeci, F Santorelli, G d'Amati, EA Schon, S Dimauro, M Hirano, MM Davidson. Pathogenesis of the deafness-associated A1555G mitochondrial DNA mutation. *Biochem Biophys Res Commun* 293: 521–529, 2002.

48. M Schwartz, J Vissing. Paternal inheritance of mitochondrial DNA. *N Engl J Med* 347: 576–580, 2002.

49. L Vilarinho, E Leao, C Barbot, M Santos, H Rocha, FM Santorelli. Clinical and molecular studies in three Portuguese mtDNA T8993G families. *Pediatr Neurol* 22: 29–32, 2000.

50. P Makela–Bengs, A Suomalainen, A Majander, J Rapola, H Kalimo, A Nuutila, H Pihko. Correlation between the clinical symptoms and the proportion of mitochondrial DNA carrying the nt8993 point mutation in the NARP syndrome. *Pediatr Res* 37: 634–639, 1995.

51. T Fujii, H Hattori, Y Higuchi, M Tsuji, I Mitsuyoshi. Phenotypic differences between T→C and T→G mutations at nt8993 of mitochondrial DNA in Leigh syndrome. *Pediatr Neurol* 18: 275–277, 1998.

52. M Makino, S Horai, Y Goto, I Nonaka. Mitochondrial DNA mutations in Leigh syndrome and their phylogenetic implications. *J Hum Genet* 45: 69–75, 2000.

53. R Rossignol, B Faustin, C Rocher, M Malgat, JP Mazat, T Letellier. Mitochondrial threshold effects. *Biochem J* 370: 751–762, 2003.

54. RN Lightowlers, SM Howitt, L Hatch, F Gibson, GB Cox. The proton pore in the *Escherichia coli* F_1F_0 ATP synthase: a requirement for arginine at position 210 of the a subunit. *Biochim Biophys Acta* 894: 399–406, 1987.

55. BD Cain, RD Simoni. Proton translocation by the F_1F_0 ATP synthase of *Escherichia coli*. *J Biol Chem* 264: 3292–3300, 1989.

56. FI Valiyaveetil, RH Fillingame. On the role of Arg-210 and Glu-219 of subunit a in proton translocation by the *Escherichia coli* F_1F_0 ATP synthase. *J Biol Chem* 272: 32635–32641, 1997.

57. JL Gardner, BD Cain. Amino acid substitutions in the a subunit affect the ε subunit of F_1F_0 ATP synthase from *Escherichia coli*. *Arch Biochem Biophys* 361: 302–308, 1999.

58. BD Cain, RD Simoni. Impaired proton conductivity resulting from mutations in the a subunit of F_1F_0 ATPase in *Escherichia coli*. *J Biol Chem* 261: 10043–10050, 1986.

59. JL Gardner, BD Cain. The a subunit ala-217→arg substitution affects catalytic activity of F_1F_0 ATP synthase. *Arch Biochem Biophys* 380: 201–207, 2000.

60. PE Hartzog, BD Cain. Second site suppressor mutations at glycine 218 and histidine 245 in the a subunit of F_1F_0 ATP synthase in *Escherichia coli*. *J Biol Chem* 269: 32313–32317, 1994.

61. J Marin–Garcia, R Ananthakrishnan, MJ Goldenthal, JJ Filiano, A Perez–Atayde. Mitochondrial dysfunction in skeletal muscle of children with cardiomyopathy. *Pediatrics* 103: 456–459, 1999.

62. SM Howitt, GB Cox. Second-site revertants of an arginine-210 to lysine mutation in the *a* subunit of the F_1F_0 ATP synthase from *Escherichia coli*: implications for structure. *Proc Natl Acad Sci* 89: 9799–9803, 1992.

63. PE Hartzog, BD Cain. The *a* $_{leu-207arg}$ mutation in F_1F_0 ATP synthase from *Escherichia coli*: a model for human mitochondrial disease. *J Biol Chem* 268: 12250–12252, 1993.

64. PE Hartzog, JL Gardner, BD Cain. Modeling the Leigh syndrome nt8993 T→C mutation in *Escherichia coli* F_1F_0 ATP synthase. *Int J Biochem Cell Biol* 31: 769–776, 1999.

65. I Ogilvie, RA Capaldi. Mutation of the mitochondrially encoded ATPase6 gene modeled in the ATP synthase of *Escherichia coli*. *FEBS Lett* 18: 179–182, 1999.

66. R Carrozzo, J Murray, FM Santorelli, RA Capaldi. The T9176G mutation of human mtDNA gives a fully assembled but inactive ATP synthase when modeled in *Escherichia coli*. *FEBS Lett* 486: 297–299, 1999.

67. R Carozzo, J Murray, O Capuano, A Tessa, G Chichierchia, MR Neglia, RA Capaldi, FM Santorelli. A novel mtDNA mutation in the ATPase6 gene studied by *E. coli* modeling. *Neurol Sci* 21: S893–894, 2000.

68. L Vergani, R Rossi, CH Brierly, M Hanna, IJ Holt. Introduction of heteroplasmic mitochondrial DNA (mtDNA) from a patient with NARP into two human ρ^0 cell lines is associated either with selection and maintenance of NARP mutant mtDNA or failure to maintain mtDNA. *Hum Molec Genet* 8: 1751–1755, 1999.

69. LGJ Nijtmans, NS Henderson, G Attardi, IJ Holt. Impaired ATP synthase assembly associated with a mutation in the human ATP synthase subunit 6 gene. *J Biol Chem* 276: 6755–6792, 2001.

70. G Manfredi, J Fu, J Ojaimi, JE Sadlock, JQ Kwong, J Guy, EA Schon. Rescue of a deficiency in ATP synthesis by transfer of MTATP6, a mitochondrial DNA-encoded gene, to the nucleus. *Nat Genet* 30: 394–399, 2002.

71. PL Sorgen, MR Bubb, KA McCormick, AE Edison, BD Cain. Formation of the *b* subunit dimer is necessary for the interaction with F_1-ATPase. *Biochemistry* 37: 923–932, 1998.

72. J Velours, G Arselin. The *Saccharomyces cerevisiae* ATP synthase. *J Bioenerg Biomembr* 32: 383–390, 2000.

73. X Roucou, IM Artika, RJ Devenish, P Nagley. Bioenergetic and structural consequences of allotropic expression of subunit 8 of yeast mitochondrial ATP synthase. *Eur J Biochem* 261: 444–451, 1999.

74. O Dmitriev, PC Jones, W Jiang, RH Fillingame. Structure of the membrane domain of subunit *b* of the *Escherichia coli* F_1F_0 ATP synthase. *J Biol Chem* 274: 15598–15604, 1999.

75. AW Hardy, TB Grabar, D Bhatt, BD Cain. Mutagenesis studies of the F_1F_0 ATP synthase *b* subunit membrane domain. *J Bioenerg Biomembr* 35: 389–398, 2003.

76. DT McLachlin, AM Coveny, SM Clark, SD Dunn. Site-directed cross-linking of *b* to the α, β and *a* subunits of the *Escherichia coli* ATP synthase. *J Biol Chem* 275: 17571–17577, 2000.

77. PC Jones, J Hermolin, W Jiang, RH Fillingame. Insights into the rotary catalytic mechanism of F_1F_0 ATP synthase from the cross-linking of subunits *b* and *c* in the *Escherichia coli* enzyme. *J Biol Chem* 275: 31340–31346, 2000.

78. PL Sorgen, TL Caviston, RC Perry, BD Cain. Deletions in the second stalk of F_1F_0 ATP synthase in *Escherichia coli*. *J Biol Chem* 273: 27873–27878, 1998.

79. PL Sorgen, MR Bubb, BD Cain. Lengthening the second stalk of F_1F_0 ATP synthase in *Escherichia coli*. *J Biol Chem* 274: 36261–36266, 1999.

80. TB Grabar, BD Cain. Integration of *b* Subunits of unequal lengths into F_1F_0 ATP synthase. *J Biol Chem* 278: 34751–34756, 2003.

81. TL Caviston, CJ Ketchum, PL Sorgen, RK Nakamoto, BD Cain. Identification of an uncoupling mutation affecting the b subunit of F_1F_0 ATP synthase in *Escherichia coli*. *FEBS Lett* 429: 201–206, 1998.

82. JC Long, J DeLeon–Rangel, SB Vik. Characterization of the first cytoplasmic loop of subunit a of the *Escherichia coli* ATP synthase by surface labeling, cross-linking, and mutagenesis. *J Biol Chem* 277: 27288–27293, 2002.

83. GI Belogrudov, JM Tomich, Y Hatefi. ATP synthase complex. *J Biol Chem* 270: 2053–2060, 1995.

84. KA McCormick, BD Cain. Targeted mutagenesis of the b subunit of F_1F_0 ATP synthase in *Escherichia coli* glu-77 through gln-85. *J Bacteriol* 173: 7240–7248, 1991.

85. KA McCormick, G Deckers–Hebestreit, K Altendorf, BD Cain. Characterization of mutations in the b subunit of F_1F_0 ATP synthase in *Escherichia coli*. *J Biol Chem* 268: 24683–24691, 1993.

86. PA Del Rizzo, Y Bi, SD Dunn, BH Shilton. The "second stalk" of *Escherichia coli* ATP synthase: structure of the isolated dimerization domain. *Biochemistry* 41: 6875–6884, 2002.

87. DT McLachlin, JA Bestard, SD Dunn. The b and δ subunits of the *Escherichia coli* ATP synthase interact via residues in their C-terminal regions. *J Biol Chem* 273: 15162–15168, 1998.

88. M Takeyama, T Noumi, M Maeda, M Futai. F_0 portion of *Escherichia coli* H⁺-ATP synthase. Carboxyl terminal region of the b subunit is essential for assembly of functional F_0. *J Biol Chem* 263: 16106–16112, 1988.

11

Cybrids in the Study of Animal Mitochondrial Genetics and Pathology

IAN A. TROUNCE AND CARL A. PINKERT

CONTENTS

INTRODUCTION

Margaret Nass first determined that threads of DNA existed in animal mitochondria (1,2) and, around the same time, Gibor demonstrated DNA in chloroplasts (3,4); these events began the resolution of mysteries surrounding ideas of "cytoplasmic inheritance" arising from careful observations made with plants 50 years earlier. Ten years following the discoveries of organelle genomes, a new term in the lexicon of somatic cell genetics, "cybrid," signaled the arrival of a powerful new experimental approach (5). This technique remains a central tool for the investigation of mtDNA mutations in mitochondrial biogenesis and disease 30 years later.

Cybrids, or cytoplasmic hybrids, are produced by first treating cells with cytochalasin B before subjecting the treated cells to a centrifugal force, as attached cells or in suspension. The dense nuclei are extruded, leaving membrane-bound cytoplasts containing cell cytoplasm and organelles, including mitochondria. These cytoplasts are then fused with a donor cell line (originally using Sendai virus, but now using poly-ethylene-glycol or electrofusion) and transformant clones or cybrids selected — with drug-resistance markers, as with CAP^R, or by selecting for respiratory competence with ρ^0 cell fusions as discussed later.

The essential feature of cybrids is their ability to distinguish effects of transferred organelle genes from those of a controlled nuclear background. Mitochondria, and in particular the assembly of the multimeric oxidative phosphorylation (OXPHOS) complexes, depend on the coordinated assembly of nuclear- and mtDNA-encoded proteins so that mutations in either genome can result in mitochondrial pathology (see references 6 through 8). Together with classical genetic studies, cybrid experiments have contributed greatly to understanding of mtDNA mutations in human disease, the subject of the present review. Mitochondrial pathology and biogenesis studies are now entering a new era, in which nuclear and mtDNA gene interactions in polygenic diseases will increasingly be the focus. For such studies, we can expect cybrids to continue to make important contributions and to help explore pathogenic mechanisms yet to be defined, such as a greater understanding of the role of normal and abnormal OXPHOS complexes in programmed cell death.

HISTORICAL OVERVIEW

Early Studies with Plants and Simple Eukaryotes

Experimental systems investigating organelle genetics in plants were in advance of those with animals until the cybrid technique was developed in the 1970s. Reciprocal crosses of different plant strains or even species were used in the first half of the 20th century, leading to the development of the idea of extranuclear, or cytoplasmic, inheritance. Ephrussi (9) and Michaelis (10) were among the first to bring these ideas into clear focus, following from earlier work of several German geneticists with various plants.

When leaf variegation patterns were used as phenotypic markers of plastid inheritance, conflicting ideas of cytoplasmic inheritance emerged. Therefore, although Lehman (11) and others favored the idea of nuclear gene copies called "plasmagenes" somehow operating in the cytoplasm, Michaelis supported Renner's argument that the most likely explanation was independently assorting genetic elements in the cytoplasm (12). Renner's work in turn owed much to the earlier work from Baur (13), whom some view as the originator of the theory of plastid inheritance (see Hagemann (14)).

This stimulated work with more experimentally amenable organisms and began a productive era of research using yeast, paramecium, and other simple eukaryotes. Once the existence of mtDNA and (soon after) chloroplast DNA was established, Slonimsky's group followed the pioneering steps of Ephrussi, Michaelis, and others (9,10), using yeast. His group was the first to show that gross changes in mtDNA buoyant densities (which later were shown to correspond to large deletions) segregated with various "petite" mutant growth phenotypes (16,17). This group also went on to discover drug-selectable, cytoplasmic yeast mutants resistant to the antibiotics chloramphenicol, erythromycin, and spiramycin (18). Chloramphenicol resistance (CAP[R]) was also used in genetic crosses of the ciliate *Tetrahymena pyriformis* around this time, providing evidence for cytoplasmic inheritance in metazoans (19).

Early Mammalian Cell Cybrids: CAP[R] and Others

The early yeast mtDNA studies preempted the first description of a mammalian cell culture mtDNA mutant in human (HeLa) cells, also resulting in resistance to the antibiotic chloramphenicol (20). Eisenstadt's group coined the term cybrid. They were the first to demonstrate cosegregation of this phenotype with mtDNA markers in a mouse cell line (5). Several other drug-resistant phenotypes were identified in the 1970s and 1980s, mostly in mouse cells, including resistance to the complex III inhibitors antimycin and myxathiozol (21) and, later, to the complex I inhibitor rotenone (22,23).

The development of accessible and reliable DNA sequencing in the late 1970s led to the identification of single base substitutions in the 16S rRNA gene of the mtDNA of independently derived yeast, mouse, and human CAP[R] cell lines (24–27). These were followed by identification of cytochrome *b* mutants (21) and, more recently, ND5, ND6, and COI mutants (Table 11.1). Mammalian mitochondrial genetics had begun.

ρ^0 Cells: Freedom from Drug-Selectable Markers

At around the time that the first disease-causing mtDNA mutations were identified (see later text), King and Attardi (33) described the successful isolation of a human cell line without mtDNA (ρ^0 cells). Employing an approach also first used with yeast (15) and then with avian cells (34), they

Table 11.1 Reported mtDNA Mutants in Cultured Mouse Cells

Locus	Nucleotide Change	Amino Acid Change	OXPHOS Phenotype	Ref.
16S	T2432C	—	Moderate complex I,	27
	A2381T	—	III, and IV	25
			defects	28
				29
ATP6/16S	T8563A	Val→Glu	NR	30
	C2380T	—		
Cytb	G14830A	Glu231→Asp	NR	31
(5 mutants)	G14251T	Gly38→Val		21
	G14563C	Gly142→Ala		
	C14578T	Thr147→Met		
	A15020T	Leu294→Phe		
ND6	13879–13884 Cins	Frameshift	Severe complex I defect	22
ND5	C12081A	Frameshift	Severe complex I defect	23
COI	C6063A	Leu246→Ile	~50% Complex IV	32
(2 mutants)	T6589C	Val421→Ala	defect	

Note: NR = not reported.

incubated the cells with low levels of the drug ethidium bromide, which intercalates DNA. The drug selectively inhibits the γ-DNA polymerase responsible for mtDNA replication; with ongoing cell division, the mtDNAs are "diluted" to vanishing quantities and clones can be isolated without detectable organelle genomes. King and Attardi (33) also discovered the absolute requirement for pyruvate gained by these cells, and confirmed the previous observation from Desjardins et al. that mtDNA-less cells also required added uridine for growth (34). This allowed a selection regime to be used after cytoplast-ρ^0 cell fusion so that unfused ρ^0 cells could be eliminated and cybrids selected with the use of an appropriate nuclear drug-resistant marker (33).

With these advances, researchers were able for the first time to introduce any mitochondria into ρ^0 cells, including those from patients with putative mtDNA disease, without the need for selectable drug-resistant markers. As such, the emerging field of mitochondrial medicine was poised to dissect and characterize specific mitochondrial mutations.

CYBRID MODELING OF PATHOGENIC MTDNA MUTATIONS ASSOCIATED WITH HUMAN DISEASE: SUCCESSES AND LIMITATIONS

From 1988 to 1990, five different mtDNA mutations associated with the major mitochondrial disease syndromes were discovered. Thus, single base substitution mutations were identified that segregated with Leber's hereditary optic neuropathy (LHON; 35); mitochondrial encephalomyopathy with lactic acidosis and stroke-like episodes (MELAS; 36); myoclonic epilepsy with ragged red fibers (MERRF; 37); large heteroplasmic mtDNA deletions in chronic progressive external ophthalmoplegia (CPEO; 38); and a single base substitution in the ATP6 gene in neurogenic weakness with ataxia and retinitis pigmentosa (NARP; 39).

The new ρ^0 cell cybrid technique was applied to these and other mtDNA mutations, quickly establishing OXPHOS phenotypes that were transferred with the different mutations. Chomyn et al. showed that the MERRF mutation, a single base substitution in the mtDNA tRNA[Lys] gene, segregated with a severe mitochondrial protein synthesis defect (40). Interestingly, their paper also showed that a heteroplasmic mtDNA donor sample from a patient could be used to produce cybrids with normal or defective protein synthesis and segregating with wild-type or mutant mtDNA cybrids, thus suggesting the mtDNA heteroplasmy was largely intercellular. This observation was repeated in several later studies with different mtDNA mutants. However, other data suggest largely intracellular heteroplasmy, so this remains a controversial topic to which we will briefly return later.

Another controversy surrounds the pathogenic mechanism associated with the MELAS tRNA[Leu(UUR)] mutation. *In vitro* studies suggested that the mutation altered the ratio of rRNA to mRNA transcripts by altering the transcriptional termination sequence located in the tRNA gene (41). Cybrid studies did not support this — showing no alteration in the transcript ratios, but, again, a severe mitochondrial protein synthesis defect when the level of mutation was high (42,43). Another report found that, in cybrids made by fusing different ρ^0 cells (i.e., different nuclear backgrounds) and enucleated patient cells harboring the heteroplasmic MELAS mutation, the different subclones analyzed showed segregation to the wild-type or mutant mtDNAs, depending on the nucleus (44).

The most commonly encountered missense mutations associated with mitochondrial disease are the LHON (at nt 3460 in the ND1 gene; 11778 in the ND4 gene; and 14484 in the ND6 gene) and Leigh's disease/NARP mutants most commonly found at nucleotide 8993 in the ATPase6 gene. The nt 8993 ATP6 mutation was shown in cybrid studies to segregate with a moderate respiration defect and a lowered ADP:O ratio (45). The LHON mutants have been more difficult to define or assign biochemical phenotypes. Jun et al. found complex I activity to be moderately reduced when the 14459 mutation associated with LHON + dystonia was transferred into lymphoblastoid cybrids (46). Others have found that nuclear background is again important (47), with different cybrid nuclear backgrounds affecting the expression of the complex I defect associated with the nt 3460 mutation. Comparison of the three "primary" LHON mutants in the same cybrid nucleus also showed variable defects of complex I, with the 3460 mutant exhibiting a greater defect than the nt 11778 and the 14484 mutants (48).

In summary, cybrid studies have greatly advanced the understanding of mtDNA disease genetics, but have also left major gaps to fill. Clues that nuclear gene effects can be important modulators of OXPHOS phenotypes have been gained. As new information on the role of the mitochondrion in programmed cell death pathways accumulates, new assays for cellular phenotypes will emerge beyond the simplistic assays of respiratory chain complex activity mostly used to date. Although the OXPHOS assays will continue to be a vital part of such characterization, wider cellular effects of mutations will need to be discovered before one can begin to understand the still baffling and mercurial clinical features of human mitochondrial diseases. An example of such new cellular phenotypes may be some recent reports using LHON cybrids, in which the mtDNA mutant cells are shown to be more sensitized to enter programmed cell death via mitochondrial pathways (49,50).

MOUSE CELL MTDNA MUTANTS AND CYBRIDS

In nearly 30 years, only a dozen different mouse cell mtDNA mutants have been described (Table 11.1). These have contributed important insights into the function of the affected OXPHOS complexes. As discussed elsewhere in this volume (Chapter 13), such mutants also represent a resource for production of transmitochondrial mice via a cybrid route. When investigated, most of the mutants described have moderate to severe OXPHOS phenotypes, which may preclude their use in mouse modeling studies. The limited mouse models produced to date include the use of one of the CAP^R mouse cell lines, which have moderate protein synthesis defects reflected in partial deficiencies of complexes I, III, and IV (29,51). Because the homoplasmic mice produced by Sligh et al. (51) showed a neonatal lethal phenotype, our expectation is that only mutants with mild OXPHOS phenotypes will be viable and, therefore, most informative.

The report of Sligh et al. is also significant for showing that mouse embryonic stem cell cybrids can maintain their pluripotency, even when treated with rhodamine 6-G prior to cytoplast fusion (51). The rhodamine treatment destroys the mitochondria and had been shown to prevent transmission of the mtDNAs of treated cells when fused to donor cytoplasts (52). This has opened the door for ES cell cybrid approaches to modeling mtDNA defects in mice (see Chapter 13).

Mouse mtDNA mutants with proven phenotypes identified in living mouse strains are completely lacking. One interesting variant has been described for the diabetes-prone BHE/Cdb rat, which was found to harbor a pair of point mutations in the ATP6 gene compared with Sprague–Dawley rats (see Berdanier (53) and Mathews (54)). A relative lack of rat ρ^0 cell lines has meant this mutation has yet to be investigated in a cybrid system. Until breakthrough technologies allow mtDNA transformation, a greater effort should be put into isolating new mouse cell mtDNA mutants for use in modeling.

Xenomitochondrial Cybrids

One approach to produce mtDNA-based defects without mutants is by creating xenomitochondrial cybrids. Interspecific transfer of mitochondria and mtDNAs was presaged by the pioneering work of the German plant geneticists (mentioned earlier) who investigated plastid compatibility between species. This line of investigation has continued to the present, representing an important experimental system in plant organelle genetics (see Gillham (55)). Early cybrid and hybrid studies using interspecific crosses of mammalian cells quickly determined that the mtDNAs segregated with the nuclear genes of the same species (56).

Studies with mice first explored the idea of introducing mtDNAs into interspecific nuclear backgrounds by backcrossing (57,58). Although limited to species in which interbreeding results in fertile offspring, these studies are interesting for demonstrating that such transfers can produce viable offspring. Some success has been reported for manipulating the germ-line mtDNAs in Drosophila, and subsequent breeding of heteroplasmic flies showed complete segregation of the endogenous mtDNA (59). Cybrids represent a cleaner experimental system because the cross is achieved in a single generation, with the advantage that no doubt about the nuclear gene complement can exist.

Because effects of endogenous mtDNA in the nuclear donor cell could be removed, ρ^0 cells presented an opportunity to revisit interspecies mitochondrial transfer. Kenyon and Moraes fused enucleated primate cells with human ρ^0 cells, selected for respiratory competent cells, and succeeded in producing xenocybrids (60). They showed that, with increased evolutionary divergence, a barrier was reached whereby the foreign mtDNA could no longer be maintained, so cybrids were not viable. Thus, chimpanzee and gorilla mtDNAs could be replicated and transcribed in human cells, but orangutan and more divergent primate mtDNAs could not. Interestingly, the OXPHOS of the cybrids showed defective complex I activity with the other respiratory chain complexes relatively preserved (60).

In a similar way, mouse ρ^0 cells were independently shown to be capable of maintaining other murid mtDNAs, including the closely related *Mus spretus* and the more distant mouse cousin the Norway rat (*Rattus norvegicus*) (61–63). Although the *Mus spretus* xenocybrid exhibited normal OXPHOS, the Rattus xenocybrid had severe combined respiratory chain defects. Murid rodents have more extant species than any other mammalian family, so a large number of potential xenocybrids can be made, although the limited number of described cell lines necessitates creation or procurement of new primary cell lines. Using this approach, McKenzie et al. (64,65) showed that species of intermediate divergence compared with Rattus could create graded respiratory chain defects in xenocybrids. Mild complex I and IV defects emerged in intermediate divergence cybrids, but a severe complex III defect was evident in the most divergent xenocybrids (64).

Like the primate xenocybrids, mouse cells also display limits to the evolutionary divergence of mtDNA donor species. Hamster mtDNAs, divergent by around 16 mybp, could not be maintained in cybrids with the mouse nucleus (61). The ρ^0 cell xenocybrids are selected by virtue of at least some recovery of respiratory chain function. These experiments cannot clearly distinguish whether the evolutionary barrier to xenocybrid creation results from total loss of respiratory chain function due to mismatching of nuclear and mtDNA subunits or from loss of mtDNA replication. This could be further addressed by transfection of key nuclear mtDNA replication factors (such as Tfam) from the mtDNA donor species. From the mouse xenocybrid studies, it is nevertheless clear that OXPHOS dysfunction will precede such replicative barriers in interspecific crosses.

SOME BURNING QUESTIONS FOR CYBRIDS

Complementation of Heteroplasmic mtDNA Mutants

Cybrid studies have provided contradictory evidence for mixing of organelles. Oliver and Wallace presented the first evidence that heteroplasmic mtDNAs could complement each other, implying fusing of organelles and mixing of mtDNAs and mitoribosomes (66). This conclusion was based on the observation that, in heteroplasmic cybrids, a CAP-sensitive marker polypeptide continued to be translated in the presence of CAP (66). Hayashi's group has been a proponent of such intramitochondrial complementation and has provided evidence in a mouse model (67) and in cultured cells (67,68). In both cases, heteroplasmic mtDNA deletions were used as a model system. When present in cells in very high levels, the deleted mtDNA prevented any mitochondrial translation and cells were cytochrome oxidase (COX) negative.

Newly created cybrids, or cells in mice carrying heteroplasmic deletions, always showed a uniform staining for COX, suggesting that single organelles had extensively fused (67,68). However, another study does not support such extensive mixing. Using cybrids carrying two different homoplasmic mtDNA point mutations, Enriquez et al. (69) found a level of transcomplementation of only around 1% of cybrids produced. This basic issue in mitochondrial genetics may have consequences for the expression of mutants in tissues and deserves further attention to better address whether different mutations or cell types will alter the rate of mitochondrial fusion and mixing of mtDNAs. With the recent emergence of real-time video florescence microscopy, cybrids could be investigated for such complementation immediately following cytoplast fusion.

mtDNA Contributions to Polygenic Disorders and Tissue-Specific Effects

In the classic mtDNA diseases, clear matrilineal inheritance, albeit with variable penetrance, strongly supports the disease link. Controversial claims have also been made over the past decade that mtDNA variants may also be contributors to the commonest age-related neurodegenerative disorders, especially Parkinson's disease (PD) and Alzheimer's disease (AD). Several cybrid reports have provided evidence for an association of mild OXPHOS defects and these diseases, but these reports await confirmation from other labs. Some support has been gained for the idea that at least a portion of PD patients may have mild, mtDNA-encoded complex I defects that can be transferred in cybrids (70,71). Adding to the controversy is the lack, to date, of identification of specific disease-associated mtDNA variants in PD and AD.

Such identification is extremely difficult when faced with the large degree of polymorphism encountered in mtDNA from different humans. As the number of fully sequenced individual mtDNAs grows, this big question may have begun to be answered. Wallace's group has recently published the largest of such analyses to date; this has provided some interesting clues (72). Populations from higher latitudes were found to exhibit higher rates of substitutions of more highly conserved amino acids, which may correlate with increased susceptibility to energetic diseases (72).

Why different tissues are adversely affected by different mtDNA mutations remains another enduring mystery of mitochondrial disease. One route to address this question has been to isolate ρ^0 cells from different cell types, including neuronal cell lines (70,73), although definitive insights into cell-specific pathogenesis are yet to emerge from the use of such lines. As mentioned earlier in relation to mtDNA disease cybrids, further OXPHOS phenotypes will probably need to be established before cybrids will be useful in determining mtDNA effects in polygenic disorders.

Unknown mtDNA and OXPHOS Functions

The discovery of the second role of the respiratory chain electron carrier, cytochrome c, as a messenger in programmed cell death signaling (74) was a startling reminder that nothing is what it seems in well-established biochemical pathways. We look forward to further surprises, especially with regard to the role of the OXPHOS complexes in mediating aspects of the cell death program. Further mysteries may still be encoded in the mtDNA. The identification of a potential open reading frame in the 16S rRNA gene, which is identical to the recently discovered nuclear-encoded neuronal survival factor "humanin" (75) is an intriguing example.

It is likely that cybrid studies will continue to represent an important tool for defining pathogenic effects of organellar genes, as well as in ongoing modeling studies in mice. Even when transformation techniques finally catch up with mtDNA, cybrids will continue to have a role in allowing ease of transfer of mitochondria with altered mtDNAs between cells.

REFERENCES

1. MM Nass, S Nass. Fibrous structures within the matrix of developing chick embryo mitochondria. *Exp Cell Res* 26:424–427; 1962.

2. MM Nass, S Nass. Intramitochondrial fibers with DNA characteristics. I. Fixation and electron staining reactions. *J Cell Biol* 19:593–611; 1963.

3. A Gibor, M Izawa. The DNA content of the chloroplasts of Acetabularia. *Proc Natl Acad Sci* (USA) 50:1164–1169; 1963.

4. A Gibor, S Granick. Plastids and mitochondria: inheritable systems. *Science* 145:890–897; 1964.

5. CL Bunn, DC Wallace, JM Eisenstadt. Cytoplasmic inheritance of chloramphenicol resistance in mouse tissue culture cells. *Proc Natl Acad Sci* (USA) 71:1681–1685; 1974.

6. DC Wallace. Mitochondrial diseases in man and mouse. *Science* 283:1482–1488; 1999.

7. J Smeitink, L van den Heuvel, S DiMauro. The genetics and pathology of oxidative phosphorylation. *Nat Rev Genet* 2:342–352; 2001.

8. PF Chinnery, EA Schon. Mitochondria. *J Neurol Neurosurg Psychiatry* 74:1188–1199; 2003.

9. B Ephrussi. The interplay of heredity and environment in the synthesis of respiratory enzymes in yeast. *Harvey Lect* 46:45–67; 1950.

10. P Michaelis. Cytoplasmic inheritance in *Epilobium* and its theoretical significance. *Adv Genet* 6:287–401; 1954.

11. G Lehman. Uber reziproke Bastarde zwischen *Epilobium roseum* und *parviflorum*. *Z Botan* 10:497–511; 1918.

12. O Renner. *Die Artbastarde bei Pflanzen*. Handb Vererbungsw II 1929; 161 pp.

13. E Baur. Das Wesen und die Erblichkeitsverhältnisse der "Varietates albomarginatae hort" von *Pelargonium zonale*. *Z Abstamm u Vererbungsl* 1:330–351; 1909.

14. R Hagemann. Erwin Baur or Carl Correns: who really created the theory of plastid inheritance? *J Hered* 91:435–440; 2000.

15. PP Slonimski, G Perrodin, JH Croft. Ethidium bromide induced mutation of yeast mitochondria: complete transformation of cells into respiratory deficient nonchromosomal "petites". *Biochem Biophys Res Commun* 30:232–239; 1968.

16. JC Mounolou, H Jakob, PP Slonimsky. Mitochondrial DNA from petite mutants: specific changes in buoyant density corresponding to different cytoplasmic mutations. *Biochem Biophys Res Comm* 24:218–224; 1966.

17. G Bernadi, M Faures, G Piperno, PP Slonimsky. Mitochondrial DNAs from respiratory-sufficient and cytoplasmic respiratory-deficient mutant yeast. *J Mol Biol* 48:23–42; 1970.

18. LA Grivell, P Netter, P Borst, PP Slonimski. Mitochondrial antibiotic resistance in yeast: ribosomal mutants resistant to chloramphenicol, erythromycin and spiramycin. *Biochim Biophys Acta* 312:358–367; 1973.

19. CT Roberts, E Orias. Cytoplasmic inheritance of chloramphenicol resistance in tetrahymena. *Genetics* 73:259–272; 1973.

20. CM Spolsky, JM Eisenstadt. Chloramphenicol-resistant mutants of human HeLa cells. *FEBS Lett* 25:319–324; 1972.

21. N Howell, K Gilbert. Mutational analysis of the mouse mitochondrial cytochrome b gene. *J Mol Biol* 203:607–618; 1988.

22. Y Bai, G Attardi. The mtDNA-encoded ND6 subunit of mitochondrial NADH dehydrogenase is essential for the assembly of the membrane arm and the respiratory function of the enzyme. *EMBO J* 17:4848–4858; 1998.

23. Y Bai, RM Shakeley, G Attardi. Tight control of respiration by NADH dehydrogenase ND5 subunit gene expression in mouse mitochondria. *Mol Cell Biol* 20:805–815; 2000.

24. B Dujon. Sequence of the intron and flanking exons of the mitochondrial 21S rRNA gene of yeast strains having different alleles at the omega and rib-1 loci. *Cell* 20:185–197; 1980.

25. SE Kearsey, IW Craig. Altered ribosomal RNA genes in mitochondria from mammalian cells with chloramphenicol resistance. *Nature* 290:607–608; 1981.

26. H Blanc, CW Adams, DC Wallace. Different nucleotide changes in the large rRNA gene of the mitochondrial DNA confer chloramphenicol resistance on two human cell lines. *Nucleic Acids Res* 9:5785–5795; 1981.

27. H Blanc, CT Wright, MJ Bibb, DC Wallace, DA Clayton. Mitochondrial DNA of chloramphenicol resistant mouse cells contains a single nucleotide change in the region encoding the 3′ end of the large ribosomal RNA. *Proc Natl Acad Sci* (USA) 78:3789–3793; 1981.

28. N Howell, A Lee. Sequence analysis of mouse mitochondrial chloramphenicol-resistant mutants. *Somat Cell Mol Genet* 15:237–244; 1989.

29. N Howell, MS Nalty. Mitochondrial chloramphenicol mutants can have deficiencies in energy metabolism. *Somat Cell Mol Genet* 14:185–193; 1987.

30. EF Slott, RO Shade, RA Lansman. Sequence analysis of mitochondrial DNA in a mouse cell line resistant to chloramphenicol and oligomycin. *Mol Cell Biol* 3:1694–1702; 1983.

31. N Howell, J Appel, JP Cook, B Howell, WW Hauswirth. The molecular basis of inhibitor resistance in a mammalian mitochondrial cytochrome b mutant. *J Biol Chem* 262:2411–2414; 1987.

32. R Acín–Pérez, MP Bayona–Bafaluy, M Bueno, C Machicado, P Fernández–Silva, A Pérez–Martos, J Montoya, MJ López–Pérez, J Sancho, JA Enríquez. An intragenic suppressor in the cytochrome c oxidase I gene of mouse mitochondrial DNA. *Hum Molec Genet* 12:329–339; 2003.

33. MP King, G Attardi. Human cells lacking mtDNA: repopulation with exogenous mitochondria by complementation. *Science* 246:500–503; 1989.

34. P Desjardins, E Frost, R Morais. Ethidium bromide-induced loss of mitochondrial DNA from primary chicken embryo fibroblasts. *Mol Cell Biol* 5:1163–1169; 1985.

35. DC Wallace, G Singh, MT Lott, JA Hodge, TG Schurr, AM Lezza, LJ Elsas, EK Nikoskelainen. Mitochondrial DNA mutation associated with Leber's hereditary optic neuropathy. *Science* 242:1427–1430; 1988.

36. Y Goto, I Nonaka, S Horai. A mutation in the tRNA[Leu(UUR)] gene associated with the MELAS subgroup of mitochondrial encephalomyopathies. *Nature* 348:651–653; 1990.

37. JM Shoffner, MT Lott, AM Lezza, P Seibel, SW Ballinger, DC Wallace. Myoclonic epilepsy and ragged-red fiber disease (MERRF) is associated with a mitochondrial DNA tRNA(Lys) mutation. *Cell* 61:931–937; 1990.

38. IJ Holt, AE Harding, JA Morgan–Hughes. Deletions of muscle mitochondrial DNA in patients with mitochondrial myopathies. *Nature* 331:717–719; 1988.

39. IJ Holt, AE Harding, RK Petty, JA Morgan–Hughes. A new mitochondrial disease associated with mitochondrial DNA heteroplasmy. *Am J Hum Genet* 46:428–433; 1990.

40. A Chomyn, G Meola, N Bresolin, ST Lai, G Scarlato, G Attardi. *In vitro* genetic transfer of protein synthesis and respiration defects to mitochondrial DNA-less cells with myopathy-patient mitochondria. *Mol Cell Biol* 11:2236–2244; 1991.

41. JF Hess, MA Parisi , JL Bennett, DA Clayton. Impairment of mitochondrial transcription termination by a point mutation associated with the MELAS subgroup of mitochondrial encephalomyopathies. *Nature* 351:236–239; 1991.

42. A Chomyn, A Martinuzzi, M Yoneda, A Daga, O Hurko, D Johns, ST Lai, I Nonaka, C Angelini, G Attardi. MELAS mutation in mtDNA binding site for transcription termination factor causes defects in protein synthesis and in respiration but no change in levels of upstream and downstream mature transcripts. *Proc Natl Acad Sci* (USA) 89:4221–4225; 1992.

43. MP King, Y Koga, M Davidson, EA Schon. Defects in mitochondrial protein synthesis and respiratory chain activity segregate with the tRNA(Leu(UUR)) mutation associated with mitochondrial myopathy, encephalopathy, lactic acidosis, and strokelike episodes. *Mol Cell Biol* 12:480–490; 1992.

44. DR Dunbar, PA Moonie, HT Jacobs, IJ Holt. Different cellular backgrounds confer a marked advantage to either mutant or wild-type mitochondrial genomes. *Proc Natl Acad Sci* (USA) 92:6562–6566; 1995.

45. IA Trounce, S Neill, DC Wallace. Cytoplasmic transfer of the mtDNA nt 8993 T→G (ATP6) point mutation associated with Leigh syndrome into mtDNA-less cells demonstrates cosegregation with a decrease in state III respiration and ADP/O ratio. *Proc Natl Acad Sci* (USA) 91:8334–8338; 1994.

46. AS Jun, IA Trounce, MD Brown, JM Shoffner, DC Wallace. Use of transmitochondrial cybrids to assign a complex I defect to the mitochondrial DNA-encoded NADH dehydrogenase subunit 6 gene mutation at nucleotide pair 14459 that causes Leber hereditary optic neuropathy and dystonia. *Mol Cell Biol* 16:771–777; 1996.

47. HR Cock, SJ Tabrizi, JM Cooper, AH Schapira. The influence of nuclear background on the biochemical expression of 3460 Leber's hereditary optic neuropathy. *Ann Neurol* 44:187–193; 1998.

48. MD Brown, IA Trounce, AS Jun, JC Allen, DC Wallace. Functional analysis of lymphoblast and cybrid mitochondria containing the 3460, 11778, or 14484 Leber's hereditary optic neuropathy mitochondrial DNA mutation. *J Biol Chem* 275:39831–39836; 2000.

49. SR Danielson, A Wong, V Carelli, A Martinuzzi, AH Schapira, GA Cortopassi. Cells bearing mutations causing Leber's hereditary optic neuropathy are sensitized to Fas-induced apoptosis. *J Biol Chem* 277:5810–5815; 2002.

50. A Ghelli, C Zanna, AM Porcelli, AH Schapira, A Martinuzzi, V Carelli, M Rugolo. Leber's hereditary optic neuropathy (LHON) pathogenic mutations induce mitochondrial-dependent apoptotic death in transmitochondrial cells incubated with galactose medium. *J Biol Chem* 278:4145–4150; 2003.

51. JE Sligh, SE Levy, KG Waymire, P Allard, DL Dillehay, S Nusinowitz, JR Heckenlively, GR MacGregor, DC Wallace. Maternal germ-line transmission of mutant mtDNAs from embryonic stem cell-derived chimeric mice. *Proc Natl Acad Sci* (USA) 97:14461–14466; 2000.

52. IA Trounce, DC Wallace. Production of transmitochondrial mouse cell lines by cybrid rescue of rhodamine-6G pretreated L-cells. *Somat Cell Mol Genet* 22:81–85; 1996.

53. CD Berdanier. Diabetes and nutrition: the mitochondrial part. *J Nutr* 131:344S–533S; 2001.

54. CE Mathews, RA McGraw, R Dean, CD Berdanier. Inheritance of a mitochondrial DNA defect and impaired glucose tolerance in BHE/Cdb rats. *Diabetologia* 42:35–40; 1999.

55. NW Gillham. *Organelle Genes and Genomes*. Oxford University Press, New York; 1994.

56. DC Wallace, Y Pollack, CL Bunn, JM Eisenstadt. Cytoplasmic inheritance in mammalian tissue culture cells. *In Vitro* 12:758–776; 1976.

57. U Gyllensten, D Wharton, AC Wilson. Maternal inheritance of mitochondrial DNA during backcrossing of two species of mice. *J Hered* 76:321–324; 1985.

58. Y Nagao, Y Totsuka, Y Atomi, H Kaneda, KF Lindahl, H Imai, H Yonekawa. Decreased physical performance of congenic mice with mismatch between the nuclear and the mitochondrial genome. *Genes Genet Syst* 73:21–27; 1998.

59. Y Niki, SI Chigusa, ET Matsuura. Complete replacement of mitochondrial DNA in Drosophila. *Nature* 341:551–552; 1989.

60. L Kenyon, CT Moraes. Expanding the functional human mitochondrial DNA database by the establishment of primate xenomitochondrial cybrids. *Proc Natl Acad Sci* (USA) 94:9131–9135; 1997.

61. R Dey, A Barrientos, CT Moraes. Functional constraints of nuclear–mitochondrial DNA interactions in xenomitochondrial rodent cell lines. *J Biol Chem* 275:31520–31527; 2000.

62. M McKenzie, IA Trounce. Expression of *Rattus norvegicus* mtDNA in *Mus musculus* cells results in multiple respiratory chain defects. *J Biol Chem* 275:31514–31519; 2000.

63. M Yamaoka, K Isobe, H Shitara, H Yonekawa, S Miyabayashi, JL Hayashi. Complete repopulation of mouse mitochondrial DNA-less cells with rat mitochondrial DNA restores mitochondrial translation but not mitochondrial respiratory function. *Genetics* 155:301–307; 2000.

64. M McKenzie, M Chiotis, CA Pinkert, IA Trounce. Functional respiratory chain analyses in murid xenomitochondrial cybrids expose coevolutionary constraints of cytochrome b and nuclear subunits of complex III. *Mol Biol Evol* 20:1117–1124; 2003.

65. M McKenzie, IA Trounce, CA Cassar, CA Pinkert. Production of homoplasmic xenomitochondrial mice. *Proc Natl Acad Sci (USA)* 101:1685–1690; 2004.

66. NA Oliver, DC Wallace. Assignment of two mitochondrially synthesized polypeptides to human mitochondrial DNA and their use in the study of intracellular mitochondrial interaction. *Mol Cell Biol* 2:30–41; 1982.

67. K Nakada, K Inoue, T Ono, K Isobe, A Ogura, YI Goto, I Nonaka, JI Hayashi. Intermitochondrial complementation: mitochondria-specific system preventing mice from expression of disease phenotypes by mutant mtDNA. *Nat Med* 7:934–940; 2001.

68. J Hayashi, M Takemitsu, Y Goto, I Nonaka. Human mitochondria and mitochondrial genome function as a single dynamic cellular unit. *J Cell Biol* 125:43–50; 1994.

69. JA Enriquez, J Cabezas–Herrera, MP Bayona–Bafaluy, G Attardi. Very rare complementation between mitochondria carrying different mitochondrial DNA mutations points to intrinsic genetic autonomy of the organelles in cultured human cells. *J Biol Chem* 275:11207–11215; 2000.

70. RH Swerdlow, JK Parks, SW Miller, JB Tuttle, PA Trimmer, JP Sheehan, JP Bennett, RE Davis, WD Parker. Origin and functional consequences of the complex I defect in Parkinson's disease. *Ann Neurol* 40:663–671; 1996.

71. M Gu, JM Cooper, JW Taanman, AH Schapira. Mitochondrial DNA transmission of the mitochondrial defect in Parkinson's disease. *Ann Neurol* 44:177–186; 1998.

72. E Ruiz–Pesini, D Mishmar, M Brandon, V Procaccio, DC Wallace. Effects of purifying and adaptive selection on regional variation in human mtDNA. *Science* 303:223–226; 2004.

73. R Fukuyama, A Nakayama, T Nakase, H Toba, T Mukainaka, H Sakaguchi, T Saiwaki, H Sakurai, M Wada, S Fushiki. A newly established neuronal rho-0 cell line highly susceptible to oxidative stress accumulates iron and other metals. Relevance to the origin of metal ion deposits in brains with neurodegenerative disorders. *J Biol Chem* 277:41455–41462; 2002.

74. X Liu, CN Kim, J Yang, R Jemmerson, X Wang. Induction of apoptotic program in cell-free extracts: requirement for dATP and cytochrome c. *Cell* 86:147–157; 1996.

75. B Guo, D Zhai, E Cabezas, K Welsh, S Nouraini, AC Satterthwait, JC Reed. Humanin peptide suppresses apoptosis by interfering with Bax activation. *Nature* 423:456–461; 2003.

12

Animal Modeling: From Transgenesis to Transmitochondrial Models

CARL A. PINKERT AND IAN A. TROUNCE

CONTENTS

INTRODUCTION

Historical Overview

The field of mitochondrial medicine is still in its infancy. The first published accounts of diseases caused by mutations of mtDNA were reported in 1988; now, scores of point mutations and rearrangements of the mitochondrial genome are known to be the underlying causes of various degenerative disorders (1–7; see also MITOMAP at http://www.mitomap.org). The Mitochondrial Research Society estimates that more than 50 million adults in the U.S. suffer from diseases in which mitochondrial dysfunction is involved and that mitochondrial dysfunction is found in a broad spectrum of diseases — from diabetes and infertility to cancer and age-related neurodegenerative disorders (see http://www.mitoresearch.org). Additionally, mutations that affect mitochondrial function have a dramatic impact upon tissues with high cellular energy requirements in the central nervous system, cardiac and skeletal muscle, and various endocrine organs (5).

In mammals, mitochondrial genetics are inherited in matrilineal fashion. Furthermore, it has been postulated that a developmental bottleneck exists in female germ cell development, whereby a small number of mitochondria give rise to the ~10^5 mitochondria that populate the mature ovum (8–10). Through the late 1990s, *in vivo* animal models of mtDNA mutation and human disease were virtually nonexistent because techniques for modifying large numbers of mitochondria in early embryos were far more complex than comparable modifications of nuclear-encoded genes.

As more is learned about mitochondrial dysfunction and developmental consequences of mitochondrial mutation, creation of animal models will be of critical importance not only in the study of mitochondrial dynamics and function but also in dissection of mtDNA-based disease pathogenesis. Currently, the term "mitochondrial disease" is used mostly to describe diseases caused by mutations in mitochondrial DNA. However, defining mitochondrial disease is probably more accurate in the broader context of recognizing that mitochondrial diseases are those with a mitochondrial component, genetically or environmentally influenced, that produce cell and organ dysfunction or failure under acute and chronic stress.

In Vitro Technologies

Manipulation of the mitochondrial genome *in vitro* was technically feasible by the mid-1980s. Human and murine mitochondrial genomes were cloned (see Clayton (2); full-length cloning of the human mtDNA in bacteria has not been possible due to "poison" sequences in the D-loop; see Bigger et al. (11)) and sequenced and are available as constituents of plasmid constructs. The differences between nuclear and mitochondrial codon assignments were identified, and the general features associated with mitochondrial targeting of nuclear-encoded mitochondrial proteins were described (12–15).

In order to employ genetically engineered mitochondria or mitochondrial genes for *in vivo* gene transfer or for gene therapy, efficient methods to introduce foreign or altered mitochondrial genomes into somatic cells or germ cells must be identified. In this regard, pioneering work in the development of cybrid technology — first, using mtDNA mutants resistant to chloramphenicol (16) and then using mtDNA-less (ρ^0) cells — made the establishment of transmitochondrial mouse models inevitable (17).

A cybrid, by definition, can harbor more than one type of mtDNA. This coexistence of more than one form of mtDNA within a single cell or within cells of a tissue, organ, or individual organism is referred to as "heteroplasmy" (in contrast to maintenance of a single population of mtDNA genomes, which is referred to as "homoplasmy"). Early cybrid experiments have demonstrated that multiple mitochondrial genotypes can be maintained within cells and that selective segregation and amplification of one genotype often occur.

SPONTANEOUS AND INDUCED MODELS

In developing a rat model for diabetes mellitus, the BHE rat was identified early as a unique nonobese model that appeared inconsistent with expected disease phenotype

(18–20). In contrast to the BHE rat, most rodents that spontaneously develop non-insulin-dependent diabetes mellitus are obese. However, the BHE rat develops abnormal glucose tolerance and is lipemic with fatty liver, but is not obese (20). In turn, the lipemic/glycemic BHE/Cdb rat studied throughout the 1990s was first characterized with two homoplasmic and three heteroplasmic mutations in the mitochondrial ATPase 6 gene, causing impairment of oxidative phosphorylation (OXPHOS) and glucose intolerance; the specific mutations have been selectively segregated and studied (21,22).

The BHE/Cdb rat and related models paved the way for current hypotheses related to nutritional regulation of mitochondrial function in mammals. As late as 1996, animal models of heritable encephalomyopathy were unavailable (23). Chronic administration of compounds such as diphenylene iodonium to rats did illustrate similarities to known myopathies and germanium dioxide administration to rats produced hallmarks of mitochondrial disorders, including ragged-red fibers accumulation with decreased cytochrome c oxidase activity (24,25).

A number of characterized spontaneously identified mouse strains in the literature have been useful in furthering understanding of mitochondrial dynamics and in response to environmental toxicants. The copper-deficient brindled (mottled-brindled, Mobr or mo[br]; now Atp7a[Mo–br]) mouse is normally considered a model of Menkes disease with decreased cytochrome oxidase enzymatic activity (26). Cytochrome c oxidase and copper–zinc superoxide dismutase were compared in copper-deficient brindled and blotchy (mottled-blotchy, Moblo or Atp7a[Mo–blo]; also an osteoarthritis model) mouse mutants and normal mice (27). In brindled and blotchy mutants, enzyme deficiencies were identified in brain, heart, and skeletal muscle. However, cytochrome c oxidase was more severely affected than was superoxide dismutase. The deficiency was correlated with decreased copper concentration; however, enzyme activity was normal in liver, kidney, and lung, despite abnormal copper concentrations in these tissues.

In addition, these results were comparable to nutritionally copper-deficient mice in which decreased enzyme activity was marked in brain, heart, and skeletal muscle. Interestingly, exogenous copper administration increased cytochrome c oxidase activity in all deficient tissues of brindled mice, but only in brain and heart from blotchy mice. However, skeletal-muscle cytochrome c oxidase in blotchy mutants did not respond to copper injection indicative of differential developmental pathways.

In 1998, identifying the lack of existing animal models of mitochondrial disease, normal human myoblast cells were injected into muscle of SCID mice postnecrosis induction in order to identify whether normal human cells would functionally repopulate the impaired mouse tissue (28). Postinjection, regenerated muscle fibers expressing human $-spectrin and human cytochrome c oxidase subunit II were observed, illustrating a successful and localized reconstitution of human myoblast-derived mitochondria. Neurodegeneration modeling and mitochondrial dysfunction were also identified following chronic exposure of mice to MPTP⁺ and rats to rotenone — both complex I inhibitors (29,30). Here, specific dopaminergic neuronal inhibition was observed resembling Parkinson's disease phenotype in humans reflective of epidemiological data related to pesticide exposure.

In the aggregate, a large body of data has accumulated over the years that readily reflected mitochondrial dysfunction in animal models, and many of these models have been molecularly characterized after their initial description in the literature. Demonstration of methodology to tailor mitochondrial genetics and mitochondrial:nuclear dynamics was the next logical progression in animal modeling efforts targeting mitochondrial biology.

TRANSGENIC MODELS

Nuclear Gene Modifications

As described in a number of recent reports (see Wallace (5)), nuclear-encoded genes and knock-out modeling have been informative in identifying novel models in mitochondrial disease pathogenesis as well as critical pathways associated with mitochondrial function. The host of characterized models includes those associated with ablation of the genes encoding adenine nucleotide translocator-1 (Ant1); manganese superoxide dismutase (MnSOD); mitochondrial transcription factor A (Tfam); glutathionine peroxidase (GPx1); and uncoupling proteins 1, 2, and 3.

As examples of the pathophysiology of knocking out key regulatory enzymes, Ant1-deficient mouse models exhibited severe mitochondrial myopathies. MnSOD-deficient mice included two different models that produced cardiomyopathy, neuronal degeneration, and liver dysfunction. Tfam-deficient mice exhibited a reduction in complex I and complex IV activity. GPx1-deficient mice were viable but displayed growth retardation and liver-specific but not heart pathology. Lastly, UCP1-, UCP2-, or UCP3-deficient mice illustrated the nature of compensatory function when one or two proteins were ablated, with phenotype ranging from cold sensitivity to an increase in reactive oxygen species generation. With the interaction of literally a few thousand nuclear genes critical to metabolic function and normal mitochondrial activity, it is not surprising to see the large body of literature developing from nuclear gene overexpression, knock-out, knock-in, and conditional models.

Modifying the Mitochondrial Genome

With initial characterization of nuclear gene-encoded models that affected mitochondrial function, our search for a greater understanding of mitochondrial biology would eventually lead us to develop methodology for mitochondria and mitochondrial gene transfer. As an initial endeavor, efficient methods for introduction of foreign or modified mtDNA, or of intact foreign mitochondrial genomes, into somatic or germ cells would be needed.

First Transmitochondrial Mice

Early attempts to create transmitochondrial strains of mammalian species by introduction of foreign mitochondria into germ cells were not successful at generating heteroplasmic mice, but they did illustrate important considerations in development of a microinjection methodology (31). In experiments designed to mimic the transfer of mitochondria from spermatozoa during mammalian fertilization, introduction of approximately 120 intact mitochondria into fertilized murine zygotes did not result in detectable levels of foreign mtDNA in offspring. Because these early experiments employed detection techniques that were less sensitive than PCR-based methods (e.g., southern blot hybridization analyses) and because of the vast discrepancy in the number of foreign mitochondria microinjected vs. the number known to be present in the average mouse ovum, it was quite possible that heteroplasmic or "transmitochondrial" offspring were produced in these experiments but remained unidentified. However, similar to sperm mitochondria programming for destruction, mitochondrial transfer efforts may have reflected a biological cascade of events critical to the fertilization/post-fertilization period (10,32).

Techniques to create heteroplasmic mice then progressed via cytoplast fusion (33,34) and by embryonic karyoplast transplantation (35). In these experiments, rapid segregation of mtDNA was possible within maternal lineages; however, specific manipulations were not readily controllable or quantifiable in first-generation animals. Generally, the recipient embryo appeared to dominate in terms of mitochondrial survival: "The mechanisms for eliminating introduced mitochondria in eukaryotic organisms are extraordinarily diverse and mysterious, but the bottom line is that any hint of heteroplasmy seems to be highly nonadaptive and is rapidly eliminated by selection processes" (36).

Interestingly, early reports on development of cloned animals by nuclear transfer resulted in conflicting consequences when retrospective studies on mitochondrial transmission were reported (37–40). Similarly, using a human *in vitro* fertilization protocol, two heteroplasmic children may have been inadvertently created (41,42). As such, research independent of targeted mitochondrial genomic modifications may also help unlock mechanisms underlying the dynamics related to persistence of foreign mitochondria and maintenance of heteroplasmy in various cloning protocols. Indeed, specific culture-related conditions that may influence the prevalence or development of heteroplasmy in these techniques provide one example to explain the range of results (heteroplasmy to homoplasmy) observed in nuclear transfer experimentation (43).

For *in vivo* modeling, a number of laboratories have reported on methodologies used to create transmitochondrial mouse models (7,44–48). All of these reports illustrated various aspects of *in vivo* modeling of mitochondrial dynamics and human disease. Interestingly, in some of the reports in which developmental consequences of the genetic manipulations were observed, an aberrant or unexpected phenotype appeared to be the rule (47,48). In contrast to existing karyoplast and cytoplast methods, our efforts to devise a direct mitochondria-transfer technique offered a number of additional advantages (Figure 12.1). Of greatest interest in initial efforts was the possibility of *in vitro* engineering of isolated mitochondria for the production of heteroplasmic mice (49).

mtDNA Injection vs. ES Cell-Derived Models

Our current modeling systems were designed to effect an efficient method to introduce mutant or foreign mitochondrial populations into the mouse germline. Initially, following cytoplast injection and *in vivo* fusion studies (50), direct mitochondrial injection into zygotes demonstrated that germline transmission of heteroplasmy was achievable (7,44). Unfortunately, these early studies did not result in high levels of heteroplasmy in founders or offspring, perhaps reflecting the somatic cell origin of the transferred mitochondria. Subsequently, ES cell transfer into morula- or blastocyst-stage embryos has proven more effective in establishing high percentages of heteroplasmy similar to reports from other laboratories, allowing us to determine more readily the fate of these foreign mitochondria in transmitochondrial mouse offspring during embryonic development and postnatal life (e.g., phenotypic changes associated with aging).

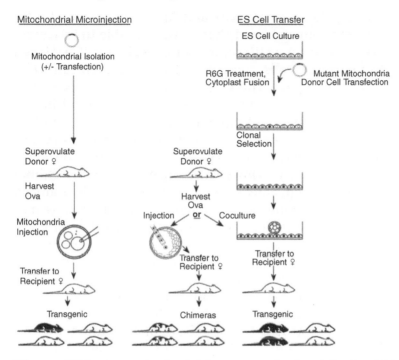

Figure 12.1 Transmitochondrial mouse production. Mitochondrial injection and ES cell transfer technologies represent two methods for producing transmitochondria derived from traditional transgenic animal modeling (72). A principal difference in the two techniques is the *in vitro* culture step allowing propagation of targeted clones using immortalized ES cells (right). Left panel: for microinjection of intact mitochondria, the mitochondrial preparation is injected into the cytoplasm of the pronuclear zygotes causing expansion of the vitelline membrane (intracellular expansion) with the occasional appearance of the extranuclear vacuole as depicted. In general, if successful, mitochondria injection results in heteroplasmic cells and consequently transgenic (or transmitochondrial) mice (depicted by the black mouse, which would harbor mutant and wild-type mitochondrial genomes). Right panel: after clonal selection of transfected ES cells, one of two additional techniques can be employed for ES cell transfer. Ova are harvested between the eight-cell and blastocyst stages. R6G-treated and -transfected ES cells are injected directly into a host blastocyst (injection), or cocultured with eight-cell to morula stage ova, so that transfected ES cells are preferentially incorporated into the inner cell mass of the developing embryo (coculture). With blastocyst injection, transgenic offspring are termed chimeric because some of their cells are derived from the host blastocyst and some from transfected ES cells (denoted by white mice with black patches). Using coculture and tetraploid embryos, one can obtain founder mice derived completely from the transfected ES cells (black mice). (From C.A. Pinkert and I.A. Trounce, *Methods*, 26:348–357, 2002. (72) With permission.)

Similarly, Levy et al. (45) and Marchington et al. (46) independently demonstrated that it was possible to fuse cytoplasts prepared from a chloramphenicol-resistant (CAP^R) cell line with mouse ES cells and then introduce the mutant cells into mouse blastocysts. The CAP^R mutation was expected to impart a respiratory deficiency in resultant animals; although not initially identified, anticipated phenotype and germ-line competence in founders was later demonstrated (45,48). Unexpectedly, chimeric founders exhibited ocular abnormalities including cataracts. However, offspring of founder chimeras exhibited heteroplasmy or homoplasmy for the introduced mutation, with striking developmental abnormalities (ranging from severe growth retardation and perinatal or *in utero* lethality).

An important innovation was in the use of rhodamine-6G (R6G) to limit the transmission of endogenous ES cell mtDNAs — something that had been previously demonstrated only for somatic cells (51). The advantage of this approach was illustrated by the failure of Marchington et al. to obtain homoplasmic ES cell cybrids or mice using chloramphenicol selection alone (46). Although these initial studies did not involve a synthesized/engineered mitochondrial genome per se, they illustrated that such modeling would be feasible as a critical component in developing targeted animal models of human disease.

In the absence of methods to produce targeted mtDNA mutants, the current limitation to the production of mtDNA mutant mice is the lack of suitable mouse cell mutants. A handful of mutants has been described over the past 20 years, all produced by *in vitro* mutagenesis followed by selection of drug-resistant mtDNA mutants. Although some of these may prove interesting if successfully introduced into mice, the findings of Sligh et al. sound a warning (48). The relatively mild impairment of respiration consequent to the CAP^R mutation as defined by polarographic measurements (45), surprisingly, resulted in the *in utero* or perinatal death of animals with high levels of the mutation. This suggested that mutations with mild respiratory chain impairment as measured *in vitro* would likely prove the most successful in modeling human disorders. Data on the respiratory chain function are lacking for many of the mouse cell mutants described, but when available, they suggest that these mutants are severe (e.g., Howell (52) and Bai and Attardi (53)).

Xenomitochondrial Mice

Other approaches for creation of mice with modified mitochondrial backgrounds included the introduction of species-specific mitochondria into zygotes or development of xenomitochondrial cybrids and creation of chimeric mice. We first used the injection of *Mus spretus* mitochondria into mouse (*Mus musculus* domesticus) zygotes (7). Subsequently, a different approach based on interprimate transfer studies (54) further illustrated that mouse cells could be repopulated with Rattus mtDNA, resulting in a severe respiratory impairment (55,56).

The pioneering xenocybrid studies could be exploited utilizing xenocybrids from several mouse species to evaluate a graded respiratory impairment *in vivo* in mouse models (57; see Chapter 11). Taking advantage of the enormous evolutionary diversity of Muridae species, one objective was to create a series of cybrids with increasing levels of respiratory chain impairment resulting from the presence of mismatched nuclear-mtDNA subunits. Modeling of the defects was first done in a ρ^0 cybrid system; then, lineages in which phenotype was observed (here, a mild respiratory impairment) could be readily transferred into mouse female ES cells for production of chimeric mice.

Introduction of Mutant mtDNA into Mitochondria

A primary focus for our work has included production of engineered mutations of mtDNA genomes and subsequent transduction or transfection into isolated mitochondria *in vitro* prior to transfer into mouse embryos. A key objective in this regard would be creation of transmitochondrial mouse models of human diseases. At present, outside of recapitulating a mitochondrial genomic deletion mutant (58), methods for the experimental manipulation of mtDNA genomes in a precise, directed fashion have yet to be described. Obviously, *in vitro* production of human cell lines containing mutant mtDNA was achieved for a number of mtDNA-related diseases by introduction of isolated patient mitochondria into ρ^0 cells (59–62).

Methods for the delivery of complete foreign mtDNA genomes into intact mitochondria are presently nonexistent. The transfection of DNA into mitochondria presents some very formidable challenges. As a prime example, the outer and inner mitochondrial membranes must be traversed; furthermore, their protein and lipid compositions are very different from the plasma membrane.

To an extent, the protein and lipid compositions of mitochondrial membranes are known (63–65; reviewed by Ellis and Reid (66)); however, questions still remain. Although *in vitro* fusion of inner mitochondrial membrane vesicles was reported, there have been no reports of fusion of outer membrane vesicles or of whole mitochondria (67). Also, the basic mechanisms for mitochondrial membrane lipid addition have yet to be elucidated (i.e., how newly synthesized lipid is targeted to mitochondrial membranes). Such information would be very useful in attempts to design liposome-mediated DNA delivery systems.

Using mitochondrial targeting peptides covalently attached to DNA molecules, Seibel et al. demonstrated internalization of up to 322-bp DNA fragment into intact mitochondria via a protein import pathway (68). However, delivery of mitochondrial-specific DNA sequences and subsequent integration into the host mitochondrial genome, although yet to be demonstrated, may not be possible if mechanisms for DNA recombination within mitochondria are lacking.

A mouse model of human mtDNA deletion syndromes may be the most straightforward to construct because the steps required are attainable with existing technologies. Currently, the greatest obstacle to surmount involves transfecting viable mitochondria with mutant mtDNA genomes (58,69). It may be that electroporation protocols may not work for full-length mtDNA molecules with introduced point mutations; therefore, models for many common human mtDNA-based diseases (e.g., LHON, MELAS, MERRF) may not be possible using this technique.

However, creation of a deletion mutation that resulted in a closed circular mtDNA molecule less than 7.2 kb (while still retaining O_H and O_L) would allow for successful electroporation into isolated mitochondria followed by transfer of viable mitochondria into recipient zygotes by the techniques described earlier. Using long-range, high-fidelity PCR and standard molecular cloning protocols, we previously produced a 6.5-kb mouse mtDNA deletion construct, designated mt-del, and introduced this construct into isolated mitochondria by electroporation. The resultant transfected mitochondria were shown to be intact and coupled by closed-chamber respirometry, according to respiratory control ratio (RCR) and P/O ratio parameters. However, when transfected mitochondria were introduced into zygotes, they were not detectable in live-born offspring (58 and unpublished data).

Use of Transfected-D° Cells as Intermediate Mitochondrial Carriers in the Production of Mouse Models

At this juncture, the major roadblock to using mitochondrial microinjection would be optimization of the transfection methodology with demonstration of our ability to produce heteroplasmic mice. Indeed, the risks inherent in this objective were significant. Therefore, an intermediate culture system could facilitate this critical aim. Here, we would take advantage of breakthroughs in ES cell-based technologies and the work of Levy et al., in which ES cells were used to transfer a foreign mitochondrial genome into mouse blastocysts (45).

In developing *in vitro* models of human mitochondrial disorders, immortalized human and mouse D° cell lines were developed (17,53,70). By convention, D° cells are devoid of any mitochondrial DNA and require uridine and pyruvate in culture media for survival (17). Thus, immortalized D∞ cells provide an excellent vehicle for propagation of our mtDNA-del containing mitochondria. With transfer of transfected mitochondria (by microinjection or cytoplast fusion), selection pressure could be exerted by removal of uridine and pyruvate from the culture medium. Thus, a means of selection could be employed prior to clonal expansion and subsequent analyses for mitochondrial mutants. In the case of the mtDNA deletion constructs in which complementation of respiratory chain activity would not be expected, the selection would not be used.

By isolation of many clones following microinjection of transfected mitochondria (71), clones with the construct could be identified by PCR analysis. Such methodology

- Allows for enhanced survival rates of transfected mitochondria and enhanced concentration of specifically modified mitochondria
- Acts as a culture system to propagate mitochondria
- Allows for controlled culture conditions to facilitate mitochondrial engineering

An important advantage of this intermediate culture system is that heteroplasmic constructs can be produced at will by simply mixing cells of interest in the enucleation step, then fusing the mixed cytoplasts (in different ratios, if desired) and genotyping many cybrid clones to select heteroplasmic clones. These clones can then be used in fusion with the R6G-treated ES cells as recently described (57).

SUMMARY/FUTURE DIRECTION

Through the early 1990s, various early attempts to create transmitochondrial strains of mammalian species by introduction of foreign mitochondria into germ cells were largely unsuccessful. A number of constraints have been identified or postulated — from perturbations of biological pathways to mechanistic aspects of the specific protocols used. Since 1997, a number of laboratories have reported on methodologies used to create transmitochondrial animals.

To date, methods for mitochondria isolation and interspecific transfer of mitochondria have been reported in laboratory and domestic animal models (35,55,72,73). Interestingly, early reports on development of cloned animals by nuclear transfer resulted in conflicting consequences when retrospective studies on mitochondrial transmission were reported (37–41,72). Indeed, depending upon the specific methodology employed for nuclear transfer and cytoplasm/ooplasm transfer to rescue embryos, additional models of heteroplasmy may have been characterized as a consequence of mitochondrial dysfunction. As such, research dependent on and independent of targeted mitochondrial genomic modifications may also help unlock mechanisms underlying the dynamics related to persistence of foreign mitochondria and maintenance of heteroplasmy in various cloning protocols. In contrast, current ES cell approaches in mice hold significant promise in targeted modification of mitochondrial genes and in the study of nuclear–mitochondrial crosstalk.

REFERENCES

1. DC Wallace, X Zheng, MT Lott, JM Shoffner, JA Hodge, RI Kelley, CM Epstein, LC Hopkins. Familial mitochondrial encephalomyopathy (MERRF): genetic, pathophysiological, and biochemical characterization of a mitochondrial DNA disease. *Cell* 55:601–610; 1988.

2. DA Clayton. Replication and transcription of vertebrate mitochondrial DNA. *Annu Rev Cell Biol* 7:453–478; 1991.

3. DA Clayton. Structure and function of the mitochondrial genome. *J Inher Metab Dis* 15:439–447; 1992.

4. JJ Lemasters, AL Nieminen. *Mitochondria in Pathogenesis.* Kluwer Academic/Plenum Publishers, New York; 2001.

5. DC Wallace. Mouse models for mitochondrial disease. *Am J Med Genet* 106:71–93; 2001.

6. WC Copeland. *Mitochondrial DNA: Methods and Protocols. Methods in Molecular Biology.* Humana Press, Totowa, NJ; 2002.

7. CA Pinkert, MH Irwin, LW Johnson, RJ Moffatt. Mitochondria transfer into mouse ova by microinjection. *Transgenic Res* 6:379–383; 1997.

8. L Piko, L Matsumoto, Number of mitochondria and some properties of mitochondrial DNA in the mouse egg. *Dev Biol* 49:1–10; 1976.

9. DR Marchington, GM Hartshorne, D Barlow, J Poulton. Homopolymeric tract heteroplasmy in mtDNA from tissues and single oocytes: support for a genetic bottleneck. *Am J Hum Genet* 60:408–416; 1997.

10. J Cummins. Mitochondrial DNA in mammalian reproduction. *Rev Reprod* 3:72–82; 1998.

11. B Bigger, O Tolmachov, JM Collombet, C Coutelle. Introduction of chloramphenicol resistance into the modified mouse mitochondrial genome: cloning of unstable sequences by passage through yeast. *Anal Biochem* 277:236–242; 2000.

12. MJ Bibb, RA Van Etten, CT Wright, MW Walberg, DA Clayton. Sequence and gene organization of mouse mitochondrial DNA. *Cell* 26:167–180; 1981.

13. DP Tapper, RA Van Etten, DA Clayton. Isolation of the mammalian mitochondrial DNA and RNA and cloning of the mitochondrial genome. *Meth Enzymol* 97:426–434; 1983.

14. N Pfanner, W Neupert. A mitochondrial machinery for membrane translocation of precursor proteins. *Biochem Soc Trans* 18:513 515; 1990.

15. N Pfanner, W Neupert. The mitochondrial protein import apparatus. *Annu Rev Biochem* 59:331–353; 1990.

16. DC Wallace, CL Bunn, JM Eisenstadt. Cytoplasmic transfer of chloramphenicol resistance in human tissue culture cells. *J Cell Biol* 67:174–188; 1975.

17. MP King, G Attardi. Human cells lacking mtDNA: repopulation with exogenous mitochondria by complementation. *Science* 246:500–503; 1989.

18. MK Lakshmanan, CD Berdanier, RL Veech. Comparative studies on lipogenesis and cholesterogenesis in lipemic BHE rats and normal Wistar rats. *Arch Biochem Biophys* 183:355–360; 1977.

19. CD Berdanier. Rat strain differences in gluconeogenesis by isolated hepatocytes. *Proc Soc Exp Biol Med* 169:74–79; 1982.

20. CD Berdanier. The BHE rat: an animal model for the study of non-insulin-dependent diabetes mellitus. *FASEB J* 5:2139–2144; 1991.

21. CE Mathews, RA McGraw, R Dean, CD Berdanier. Inheritance of a mitochondrial DNA defect and impaired glucose tolerance in BHE/Cdb rats. *Diabetologia* 42:35–41; 1999.

22. HB Everts, CD Berdanier. Nutrient-gene interactions in mitochondrial function: vitamin A needs are increased in BHE/Cdb rats. *IUBMB Life* 53:289–294; 2002.

23. I Tracey, JF Dunn, GK Radda. A ^{31}P-magnetic resonance spectroscopy and biochemical study of the mo^{vbr} mouse: potential model for the mitochondrial encephalomyopathies. *Muscle Nerve* 20:1352–1359; 1997.

24. I Higuchi, K Takahashi, K Nakahara, S Izumo, M Nakagawa, M Osame. Experimental germanium myopathy. *Acta Neuropathol* 82:55–59; 1991.

25. JM Cooper, DJ Hayes, RA Challiss, JA Morgan–Hughes, JB Clark. Treatment of experimental NADH ubiquinone reductase deficiency with menadione. *Brain* 115:991–1000; 1992.

26. JH Menkes, M Alter, GK Steigleder, DR Weakley, JH Sung. A sex-linked recessive disorder with retardation of growth, peculiar hair, and focal cerebral and cerebellar degeneration. *Pediatrics* 29:764–779; 1962.

27. M Phillips, J Camakaris, DM Danks. Comparisons of copper deficiency states in the murine mutants blotchy and brindled. Changes in copper-dependent enzyme activity in 13-day old mice. *Biochem J* 238:177–183; 1986.

28. KM Clark, DJ Watt, RN Lightowlers, MA Johnson, JB Relvas, JW Taanman, DM Turnbull. SCID mice containing muscle with human mitochondrial DNA mutations. An animal model for mitochondrial DNA defects. *J Clin Invest* 15:2090–2095; 1998.

29. RE Heikkila, L Manzino, FS Cabbat, RC Duvoisin. Protection against the dopaminergic neurotoxicity of 1-methyl-4-phenyl-1,2,5,6-tetrahydropyridine by monoamine oxidase inhibitors. *Nature* 10:467–469; 1984.

30. R Betarbet, TB Sherer, G MacKenzie, M Garcia–Osuna, AV Panov, JT Greenamyre. Chronic systemic pesticide exposure reproduces features of Parkinson's disease. *Nat Neurosci* 3:1301–1306; 2000.

31. KM Ebert, A Alcivar, B Liem, R Goggins, NB Hecht. Mouse zygotes injected with mitochondria develop normally but the exogenous mitochondria are not detectable in the progeny. *Mol Reprod Dev* 1:156–163; 1989.

32. JM Cummins, H Kishikawa, D Mehmet, R Yanagimachi. Fate of genetically marked mitochondria DNA from spermatocytes microinjected into mouse zygotes. *Zygote* 7:151–156; 1999.

33. JP Jenuth, AC Peterson, K Fu, EA Shoubridge. Random genetic drift in the female germline explains the rapid segregation of mammalian mitochondrial DNA. *Nature Genet* 14:146–151; 1996.

34. JP Jenuth, AC Peterson, EA Shoubridge. Tissue-specific selection for different mtDNA genotypes in heteroplasmic mice. *Nature Genet* 16:93–95; 1997.

35. FV Meirelles, IC Smith. Mitochondrial genotype segregation in a mouse heteroplasmic lineage produced by embryonic karyoplast transplantation. *Genetics* 145:445–451; 1997.

36. J Cummins. Elimination of the sperm mitochondrial DNA. *Embryo Mail News Commun* #1060, 30 September, 1999.

37. MJ Evans, C Gurer, JD Loike, I Wilmut, AE Schnieke, EA Schon. Mitochondrial DNA genotypes in nuclear transfer-derived cloned sheep. *Nature Genet* 23:90–93; 1999.

38. S Hiendleder, SM Schmutz, G Erhardt, RD Green, Y Plante. Transmitochondrial differences and varying levels of heteroplasmy in nuclear transfer cloned cattle. *Mol Reprod Dev* 54:24–31; 1999.

39. K Takeda, S Takahashi, A Onishi, Y Goto, A Miyazawa, H Imai. Dominant distribution of mitochondrial DNA from recipient oocytes in bovine embryos and offspring after nuclear transfer. *J Reprod Fertil* 116:253–259; 1999.

40. R Steinborn, P Schinogl, V Zakhartchenko, R Achmann, W Schernthaner, M Stojkovic, E Wolf, M Muller, G Brem. Mitochondrial DNA heteroplasmy in cloned cattle produced by fetal and adult cell cloning. *Nature Genet* 25:255–257; 2000.

41. J Cohen, R Scott, T Schimmel, J Levron, S Willadsen. Birth of infant after transfer of anucleate donor oocyte cytoplasm into recipient eggs. *Lancet* 350:186–187; 1997.

42. JA Barritt, CA Brenner, H Malter, J Cohen. Mitochondria in human offspring derived from ooplasmic transplantation. *Hum Reprod* 16:513–516; 2001.

43. K Takeda, S Akagi, S Takahashi, A Onishi, H Hanada, CA Pinkert. Mitochondrial activity in response to serum starvation in bovine (Bos taurus) cell culture. *Cloning Stem Cells* 4:223–230; 2002.

44. MH Irwin, LW Johnson, CA Pinkert. Isolation and microinjection of somatic cell-derived mitochondria and germline heteroplasmy in transmitochondrial mice. *Transgenic Res* 8:119–123; 1999.

45. SE Levy, KG Waymire, YL Kim, GR MacGregor, DC Wallace. Transfer of chloramphenicol-resistant mitochondrial DNA into the chimeric mouse. *Transgenic Res* 8:137–145; 1999.

46. DR Marchington, D Barlow, J Poulton. Transmitochondrial mice carrying resistance to chloramphenicol on mitochondrial DNA: developing the first mouse model of mitochondrial DNA disease. *Nature Med* 5:957–960; 1999.

47. K Inoue, K Nakada, A Ogura, K Isobe, Y-I Goto, I Nonaka, J-I Hayashi. Generation of mice with mitochondrial dysfunction by introducing mouse mtDNA carrying a deletion into zygotes. *Nature Genet* 26:176–181; 2000.

48. JE Sligh, SE Levy, KG Waymire, P Allard, DL Dillehay, JR Heckenlively, GR MacGregor, DC Wallace. Maternal germ-line transmission of mutant mtDNAs from embryonic stem cell-derived chimeric mice. *Proc Nat Acad Sci USA* 97:14461–14466; 2000.

49. MH Irwin, V Parrino, CA Pinkert. Construction of a mutated mtDNA genome and transfection into isolated mitochondria by electroporarion. *Adv Reprod* 5:59–66; 2001.

50. PJ Laipis. Construction of heteroplasmic mice containing two mitochondrial DNA genotypes by micromanipulation of single-cell embryos. *Methods Enzymol* 264:345–357; 1996.

51. IA Trounce, DC Wallace. Production of transmitochondrial mouse cell lines by cybrid rescue of rhodamine-6G pretreated L cells. *Somatic Cell Mol Genet* 22:81–85; 1996.

52. N Howell. Glycine 231 residue of the mouse mitochondrial protonmotive cytochrome b: mutation to aspartic acid deranges electron transport. *Biochemistry* 29:8970–8977; 1990.

53. Y Bai, G Attardi. The mtDNA-encoded ND6 subunit of mitochondrial NADH dehydrogenase is essential for the assembly of the membrane arm and the respiratory function of the enzyme. *EMBO J* 17:4848–4858; 1998.

54. L Kenyon, CT Moraes. Expanding the functional human mitochondrial DNA database by the establishment of primate xenomitochondrial cybrids. *Proc Natl Acad Sci USA* 94:9131–9135; 1997.

55. M McKenzie, IA Trounce. Expression of *Rattus norvegicus* mtDNA in *Mus musculus* cells results in multiple respiratory chain defects. *J Biol Chem* 275:31514–31519; 2000.

56. R Dey, A Barrientos, CT Moraes. Functional constraints of nuclear-mitochondrial DNA interactions in xenomitochondrial rodent cell lines. *J Biol Chem* 275:31520–31527; 2000.

57. M McKenzie, IA Trounce, CA Cassar, CA Pinkert. Production of homoplasmic xenomitochondrial mice. *Proc Nat Acad Sci USA* 101:1685–1690; 2004.

58. MH Irwin, V Parrino, CA Pinkert. Construction of a mutated mtDNA genome and transfection into isolated mitochondria by electroporation. *Adv Reprod* 5:59–66; 2001.

59. A Chomyn, G Meola, N Bresolin, ST Lai, G Scarlato, G Attardi. *In vitro* genetic transfer of protein synthesis and respiratory defects to mitochondrial DNA-less cells with myopathy-patient mitochondria. *Mol Cell Biol* 11:2236–2244; 1991.

60. J-I Hayashi, S Ohta, A Kikuchi, M Takemitsu, Y-I Goto, I Nonaka. Introduction of disease-related mitochondrial DNA deletions into HeLa cells lacking mitochondrial DNA results in mitochondrial dysfunction. *Proc Natl Acad Sci USA* 88:10614–10618; 1991.

61. J-I Hayashi, S Ohta, D Takai, S Miyabayashi, R Sakuta, Y-I Goto, I Nonaka. Accumulation of mtDNA with a mutation at position 3271 in tRNA$^{Leu(UUR)}$ gene introduced from a MELAS patient to HeLa cells lacking mtDNA results in a progressive inhibition of mitochondrial respiratory function. *Biochem Biophys Res Commun* 197:1049–1055; 1993.

62. I Trounce, S Neill, DC Wallace. Cytoplasmic transfer of the mitochondrial DNA 8993T→G (ATP6) point mutation associated with Leigh syndrome into mtDNA-less cells demonstrates co-segregation of decreased state III respiration and ADP/O ratio. *Proc Natl Acad Sci USA* 91:8334–8338; 1994.

63. R Hovius, J Thijssen, P van der Linden, K Nicolay, B De Kruijff. Phospholipid asymmetry of the outer membrane of rat liver mitochondria. Evidence for the presence of cardiolipin on the outside of the outer membrane. *FEBS Lett* 330:71–76; 1993.

64. K Adachi, T Matsuhashi, Y Nishizawa, J Usukura, M Momota, J Popinigis, T Wakabayashi. Further studies on physicochemical properties of mitochondrial membranes during formation of megamitochondria in the rat liver by hydrazine. *Exp Mol Path* 61:134–151; 1994.

65. AI de Kroon, D Dolis, A Mayer, R Lill, B De Kruijff. Phospholipid composition of highly purified mitochondrial outer membranes of rat liver and *Neurospora crassa*. Is cardiolipin present in the mitochondrial outer membrane? *Biochim Biophys Acta* 1325:108–116; 1997.

66. EM Ellis, GA Reid. Assembly of mitochondrial membranes. *Sub-Cell Biochem* 22:151–181; 1994.

67. CR Hackenbrock, B Chazotte. Lipid enrichment and fusion of mitochondrial inner membranes. *Methods Enzymol* 125:35–45; 1986.

68. P Seibel, J Trappe, G Villani, T Klopstock, S Papa, H Reichmann. Transfection of mitochondria: strategy towards a gene therapy of mitochondrial DNA diseases. *Nucleic Acids Res* 11:10–17; 1995.

69. JM Collombet, VC Wheeler, F Vogel, C Coutelle. Introduction of plasmid DNA into isolated mitochondria by electroporation. A novel approach toward gene correction for mitochondrial disorders. *J Biol Chem* 272:5342–5347; 1997.

70. IA Trounce, J Schmiedel, HC Yen, S Hosseini, MD Brown, JJ Olson, DC Wallace. Cloning of neuronal mtDNA variants in cultured cells by synaptosome fusion with mtDNA-less cells. *Nucleic Acids Res* 28:2164–2170; 2000.

71. MP King, G Attardi. Injection of mitochondria into human cells leads to rapid replacement of the endogenous mitochondrial DNA. *Cell* 52:811–819; 1988.

72. CA Pinkert, IA Trounce. Production of transmitochondrial mice. *Methods* 26:348–357; 2002.

73. S Hiendleder, V Zakhartchenko, H Wenigerkind, HD Reichenbach, K Bruggerhoff, K Prelle, G Brem, M Stojkovic, E Wolf. Heteroplasmy in bovine fetuses produced by intra- and inter-subspecific somatic cell nuclear transfer: neutral segregation of nuclear donor mitochondrial DNA in various tissues and evidence for recipient cow mitochondria in fetal blood. *Biol Reprod* 68:159–166; 2003.

13

BHE/Cdb Rats as Tools to Study Mitochondrial Gene Expression

CAROLYN D. BERDANIER AND
MOON-JEONG CHANG KIM

CONTENTS

THE BHE/CDB RAT

The BHE/Cdb rat strain was developed as an inbred strain designed for the study of maturity onset diabetes.* The original strain, the BHE strain, was developed through a cross of the Osborne Mendel strain and the now nonexistent Pennsylvania State College strain. Scientists at the Bureau of Home Economics (thus the BHE name for the strain) were part of the USDA research facility at Beltsville, Maryland.

The first reported use of these rats appeared in the 1960s. It was noted that the rats were highly variable, although some of them were hyperglycemic and developed fatty livers with high fasting blood lipids (1–3). Some rats also developed hydronephrosis and others developed nephropathy without the hydronephrosis (4–6). Some became obese and others remained of normal weight (3,6). Some were very inactive; others had a normal voluntary activity pattern (6). Coat color varied: all black; all brown; agouti; black and white; brown and white; and all white. The metabolic heterogeneity made it hard to use these rats to test specific hypotheses about the role of diet ingredients in determining metabolic response (7–12).

Thus, in 1975, it was decided to reduce the heterogeneity through selective breeding. Selection pressure was placed on the lipemic–hyperglycemic characteristic, associated fatty liver, and a form of diabetic nephropathy. Only black and white coat color was allowed.

---

*The terminology used for these rats changed as the degree of homogeneity changed. When originally produced, the progeny of the closed colony were identified as BHE rats from this colony. As the colony began to become more uniform, it was identified as BHE/Cdb. Finally, after 86 generations, full homogeneity was achieved and the rats were then identified as BHE/Cdb rats.

COLONY DEVELOPMENT

The protocol for the selective breeding was as follows. Randomly selected 100-day-old adult male and female rats (24 pairs in total) were tested for their fasting glucose and insulin levels. Those whose blood glucose levels were high in the face of elevated insulin levels were selected as the parent generation. Ten pairs of animals were used initially. The parents were randomly bred and their progeny then screened for fasting insulin levels at 50 days of age. Those with elevated insulin levels were then used for breeding generation 2. Generations 3 through 5 were selected in the same manner.

The rats were bred at 100 days of age. Their progeny were kept until tested; if they had elevated insulin levels, they were kept for future breeding. The fifth generation began to be useful with respect to a reduction in variability. However, it actually took several more generations before uniformity in growth rate and metabolic response to purified diets occurred. At this time, a somewhat different breeding strategy was developed: the males were back-crossed to their female progeny to strengthen the hyperglycemic characteristic. However, because the attempt was to put selection pressure on the glycemic characteristic that appeared at mid-life (300 days of age), it was necessary to abandon the 50-day screening procedure. Instead, breeding pairs were kept until 300 days of age and, if they developed hyperglycemia, their progeny were kept in the breeding pool. Progeny of normoglycemic breeders were discarded.

For several generations, full sib matings were used — again, to strengthen the hyperglycemic trait. At about the 15th generation, too many progeny were being saved on the chance that they would produce the desired traits. To reduce the number of animals housed, we reduced the number of male breeders; one male was used for two females, rather than the one-on-one plan used until then. Because glycemic females were more likely to produce glycemic progeny than glycemic male breeders, the female lines were carefully tracked and interbred for production purposes to produce a uniform animal. This breeding plan was followed for the next 70 generations.

The female lines were interbred at each generation; full sib matings were avoided. Two females and two males out of each first litter were kept for future breeding. Both females were bred when experimental animals were needed, but only one was bred if only the next generation of breeding stock was needed. All breeders were kept and tested for glucose tolerance at 300 days of age. The glucose tolerance test consisted of sampling tail blood before and at 30, 60, and 120 min after 1 g glucose/100 g body weight as a 25% glucose solution gavaged into a 16-h starved rat. If normal glucose tolerance was detected in the retired breeder, the progeny of that female were discarded and that female line deleted from the breeding pool. By the time the colony was given to the NIH Animal Resources Center in 2001, the BHE/Cdb strain was uniform and fully inbred. It was then in its 98th generation as a closed colony. BHE/Cdb animals can be obtained through Dr. Carl Hanson, NIH Animal Resources Center, Bethesda, Maryland.

Colony maintenance was difficult for several reasons:

- The females tended to develop gestational diabetes. This was especially apparent when purified diets were fed (13–15). If this occurred, the females would resorb some or all of their litter. In some instances, abnormally long gestation times would occur and sometimes very large pups were born. Sometimes only one or two pups survived to weaning. The occurrence of gestational diabetes did not have long-lasting effects on the progeny in one study (16).
- Rats of this strain needed more micronutrients than could be provided by the stock diets normally used for rat colonies. A normal stock diet provides all of the nutrients deemed essential for rats by the National Academy of Sciences (17). However, this diet would not maintain the colony. It was necessary to feed a vitamin- and mineral-enriched stock diet for normal maintenance in addition to that needed for

growth and reproduction. Three times normal vitamin A was required for optimal maintenance (18); ten times normal vitamin E was required (19); and twice normal amounts of biotin, pantothenic acid, and pyridoxine were needed (unpublished observations). Scientists at the USDA Beltsville facility reported that the biotin need of the parent BHE strain was also increased (20). The energy requirement was similar to that of normal rats (21), as was the protein requirement.

- In contrast to the commercially viable Sprague–Dawley or Wistar rats, BHE/Cdb females needed a 3-week rest period between weaning and rebreeding. The litter size varied between six and ten pups. Litters of 12 to 14 pups were rare. Normal Sprague–Dawley or Wistar rats can be bred every 3 weeks and successfully rear 12 to 16 pups each time. The relatively poor reproductive performance limited the number of rats that could be produced for laboratory work.

- BHE/Cdb rats developed diabetic nephropathy as they aged; this was apparent in rats as young as 200 days of age (22–29). This pathology was a rodent variety of the disease and somewhat different from that which develops in humans. The diabetic nephropathy responded to diet in that high-fat diets (27,28), diets with menhaden oil as the fat source (23–29), or high-protein diets (22) seemed to accelerate the time course of disease development. In a study of the effects of lifelong ingestion of 9% menhaden oil, corn oil, or beef tallow plus 1% corn oil (total fat = 10% of the diet by weight), it was found that free radicals accumulated in the kidneys of these animals and preceded the renal disease development (25,27). Rats fed menhaden oil died sooner than rats fed corn oil, and those fed a beef tallow diet lived the longest. The renal disease was independent of the cholesterol content of the diet and was not related to the arginine content of the diet (26,29).

THE BHE/CDB RAT AS A MODEL FOR THE
NONOBESE TYPE 2 DIABETIC HUMANS

As a model of maturity onset diabetes mellitus, the BHE/Cdb rat was considered useful because few of the animal models of diabetes develop the disease at maturity and few could be shown to have the time course of their disease influenced by diet (9–12,28–31). This rat was a unique model for diabetes because, although it had a fatty liver, it was not obese. Genetically obese animals such as the Zucker fatty rat and the obese (ob/ob) mouse likewise have a fatty liver; however, they are obese regardless of diet and many develop a form of diabetes that was related to body fatness. In this respect, these animals model the human "diabesity" syndrome (33,34).

The BHE/Cdb rat has a brief period of hyperinsulinemia early in life (35), but this trait soon disappears and the islet cells become increasingly less responsive to a glucose signal (36). The fatty liver is especially prominent when this rat is fed a diet rich in sugar or sugar and saturated fat (9,29–31,37). Even if the animal is fed a standard stock diet, hepatic lipogenesis and cholesterogenesis are twice that of a normal rat; of course, this increase in hepatic lipogenesis explains its high blood lipid (38). Interestingly, the elevated lipogenesis is not accompanied by excess body fat stores because these rats also have a tenfold increase in lipolysis (39). The obesity seen in other models of diabetes does not develop because of an active and futile energy-wasting lipogenesis–lipolysis cycle.

Studies of oxidative phosphorylation (OXPHOS) in these animals revealed a 20% reduction in ATP synthesis efficiency compared to normal rats (40). Efficiency could be further reduced by feeding a sugar-rich diet (41–43) or a saturated-fat diet (44–46). Efficiency could be increased if a very unsaturated-fat (menhaden oil) diet was fed (45,46). Efficiency decreased as the animals aged (47). These dietary treatments affected the fatty acid composition of the inner membrane and this in turn affected the membrane's fluidity.

An increase in fluidity facilitates the motion of the ATPase, which moves within the inner membrane as it synthesizes ATP (48). Any hindrance or enhancement of this motion will have effects on ATP synthesis. The observations of a diet effect on ATP synthesis efficiency paralleled observations that, with the feeding of an unsaturated fat, the percentage of fat in the liver fell. When rats were fed a very saturated-fat (hydrogenated coconut oil) diet, fat percentage rose and ATP synthesis efficiency fell dramatically.

Corresponding changes in glucose homeostasis were observed with the previously described dietary manipulations. Gluconeogenesis was more active in BHE/Cdb rats compared to control rats (49); when a high-sucrose diet was fed, this gluconeogenesis was further increased (50), as it was when coconut oil was fed (51).

Tracer studies using tritiated and ^{14}C-labeled glucose, tritiated water, and ^{14}C-labeled alanine confirmed that the saturation of the fat and the source of carbohydrate affected not only gluconeogenesis but also glucose turnover and glycogen synthesis and degradation (52,53). In all instances when diet had an effect on OXPHOS, corresponding effects on glucose homeostasis were found. That is, if mitochondria were less well coupled, liver fat rose, gluconeogenesis rose, lipogenesis rose, glycogen breakdown fell, and glycogen synthesis and stores rose. The irreversible glucose oxidation (glucose oxidized to CO_2) fell. These changes are typical features of glucose dysregulation or non-insulin-dependent diabetes mellitus. These strain differences in metabolism are summarized in Table 13.1.

THYROID HORMONE AFFECTS THE METABOLIC FEATURES OF THE BHE/CDB RAT

In another study, treatment with thyroid hormone resulted in an increase in respiratory rate with no effect on ATP synthesis when fatty acid substrates, but not citric acid cycle substrates, were used (54). In addition, the α-glycerophosphate and malate–aspartate shuttles, the Mg-ATPase (the F_1 portion of F_1F_0 ATPase), and gluconeogenesis were also responsive to thyroid stimulation (54–56). Again, dietary fat

Table 13.1 Metabolic Characteristics of
BHE/Cdb Rats Compared to Rats from a
Control[a] Strain

Feature	Difference
Glucose oxidation	↓
Fatty acid synthesis	↑
Lipolysis	↑
Glucose-stimulated insulin release	↓
Cholesterol synthesis	↑
Cholesterol degradation	↑
ATP synthesis	↓
Gluconeogenesis	↑
Glycogen stores	↑
Hepatic lipid	↑
Average life span	↓

[a] Originally, rats of the Wistar strain were used as
the control rats. However, because these rats
became difficult to obtain as SFP rats, the Sprague–Dawley strain was substituted. Comparison of
these two strains revealed very few strain differences relevant to glycemia and lipemia.

could modify the response to thyroid hormone and vice versa
(45,53,56).

In normal rats, stimulation of respiration with thyroid
hormone treatment (within the normal range of thyroid hormone levels, not toxic levels) usually results in an increase in
ATP synthesis (57–59). However, in BHE/Cdb rats, thyroid
hormone treatment stimulated respiration only when fatty
acid substrates were used to support OXPHOS — not when
citric acid cycle intermediates were used. In addition, even
with fatty acid substrates, thyroid treatment did not result
in an increase in ATP synthesis. Even though the shuttles
were increased and a response to thyroid hormone could be
demonstrated in the MgATPase, an increase in ATP synthesis
could not be induced. This suggested that the BHE/Cdb rat
had a problem with the F_0 ATPase. It was not inducible by
thyroid hormone.

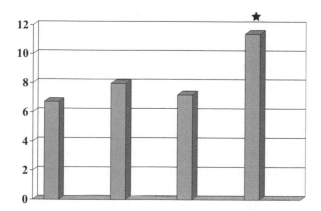

Figure 13.1 Hepatic fatty acid synthesis (mmol tritiated fatty acids/g liver) in BHE/Cdb rats untreated or treated with DCMU (columns 1 and 2) and in Sprague Dawley rats untreated or treated with DCMU (columns 3 and 4).

To test whether the excess lipogenic character of the BHE/Cdb rat was due to decreased ATP synthesis efficiency, i.e., a reduction in the coupling of respiration to ATP synthesis, BHE/Cdb and Sprague–Dawley rats were fed a diet containing a low level of a mild uncoupler, DCMU. The reasoning was that enhanced hepatic lipogenesis was causally related to mild uncoupling. It was hypothesized that DCMU-induced uncoupling would result in an increase in *de novo* lipogenesis in the control rat and would have an additive effect (to the mitochondrial problem) in the BHE/Cdb rat. Figure 13.1 shows the results of this study (60). Clearly, DCMU uncoupling resulted in an increase in lipogenesis in the normal rat.

GENETIC STUDIES

Once this problem was identified, the next task was to determine whether a genetic problem could exist in this rat. This was an inbred strain and, through tracking the trait of maturity onset impaired glucose tolerance, it was known that the traits of nonobese-abnormal glucose tolerance and pancreatic dysfunction were maternally inherited. It was hypothesized that, because of the maternal inheritance of these traits, the problem was probably due to a mutation in the mitochondrial genome in the region of the two F_0 ATPase mitochondrially encoded genes — the ATPase 6 and 8 genes. The ATPase 6 encodes subunit a of the F_0 ATPase and overlaps the ATPase 8 gene.

Sequencing this region of the genome (61) was then begun. Two homoplasmic base substitutions in the ATPase 6 gene were found: one at bp 8204 and the other at bp 8289. Other substitutions were also found; three were silent because they resulted in a codon that would have translated to the same amino acid as the codon in a normal rat. Two more were heteroplasmic with a low level of the substituted codon present. This heteroplasmy likely would have no effect on the function of the gene product.

One of the substitutions in the ATPase 6 gene occurred in the region that translates into the part of subunit a that forms the proton channel. The other translates into the part of subunit a forming a hinge in the molecule that seems to be part of its flexibility. That is, it is in a region important to the movement of the F_0 within the membrane. The first mutation could explain the reduced ATP synthesizing efficiency and the second explain why changes in inner membrane fluidity had an effect on ATP synthesis. Obviously, a change in the mobility of the complex would affect its motion within the membrane and, because motion is important to ATP synthesis, the diet effects would be explained.

A crossbreeding experiment showed that the mutations and the phenotype of impaired glucose tolerance and impaired mitochondrial function were maternally inherited traits (62,63). Progeny of female BHE/Cdb rats mated to Sprague–Dawley males had the same base sequence as full-bred BHE/Cdb rats; progeny of female Sprague–Dawley rats bred to BHE/Cdb males had the same base sequence as the full-bred Sprague–Dawley rats. This meant that future experiments using the BHE/Cdb rat did not need to use two control groups: cross-bred and full-bred Sprague–Dawley rats. This assumes that all of the strain differences in metabolism can be attributed to the difference in the mitochondrial genome.

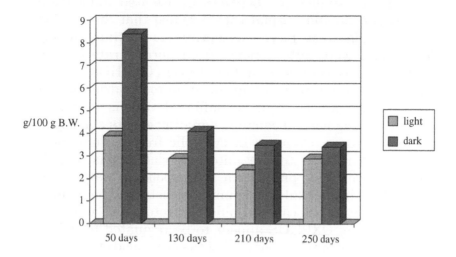

Figure 13.2   Food intake of BHE/Cdb rats given as g food consumed per 100 g body weight. The food offered was an unpurified ingredient stock diet. Note that the proportion of food consumed during the light hours increased as the animals aged. The differences in food consumed as a percentage of the total was significant.

Base substitutions were found in all the tissues examined in the BHE/Cdb rat (62), so how did these rats manage their metabolism to ensure their survival and reproduction? A careful study of their eating patterns showed that as the rats aged, they shifted their eating pattern from one of primarily night-time (dark phase of the lighting schedule) eating typical of the usual rat to one in which the rat ate throughout the 24 h (64). The total amount of food consumed was not different from that of normal control rats, but the eating pattern as well as the RQ (respiratory quotient) shifted (Figure 13.2). The composition of the diet affected this shift as well. Rats fed a high-fat diet shifted their rhythms earlier than did rats fed a stock diet or a purified diet with the same proximate composition as the stock diet. The suspicion is that, by shifting the eating pattern, the rat is able to meter out the substrates from food that need to be metabolized.

Further studies of OXPHOS by isolated mitochondria from BHE/Cdb and normal rats revealed that the BHE/Cdb mitochondria were more susceptible to calcium-mediated death than were mitochondria from normal rats (65). The calcium ion activates several key steps in OXPHOS and actually serves as a central integrator of mitochondrial function. Mitochondria are thought to buffer elevations in cytosolic calcium during normal calcium signaling events and atypical calcium burdens. Changes in cytosolic free calcium concentration are relayed to the mitochondrial matrix, enabling mitochondrial ATP production to respond to varying ATP demands of intracellular calcium homeostasis. The β-subunit of the F_1ATPase is a calcium-binding protein (66) and several components of OXPHOS appear to be regulated by calcium. In addition, several components of the citric acid cycle respond to calcium. Calcium influx into and efflux out of the mitochondrial compartment occurs via several mechanisms: a uniporter and a sodium-dependent and sodium-independent efflux mechanism (67–72).

Kimura and Rasmussen (68) reported that the administration of dexamethasone to rats markedly diminished the initial rate and maximal extent of substrate-dependent calcium uptake and enhanced the release of calcium by hepatic mitochondria. These mitochondria retained calcium until the total ATP (synthesized by OXPHOS) reached a critical level and was released into the cytoplasm. When the ATP content fell below 5 to 7 mol ATP/mg protein, the mitochondria quickly released the calcium. Thus, calcium uptake and release followed the pattern of ATP synthesis and release.

Compton et al. showed that calcium efflux from mitochondria was sodium dependent (69). Kowaltowski et al. reported that agents that uncouple or inhibit OXPHOS also affect calcium efflux (70). Increased calcium binding by the F_0 ATPase subunits causes a reduction in ATP formation. Such calcium-binding effects are consistent with a pH-dependent gating mechanism for the control of H^+ ion flux across the opening of the H^+ channel. In mammalian systems, an impairment of calcium egress should have effects on respiration and ATP synthesis.

Luvissetto et al. reported that hyperthyroidism activates respiration with a loss in thermodynamic control due to increased leak (73). We suspect that the mutation in the ATPase 6 gene in the BHE/Cdb rat also results in an increased leak rate and thus affects the membrane potential. This probably accounts for the reduction in ATPase synthesis efficiency reported many years ago (40).

Treatment with thyroid hormone compounds the genetically determined reduction in OXPHOS efficiency. Actually, when mitochondria from hyperthyroid BHE/Cdb rats were compared to mitochondria from control rats or non-thyroid-treated BHE/Cdb rats, the thyroid treatment had an additive effect on the impairment of OXPHOS, with rising levels of calcium in the incubation medium (65). In other words it took less calcium to shut down OXPHOS in mitochondria from the thyroid-treated BHE/Cdb rats than in mitochondria from untreated rats. Normal rats treated with thyroid hormone had the expected rise in respiration and ATP synthesis. This is consistent with studies of mitochondria from humans with mitochondrial disease (74).

Mitochondria from BHE/Cdb rats were also more susceptible to oligomycin inhibition of OXPHOS than mitochondria from control rats (75). Oligomycin blocks proton conductance, primarily through its binding to the oligomycin-sensitive protein found in the stalk that connects the F_1 portion of the F_1F_0 ATPase to the F_0 portion. Oligomycin dose–response curves of mitochondria from BHE/Cdb and control rats showed a shift to the left with increasing amounts of oligomycin for the BHE/Cdb mitochondria.

These observations are consistent with studies of Vazquez–Memiji et al. (76), who studied fibroblasts from patients with a T8993G or a T8993C point mutation in the ATPase 6 gene. The human mutation is within 30 bp of the mutation found in BHE/Cdb rats. Humans and rats become diabetic with these base substitutions. In the human and rat mutation is an inferred change in polarity of one of the amino acids that protrude out into the proton channel. In the human with the T8993G mutation, this is a substitution of arginine for leucine. In the rat, it is the substitution of asparagine for aspartate. In both species, ATP synthesis is impaired and sensitivity to oligomycin inhibition of OXPHOS is increased.

LONGEVITY STUDIES

While the sequencing of the ATPase gene was performed and the responsiveness of the mitochondria to various treatments was studied, a series of longevity studies were conducted (23–29). These studies were designed to determine how and whether diet composition could affect longevity and determine cause of death. The results were not consistent with the earlier short-term studies of the effects of saturated or unsaturated fats on OXPHOS (41–46, 50–56).

Rats fed menhaden oil were shorter lived than rats fed a saturated fat such as beef tallow (23); in fact, rats fed an egg-rich diet were even longer lived than those fed a beef tallow diet (28). Thus, although short-term menhaden oil feeding enhanced OXPHOS and improved glucose metabolism (45,77,78), long-term feeding was quite detrimental (23,26). Conversely, the egg-rich diet had some rather remarkable effects on glucose tolerance (delayed its appearance) and OXPHOS (improved efficiency) and lengthened the life span. What could explain these remarkable effects? Although a number of studies were designed to determine possible mechanisms, none were as exciting as those that tested the hypothesis that the egg effect was due to its vitamin A content (18,29,79).

VITAMIN A

A feeding study that used graded levels of dietary vitamin A fed to deficient rats revealed that BHE/Cdb rats had optimal OXPHOS at an intake three times that recommended for normal rats (18). This intake level resulted in an increase in ATPase 6 gene product (F_0 ATPase subunit a) (79). In addition, vitamin A restoration to vitamin A-depleted BHE/Cdb rats resulted in an increase in mitochondrial number and increase in mitochondrial TFA. This was the first report of a nutrient having an effect on mitochondrial function via an effect on mitochondrial gene transcription and translation. The presence of a retinoic acid-binding protein in the matrix confirmed the supposition that retinoic acid stimulated gene transcription in this compartment. Studies using primary cultures of hepatocytes showed that retinoic acid at a level of 10^{-9} elicited a maximal ATPase 6 mRNA and ATPase 6 gene product (79).

Other factors have been shown to up-regulate mitochondrial gene expression as well; among these are thyroid hormone, dehydroepiandrosterone, and retinoic acid (80). Zinc also increases the transcription of the ATPase gene (unpublished observations). Others have also shown a role for retinoic acid in mitochondrial transcription (81,82). Retinoic acid has been shown to up-regulate NADH dehydrogenase subunit 5 mRNA, as well as cytochrome c oxidase subunit I and 16 S r RNA. In addition, retinoid X receptor " knockout mice were shown to have alterations in mitochondrial gene expression (83).

Retinoic acid works in two ways to increase mitochondrial transcription. One is through increasing mitochondrial TFA gene expression that, in turn, would increase the steady state levels of this protein in the matrix. Because mitogenesis is linked to mtTFA, it would also result in an increase in mitochondrial number. Studies with the BHE/Cdb rats have shown that this occurs. OXPHOS was optimized, as was the level of mitochondrial TFA; mitochondrial number was increased (79).

The conclusions drawn from these studies were that these rats needed more dietary vitamin A to maintain normal OXPHOS function because, without supplemental amounts, the performance of the mitochondria was inadequate to sustain normal metabolism. Because their mitochondrial mutation affected OXPHOS modestly, this could be overcome with an increase in the number of mitochondria as well as an increase in the amount of mitochondrial gene products. This seems to depend on the increased expression of mitochondrial and nuclear genes — probably coordinated through an effect of retinoic acid on mt TFA expression.

SUMMING UP

As discussed in Chapter 8, mitochondrial mutation has been found to be associated with diabetes mellitus. Clearly our studies of the BHE/Cdb rats show that subtle differences in the mitochondrial genome are associated with subtle differences in OXPHOS; these, in turn, are related to changes in the regulation of glucose homeostasis. Because the islet cell (as well as the renal cell) is so dependent on its ATP supply, any shortfall in ATP production by the mitochondria in those cells will result in eventual failure. Thus, with age, the pancreatic islet cell becomes less responsive to glucose as a signal for insulin release and the renal cell progressively fails as part of the filtration system.

Both of these age-related changes have been documented in the BHE/Cdb rat, making this strain quite valuable for the study of mitochondrial diabetes. Like the human with this disorder, it is not excessively fat, yet has a fatty liver. Diabetes mellitus is frequently observed in patients with more serious disease due to mitochondrial mutation (84–91). In these patients, epilepsy, myocardial degeneration, neuromuscular malfunction, lactic acidosis, retinitis pigmentosa, deafness, and early death are conditions of greater concern than the progressive loss in the regulation of glucose metabolism due to mitochondrial gene mutation.

Mitochondrial base substitutions, deletions, and rearrangements have been reported to be associated with diabetes (89–91) and some scientists have estimated that as many as 9% (or as few as 1%) of the population with diabetes may have the disease because of one or more of these mutations. As epidemiologists examine populations throughout the world, better estimates of those with mitochondrial diabetes will emerge. In the meantime, scientists wishing to study the disorder in small animals might find the BHE/Cdb rat quite useful.

REFERENCES

1. M Adams. Diet as a factor in length of life and in structure and composition of tissues of the rat with aging. Home Economics Research Report No 24. USDA Washington, D.C. 1964 108 pp.

2. DD Taylor, ES Conroy, EM Schuster, M Adams. Influence of dietary carbohydrates on liver content and on serum lipids in relation to age and strain of rat. *J Nutr* 91:275–282, 1967.

3. AMA Durand, M Fisher, M Adams. Histology in rats as influenced by age and diet. *Arch Pathol* 77:268–277, 1964.

4. AMA Durrand, M Fisher, M Adams. The influence of type of dietary carbohydrate: effects of histological findings in two strains of rats. *Arch Pathol* 85:318–327, 1968.

5. J Dupont, AA Spindler, MM Mathias. Dietary fat saturation and cholesterol and fatty acid synthesis in aging rats. *Age* 1:93–99, 1968.

6. MW Marshall, AMA Durrand, M Adams. Different characteristics of rat strains: lipid metabolism and response to diet. In: *Defining the Laboratory Animal*, National Academy of Science, Washington, D.C. 1971, 383–412.

7. MLW Chang, EM Schuster, JA Lee, C Snodgrass, DA Benton. Effects of diet, dietary regimens, and strain differences on some enzyme activities in rat tissues. *J Nutr* 96:368–374, 1968.

8. MLW Chang, LA Lee, N Simmons. Rat strain differences in the utilization of glucose-U-C14 and acetate-C14 for fat synthesis. *Proc Soc Exp Biol Med* 138:742–747, 1971.

9. CD Berdanier. Metabolic characteristics of the carbohydrate sensitive BHE rat. *J Nutr* 104:1246–1256, 1974.

10. CD Berdanier. The BHE strain of rat: an example of the role of inheritance in determining metabolic controls. *Fed Proc* 35:2295–2299, 1975.

11. PB Moser, CD Berdanier. Effect of early sucrose feeding on the metabolic patterns of mature rats. *J Nutr* 104:687–694, 1974.

12. CD Berdanier. Effect of maternal sucrose intake on the metabolic patterns of mature rat progeny. *Am J Clin Nutr* 28:1416–1421, 1975.

13. CD Berdanier, JM Bue, DB Hausman. Glucose tolerance in female BHE rats as a model for gestational diabetes. In: *Lessons from Animal Diabetes* (E Shafrir, AE Renold, Eds.) J Libby, London 1986, 427–431.

14. JM Bue, DB Hausman, CD Berdanier. Gestational diabetes in the BHE rat: influence of dietary fat. *Am J Obst Gynecol* 161:234–240, 1989.

15. K Wickwire, D Hathcock–Rice, CD Berdanier. Impaired glucose tolerance in pregnant BHE/Cdb rats is attenuated by feeding menhaden oil but reproduction is impaired. *Int J Diab* 4:5–15, 1996.

16. CE Mathews, K Wickwire, D Hathcock–Rice, CD Berdanier. Maternal diet has little effect on the glucose tolerance and age changes in enzyme activity in the male progeny of gestationally diabetic rats. *Int J Diab* 4:17–27, 1996.

17. *Nutrient Requirements of Laboratory Animals* 4th ed. National Academic Press Washington, D.C. 1995, 11–79.

18. HB Everts, CD Berdanier. Nutrient-gene interactions in mitochondrial function: vitamin A need in two strains of rats. *IUMBM Life* 53:289–294, 2002.

19. MJ Kullen, CD Berdanier. Influence of fish oil feeding on the vitamin E requirement of BHE/Cdb rats. *Biochem Arch* 8:247–257, 1992.

20. CD Berdanier, MW Marshall. Biotin intake and insulin response in adult rats. *Nutr Rep Int* 3:383–388, 1971.

21. S Krishnamachar, CD Berdanier, NL Canolty. Protein and energy utilization in female BHE rats. *Nutr Rep Int* 33:791–799, 1986.

22. J Noll–Herndon, CD Berdanier, WA Crowell. Influence of dietary casein and sucrose levels on urea cycle enzyme activities and renal histology in young BHE rats. *Nutr Rep Int* 32:403–412, 1986.

23. CD Berdanier, B Johnson, DK Hartle, W Crowell. Life span is shortened in BHE/Cdb rats fed a diet containing 9% menhaden oil and 1% corn oil. *J. Nutr* 122:1309–1317, 1992.

24. K Fowler, WA Crowell, CD Berdanier. Early renal disease in BHE/Cdb rats is less in rats fed beef tallow than in rats fed menhaden oil. *Proc Soc Exp Biol Med* 203:163–171, 1993.

25. K Wickwire, K Kras, C Gunnett, D Hartle, CD Berdanier. Menhaden oil feeding increases potential for renal free radical production in the BHE/Cdb rat. *Proc Soc Exp Biol Med* 209:397–402, 1995.

26. K Wickwire, M Porter, CD Berdanier. Differential hepatic and renal cholesterol levels in diabetes-prone BHE/Cdb rats fed menhaden oil or beef tallow. *Proc Soc Exp Biol Med* 214:346–351, 1997.

27. CD Berdanier, K Kras, K Wickwire, DG Hall, C Gunnett, D Hartle. Progressive glucose intolerance and renal disease in aging rats. *Int J Diabetes* 5:27–38, 1997.

28. CD Berdanier, K Kras, K Wickwire, DG Hall. Whole egg diet delays the age related impaired glucose tolerance of BHE/Cdb rats. *Proc Soc Exp Biol Med* 219:28–36, 1998.

29. T Jia, K Wickwire, CE Mathews, CD Berdanier. Neither the cholesterol nor arginine content of the whole egg diet explains its beneficial effect on glucose homeostasis in BHE/Cdb rats. *J Nutr Biochem* 9:170–177, 1998.

30. CD Berdanier. The BHE rat: an animal model for the study of non insulin dependent diabetes mellitus. *FASEB J* 5:2139–2144, 1991.

31. CD Berdanier. The BHE/Cdb rat — a model for NIDDM. *ILAR News* 33:58–62, 1992.

32. CD Berdanier. NIDDM in the nonobese BHE/Cdb rat. In: *Lessons from Animal Diabetes* (E. Shafrir, Ed.) Smith Gordon, London, 1994, 231–246.

33. BB Kahn, JS Flier. Obesity and insulin resistance. *J Clin Invest* 106:473–481, 2000.

34. LE Wagenknecht, CD Langefeld, AL Scherzinger, JM Norris, SM Haffner, MF Saad, RM Bergman. The insulin sensitivity, insulin secretion and abdominal fat. The insulin resistance atherosclerosis study (IRAS) family study. *Diabetes* 52:2490–2496, 2003.

35. CD Berdanier. Metabolic abnormalities in BHE rats. *Diabetologia* 10:691–695, 1974.

36. Y Liang, S Bonner–Weir, Y-J Wu, CD Berdanier, DK Berner, S Efrat, FM Matschinsky. *In situ* glucose uptake and glucokinase activity of pancreatic islets in diabetic and obese rodents. *J Clin Invest* 93:2473–2481, 1994.

37. CD Berdanier, AR Thomson, DJ Bouillon, R Wander. Relevance of hepatic mitochondrial activity to hepatic lipogenesis in genetically difference strains of rats. In: *Metabolic Effects of Utilizable Dietary Carbohydrate* (S. Reiser, Ed.) Marcel Dekker, Inc., New York, 1982, 71–117.

38. MK Lakshmanan, CD Berdanier, RL Veech. Comparative studies on lipogenesis and cholerogenesis in lipemic BHE and normal Wistar rats. *Arch Biochem* 183:355–360, 1977.

39. CD Berdanier, RB Tobin, V DeVore, R Wurdeman. Studies on the metabolism of glycerol by hyperlipemic and normolipemic rats. *Proc Soc Exp Biol Med* 157:5–11, 1978.

40. CD Berdanier, AR Thomson. Comparative studies on mitochondrial respiration in four strains of rats. *Comp Biochem Physiol* 85A:725–727, 1986.

41. RH McCusker, OE Deaver, CD Berdanier. Effect of sucrose or starch feeding on the hepatic mitochondrial activity of BHE and Wistar rats. *J Nutr* 113:1327–1334, 1983.

42. DJ Bouillon, CD Berdanier. Effect of maternal carbohydrate intake on mitochondrial activity and on lipogenesis by young and mature progeny. *J Nutr* 113:2205–2216, 1983.

43. RC Wander, CD Berdanier. Effects of dietary carbohydrate on mitochondrial composition and function in two strains of rats. *J Nutr* 115:190–199, 1985.

44. OE Deaver, RC Wander, RH McCusker, CD Berdanier. Diet effects on membrane phospholipid fatty acids and mitochondrial function in BHE rats. *J Nutr* 116:1148–1155, 1986.

45. MJC Kim, CD Berdanier. Influence of menhaden oil on mitochondrial respiration in BHE rats. *Proc Soc Exp Biol Med* 192:172–176, 1989.

46. MJC Kim, CD Berdanier. Nutrient-gene interactions determine mitochondrial function: effect of dietary fat. *FASEB J* 12:243–248, 1998.

47. CD Berdanier, S McNamara. Aging and mitochondrial activity in BHE and Wistar rats. *Exp Gerontol* 15:519–525, 1980.

48. PL Pederson. Frontiers in ATP synthase research: understanding the relationship between subunit movements and ATP synthesis. *J Bioenergetics Biomembranes* 28:389–395, 1996.

49. CD Berdanier. Rat strain differences in gluconeogenesis by isolated hepatocytes. *Proc Soc Exp Biol Med* 169:74–79, 1982.

50. JHY Park, CD Berdanier, OE Deaver, B Szepesi. Effects of dietary carbohydrate on hepatic gluconeogenesis in BHE rats. *J Nutr* 116:1193–1203, 1986.

51. RC Wander, CD Berdanier. Effects of type of dietary fat and carbohydrate on gluconeogenesis in isolated hepatocytes from BHE rats. *J Nutr* 116:1156–1164, 1986.

52. MJC Kim, JS Pan, CD Berdanier. Glucose turnover in BHE rats fed EFA deficient hydrogenated coconut oil. *Diabetes Res* 13:43–47, 1990.

53. MJC Kim, JS Pan, CD Berdanier. Glucose homeostasis in thyroxine treated BHE/Cdb rats fed corn oil or hydrogenated coconut oil. *J Nutr Biochem* 4:20–26, 1993.

54. CD Berdanier, D Shubeck. Effects of thyroid hormone on mitochondrial activity of lipemic BHE rats. *Proc Soc Exp Biol Med* 166:348–354, 1981.

55. CD Berdanier. Effects of thyroid hormone on the gluconeogenic capacity of lipemic BHE rats. *Proc Soc Exp Biol Med* 172:187–193, 1983.

56. CD Berdanier, MJC Kim. Hyperthyroidism does not induce an increase in mitochondrial respiration. *J Nutr Biochem* 4:10–19, 1993.

57. AJ Verhoeven, P Kamer, AK Groen, JM Tager. Effects of thyroid hormone on mitochondrial oxidative phosphorylation. *Biochem J* 226:183–192, 1985.

58. S Soboll, C Horst, H Hummereich, JP Schumacher, H Seitz. Mitochondrial metabolism in different thyroid states. *Biochem J* 281:171–173, 1992.

59. RP Hafner, GC Brown, MD Brand. Thyroid hormone control of state 3 respiration in isolated rat liver mitochondria. *Biochem J* 265:731–734, 1990.

60. K Fowler, CD Berdanier. Low dose intakes of mitochondrial uncouplers differentially affects hepatic lipogenesis in BHE/Cdb and Sprague–Dawley rats. *Biochem Arch* 7:269–274, 1991.

61. CE Mathews, RA McGraw, CD Berdanier. A point mutation in the mitochondrial DNA of diabetes-prone BHE/Cdb rats. *FASEB J* 9:1638–1642, 1995.

62. CE Mathews, RA McGraw, R Dean, CD Berdanier. Inheritance of a mitochondrial DNA defect and impaired glucose tolerance in BHE/Cdb rats. *Diabetologia* 42:35–41, 1999.

63. CD Berdanier, HB Everts, C Hermoyian, CE Mathews. Role of vitamin A in mitochondrial gene expression. *Diab Res Clin Prac* 54:S11–S27, 2001.

64. CE Mathews, K Wickwire, WP Flatt, CD Berdanier. Attenuation of circadian rhythms of food intake and respiration in aging diabetes prone BHE/Cdb rats. *Am J Physiol* 279:R230–238, 2000.

65. SB Kim, CD Berdanier. Further studies of diabetes-prone BHE/Cdb rats: increased sensitivity to calcium ion suppression of oxidative phosphorylation. *J Nutr Biochem* 10:31–36, 1999.

66. M Hubbard, NJ McHugh. Mitochondrial ATP synthase F_1-$-subunit is a calcium binding protein. *FEBS Lett* 391:323–329, 1996.

67. KK Gunter, TE Gunter. Transport of calcium by mitochondria. *J Bioenergetics Biomembranes* 26:471–485, 1994.

68. S Kimura, H Rasmussen. Adrenal glucocorticoids, adenine nucleotide translocation and mitochondrial calcium accumulation. *J Biol Chem* 252:1217–1225, 1977.

69. M Compton, R Moser, H Ludi, E Carafoli. Interrelations between the transport of sodium and calcium in mitochondria of various mammalian tissues. *Eur J Biochem* 82:25–31, 1978.

70. AJ Kowaltowski, RF Castilho, AE Vercisi. Opening of the mito-chondrial permeability transition pore by uncoupling or inor-ganic phosphate in the presence of Ca^{++} is dependent on mitochondrial-generated reactive oxygen species. *FEBS Lett* 378:150–152, 1996.

71. WE Thomas, A Crespo–Armas, J Mowbray. Influence of nano-molar calcium ions and physiological levels of thyroid hormone on oxidative phosphorylation in rat liver mitochondria. *Biochem J* 247:315–320, 1996.

72. SD Zakharov, X Li, TP Red'ko, RA Dilley. Calcium binding to subunit *c* of *E. coli* ATPsynthase and possible implications of energy coupling. *J Bioenergetics Biomembranes* 28:483–494, 1996.

73. S Luvissetto, I Schmehl, E Conti, E Intravaia, GF Azzone. Activation of respiration and loss of thermodynamic control in hyperthyroidism. Is it due increased slipping in mitochondrial proton pumps? *FEBS Lett* 291:117–120, 1991.

74. A Wong. Mitochondrial DNA mutations confer cellular sensitiv-ity to oxidant stress that is partially rescued by calcium deple-tion and cyclosporin A. *Biochem Biophys Res Commun* 239:139–145, 1997.

75. SB Kim, CD Berdanier. Oligomycin sensitivity of mitochondrial F_1F_0-ATPase in diabetes-prone BHE/Cdb rats. *Am J Physiol* 277:E702–E707, 1999.

76. ME Vazquez–Memije, S Shanske, FM Sanlorelli, P Kranz–Eble, DC DeVivo, S DiMauro. Comparative biochemical studies of ATPase in cells from patients with T8993G or T8993C mito-chondrial mutations. *J Infect Dis* 21:829–836, 1998.

77. MJ Kullen, LA Berdanier, R Dean, CD Berdanier. Gluconeogen-esis is less active in BHE/Cdb rats fed menhaden oil than in rats fed beef tallow. *Biochem Arch* 13:75–85, 1997.

78. JS Pan, CD Berdanier. Thyroxine effects on parameters of glu-cose turnover in BHE rats fed menhaden oil. *J Nutr Biochem* 2:262–266, 1991.

79. HB Everts, DO Claassen, CL Hermoyian, CD Berdanier. Nutri-ent-gene interactions: dietary vitamin A and mitochondrial gene expression. *IUBMB Life* 53:295–301, 2002.

80. CL Hermoyian. Hepatocyte mitochondrial gene expression is influenced by dehydroepiandrosteron (DHEA), retinoic acid, and thyroid hormone (T3) in Sprague–Dawley and BHE/Cdb rats. M.S. thesis, University of Georgia, 2000.

81. G Li, Y Liu, SS Tsang. Expression of a retinoic acid inducible mitochondrial ND 5 gene is regulated by cell density in bovine papilloma virus DNA-transformed mouse C127 cells but not in revertant cells. *Int J Oncol* 5:301–307, 1994.

82. IC Gaemers, AMN Van Pelt, APM Themmen, DG De Rooij. Isolation and characterization of all-trans retinoic acid-responsive genes in the rat testis. *Mol Reprod Dev* 50:1–6, 1998.

83. SJ Ruff, DE Ong. Cellular retinoic acid binding protein is associated with mitochondria. *FEBS Lett* 487:282–286, 2000.

84. DC Wallace. Diseases of the mtDNA. *Annu Rev Biochem* 61:1175–1212, 1992.

85. JM Shoffner, DC Wallace. Oxidative phosphorylation disease and mitochondrial mutations: diagnosis and treatment. *Annu Rev Nutr* 14:535–568, 1994.

86. P Lestienne. Mitochondrial DNA mutations in human diseases: a review. *Biochemistry* 74:123–130, 1992.

87. EA Schon, MH Grossman. Mitochondrial diseases: genetics. *Biofactors* 7:191–195, 1998.

88. EA Shoubridge. Mitochondrial DNA diseases: histological and cellular studies. *J Bioenergetics Biomembranes* 26:301B–310B, 1994.

89. KD Gerbitz, K Gempel, D Brdiczka. Mitochondria and diabetes. Genetic, biochemical and clinical implications of the cellular energy circuit. *Diabetes* 45:113–126, 1996.

90. M Odowara. Involvement of mitochondrial gene abnormalities in the pathogenesis of diabetes mellitus. *Ann NY Acad Sci* 865:722–781, 1996.

91. CE Mathews, CD Berdanier. Non-insulin dependent diabetes mellitus as a mitochondrial genomic disease. *Proc Soc Exp Biol Med* 219:97–108, 1998.

Appendix 1

Resources for Methods to Examine Mitochondrial Function

H.U. Bergmeyer (editor). *Methods of Enzymatic Analysis*, Academic Press. This is a multivolume book designed to provide the details of many methods including those relevant to mitochondria. The methods are described in detail and this book is updated periodically.

S. Fleischer and L. Packer. Vol. LVI *Biomembranes* Part G *Bioenergetics, Biogenesis of Mitochondria, Organization and Transport*. 1979 in the series, *Methods in Enzymology* (SP Colowick and NO Kaplan, editors) Academic Press. *Methods in Enzymology* is an annual series. Each volume is dedicated to a specific methods area. In this particular volume, the details of mitochondrial isolation are given. There are few places to find such detailed methods of examining mitochondrial function and this volume gathers many of them together. Although it was published in 1979, the methods are still useful and in use throughout the world by persons interested in mitochondria. This book is valuable for its descriptions of how to prepare mitochondria and so forth; however, other volumes in this series can provide additional useful methods.

For example, a method to assess the activity of mito-chondrial superoxide dismutase can be found in volume 53 of the 1978 book in this series.

O.H. Lowry and J.V. Passonneau (1972). *A Flexible System of Enzymatic Analysis*, Academic Press. Here is another older book that is an invaluable aid to the mitochondriologist. This slim volume details the methods for substrate analysis as well as metabolic flux. The methods are useful and easily validated.

J. Sambrook, E.F. Fritsch, and T. Maniatis (1989 and subsequent revisions). *Molecular Cloning, A Laboratory Manual*. Cold Spring Harbor Press. This manual deals with the common tools of molecular biology. The methods are given in great detail and are periodically updated. Molecular biology is a very rapidly moving field. Many of the newer methods can be obtained from manufacturers of kits for specific assays. Beware of these kits. Some are truly excellent but others are not. Kits for westerns, northerns, and Southerns are available and, if the user has the appropriate gene or gene product with which to work, these kits can save considerable time and effort.

F.M. Ausubel, R. Brent, R.E. Kingston, D.D. Moore, J.G. Seidman, J.A. Smith, and K. Struhl (1998) *Current Protocols in Molecular Biology*. John Wiley & Sons, New York. This is another of the multiple volume book sets. It is very well written with clear and detailed methods for molecular biology as well as protein analysis, immunology, informatics, and so forth. It is periodically updated with new and valuable methods. The book comes in a loose-leaf notebook style allowing owners to remove outdated pages and replace them with improved pages. Users can also insert their own pages with additional information.

Appendix 2

Human Mitochondrial Mutations[a] and Phenotypes[b]

Mutation	Nucleotide	Gene	Phenotype	Ref.
		Base Substitutions		
T→C	618	tRNA[phe]	Myopathy	1
A→G	1555	12S rRNA	Aminoglycoside-induced deafness	2
G→A	1642	tRNA[val]	MELAS	3
A→G	3243	tRNA[leu(UUR)]	MELAS/PEO/deafness/diabetes	4–7
A→T	3243	tRNA[leu(UUR)]	Encephalomyopathy	8
T→C	3250	tRNA[leu(UUR)]	Myopathy	9
A→G	3251	tRNA[leu(UUR)]	PEO/myopathy	10
A→G	3252	tRNA[leu(UUR)]	MELAS	11,12
C→T	3256	tRNA[leu(UUR)]	Multisystem/PEO	13,14
A→G	3260	tRNA[leu(UUR)]	Cardiomyopathy/myopathy	15
T→C	3271	tRNA[leu(UUR)]	MELAS	16
T→C	3291	tRNA[leu(UUR)]	MELAS	17
A→G	3302	tRNA[leu(UUR)]	Myopathy	18
C→T	3303	tRNA[leu(UUR)]	Cardiomyopathy	19
T→C	3394	ND1	LHON	20,21
A→G	3426	ND1	Diabetes	22
G→A	3460	ND1	LHON	23
C→T	4024	ND1	LHON	24
T→C	4160	ND1	LHON	20,25
T→C	4216	ND1	LHON	26
A→G	4269	tRNA[ile]	Encephalomyopathy/cardiomyopathy	27

(continued)

Human Mitochondrial Mutations[a] and Phenotypes[b] (Continued)

Mutation	Nucleotide	Gene	Phenotype	Ref.
T→C	4285	tRNA[ile]	PEO	28
A→G	4295	tRNA[ile]	Hypertrophic cardiomyopathy	29
A→G	4300	tRNA[ile]	Cardiomyopathy	30
C→T	4320	tRNA[ile]	Hypertrophic cardiomyopathy	31
A→G	4336	tRNA[glu]	↑Risk, Alzheimer's and Parkinson's diseases	32
A→G	4732	ND2	LHON	24
A→G	4917	ND2	LHON	33
G→A	5244	ND2	LHON	21
G→A	5549	tRNA[trp]	Chorea/ encephalomyopathy	34
G→A	5703	tRNA[asn]	Myopathy/PEO	11
T→C	5814	tRNA[cys]	Encephalopathy	35
G→A	5877	tRNA[tyr]	Myopathy	36
G→A	7444	COX I	LHON	21
A→G	7445	tRNA[ser(UCN)]	Deafness	37,38
T→C	7512	tRNA[ser(UCN)]	MERRF/MELAS	39
A→G	7543	tRNA[asp]	Myoclonus	40
A→G	8344	tRNA[lys]	MERRF	41,42
T→C	8356	tRNA[lys]	MERRF	42
G→A	8363	tRNA[lys]	MERRF/deafness/ cardiopathy	43
T→C	8851	ATPase 6	FBSN	44
T→G	8893	ATPase 6	NARP/MILS	45–48
T→C	8893	ATPase 6	NARP/MILS	46,47
T→C	9176	ATPase 6	FBSN	47
T→C	9957	COX III	MELAS	49
T→C	9997	tRNA[gly]	Cardiomyopathy	50
A→G	10398	ND3	LHON	24
A→G	11084	ND4	MELAS	51
A→G	11696	ND4	LHON/dystonia	52
G→A	11778	ND4	LHON	53–55
G→A	12301	tRNA[leu(CUN)]	Sideroblastic anemia	56
A→G	12308	tRNA[leu(CUN)]	Deafness, diabetes	57
G→A	12315	tRNA[leu(CUN)]	Encephalomyopathy	58
A→G	12320	tRNA[leu(CUN)]	Myopathy	59,60
T→C	12811	ND5	LHON	24
A→G	13637	ND5	LHON	24
G→A	13708	ND5	LHON	24,25

(continued)

Human Mitochondrial Mutations[a] and Phenotypes[b] (Continued)

Mutation	Nucleotide	Gene	Phenotype	Ref.
C→T	13967	ND5	LHON	24
G→A	14459	ND6	LHON/dystonia	61
T→C	14484	ND6	LHON	53,60
T→C	14709	tRNA[glu]	Encephalomyopathy	62
C→T	15904	tRNA[thr]	PEO	56
G→A	15257	cyt b	LHON	20,63
C→A	15452	cyt b	LHON	64
G→A	15812	cyt b	LHON	65
G→A	15915	tRNA[thr]	Encephalomyopathy	66
A→G	15923	tRNA[thr]	Fatal infantile resp. def.	67
C→T	15990	tRNA[pro]	Myopathy	68
		Insertion		
Insert C	7471	tRNA[ser(UCN)]	Deafness/myoclonus	69
		Deletions		
2.461 kb	10368–12828	ND1, tRNA[arg,his,ser,leu]	Pearson's syndrome	70
4.977 kb	8470–13446	ND1,COXII, ATPase6	Aging skin	71
4.978 kb	8482–13460	COXII	KSS	72
7.4 kb	D-loop through ATPase6		Diabetes	73
2 bp	9204–9205	ATPase6	Transient lactic acidosis	74
		Duplications		
10 kb	6130–15056	COX1	KSS	75
260 bp	D-loop	—	No symptoms	76

[a] The human mitochondrial genome is a 16,569-bp circle of double-stranded DNA containing 37 genes: 13 components of oxidative phosphorylation plus the tRNAs and ribosomes needed for *in situ* protein synthesis.

[b] Abbreviations used: LHON, Leber's hereditary optic neuropathy; MERRF, myoclonus epilepsy with ragged red fibers; MELAS, mitochondrial encephalomyopathy with lactic acidosis and stroke-like episodes; COX, cytochrome oxidase; PEO, progressive external ophthalmoplegia; MILS, maternally inherited Leigh's syndrome; FBSN, familial bilateral striatal necrosis; KSS, Kearns–Sayre syndrome. The degree of heteroplasmy determines the severity of symptoms.

REFERENCES

1. S Kleinle, V Schneider, P Moosman, S Brandner, S Krahenbuhl, S Liechti–Gallati. *Biochem Biophys Res Commun* 247: 112–115; 1998.

2. TR Presant, JV Agapian, MC Bohlman, X Bu, S Oztas, W-Q Qui, KS Arnos, GA Cortopassi, L Jaber, JI Rotter, M Shohat, N Fischel–Ghodsian. *Nat Genet* 4: 289–294; 1993.

3. RW Taylor, PF Chinnery, F Haldane, AAM Morris, LA Bindoff, J Wilson, DM Turnbull. *Ann Neurol* 40: 459–462; 1996.

4. Y Goto, I Nonaka, S Horai. *Nature* 348: 651–653; 1990.

5. JMW van der Ouweland, HHPJ Lemkes, W Ruitenbeck. *Nat Genet* 1: 368–371; 1992.

6. ML Smith, X-Y Hua, DL Marsden, D Liu, NG Kennaway, K-Y Ngo, RH Haas. *J Clin Endo Metab* 82: 2826–2831; 1997.

7. CC Huang, R-S Chen, C-M Chen, H-S Wang, CC Lee, CY Pang, H-S Hsu, H-C Lee, Y-H Wei. *J Neurol* 57: 586–589; 1994.

8. A Shaag, A Saada, A Steinberg, P Navon, ON Elpeleg. *Biochem Biophys Res Commun* 233: 637–639; 1997.

9. Y Goto, M Tojo, J Tohyama, S Horai, I Nonaki. *Ann Neurol* 31: 672–675; 1992.

10. MG Sweeney, S Bundey, M Brockington, JR Poulton, JB Weiner, AE Harding. *Quarterly J Med* 86: 709–713; 1993.

11. KJ Morten, JM Cooper, GK Brown, BD Lake, D Pike, J Poulton. *Hum Mol Genet* 2: 2081–2087; 1993.

12. EA Schon, V Kega, M Davidson, CT Moraes, MP King. *Biochem Biophys Acta* 1101: 206–209; 1992.

13. CT Moraes, F Ciacci, E Bonilla, C Jansen, M Hirano, N Rao, RE Lovelce, LP Rowland, EA Schon, S DiMauro. *J Clin Invest* 92: 2906–2915; 1993.

14. W Sato, K Hayasaka, Y Shoji, T Takahashi, G Takada, M Saito, O Fukawa, E Wachi. *Biochem Mol Biol Int* 33: 1055–1061; 1994.

15. M Zevianni, C Gellera, C Antozzi, M Rimoldi, L Morandi, F Villani, V Tiranti, S DiDonato. *Lancet* 338: 143–147; 1991.

16. Y Goto, I Nonaka, S Horai. *Biochem Biophys Acta* 1097: 238–240; 1991.

17. Y Goto, K Tsugane, Y Tanabe, I Nonaka, S Horai. *Biochem Biophys Res Commun* 202: 1624–1630; 1994.

18. LA Bindoff, N Howell, J Poulton, DA McCullough, KJ Morten, RN Lightowlers, DM Turnbull, K Weber. *J Biol Chem* 268: 19559–19564; 1993.

19. G Silvestri, FM Santorelli, S Shancke, CB Whitley, LA Schimmenti, SA Smith, S DiMauro. *Hum Mutat* 3: 37–43; 1994.

20. AE Harding, MG Sweeney. In: *Mitochondrial Disorders in Neurology* (AHV Schapira and S DiMauro, Eds.) Butterworth Heineman, Oxford, 181–198; 1994.

21. MD Brown, AS Voljavec, MT Lott, I MacDonald, DC Wallace. *FASEB J* 6: 2791–2799; 1992.

22. CS Shin, SK Kim, KS Park, WB Kim, SY Kim, BY Cho, HK Lee, CS Koh, CH Shin, JB Lee. *Endocrinol J* 45: 105–110; 1998.

23. K Huoponen, J Vilkki, P Aula, EK Nikoskelain, ML Savontaus. *Am J Hum Genet* 48: 1147–1153; 1991.

24. DC Wallace. *Annu Rev Biochem* 61: 1175–1212; 1992.

25. N Howell, L Kubacka, M Xu, DA McCullough. *Am J Hum Genet* 48: 935–942; 1991.

26. D Mackey, N Howell. *Am J Hum Genet* 51: 1218–1228; 1992.

27. M Taniike, H Fukushima, I Yanagihara, H Tsukamoto, J Tanaka, H Fujimura, T Nagai, T Sano, K Yamaoka, K Inui, S Okada. *Biochem Biophys Res Commun* 186: 47–53; 1992.

28. G Silvesti, S Servidei, M Rana, E Ricci, A Spinazzola, E Paris, P Tonali. *Biochem Biophys Res Commun* 220: 623–627; 1996.

29. F Merante, T Myint, I Tein, L Benson, BH Robinson. *Hum Mutat* 8: 216–222; 1996.

30. C Casali, FM Santorelli, G D'Amati, P Bernucci, L DeBiase, S DiMauro. *Biochem Biophys Res Commun* 213: 588–593; 1995.

31. FM Santorelli, SC Mak, M Vazquez–Acevedo, A Gonzalez–Astiazaran, C Ridaura–Sanz, D Gonzalez–Halpen, S DiMauro. *Biochem Biophys Res Commun* 216: 835–840; 1995.

32. MD Brown, DC Wallace. *J Bioenergetics Biomembranes* 26: 273–289; 1994.

33. DR Johns, J Berman. *Biochem Biophys Res Commun* 174: 1324–1330; 1991.

34. I Nelson, MG Hanna, N Alsanjari, F Scaravilli, JA Morgan–Hughes, AE Harding. *Ann Neurol* 37: 400–403; 1995.

35. G Manfredi, EA Schon, E Bonilla, CT Moraes, S Shanske, S DiMauro. *Hum Mutat* 7: 158–163; 1996.

36. K Sahashi, T Ibi, R Ohno, M Tanaka, M Tashiro, K Marui, N Nakao, T Ozawa. *Muscle Nerve* Suppl 1: S139; 1994.

37. FM Reid, GA Vernham, HT Jacobs. *Hum Mutat* 3: 243–247; 1994.

38. SJ Hyslop, AM James, M Maw, N Fischel–Ghodsian, MP Murphy. *Biochem Mol Biol Int* 42: 567–575; 1997.

39. M Nakamura, S Nakano, Y Goto, M Ozawa, Y Nagahama, H Fukuyama, I Akiguchi, R Kaji, J Kimura. *Biochem Biophys Res Comm* 214: 86–93; 1995.

40. M El-Schahawi, FM Santorelli, E Malkin, S Shanske, S DiMauro. *J Mol Med* 73: B37; 1995.

41. JM Shoffner, MT Lott, AMS Lezza, P Seibel, SW Ballinger, DC Wallace. *Cell* 61: 931–937; 1990.

42. G Silvestri, CT Moraes, S Shanske, SJ Oh, S DiMauro. *Am J Hum Genet* 51: 1213–1217; 1992.

43. FM Santorelli, SC Mak, M El-Schahawi, C Casali, S Shanske, TZ Baram, RE Madrid, S DiMauro. *Am J Hum Genet* 58: 933–939; 1996.

44. L De Meirleir, S Seneca, W Lissens, E Schoentjes, B Desprechins. *Pediatr Neurol* 13: 242–246; 1995.

45. IJ Holt, AE Harding, RKH Petty, J Morgan–Hughes. *Am J Hum Genet* 46: 428–433; 1990.

46. Y Tatuch, J Christodoulou, A Feigenbaum. *Am J Hum Genet* 50: 852–858; 1992.

47. DD de Vries, BGM van Engelen, FJM Gabreels, W Ruitenbeek, BA van Oost. *Ann Neurol* 34: 410–412; 1993.

48. IA Trounce, S Neill, DC Wallace. *Proc Nat Acad Sci USA* 91: 8334–8338; 1994.

49. D Thyagarajan, S Shanske, M Vasquez–Memiji, D DeVivo, S DiMauro. *Ann Neurol* 38: 468–472; 1995.

50. G Manfredi, EA Schon, CT Moraes, E Bonilla, G Berry, JT Sladky, S DiMauro. *Neuromusc Disord* 5: 391–398; 1995.

51. F Merante, I Tein, L Benson, BH Robinson. *Am J Hum Genet* 55: 437–446; 1994.

52. P Letrit, AS Noer, MJB Jean–Francois, R Kapsa, X Dennet, D Thyagarajan, K Lethlean, E Byrne, S Marzuki. *Am J Hum Genet* 51: 457–468; 1992.

53. DD de Vries, LN Went, GW Bruyn, HR Scholte, RMW Hofstra, PA Bolhuis, BA van Oost. *Am J Hum Genet* 58: 703–711; 1996.

54. DC Wallace, G Singh, MT Lott, JA Hodge, TG Schurr, AMS Lezza, LJ Elsas, EK Nikoskelainen. *Science* 242: 1427–1430; 1988.

55. Y-H Wei, T-C Yen, C-Y Pang, M-Y Yen. *Clin Biotech* 3: 243–247; 1991.

56. H Sudoyo, M Sitepu, S Malik, HD Poesponegoro, S Marzuki. *Hum Mutat* 1: S271–S274; 1998.

57. N Gattermann, S Retzlaff, Y-L Wang, M Berneburg, J Heinisch, M Wlaschek, C Aul, W Schneider. *Br J Haematol* 93: 845–855; 1996.

58. K Fu, R Hartlen, T Johns, A Genge, G Karpati, EA Shoubridge. *Hum Mol Genet* 5: 1835–1840; 1996.

59. K Weber, JN Wilson, L Taylor, E Brierley, MA Johnson, DM Turnbull, LA Bindoff. *Am J Hum Genet* 60: 373–380; 1997.

60. AS Jun, MD Brown, DC Wallace. *Proc Nat Acad Sci USA* 91: 6206–6210; 1994.

61. DR Johns, MJ Neufeld, RD Park. *Biochem Biophys Res Commun* 223: 496–501; 1992.

62. MG Hanna, I Nelson, MG Sweeney, DA Wicks, JA Morgan–Hughes, AE Harding. *Am J Hum Genet* 56: 1026–1033; 1995.

63. J Marin–Garcia, Y Hu, R Ananthakrushnan, ME Pierpont, GL Pierpont, MJ Goldenthal. *Biochem Mol Biol Int* 40: 487–495; 1996.

64. DR Johns, MJ Neufeld. *Biochem Biophys Res Commun* 187: 1551–1557; 1991.

65. JMW van den Ouweland, GJ Bruining, D Lindhout, J-M Wit, BFE Veldhuyzen, JA Maassen. *Nucleic Acids Res* 20: 679–682; 1992.

66. I Nishino, A Seki, Y Maegaki, K Takeshita, S Horai, I Nonaka, I Goto. *Am J Hum Genet* 59: A400; 1996.

67. KL Yoon, JR Aprille, SG Ernst. *Biochem Biophys Res Commun* 176: 1112–1115; 1991.

68. CT Moraes, F Ciavcci, E Bonilla, V Ionasescu, EA Schon, S DiMauro. *Nat Genet* 4: 284–288; 1993.

69. V Tiranti, P Chariot, F Carella, A Toscano, P Sioliveri, P Girlanda, F Carrara, GM Fratta, FM Reid, C Mariotti, M Zeviani. *Hum Mol Genet* 4: 1421–1427; 1995.

70. R Kappa, GN Thompson, DR Thorburn, HHM Dahl, S Marzuki, E Byrne, RB Blok. *J Inher Metab Dis* 17: 521–526; 1994.

71. JH Yang, HC Lee, KJ Lin, Y-H Wei. *Arch Dermatol Res* 286: 386–390; 1994.

72. N Takeda. *Mol Clin Biochem* 176: 287–290; 1997.

73. S Mita, B Schmidt, EA Schon, S DiMauro, E Bonilla. *Proc Nat Acad Sci USA* 86: 9509–9513; 1989.

74. S Seneca, M Abramowicz, W Lissens, MF Muller, E Vamos, L de Meirleir. *J Inher Metab Dis* 19: 115–118; 1996.

75. J Poulton. *J Inher Metab Dis* 15: 487–498; 1992.

76. H Has, G Manfredi, CT Moreas. *Am J Hum Genet* 60: 1363–1372; 1997.

Index